Acta Numerica 1995

Acta
Numerica
1995

CAMBRIDGE
UNIVERSITY PRESS

CAMBRIDGE UNIVERSITY PRESS
Cambridge, New York, Melbourne, Madrid, Cape Town, Singapore,
São Paulo, Delhi, Dubai, Tokyo, Mexico City

Cambridge University Press
The Edinburgh Building, Cambridge CB2 8RU, UK

Published in the United States of America by Cambridge University Press, New York

www.cambridge.org
Information on this title: www.cambridge.org/9780521157582

First published 1995
First paperback edition 2010

A catalogue record for this publication is available from the British Library

ISBN 978-0-521-48255-4 Hardback
ISBN 978-0-521-15758-2 Paperback

Contents

Acta Numerica (1995), *pp.* 1–51

Sequential Quadratic Programming *

Paul T. Boggs

Applied and Computational Mathematics Division
National Institute of Standards and Technology
Gaithersburg, Maryland 20899
USA
E-mail: boggs@cam.nist.gov

Jon W. Tolle

Departments of Mathematics and Operations Research
University of North Carolina
Chapel Hill, North Carolina 27599
USA
E-mail: tolle@math.unc.edu

CONTENTS

1. Introduction

Since its popularization in the late 1970s, Sequential Quadratic Programming (SQP) has arguably become the most successful method for solving nonlinearly constrained optimization problems. As with most optimization methods, SQP is not a single algorithm, but rather a conceptual method from which numerous specific algorithms have evolved. Backed by a solid theoretical and computational foundation, both commercial and public-domain SQP algorithms have been developed and used to solve a remarkably large set of important practical problems. Recently large-scale versions have been devised and tested with promising results.

In this paper we examine the underlying ideas of the SQP method and the theory that establishes it as a framework from which effective algorithms can

---

* Contribution of the National Institute of Standards and Technology and not subject to copyright in the United States.

be derived. In the process we will describe the most popular manifestations of the method, discuss their theoretical properties and comment on their practical implementations.

The nonlinear programming problem to be solved is

$$\text{minimize} \quad f(\boldsymbol{x})$$
$$\text{subject to:} \quad \begin{aligned} \boldsymbol{h}(\boldsymbol{x}) &= \boldsymbol{0}, \\ \boldsymbol{g}(\boldsymbol{x}) &\le \boldsymbol{0}, \end{aligned} \qquad \text{(NLP)}$$

where $f\colon \mathcal{R}^n \to \mathcal{R}$, $\boldsymbol{h}\colon \mathcal{R}^n \to \mathcal{R}^m$, and $\boldsymbol{g}\colon \mathcal{R}^n \to \mathcal{R}^p$. Such problems arise in a variety of applications in science, engineering, industry and management. In the form (NLP) the problem is quite general; it includes as special cases linear and quadratic programs in which the constraint functions, $\boldsymbol{h}$ and $\boldsymbol{g}$, are affine and f is linear or quadratic. While these problems are important and numerous, the great strength of the SQP method is its ability to solve problems with nonlinear constraints. For this reason it is assumed that (NLP) contains at least one nonlinear constraint function.

The basic idea of SQP is to model (NLP) at a given approximate solution, say $\boldsymbol{x}^k$, by a quadratic programming subproblem, and then to use the solution to this subproblem to construct a better approximation $\boldsymbol{x}^{k+1}$. This process is iterated to create a sequence of approximations that, it is hoped, will converge to a solution $\boldsymbol{x}^*$. Perhaps the key to understanding the performance and theory of SQP is the fact that, with an appropriate choice of quadratic subproblem, the method can be viewed as the natural extension of Newton and quasi-Newton methods to the constrained optimization setting. Thus one would expect SQP methods to share the characteristics of Newton-like methods, namely, rapid convergence when the iterates are close to the solution but possible erratic behaviour that needs to be carefully controlled when the iterates are far from a solution. While this correspondence is valid in general, the presence of constraints makes both the analysis and implementation of SQP methods significantly more complex.

Two additional properties of the SQP method should be pointed out. First, SQP is not a feasible-point method; that is, neither the initial point nor any of the subsequent iterates need be *feasible* (a feasible point satisfies all of the constraints of (NLP)). This is a major advantage since finding a feasible point when there are nonlinear constraints may be nearly as hard as solving (NLP) itself. SQP methods can be easily modified so that linear constraints, including simple bounds, are always satisfied. Second, the success of the SQP methods depends on the existence of rapid and accurate algorithms for solving quadratic programs. Fortunately, quadratic programs are easy to solve in the sense that there are good procedures for their solution. Indeed, when there are only equality constraints the solution to a quadratic program reduces to the solution of a linear system of equations. When there are inequality constraints a sequence of systems may have to be solved.

A comprehensive theoretical foundation does not guarantee that a proposed algorithm will be effective in practice. In real-life problems hypotheses may not be satisfied, certain constants and bounds may not be computable, matrices may be numerically singular, or the scaling may be poor. A successful algorithm needs adaptive safeguards that deal with these pathologies. The algorithmic details to overcome such difficulties, as well as more mundane questions – how to choose parameters, how to recognize convergence and how to carry out the numerical linear algebra – are lumped under the term 'implementation'. While a detailed treatment of this topic is not possible here, we will take care to point out questions of implementation that pertain specifically to the SQP method.

This survey is arranged as follows. In Section 2 we state the basic SQP method along with the assumptions about (NLP) that will hold throughout the paper. We also make some necessary remarks about notation and terminology. Section 3 treats *local convergence,* that is, behaviour of the iterates when they are close to the solution. Rates of convergence are provided both in general terms and for some specific SQP algorithms. The goal is not to present the strongest results but to establish the relation between Newton's method and SQP, to delineate the kinds of quadratic models that will yield satisfactory theoretical results and to place current variations of the SQP method within this scheme.

The term *global* is used in two different contexts in nonlinear optimization and is often the source of confusion. An algorithm is said to be *globally convergent* if, under suitable conditions, it will converge to some local solution from any remote starting point. Nonlinear optimization problems can have multiple local solutions; the *global solution* is that local solution corresponding to the least value of f. SQP methods, like Newton's method and steepest descent, are only guaranteed to find a local solution of (NLP); they should not be confused with algorithms for finding the global solution, which are of an entirely different flavour.

To establish global convergence for constrained optimization algorithms, a way of measuring progress towards a solution is needed. For SQP this is done by constructing a *merit function,* a reduction in which implies that an acceptable step has been taken. In Section 4, two standard merit functions are defined and their advantages and disadvantages for forcing global convergence are considered. In Section 5, the problems of the transition from global to local convergence are discussed. In Sections 4 and 5 the emphasis is on line-search methods. Because trust region methods, which have been found to be effective in unconstrained optimization, have been extended to fit into the SQP framework, a brief description of this approach is given in

Section 6. Section 7 is devoted to implementation issues, including those associated with large-scale problems.

To avoid interrupting the flow of the presentation, comments on related results and references to the literature are provided at the end of each section. The number of papers on the SQP method and its variants is large and space prohibits us from compiling a complete list of references; we have tried to give enough references to each topic to direct the interested reader to further material.

1.1. Notes and References

The earliest reference to SQP-type algorithms seems to have been in the PhD thesis of Wilson (1963) at Harvard University, in which he proposed the method we call in Section 3 the Newton–SQP algorithm. The development of the secant or variable-metric algorithms for unconstrained optimization in the late 1960s and early 1970s naturally led to the extension of these methods to the constrained problem. The initial work on these methods was done by Mangasarian and his students at the University of Wisconsin. Garcia-Palomares and Mangasarian (1976) investigated an SQP-type algorithm in which the entire Hessian matrix of the Lagrangian, that is, the matrix of second derivatives with respect to both to x and the multipliers, was updated at each step. Shortly thereafter, Han (1976, 1977) provided further impetus for the study of SQP methods. In the first paper Han gave local convergence and rate of convergence theorems for the PSB- and BFGS-SQP algorithms for the inequality-constrained problem and, in the second, employed the ℓ_1 merit function to obtain a global convergence theorem in the convex case. In a series of papers presented at conferences, Powell (1977, 1978a, 1978b), Han's work was brought to the attention of the general optimization audience. From that time there has been a continuous production of research papers on the SQP method.

As noted, a significant advantage of SQP is that feasible points are not required at any stage of the process. Nevertheless, a version of SQP that always remains feasible has been developed and studied, for example, by Bonnans et al. (1992).

2. The Basic SQP Method

2.1. Assumptions and Notation

As with virtually all nonlinear problems, it is necessary to make some assumptions on the problem (NLP) that clarify the class of problems for which the algorithm can be shown to perform well. These assumptions, as well as the consequent theory of nonlinear programming, are also needed to describe the SQP algorithm itself. In this presentation we do not attempt to make the weakest assumptions possible, but rather provide what we consider reasonable assumptions under which the main ideas can be illustrated.

We make the blanket assumption that all of the functions in (NLP) are three times continuously differentiable. We denote the gradient of a scalar-valued function by ∇, for example, $\nabla f(x)$. (Here, and throughout the paper, all vectors are assumed to be column vectors; we use the superscript ^{t} to denote transpose.) For vector-valued functions we also use ∇ to denote the Jacobian of the function. For example,

$$\nabla h(x) = (\nabla h_1(x), \nabla h_2(x), \ldots, \nabla h_m(x)).$$

The Hessian of a scalar-valued function, denoted by the letter H, is defined to be the symmetric matrix whose (i,j)th component is

$$Hf(x)_{i,j} = \frac{\partial^2 f(x)}{\partial x_i \partial x_j}.$$

Where a function is defined on more than one set of variables, such as the Lagrangian function defined below, the differentiation operators ∇ and H will refer to differentiation *with respect to x only*.

A key function, one that plays a central role in all of the theory of constrained optimization, is the scalar-valued *Lagrangian function* defined by

$$\mathcal{L}(x, u, v) = f(x) + u^t h(x) + v^t g(x), \tag{2.1}$$

where $u \in \mathcal{R}^m$ and $v \in \mathcal{R}^p$ are the *multiplier vectors*. Given a vector x, the set of *active* constraints at x consists of the inequality constraints, if any, satisfied as equalities at x. We denote the index set of active constraints by

$$\mathcal{I}(x) = \{i \ : \ g_i(x) = 0\}.$$

The matrix $G(x)$ made up of the matrix $\nabla h(x)$ along with the columns $\nabla g_i(x)$, $i \in \mathcal{I}(x)$, will be important in describing the basic assumptions and carrying out the subsequent analyses. Assuming that the matrix $G(x)$ has full column rank, the null space of $G(x)^t$ defines the tangent space to the equality and active inequality constraints at x. The projection onto this tangent space can be written as

$$P(x) = I - G(x)\left(G(x)^t G(x)\right)^{-1} G(x)^t. \tag{2.2}$$

The corresponding projection onto the range space of $G(x)$ will be written

$$\mathcal{Q}(x) = I - P(x). \tag{2.3}$$

For convenience, these projection matrices evaluated at iterates x^k and at a solution x^* will be denoted by P^k, P^*, $\mathcal{Q}^k$ and $\mathcal{Q}^*$. Similarly, we will write

$$H\mathcal{L}^* = H\mathcal{L}(x^*, u^*, v^*)$$

throughout the remainder of this paper.

In this paper x^* will represent any particular local solution of (NLP). We assume that the following conditions apply to each such solution.

A1: The *first order necessary conditions* hold, that is, there exist optimal multiplier vectors u^* and $v^* \geq 0$ such that

$$\nabla \mathcal{L}(x^*, u^*, v^*) = \nabla f(x^*) + \nabla h(x^*)u^* + \nabla g(x^*)v^* = 0.$$

A2: The columns of $G(x^*)$ are linearly independent.

A3: Strict complementary slackness holds, that is,

$$g_i(x^*)\, v_i^* = 0$$

for $i = 1, \ldots, p$ and, if $g_i(x^*) = 0$, then $v_i^* > 0$.

A4: The Hessian of the Lagrangian function with respect to x is positive definite on the null space of $G(x^*)^{\mathsf{t}}$; that is,

$$d^{\mathsf{t}} H \mathcal{L}^* d > 0$$

for all $d \neq 0$ such that $G(x^*)^{\mathsf{t}} d = 0$.

The above conditions, sometimes called the *strong second order sufficient conditions*, in fact guarantee that x^* is an isolated local minimum of (NLP) and that the optimal multiplier vectors u^* and v^* are unique. It should be noted that without strong additional assumptions (NLP) may have multiple local solutions.

We use the term *critical point* to denote a feasible point that satisfies the first order necessary conditions **A1**. A critical point may or may not be a local minimum of (NLP).

In discussing convergence of SQP algorithms the asymptotic *rate of convergence* plays an important role. Three standard measures of convergence rates will be emphasized in this paper. In the definitions that follow, and throughout the paper, the norm $\|\cdot\|$ will refer to the 2-norm unless specifically noted otherwise. Other norms can be used for most of the analysis.

Definition 1 Let $\{x^k\}$ be a sequence converging to x^*. The sequence is said to converge *linearly* if there exists a positive constant $\xi < 1$ such that

$$\left\| x^{k+1} - x^* \right\| \leq \xi \left\| x^k - x^* \right\|$$

for all k sufficiently large, *superlinearly* if there exists a sequence of positive constants $\xi_k \to 0$ such that

$$\left\| x^{k+1} - x^* \right\| \leq \xi_k \left\| x^k - x^* \right\|$$

for all k sufficiently large, and *quadratically* if there exists a positive constant

ξ such that

$$\left\|x^{k+1} - x^*\right\| \le \xi \left\|x^k - x^*\right\|^2$$

for all k sufficiently large.

These rates, which measure improvement at each step, are sometimes referred to as Q-rates. We will also have occasion to state results in terms of a measure of the average rate of convergence called the R-rate. A sequence $\{x^k\}$ will be said to converge *R-linearly* if the sequence $\{\left\|x^k - x^*\right\|\}$ is bounded by a sequence that converges Q-linearly to zero. Similar definitions exist for R-superlinear and R-quadratic. There are two relations between R-rate and Q-rate convergence that are of importance for this paper. First, m-step Q-linear convergence (where $k+1$ is replaced by $k+m$ in the above definitions) implies an R-linear rate of convergence and, second, a Q-rate of convergence of a sequence of vectors implies the same R-rate (but not the same Q-rate) of convergence of its components. In the analyses that follow, rates not designated explicitly as Q- or R-rates are assumed to be Q-rates.

2.2. The Quadratic Subproblem

As suggested in the Introduction the SQP method is an iterative method in which, at a current iterate x^k, the step to the next iterate is obtained through information generated by solving a quadratic subproblem. The subproblem is assumed to reflect in some way the local properties of the original problem. The major reason for using a quadratic subproblem, that is, a problem with a quadratic objective function and linear constraints, is that such problems are relatively easy to solve and yet, in their objective function, can reflect the nonlinearities of the original problem. The technical details of solving quadratic programs will not be dealt with here, although the algorithmic issues involved in solving these quadratic subproblems are nontrivial and their resolution will affect the overall performance of the SQP algorithm. Further comments on this are made in Section 7.

A major concern in SQP methods is the choice of appropriate quadratic subproblems. At a current approximation x^k a reasonable choice for the constraints is a linearization of the actual constraints about x^k. Thus the quadratic subproblem will have the form

$$\begin{aligned}
\underset{d_x}{\text{minimize}} \quad & (\mathbf{r}^k)^\mathsf{t} d_x + \tfrac{1}{2} d_x{}^\mathsf{t} B_k \, d_x \\
\text{subject to:} \quad & \nabla h(x^k)^\mathsf{t} d_x + h(x^k) = 0, \\
& \nabla g(x^k)^\mathsf{t} d_x + g(x^k) \le 0,
\end{aligned}$$

where $d_x = x - x^k$. The vector $\mathbf{r}^k$ and the symmetric matrix B_k remain to be chosen.

The most obvious choice for the objective function in this quadratic program is the local quadratic approximation to f at x^k. That is, B_k is taken as the Hessian and r^k as the gradient of f at x^k. While this is a reasonable approximation to use if the constraints are linear, the presence of nonlinear constraints makes this choice inappropriate. For example, consider the problem

$$\text{minimize} \quad x_1 - \tfrac{1}{2}(x_2)^2$$
$$\quad x$$
$$\text{subject to: } (x_1)^2 + (x_2)^2 - 1 = 0.$$

The point $(1,0)$ is a solution satisfying **A1–A4**, but at the point $(1 + \epsilon, 0)$ the approximating quadratic program is (with d replacing d_x)

$$\text{minimize} \quad -d_1 - \tfrac{1}{2}(d_2)^2$$
$$\quad d$$
$$\text{subject to: } d_1 = -\tfrac{1}{2}\epsilon(2 + \epsilon)(1 + \epsilon)^{-1}.$$

which is unbounded regardless of how small ϵ is. Thus the algorithm that computes d breaks down for this problem.

To take nonlinearities in the constraints into account while maintaining the linearity of the constraints in the subproblem, the SQP method uses a quadratic model of the Lagrangian function as the objective. This can be justified by noting that conditions **A1–A4** imply that x^* is a local minimum for the problem

$$\text{minimize} \quad L(x, u^*, v^*)$$
$$\quad x$$
$$\text{subject to: } \quad h(x) \;=\; 0,$$
$$\qquad\qquad\qquad g(x) \;\leq\; 0.$$

Note that the constraint functions are included in the objective function for this equivalent problem. Although the optimal multipliers are not known, approximations u^k and v^k to the multipliers can be maintained as part of the iterative process. Then given a current iterate, (x^k, u^k, v^k), the quadratic Taylor-series approximation in x for the Lagrangian is

$$L(x^k, u^k, v^k) + \nabla L(x^k, u^k, v^k)^t d_x + \tfrac{1}{2} d_x{}^t H L(x^k, u^k, v^k) d_x.$$

A strong motivation for using this function as the objective function in the quadratic subproblem is that it generates iterates that are identical to those generated by Newton's method when applied to the system composed of the first order necessary condition (condition **A1**) and the constraint equations (including the active inequality constraints). This means that the resulting algorithm will have good local convergence properties. In spite of these local convergence properties there are good reasons to consider choices other than the actual Hessian of the Lagrangian, for example, approximating matrices that have properties that permit the quadratic subproblem to be solved at any x^k and the resulting algorithm to be amenable to a global convergence

analysis. Letting B_k be an approximation of $H\mathcal{L}(x^k, u^k, v^k)$, we can write
the quadratic subproblem as:

$$\begin{aligned}
\underset{d_x}{\text{minimize}} \quad & \nabla \mathcal{L}(x^k, u^k, v^k)^t d_x + \tfrac{1}{2} d_x{}^t B_k\, d_x \\
\text{subject to:} \quad & \nabla h(x^k)^t d_x + h(x^k) = 0, \\
& \nabla g(x^k)^t d_x + g(x^k) \le 0.
\end{aligned} \qquad (2.4)$$

The form of the quadratic subproblem most often found in the literature,
and the one that will be employed here, is

$$\begin{aligned}
\underset{d_x}{\text{minimize}} \quad & \nabla f(x^k)^t d_x + \tfrac{1}{2} d_x{}^t B_k\, d_x \\
\text{subject to:} \quad & \nabla h(x^k)^t d_x + h(x^k) = 0, \\
& \nabla g(x^k)^t d_x + g(x^k) \le 0.
\end{aligned} \qquad \text{(QP)}$$

These two forms are equivalent for problems with only equality constraints
since, by virtue of the linearized constraints, the term $\nabla h(x^k)^t d_x$ is constant
and the objective function becomes $\nabla f(x^k)^t d_x + \tfrac{1}{2} d_x{}^t B_k d_x$. The two subproblems are not quite equivalent in the inequality-constrained case unless
the multiplier estimate v^k is zero for all inactive linear constraints. However,
(QP) is equivalent to (2.4) for the slack-variable formulation of (NLP) given
by

$$\begin{aligned}
\underset{x,\, z}{\text{minimize}} \quad & f(x) \\
\text{subject to:} \quad & h(x) = 0, \\
& g(x) + z = 0, \\
& z \ge 0,
\end{aligned} \qquad (2.5)$$

where $z \in \mathcal{R}^p$ is the vector of slack variables. Therefore (QP) can be considered an appropriate quadratic subproblem for (NLP).

The solution d_x of (QP) can be used to generate a new iterate x^{k+1}, by
taking a step from x^k in the direction of d_x. But to continue to the next
iteration new estimates for the multipliers are needed. There are several
ways in which these can be chosen, but one obvious approach is to use the
optimal multipliers of the quadratic subproblem. (Other possibilities will be
discussed where appropriate.) Letting the optimal multipliers of (QP) be
denoted by u_{qp} and v_{qp} and setting

$$\begin{aligned}
d_u &= u_{\text{qp}} - u^k, \\
d_v &= v_{\text{qp}} - v^k,
\end{aligned}$$

allow the updates of (x, u, v) to be written in the compact form

$$\begin{aligned}
x^{k+1} &= x^k + \alpha d_x, \\
u^{k+1} &= u^k + \alpha d_u, \\
v^{k+1} &= v^k + \alpha d_v
\end{aligned} \qquad (2.6)$$

for some selection of the *steplength parameter* α. Once the new iterates
are constructed, the problem functions and derivatives are evaluated and a
prescribed choice of B_{k+1} calculated. It will be seen that the effectiveness
of the algorithm is determined in large part by this choice.

2.3. The Basic Algorithm

Since the quadratic subproblem (QP) has to be solved to generate steps in
the algorithm, the first priority in analyzing an SQP algorithm is to determine conditions that guarantee that (QP) has a solution. To have a solution
the system of constraints of (QP) must have a nonempty feasible set (i.e.
the system must be *consistent*) and the quadratic objective function should
be bounded below on that set (although a local solution can sometimes exist
without this condition). The consistency condition can be guaranteed when
x^k is in a neighborhood of x^* by virtue of Assumption **A2** but, depending
on the problem, may fail at nonlocal points. Practical means of dealing
with this possibility will be considered in Section 7. An appropriate choice
of B_k will ensure that a consistent quadratic problem will have a solution;
the discussion of this point will be a significant part of the analysis of the
next two sections.

Assuming that (QP) can be solved, the question of whether the sequence
generated by the algorithm will converge must then be resolved. As described in the Introduction, convergence properties generally are classified
as either local or global. *Local convergence* results proceed from the assumptions that the initial x-iterate is close to a solution x^* and the initial
Hessian approximation is, in an appropriate sense, close to $H\mathcal{L}^*$. Conditions
on the updating schemes that ensure that the x^k (and the B_k) stay close to
x^* (and $H\mathcal{L}^*$) are then formulated. These conditions allow one to conclude
that (QP) is a good model of (NLP) at each iteration and hence that the
system of equations defined by the first order conditions and the constraints
for (QP) are nearly the same as those for (NLP) at x^*. Local convergence
proofs can be modeled on the proof of convergence of the classical Newton's
method, which assumes $\alpha = 1$ in (2.6).

Convergence from a remote starting point is called *global convergence*. As
stated in the Introduction, to ensure global convergence the SQP algorithm
needs to be equipped with a measure of progress, a merit function ϕ, whose
reduction implies progress towards a solution. In order to guarantee that ϕ
is reduced at each step a procedure for adjusting the steplength parameter
α in (2.6) is required. Using the decrease in ϕ it can be shown that under
certain assumptions the iterates will converge to a solution (or, to be precise,
a potential solution) even if the initial x-iterate, x^0, is not close to a solution.

Local convergence theorems are based on Newton's method whereas global
convergence theorems are based on descent. Ideally an algorithm should be

such that as the iterates get close to the solution, the conditions for local convergence will be satisfied and the local convergence theory will apply without the need for the merit function. In general a global algorithm, although ultimately forcing the x-iterates to get close to x^*, does not automatically force the other local conditions (such as unit steplengths and close Hessian approximations) to hold and therefore the merit function must be used throughout the iteration process. Since there is no practical way to know when, if ever, the local convergence conditions will hold an implementation's true performance can be deduced only from numerical experience. These issues will be discussed more fully in Section 5.

With this background we can now state a basic SQP algorithm. This template indicates the general steps only, without the numerous details required for a general code.

Basic Algorithm

Given approximations (x^0, u^0, v^0), B_0, and a merit function ϕ, set $k = 0$.

1. Form and solve (QP) to obtain (d_x, d_u, d_v).
2. Choose steplength α so that

$$\phi(x^k + \alpha d_x) < \phi(x^k).$$

3. Set

$$\begin{aligned}
x^{k+1} &= x^k + \alpha d_x, \\
u^{k+1} &= u^k + \alpha d_u, \\
v^{k+1} &= v^k + \alpha d_v.
\end{aligned}$$

4. Stop if converged.
5. Compute B_{k+1}.
6. Set $k := k + 1$; go to 1.

2.4. Notes and References

The basic second order sufficient conditions as well as a description of the major theoretical ideas of finite-dimensional nonlinear constrained optimization can be found in numerous sources. See, for example, Luenberger (1984), Nash and Sofer (1995) or Gill, Murray and Wright (1981). The basic definitions for various rates of convergence and the relations among them can be found in Ortega and Rheinboldt (1970) and Dennis and Schnabel (1983).

The trust region methods discussed in Section 6 use different quadratic subproblems than those given here. Also Fukushima (1986) considers a different quadratic subproblem in order to avoid the Maratos effect discussed in Section 5.

3. Local Convergence

In this section the theory of local convergence for the most common versions of the SQP method is developed. The local convergence analysis establishes conditions under which the iterates converge to a solution and at what rate, given that the starting data (e.g. x^0, u^0, v^0, B_0) are sufficiently close to the corresponding data at a solution x^*. As will be seen, it is the method of generating the matrices B_k that will be the determining factor in obtaining the convergence results.

An SQP method can have excellent theoretical local convergence properties; quadratic, superlinear or two-step superlinear convergence of the x - iterates can be achieved by requiring the B_k to approximate HL^* in an appropriate manner. Although each of these local convergence rates is achievable in practice under certain conditions, the need for a globally convergent algorithm further restricts the choice of the B_k. Certain choices of the B_k have led to algorithms that extensive numerical experience has shown to be quite acceptable. Therefore, while there is a gap between what is theoretically possible and what has been achieved in terms of local convergence, this discrepancy should not obscure the very real progress that the SQP methods represent in solving nonlinearly constrained optimization problems.

Two important assumptions simplify the presentation. First, it will be assumed that the active inequality constraints for (NLP) at x^* are known. As will be discussed in Section 5, this assumption can be justified for many of the SQP algorithms because the problem (QP) at x^k will have the same active constraints as (NLP) at x^* when x^k is near x^*. The fact that the active constraints, and hence the inactive constraints, are correctly identified at x^k means that those inequality constraints that are inactive for (NLP) at x^* can be ignored and those that are active can be changed to equality constraints without changing the solution of (QP). Thus, under this assumption, only equality-constrained problems need be considered for the local analysis. For the remainder of this section (NLP) will be assumed to be an equality-constrained problem and the reference to inequality multiplier vectors (the $v^t g(x)$ term) will be eliminated from the Lagrangian. For reference, we rewrite the quadratic subproblem with only equality constraints:

$$\begin{aligned} \underset{d_x}{\text{minimize}} \quad & \nabla f(x^k)^t d_x + \tfrac{1}{2} d_x^t B_k d_x \\ \text{subject to:} \quad & \nabla h(x^k)^t d_x + h(x^k) = 0. \end{aligned} \qquad \text{(ECQP)}$$

The second assumption follows from the fact that the local convergence for SQP methods, as for many optimization methods, is based on Newton's method; that is, the convergence and rate-of-convergence results are obtained by demonstrating an asymptotic relation between the iterates of the method being analyzed and the classical Newton steps. Because Newton's method for a system with a nonsingular Jacobian at the solution requires that unit steps be taken to obtain rapid convergence, we assume that the merit function allows $\alpha = 1$ to be used in the update formulae (2.6). Similar results to those given in this section can be obtained if the steplengths converge to one sufficiently quickly.

For the equality-constrained problem a good initial estimate of $\boldsymbol{x}$ can be used to obtain a good estimate for the optimal multiplier vector. The first order necessary condition, **A1**, together with condition **A2**, leads to the formula

$$\boldsymbol{u}^* = -[\nabla h(\boldsymbol{x}^*)^{\mathrm{t}} A \nabla h(\boldsymbol{x}^*)]^{-1}\nabla h(\boldsymbol{x}^*)^{\mathrm{t}} A \nabla f(\boldsymbol{x}^*) \qquad (3.1)$$

for any nonsingular matrix A that is positive definite on the null space of $\nabla h(\boldsymbol{x}^*)^{\mathrm{t}}$. In particular, if A is taken to be the identity matrix, then (3.1) defines the least squares solution of the first order necessary conditions. By our smoothness assumptions, it follows that

$$\boldsymbol{u}^0 = -[\nabla h(\boldsymbol{x}^0)^{\mathrm{t}}\nabla h(\boldsymbol{x}^0)]^{-1}\nabla h(\boldsymbol{x}^0)^{\mathrm{t}}\nabla f(\boldsymbol{x}^0) \qquad (3.2)$$

can be made arbitrarily close to $\boldsymbol{u}^*$ by choosing x^0 close to $\boldsymbol{x}^*$. Consequently, no additional assumption will be made about an initial optimal multiplier estimate for the local convergence analysis.

Denoting the optimal multiplier for (ECQP) by $\boldsymbol{u}_{qp}$ we see that the first order conditions for this problem are

$$\begin{aligned} B_k\,\boldsymbol{d}_x + \nabla h(\boldsymbol{x}^k)\boldsymbol{u}_{qp} &= -\nabla f(\boldsymbol{x}^k), \\ \nabla h(\boldsymbol{x}^k)^{\mathrm{t}}\boldsymbol{d}_x &= -h(\boldsymbol{x}^k). \end{aligned}$$

If, as discussed in Section 2, we set

$$\boldsymbol{u}^{k+1} = \boldsymbol{u}_{qp} = \boldsymbol{u}^k + \boldsymbol{d}_u \qquad (3.3)$$

then the above equations become

$$\begin{aligned} B_k\,\boldsymbol{d}_x + \nabla h(\boldsymbol{x}^k)\boldsymbol{d}_u &= -\nabla \mathcal{L}(\boldsymbol{x}^k, \boldsymbol{u}^k), & (3.4) \\ \nabla h(\boldsymbol{x}^k)^{\mathrm{t}}\boldsymbol{d}_x &= -h(\boldsymbol{x}^k). & (3.5) \end{aligned}$$

We are now in a position to begin the analysis of the SQP methods.

3.1. The Newton SQP Method

The straightforward SQP method derived from setting

$$B_k = H\mathcal{L}(\boldsymbol{x}^k, \boldsymbol{u}^k)$$

will be analyzed first. Assuming that $\boldsymbol{x}^0$ is close to $\boldsymbol{x}^*$ it follows from (3.2) that $\boldsymbol{u}^0$ can be presumed to be close to $\boldsymbol{u}^*$ and hence that $H\mathcal{L}(\boldsymbol{x}^0, \boldsymbol{u}^0)$ is close to $H\mathcal{L}^*$. The local convergence for the SQP algorithm now follows from the application of Newton's method to the nonlinear system of equations obtained from the first order necessary conditions and the constraint equation:

$$\Psi(x, u) = \left[\begin{array}{c} \nabla \mathcal{L}(x, u) \\ h(x) \end{array} \right] = 0.$$

From assumptions **A1** and **A4** the Jacobian of this system at the solution,

$$J(x^*, u^*) = \left[\begin{array}{cc} H\mathcal{L}^* & \nabla h(x^*) \\ \nabla h(x^*)^t & 0 \end{array} \right], \tag{3.6}$$

is nonsingular. Therefore, the Newton iteration scheme

$$\begin{aligned} x^{k+1} &= x^k + s_x, \\ u^{k+1} &= u^k + s_u, \end{aligned}$$

where the vector $s = (s_x, s_u)$ is the solution to

$$J(x^k, u^k)\, s = -\Psi(x^k, u^k), \tag{3.7}$$

yields iterates that converge to (x^*, u^*) quadratically provided (x^0, u^0) is sufficiently close to (x^*, u^*). The equations (3.7) are identical to equations (3.4) and (3.5) with $B_k = H\mathcal{L}(x^k, u^k)$, $d_x = s_x$ and $d_u = s_u$. Consequently the iterates (x^{k+1}, u^{k+1}) are exactly those generated by the SQP algorithm. For this reason we call this version of the algorithm the (local) Newton SQP method. The basic local convergence results are summarized in the following theorem:

Theorem 1   Let x^0 be an initial estimate of the solution to (NLP) and let u^0 be given by (3.2). Suppose that the sequence of iterates $\{(x^k, u^k)\}$ is generated by

$$\begin{aligned} x^{k+1} &= x^k + d_x, \\ u^{k+1} &= u^k + d_u, \end{aligned}$$

where d_x and $u_{qp} = u^k + d_u$ are the solution and multiplier of the quadratic program (ECQP) with $B_k = H\mathcal{L}(x^k, u^k)$. Then, if $\|x^0 - x^*\|$ is sufficiently small, the sequence of iterates in (x, u)-space is well defined and converges quadratically to the pair (x^*, u^*).

It is important to emphasize that this version of the SQP algorithm always requires a unit step in the variables; there is no line search to try to determine a better point. If a line search is used (the so-called damped Newton method) then the rate of convergence of the iterates can be significantly decreased – usually to linear.

In a sense, the Newton version of the SQP algorithm can be thought of as the ideal algorithm for solving the nonlinear program: it provides rapid

convergence with no requirement for line searches and no necessity for the
introduction of additional parameters. Of course, in this form it has little
practical value. It is often difficult to choose an initial point close enough
to the true solution to guarantee that the Newton algorithm will converge
to a minimum of f. When remote from the solution $H\mathcal{L}(\boldsymbol{x}^k, \boldsymbol{u}^k)$ cannot
be assumed to be positive definite on the appropriate subspace and hence
a solution to the subproblem (ECQP) cannot be guaranteed to exist. The
value of this Newton SQP method is that it provides a standard against
which the local behavior of other versions can be measured. In fact, to obtain
superlinear convergence it is necessary and sufficient that the steps aproach
the Newton steps as the solution is neared. Despite the disadvantages, the
use of $H\mathcal{L}(\boldsymbol{x}^k, \boldsymbol{u}^k)$ often makes (ECQP) an excellent model of (NLP) and
its use, combined with some of the techniques of Sections 6 and 7, can
lead to effective algorithms. The actual computation of $H\mathcal{L}(\boldsymbol{x}^k, \boldsymbol{u}^k)$ can be
accomplished efficiently by using finite difference techniques or by automatic
differentiation methods.

3.2. Conditions for Local Convergence

Here we discuss the theoretical properties of the approximation matrices,
B_k, that are sufficient to guarantee that the local SQP method will give
Newton-like convergence results. In the following subsections the practical attempts to generate matrices that satisfy these properties will be described.

   In order to formulate conditions on B_k that will yield locally convergent
SQP algorithms, the following assumption on $H\mathcal{L}^*$ will be imposed.

A5: The matrix $H\mathcal{L}^*$ is nonsingular.

In light of **A4** most of the results that follow could be extended to cover
the case where this condition is not satisfied, but the resulting complexity
of the arguments would clutter the exposition.

   The following conditions on the matrix approximations will be referenced
in the remainder of the paper.

B1: The matrices B_k are uniformly positive definite on the null spaces of
the matrices $\nabla h(\boldsymbol{x}^k)^t$, that is, there exists a $\beta_1 > 0$ such that for each k

$$\boldsymbol{d}^t B_k \boldsymbol{d} \geq \beta_1 \|\boldsymbol{d}\|^2$$

for all $\boldsymbol{d}$ satisfying

$$\nabla h(\boldsymbol{x}^k)^t \boldsymbol{d} = 0.$$

B2: The sequence $\{B_k\}$ is uniformly bounded, that is, there exists a $\beta_2 > 0$
such that for each k

$$\|B_k\| \leq \beta_2.$$

B3: The matrices B_k have uniformly bounded inverses, that is, there exists a $\beta_3 > 0$ such that for each k, B_k^{-1} exists and

$$\left\| B_k^{-1} \right\| \leq \beta_3.$$

As the SQP methods require the solution of the quadratic subproblem at each step, it is imperative that the choice of matrices B_k make that possible. The second order sufficient conditions for a solution to (ECQP) to exist require that the matrix B_k be positive definite on the null space of $\nabla h(x^k)^{\text{t}}$ (cf. **A4**). Condition **B1** is a slight strengthening of this requirement. Assumption **A5** and the fact that the matrices B_k are to approximate the Hessian of the Lagrangian suggest that conditions **B2** and **B3** are not unreasonable.

Under assumptions **B1**–**B3** (3.4) and (3.5) have a solution provided (x^k, u^k) is sufficiently close to (x^*, u^*). Moreover, the solutions can be written in the form

$$d_u = [\nabla h(x^k)^{\text{t}} B_k^{-1} \nabla h(x^k)]^{-1} [h(x^k) - \nabla h(x^k)^{\text{t}} B_k^{-1} \nabla \mathcal{L}(x^k, u^k)] \quad (3.8)$$

and

$$d_x = -B_k^{-1} \nabla \mathcal{L}(x^k, u^{k+1}), \quad (3.9)$$

where u^{k+1} is given by (3.3). In particular (3.8) leads to the relation

$$
\begin{aligned}
u^{k+1} &= [\nabla h(x^k)^{\text{t}} B_k^{-1} \nabla h(x^k)]^{-1} [h(x^k) - \nabla h(x^k)^{\text{t}} B_k^{-1} \nabla f(x^k)] \\
&\equiv W(x^k, B_k).
\end{aligned}
$$

Setting $A = B_k^{-1}$ in (3.1) yields

$$u^* = W(x^*, B_k).$$

It can now be deduced that

$$
\begin{aligned}
u^{k+1} - u^* &= W(x^k, B_k) - W(x^*, B_k) \\
&= [\nabla h(x^*)^{\text{t}} B_k^{-1} \nabla h(x^*)]^{-1} \nabla h(x^*)^{\text{t}} B_k^{-1} (B_k - H\mathcal{L}^*)(x^k - x^*) \\
&\quad + w_k, \quad (3.10)
\end{aligned}
$$

where, by assumptions **B2** and **B3**,

$$w_k \leq \kappa \left\| x^k - x^* \right\|^2$$

for some κ independent of k. It should be noted that (3.10) can be used in conjunction with Theorem 1 to prove that the sequence $\{x^k\}$ is quadratically convergent when the Newton SQP method is used and $H\mathcal{L}^*$ is nonsingular. In that case $\{u^k\}$ converges R-quadratically.

Equations (3.9) and (3.10) and **A2** now yield

$$
\begin{aligned}
\boldsymbol{x}^{k+1} - \boldsymbol{x}^* &= \boldsymbol{x}^k - \boldsymbol{x}^* - B_k^{-1}\left[\nabla \mathcal{L}(\boldsymbol{x}^k, \boldsymbol{u}^{k+1}) - \nabla \mathcal{L}(\boldsymbol{x}^*, \boldsymbol{u}^*)\right] \\
&= B_k^{-1}\left[(B_k - H\mathcal{L}^*)(\boldsymbol{x}^k - \boldsymbol{x}^*) - \nabla h(\boldsymbol{x}^*)(\boldsymbol{u}^{k+1} - \boldsymbol{u}^*)\right] \\
&\quad + \mathcal{O}\left(\left\|\boldsymbol{x}^k - \boldsymbol{x}^*\right\|^2\right) \\
&= B_k^{-1} V_k (B_k - H\mathcal{L}^*)(\boldsymbol{x}^k - \boldsymbol{x}^*) + \mathcal{O}\left(\left\|\boldsymbol{x}^k - \boldsymbol{x}^*\right\|^2\right), \ (3.11)
\end{aligned}
$$

where

$$
V_k = I - \nabla h(\boldsymbol{x}^*)[\nabla h(\boldsymbol{x}^*)^{\mathrm{t}} B_k^{-1} \nabla h(\boldsymbol{x}^*)]^{-1}\nabla h(\boldsymbol{x}^*)^{\mathrm{t}} B_k^{-1}.
$$

The projection operator defined by (2.2), which in the equality-constrained case has the form

$$
\mathcal{P}(\boldsymbol{x}) = I - \nabla h(\boldsymbol{x})[\nabla h(\boldsymbol{x})^{\mathrm{t}}\nabla h(\boldsymbol{x})]^{-1}\nabla h(\boldsymbol{x})^{\mathrm{t}},
$$

satisfies

$$
V_k \mathcal{P}^k = V_k.
$$

Thus (3.11) leads to the inequality

$$
\begin{aligned}
\left\|\boldsymbol{x}^{k+1} - \boldsymbol{x}^*\right\| &\leq \left\|B_k^{-1}\right\| \|V_k\| \left\|\mathcal{P}^k(B_k - H\mathcal{L}^*)(\boldsymbol{x}^k - \boldsymbol{x}^*)\right\| \\
&\quad + \mathcal{O}\left(\left\|\boldsymbol{x}^k - \boldsymbol{x}^*\right\|^2\right).
\end{aligned} \tag{3.12}
$$

Using induction the above analysis can be made rigorous to yield the following local convergence theorem.

Theorem 2 *Let assumptions **B1**–**B3** hold. Then there exist positive constants ϵ and γ such that if*

$$
\left\|\boldsymbol{x}^0 - \boldsymbol{x}^*\right\| < \epsilon,
$$

$$
\left\|\boldsymbol{u}^0 - \boldsymbol{u}^*\right\| < \epsilon
$$

and

$$
\left\|\mathcal{P}^k(B_k - H\mathcal{L}^*)(\boldsymbol{x}^k - \boldsymbol{x}^*)\right\| < \gamma \left\|\boldsymbol{x}^k - \boldsymbol{x}^*\right\| \tag{3.13}
$$

for all k, then the sequences $\{\boldsymbol{x}^k\}$ and $\{(\boldsymbol{x}^k, \boldsymbol{u}^k)\}$ generated by the SQP algorithm are well defined and converge linearly to $\boldsymbol{x}^*$ and $(\boldsymbol{x}^*, \boldsymbol{u}^*)$, respectively. The sequence $\{\boldsymbol{u}^k\}$ converges R-linearly to $\boldsymbol{u}^*$.

Condition (3.13) is, in conjuction with the other hypotheses of the theorem, almost necessary for linear convergence. In fact, if the other hypotheses of the theorem hold and the sequence $\{\boldsymbol{x}^k\}$ converges linearly, then there exists a ξ such that

$$\left\|\mathcal{P}^k(B_k - H\mathcal{L}^*)(x^k - x^*)\right\| < \xi\left\|(x^k - x^*)\right\|$$

for all k. These results and those that follow indicate the crucial role that the projection of $(B_k - H\mathcal{L}^*)$ plays in the theory. The inequality (3.13) is guaranteed by either of the stronger conditions:

$$\left\|\mathcal{P}^k(B_k - H\mathcal{L}^*)\right\| \le \gamma,$$

or

$$\|(B_k - H\mathcal{L}^*)\| \le \gamma \tag{3.14}$$

which are easier to verify in practice.

In order to satisfy (3.14) it is not necessary that the approximations converge to the true Hessian but only that the growth of the difference $\|B_k - H\mathcal{L}^*\|$ be kept under some control. In the quasi-Newton theory for unconstrained optimization this can be accomplished if the approximating matrices have a property called *bounded deterioration*. This concept can be generalized from unconstrained optimization to the constrained setting in a straightforward way.

Definition 2 A sequence of matrix approximations, $\{B_k\}$, for the SQP method is said to have the *bounded deterioration* property if there exist constants α_1 and α_2 independent of k such that

$$\|B_{k+1} - H\mathcal{L}^*\| \le (1 + \alpha_1\sigma_k)\|B_k - H\mathcal{L}^*\| + \alpha_2\sigma_k, \tag{3.15}$$

where

$$\sigma_k = \max\{\left\|x^{k+1} - x^*\right\|, \left\|x^k - x^*\right\|, \left\|u^{k+1} - u^*\right\|, \left\|u^k - u^*\right\|\}.$$

It seems plausible that the bounded deterioration condition when applied to the SQP process will lead to a sequence of matrices that satisfy **B2**, **B3** and (3.13) provided the initial matrix is close to the Hessian of the Lagrangian and (x^k, u^k) is close to (x^*, u^*). Indeed this is the case, as can be shown by induction to lead to the following result.

Theorem 3 *Suppose that the sequence of iterates $\{(x^k, u^k)\}$ is generated by the SQP algorithm and the sequence $\{B_k\}$ of symmetric matrix approximations satisfies **B1** and (3.15). If $\|x^0 - x^*\|$ and $\|B_0 - H\mathcal{L}^*\|$ are sufficiently small and u^0 is given by (3.2) then the hypotheses of Theorem 2 hold.*

The linear convergence for the iterates guaranteed by the above theorem is hardly satisfactory in light of the quadratic convergence of the Newton SQP method. As would be expected, a stronger relation between the approximation matrices, B_k, and the Hessian of the Lagrangian is needed to improve the rate of convergence; however, this relation still depends only on the projection of the difference between the approximation and the true Hessian. With some effort the following theorem can be deduced as a consequence of the inequalities above.

Theorem 4   *Let assumptions* **B1–B3** *hold and let the sequence* $\{(x^k, u^k)\}$ *be generated by the SQP algorithm. Assume that* $x^k \to x^*$*. Then the sequence* $\{x^k\}$ *converges to* x^* *superlinearly if and only if the matrix approximations satisfy*

$$\lim_{k\to\infty} \frac{\left\|\mathcal{P}^k(B_k - H\mathcal{L}^*)(x^{k+1} - x^k)\right\|}{\left\|(x^{k+1} - x^k)\right\|} = 0. \qquad (3.16)$$

If this equation holds then the sequence $\{u^k\}$ *converges R-superlinearly to* u^* *and the sequence* $\{(x^k, u^k)\}$ *converges superlinearly.*

Not that this theorem requires convergence of the x-iterates; (3.16) does not appear to be enough to yield convergence by itself. A slightly different result that uses a two-sided approximation of the Hessian approximation can be obtained by writing (3.16) as

$$\lim_{k\to\infty} \left\{ \frac{\mathcal{P}^k(B_k - H\mathcal{L}^*)\mathcal{P}^k s_k}{\|s_k\|} + \frac{\mathcal{P}^k(B_k - H\mathcal{L}^*)\mathcal{Q}^k s_k}{\|s_k\|} \right\} = 0, \qquad (3.17)$$

where $s_k = (x^{k+1} - x^k)$. It can be shown that if only the first term goes to zero then a weaker form of superlinear convergence holds.

Theorem 5   *Let assumptions* **B1–B3** *hold and let the sequence* $\{(x^k, u^k)\}$ *be generated by the SQP algorithm. Assume that* $x^k \to x^*$*. Then, if*

$$\lim_{k\to\infty} \frac{\left\|\mathcal{P}^k(B_k - H\mathcal{L}^*)\mathcal{P}^k(x^{k+1} - x^k)\right\|}{\left\|(x^{k+1} - x^k)\right\|} = 0, \qquad (3.18)$$

the sequence $\{x^k\}$ *converges to* x^* *two-step superlinearly.*

If the sequence $\{x^{k+1} - x^k\}$ approaches zero in a manner tangent to the null space of the Jacobians, that is,

$$\lim_{k\to\infty} \frac{\mathcal{Q}^k(x^{k+1} - x^k)}{\left\|x^{k+1} - x^k\right\|} = 0,$$

then (3.18) implies (3.16) and superlinear convergence results. This *tangential convergence* has been observed in practice for some of the methods discussed below.

It has been difficult to find useful updating schemes for the B_k that satisfy these conditions for linear and superlinear convergence. Two basic approaches have been tried for generating good matrix approximations. In *full* Hessian approximations the matrices B_k are chosen to approximate $H\mathcal{L}^*$ while in *reduced* Hessian approximations only matrices that approximate the Hessian on the null space of the Jacobians of the constraints are computed. Each of these methods will be described in turn.

3.3. Full Hessian Approximations

An obvious choice of the matrix B_k is a finite-difference approximation of the Hessian of the Lagrangian at (x^k, u^k). It is clear from Theorems 2 and 4 that if a finite-difference method is used, the resulting sequence will be superlinearly convergent if the finite-difference step size goes to zero and will have rapid linear convergence for fixed step size. Of course using finite-difference approximations, while not requiring evaluation of second derivatives, suffers from the same global difficulties as the Newton SQP method described in the earlier subsection.

As has been seen, one way of obtaining convergence is to use a sequence of Hessian approximations that satisfies the bounded deterioration property. However, this property is not, by itself, enough to guarantee that (3.16) is satisfied and hence it does not guarantee superlinear convergence. The condition (3.16) essentially requires the component of the step generated by the Hessian approximation in the direction of the null space to converge asymptotically to the corresponding component of the step generated by the Newton SQP method. In the following a class of approximations called *secant* approximations, which satisfy a version of this condition and have the bounded deterioration property as well, is considered.

Under the smoothness assumptions of this paper the Lagrangian satisfies

$$\nabla \mathcal{L}(x^{k+1}, u^{k+1}) - \nabla \mathcal{L}(x^k, u^{k+1}) \approx H\mathcal{L}(x^{k+1}, u^{k+1})(x^{k+1} - x^k),$$

with equality if the Lagrangian is quadratic in x. As a result, it makes sense to approximate the Hessian of the Lagrangian at (x^{k+1}, u^{k+1}) by requiring B_{k+1} to satisfy

$$B_{k+1}(x^{k+1} - x^k) = \nabla \mathcal{L}(x^{k+1}, u^{k+1}) - \nabla \mathcal{L}(x^k, u^{k+1}), \qquad (3.19)$$

especially since this approximation strongly suggests that (3.16) is likely to be satisfied near the solution. Equation (3.19) is called the *secant* equation; it plays an important role in the algorithmic theory of nonlinear systems and unconstrained optimization.

A common procedure for generating the Hessian approximations that satisfy (3.19) is to compute (x^{k+1}, u^{k+1}) with a given B_k and then to update B_k according to

$$B_{k+1} = B_k + U_k,$$

where U_k is a rank-one or rank-two matrix that depends on the values of B_k, x^k, u^k, x^{k+1} and u^{k+1}. In choosing secant approximations that have the bounded deterioration property, it is natural to look to those updating schemes of this type that have been developed for unconstrained optimization.

The rank-two Powell-symmetric-Broyden (PSB) formula gives one such updating scheme. For the constrained case this update is given by

$$
\begin{aligned}
B_{k+1} \;=\; & B_k + \frac{1}{(s^t s)}[(y - B_k s)s^t + s(y - B_k s)^t] \\
& - \frac{(y - B_k s)^t s}{(s^t s)^2}\, s\, s^t,
\end{aligned}
\tag{3.20}
$$

where

$$
s = x^{k+1} - x^k \tag{3.21}
$$

and

$$
y = \nabla \mathcal{L}(x^{k+1}, u^{k+1}) - \nabla \mathcal{L}(x^k, u^{k+1}). \tag{3.22}
$$

The PSB-SQP algorithm which employs this update has been shown to have the desired local convergence properties.

Theorem 6 *Suppose that the sequence of iterates $\{(x^k, u^k)\}$ is generated by the SQP algorithm using the sequence $\{B_k\}$ of matrix approximations generated by the PSB update formulas, (3.20)–(3.22). Then, if $\|x^0 - x^*\|$ and $\|B_0 - H\mathcal{L}(x^*, u^*)\|$ are sufficiently small and u^0 is given by (3.2), the sequence $\{B_k\}$ is of bounded deterioration and the iterates (x^k, u^k) are well defined and converge superlinearly to (x^*, u^*). In addition the x-iterates converge superlinearly and the multipliers converge R-superlinearly.*

Note that there is no assumption of positive definiteness here. In fact, while **B2** and **B3** are satisfied as a result of the bounded deterioration, **B1** does not necessarily hold. Consequently, d_x and d_u are solutions of not necessarily (ECQP), but of the first order conditions (3.4) and (3.5).

As a practical method, the PSB-SQP algorithm has the advantage over the Newton SQP method of not requiring the computation of the Hessian of the Lagrangian (but, as a result, yields only superlinear rather than quadratic convergence). However, because the matrices are not necessarily positive definite it suffers from the same serious drawback; the problem (ECQP) may not have a solution if the initial starting point is not close to the solution. Consequently, it does not appear to be useful in establishing a globally convergent algorithm (see, however, Section 6). As mentioned above, solving (ECQP) requires that the matrices be positive definite on the appropriate subspace for each subproblem. One way to enforce this is to require the matrices to be positive definite.

A rank-two updating scheme that is considered to be the most effective for unconstrained problems and has useful positive definite properties is the BFGS method. The formula for this update, generalized to be applicable to the constrained problem, is

$$B_{k+1} = B_k + \frac{yy^t}{s^t y} - \frac{B_k s s^t B_k}{s^t B_k s}, \tag{3.23}$$

where s and y are as in (3.21) and (3.22). The important qualities of the matrices generated by this formula are that they have the bounded deterioration property and satisfy a hereditary positive definiteness condition. The latter states that if B_k is positive definite and

$$y^t s > 0 \tag{3.24}$$

then B_{k+1} is positive definite. If the matrix approximations are positive definite then (ECQP) is easily solved and the SQP steps are well defined. Unfortunately, because the Hessian of the Lagrangian at (x^*, u^*) need only be positive definite on a subspace, it need not be the case that (3.24) is satisfied and hence the algorithm may break down, even close to the solution. However, if the Hessian of the Lagrangian is positive definite, (3.24) is valid provided B_k is positive definite and (x^k, u^k) is close to (x^*, u^*). In this case this BFGS-SQP algorithm has the same local convergence properties as the PSB-SQP method.

Theorem 7 *Suppose that $H\mathcal{L}^*$ is positive definite and let B_0 be an initial positive definite matrix. Suppose that the sequence of iterates $\{(x^k, u^k)\}$ is generated by the SQP algorithm using the sequence of matrix approximations generated by the BFGS update (3.23). Then, if $\|x^0 - x^*\|$ and $\|B_0 - H\mathcal{L}(x^*, u^*)\|$ are sufficiently small and u^0 is given by (3.2), the sequence $\{B_k\}$ is of bounded deterioration and (3.16) is satisfied. Therefore, the iterates (x^k, u^k) converge superlinearly to the pair (x^*, u^*). In addition the x-iterates converge superlinearly and the multipliers converge R-superlinearly.*

The assumption that $H\mathcal{L}^*$ is positive definite allows B_0 to be both positive definite and close to $H\mathcal{L}^*$. The requirement that $H\mathcal{L}^*$ be positive definite is satisfied, for example, if (NLP) is convex, but cannot be assumed in general. However, because the positive definiteness of B_k permits the solution of the quadratic subproblems independently of the nearness of the iterate to the solution, the BFGS method has been the focus of vigorous research efforts to adapt it to the general case. Several of these efforts have resulted in implementable algorithms.

One scheme, which we call the Powell-SQP method, is to maintain the positive definite property of the Hessian approximations by modifying the update formula (3.23) by replacing y in (3.22) with

$$\hat{y} = \theta\, y + (1 - \theta)\, B_k\, s \qquad (3.25)$$

for some $\theta \in (0, 1]$. With this modification, the condition (3.24) can always be satisfied although the updates no longer satisfy the secant condition. Nevertheless, it has been shown that a specific choice of θ leads to a sequence, $\{x^k\}$, that converges R-superlinearly to x^* *provided that the sequence converges*. Unfortunately, no proof of local covergence has been found although algorithms based on this procedure have proven to be quite successful in practice.

A second approach is to transform the problem so that $H\mathcal{L}^*$ is positive definite and then to apply the BFGS-SQP method on the transformed problem. It is a common trick in developing algorithms for solving equality-constrained optimization problems to replace the objective function in (NLP) by the function

$$f_A(x) = f(x) + \frac{\eta}{2} \|h(x)\|^2$$

for a positive value of the parameter η. This change gives a new Lagrangian function, the so-called augmented Lagrangian

$$\mathcal{L}_A(x, u) = \mathcal{L}(x, u) + \frac{\eta}{2} \|h(x)\|^2,$$

for which (x^*, u^*) is a critical point. The Hessian of the the augmented Lagrangian at (x^*, u^*) has the form

$$H\mathcal{L}_A(x^*, u^*) = H\mathcal{L}^* + \eta \nabla h(x^*) \nabla h(x^*)^t.$$

It follows from **A4** that there exists a positive value η^* such that for $\eta \geq \eta^*$, $H\mathcal{L}_A(x^*, u^*)$ is positive definite. If the value of η^* is known then the BFGS-SQP algorithm can be applied directly to the transformed problem with $\eta \geq \eta^*$ and, by Theorem 7, a superlinearly convergent algorithm results. This version of the SQP algorithm has been extensively studied and implemented. Although adequate for local convergence theory, this augmented-BFGS method has major drawbacks that are related to the fact that it is difficult to choose an appropriate value of η. To apply Theorem 7, the value of η must be large enough to ensure that $H\mathcal{L}_A(x^*, u^*)$ is positive definite, a condition that requires a priori knowledge of x^*. If unnecessarily large values of η are used without care numerical instabilities can result. This problem is exacerbated by the fact that if the iterates are not close to the solution appropriate values of η may not exist.

Quite recently, an intriguing adaptation of the BFGS-SQP algorithm has been suggested that shows promise of leading to a resolution of the above difficulties. This method, called the SALSA-SQP method, is related to the augmented Lagrangian SQP method but differs (locally) in the sense that a precise estimate of η can be chosen independently of x^*. If

$$y_A = \nabla \mathcal{L}_A(x^{k+1}, u^{k+1}) - \nabla \mathcal{L}_A(x^k, u^{k+1})$$

then

$$y_A = y + \eta \nabla h(x^{k+1}) h(x^{k+1}),$$

where y is the vector generated by the ordinary Lagrangian in (3.22). If

$$y_A^t s > 0, \tag{3.26}$$

then the BFGS updates for the augmented Lagrangian have the hereditary positive definiteness property. A minimum value of η_k, not directly dependent on x^*, can be given that guarantees that $y_A^t s$ is 'sufficiently' positive for given values of x^k and u^{k+1} in a neighbourhood of (x^*, u^*). The (local) version of this algorithm proceeds by using the augmented Lagrangian at each iteration with the required value of η_k (which may be zero) to force the satisfaction of this condition. The B_k generated by this procedure are positive definite while $H\mathcal{L}^*$ may not be; hence, the standard bounded deterioration property is not applicable. As a result a local convergence theorem is not yet available. As in the Powell-SQP algorithm, an R-superlinear rate of convergence of the x^k to x^* has been shown, provided that the sequence is known to converge.

3.4. Reduced Hessian SQP Methods

An entirely different approach to approximating the Hessian is based on the fact that assumption **A4** requires $H\mathcal{L}^*$ to be positive definite only on a particular subspace. The *reduced Hessian* methods approximate only the portion of the Hessian matrix relevant to this subspace. The advantages of these methods are that the standard positive definite updates can be used and that the dimension of the problem is reduced to $n - m$ (possibly a significant reduction). In this section we discuss the local convergence properties of such an approach. Several versions of a reduced Hessian type of algorithm have been proposed; they differ in the ways the multiplier vectors are chosen and the way the reduced Hessian approximation is updated, in particular, in the form of y that is used (see (3.31)). A general outline of the essential features of the method is given below.

Let x^k be an iterate for which $\nabla h(x^k)$ has full rank and let Z_k and Y_k be matrices whose columns form bases for the null space of $\nabla h(x^k)^t$ and range space of $\nabla h(x^k)$, respectively. Also assume that the columns of Z_k are orthogonal. Z_k and Y_k could be obtained, for example, by a QR factorization of $\nabla h(x^k)$.

Definition 3 Let (x^k, u^k) be a given solution–multiplier pair and assume that $\nabla h(x^k)$ has full rank. The matrix

$$Z_k{}^t H \mathcal{L}(x^k, u^k) Z_k$$

is called a **reduced Hessian** for the Lagrangian function at (x^k, u^k).

The reduced Hessian is not unique; its form depends upon the choice of basis for the null space of $\nabla h(x^k)^t$. Since by **A4** a reduced Hessian at the solution is positive definite, it follows that if (x^k, u^k) is close enough to (x^*, u^*) then the reduced Hessian at (x^k, u^k) is positive definite. Decomposing the vector d_x as

$$d_x = Z_k p_Z + Y_k p_Y \tag{3.27}$$

it can be seen that the constraint equation of (ECQP) becomes

$$\nabla h(x^k)^t Y_k p_Y = -h(x^k),$$

which by virtue of **A2** can be solved to obtain

$$p_Y = -[\nabla h(x^k)^t Y_k]^{-1} h(x^k). \tag{3.28}$$

The minimization problem (ECQP) is now an unconstrained problem in $n - m$ variables given by

$$\underset{p_Z}{\text{minimize}} \quad \tfrac{1}{2} p_Z{}^t Z_k{}^t B_k Z_k p_Z + (\nabla f(x^k)^t + p_Y{}^t B_k) Z_k p_Z.$$

The matrix in this unconstrained problem is an approximation to the reduced Hessian of the Lagrangian at x^k. Rather than update B_k and then compute $Z_k{}^t B_k Z_k$, the reduced matrix itself is updated. Thus at a particular iteration, given a positive definite $(n - m) \times (n - m)$ matrix R_k, an iterate x^k and a multiplier estimate u^k, the new iterate can be found by first computing p_Y from (3.28), setting

$$p_Z = -R_k^{-1} Z_k{}^t (\nabla f(x^k) + B_k p_Y), \tag{3.29}$$

and setting $x^{k+1} = x^k + d_x$, where d_x is given by (3.27). A new multiplier, u^{k+1}, is generated and the reduced Hessian approximation is updated using the BFGS formula (3.23) with the R_k replacing B_k,

$$s = Z_k{}^t (x^{k+1} - x^k) = Z_k{}^t Z_k p_Z \tag{3.30}$$

and

$$y = Z_k{}^t [\nabla \mathcal{L}(x^k + Z_k p_Z, u^k) - \nabla \mathcal{L}(x^k, u^k)]. \tag{3.31}$$

The choices of s and y are motivated by the fact that only the reduced Hessian is being approximated.

This method does not stipulate a new multiplier iterate directly since the problem being solved at each step is unconstrained. However, the least squares solution for the first order conditions (cf. (3.2)) can be used. Generally, all that is needed is that the multipliers satisfy

$$\left\| u^k - u^* \right\| = \mathcal{O}\left(\left\| x^k - x^* \right\| \right). \tag{3.32}$$

Since $\mathcal{P}^k = Z_k Z_k{}^{\mathrm{t}}$ the approximation

$$Z_k R_k Z_k{}^{\mathrm{t}}$$

can be thought of as an approximation of

$$\mathcal{P}^k H \mathcal{L}(x^k, u^k) \mathcal{P}^k.$$

Thus since this method does not approximate

$$\mathcal{P}^k H \mathcal{L}^*$$

neither the local convergence theorem nor the superlinear-rate-of-convergence theorem, Theorems 2 and 4, follow as for full Hessian approximations. Nevertheless, the two-sided approximation of the Hessian matrix suggests that the conditions of Theorem 3.5 may hold. In fact, if it is assumed that the matrices Z_k are chosen in a smooth way, that is, so that

$$\left\| Z_k - Z(x^*) \right\| = \mathcal{O}\left(\left\| x^k - x^* \right\| \right), \tag{3.33}$$

the assumption of local convergence leads to two-step superlinear convergence.

Theorem 8   *Assume that the reduced Hessian algorithm is applied with u^k and Z_k chosen so that (3.32) and (3.33) are satisfied. If the sequence $\{x^k\}$ converges to x^* R-linearly then $\{R_k\}$ and $\{R_k^{-1}\}$ are uniformly bounded and $\{x^k\}$ converges two-step superlinearly.*

While the condition on the multiplier iterates is easy to satisfy, some care is required to ensure that (3.33) is satisfied as it has been observed that arbitrary choices of Z_k may be be discontinuous and hence invalidate the theorem.

3.5. *Notes and References*

The study of Newton's method applied to the first order necessary conditions for constrained optimization can be found in Tapia (1974) and Goodman (1985). The equivalence of Newton's method applied to the system of first order equations and feasibility conditions and the SQP step with the Lagrangian Hessian was first mentioned by Tapia (1978).

The expository paper Dennis and Moré (1977) is a good source for an introductory discussion of bounded deterioration and of updating methods satisfying the secant condition in the setting of unconstrained optimization.

The paper by Broyden, Dennis and Moré (1973) provides the basic results on the convergence properties for quasi-Newton methods.

Theorems (2) and (4) were first proven by Boggs, Tolle and Wang (1982). The latter theorem was proven under the assumption that the iterates converge linearly, generalizing the result for unconstrained optimization given in Dennis and Moré (1977). The assumption was substantially weakened by several authors; see, for example, Fontecilla, Steihaug and Tapia (1987). Theorem (5) is due to Powell (1978b). The paper by Coleman (1990) contains a good general overview of superlinear convergence theorems for SQP methods.

The (local) superlinear convergence of the PSB-SQP method was proven by Han (1976) in his first paper on SQP methods. This paper also included the proof of superlinear covergence for the positive definite updates when the Hessian of the Lagrangian is positive definite (Theorem (7)) and suggested the use of the augmenting term in the general case. The augmented method was also investigated in Schittkowski (1981) and Tapia (1977). A description of the Powell-SQP method appears in Powell (1978b) while the SALSA-SQP method is introduced in Tapia (1988) and developed further in Byrd, Tapia and Zhang (1992).

The reduced Hessian SQP method has been the subject of research by a number of authors. The presentation here follows that of Byrd and Nocedal (1991). Other important articles include Coleman and Conn (1984), Nocedal and Overton (1985), Gabay (1982), Yuan (1985) and Gilbert (1993). The problems involved in satisfying (3.33) have been addressed in Byrd and Schnabel (1986), Coleman and Sorensen (1984) and Gill et al. (1985).

Other papers of interest on local convergence for SQP methods include Schittkowski (1983), Panier and Tits (1993), Bertsekas (1980), Coleman and Feynes (1992), Glad (1979), Fukushima (1986) and Fontecilla (1988).

4. Merit Functions and Global Convergence

In the previous section we demonstrated the existence of variants of the SQP method that are rapidly locally convergent. Here we show how the merit function assures that the iterates eventually get close to a critical point; in the next section we point out some gaps in the theory that prevent a complete analysis.

A *merit function* ϕ is incorporated into an SQP algorithm for the purpose of achieving global convergence. As described in Section 2.3, a line-search procedure is used to modify the length of the step d_x so that the step from x^k to x^{k+1} reduces the value of ϕ. This reduction is taken to imply that acceptable progress towards the solution is being made.

The standard way to ensure that a reduction in ϕ indicates progress is to construct ϕ so that the solutions of (NLP) are the unconstrained minimizers of ϕ. Then it must be possible to decrease ϕ by taking a step in the direction d_x generated by solving the quadratic subproblem, that is, d_x must be a descent direction for ϕ. If this is the case, then for a sufficiently small α, $\phi(x^k + \alpha d_x)$ will be less than $\phi(x^k)$. An appropriate steplength that decreases ϕ can then be computed; for example, by a 'backtracking' procedure of trying successively smaller values of α until a suitable one is obtained.

Given that a merit function is found that has these properties and that a procedure is used to take a step that decreases the function, global convergence proofs for the resulting SQP algorithm are generally similar to those found in unconstrained optimization: they depend on proving that the limit points of the x-iterates are critical points of ϕ. These proofs rely heavily on being able to guarantee that a 'sufficient' decrease in ϕ can be achieved at each iteration.

In unconstrained minimization there is a natural merit function, namely, the objective function itself. In the constrained setting, unless the iterates are always feasible, a merit function has to balance the drive to decrease f with the need to satisfy the constraints. This balance is often controlled by a parameter in ϕ that weights a measure of the infeasibility against the value of either the objective function or the Lagrangian function. In this section, we illustrate these ideas by describing two of the most popular merit functions: a differentiable augmented Lagrangian function and the simpler, but nondifferentiable, ℓ_1 penalty function. We derive some of the important properties of these functions and discuss some of the advantages and disadvantages of each. In addition, we provide appropriate rules for choosing the steplength parameter so as to obtain the basic convergence results.

To simplify the presentation we restrict our attention at the beginning to the problem with only equality constraints. Extensions that allow inequality constraints are discussed at the ends of the sections.

As might be expected, some additional assumptions are needed if global convergence is to be achieved. In fact, the need for these assumptions raises the issue as to what exactly is meant by 'global' convergence. It is impossible to conceive of an algorithm that would converge from any starting point for every (NLP). The very nature of nonlinearity allows the possibility, for example, of perfectly satisfactory iterations following a steadily decreasing function towards achieving feasibility and a minimum at some infinite point. In order to focus attention on the methods and not the problem structure we make the following assumptions.

C1: The starting point and all succeeding iterates lie in some compact set $\mathcal{C}$.

C2: The columns of $\nabla h(x)$ are linearly independent for all $x \in \mathcal{C}$.

The first assumption is made in some guise in almost all global convergence proofs, often by making specific assumptions about the functions. The second assumption ensures that the systems of linearized constraints are consistent. An additional assumption will be made about the matrix approximations B_k, depending on the particular merit function.

4.1. Augmented Lagrangian merit functions

Our first example of a merit function is the augmented Lagrangian function. There are several versions of this function; we use the following version to illustrate the class:

$$\phi_F(x; \eta) = f(x) + h(x)^t \bar{u}(x) + \frac{\eta}{2} \|h(x)\|_2^2, \tag{4.1}$$

where η is a constant to be determined and

$$\bar{u}(x) = - \left[\nabla h(x)^t \nabla h(x)\right]^{-1} \nabla h(x)^t \nabla f(x). \tag{4.2}$$

The multipliers $\bar{u}(x)$ defined by (4.2) are the least squares estimates of the optimal multipliers based on the first order necessary conditions and hence $\bar{u}(x^*) = u^*$ (see (3.1) and (3.2)). Under the assumptions, ϕ_F and $\bar{u}$ are differentiable and ϕ_F is bounded from below on C for η sufficiently large. The following formulae will be useful in the discussion:

$$\begin{aligned} \nabla \bar{u}(x) &= -H\mathcal{L}(x, \bar{u}(x))\nabla h(x) \left[\nabla h(x)^t \nabla h(x)\right]^{-1} \\ &\quad + R_1(x), \end{aligned} \tag{4.3}$$

$$\begin{aligned} \nabla \phi_F(x; \eta) &= \nabla f(x) + \nabla h(x)\bar{u}(x) \\ &\quad + \nabla \bar{u}(x)h(x) + \eta \nabla h(x)h(x). \end{aligned} \tag{4.4}$$

The function $R_1(x)$ in (4.3) is bounded and satisfies $R_1(x^*) = 0$ if x^* satisfies the first order necessary conditions.

The following theorem establishes that the augmented Lagrangian merit function has the essential properties of a merit function for the equality-constrained nonlinear program.

Theorem 9 Assume **B1, B2, C1** and **C2** are satisfied. Then for η sufficiently large the following properties hold:

(i) $x^* \in C$ is a strong local minimum of ϕ_F if and only if x^* is a strong local minimum of (NLP) and

(ii) if x is not a critical point of (NLP) then d_x is a descent direction for ϕ_F.

The proof is rather technical, but it does illustrate the techniques that are often used in such arguments and it provides a useful intermediate result, namely, (4.10). A key idea is to decompose certain vectors according to the range- and null-space projections $\mathcal{P}$ and $\mathcal{Q}$.

Proof. If $x^* \in C$ is feasible, and satisfies the first order necessary conditions, then it follows from (4.4) that $\nabla \phi_F(x^*, \eta) = 0$. Conversely, if $x^* \in C$, $\nabla \phi_F(x^*; \eta) = 0$ and η is sufficiently large then it follows from **C1** and **C2** that $h(x^*) = 0$ and x^* satisfies **A1** for (NLP). To establish (i), the relation between $H\phi_F(x^*; \eta)$ and HL^* must be explored for such points x^*. It follows from (4.4) that

$$H\phi_F(x^*; \eta) = HL^* - Q^* HL^* - HL^* Q^* + \eta \nabla h(x^*) \nabla h(x^*)^t, \qquad (4.5)$$

where Q^* is defined by (2.3). Now let $y \in \mathcal{R}^n$, $y \neq 0$, be arbitrary. Then setting $y = Q^* y + P^* y$, (4.5) yields

$$
\begin{aligned}
y^t H\phi_F(x^*; \eta) y &= (P^* y)^t HL^* (P^* y) - (Q^* y)^t HL^* (Q^* y) \\
&\quad + \eta (Q^* y)^t \left[\nabla h(x^*) \nabla h(x^*)^t \right] (Q^* y).
\end{aligned}
\qquad (4.6)
$$

Assume that x^* is a strong local minimum of (NLP). Since $Q^* y$ is in the range space of $\nabla h(x^*)$, it follows from **A2** that there exists a constant μ such that

$$(Q^* y)^t \left[\nabla h(x^*) \nabla h(x^*)^t \right] (Q^* y) \geq \mu \| Q^* y \|^2.$$

Let the maximum eigenvalue of HL^* in absolute value be σ_{max} and let σ_{min} be the minimum eigenvalue of $P^* HL^* P^*$, which by **A4** is positive. Defining

$$\epsilon = \frac{\| Q^* y \|}{\| y \|},$$

which implies

$$\frac{\| P^* y \|^2}{\| y \|^2} = 1 - \epsilon^2,$$

and dividing both sides of (4.6) by $\| y \|^2$ gives

$$\frac{y^t H\phi_F(x^*; \eta) y}{\| y \|^2} \geq \sigma_{min} + (\eta\mu - \sigma_{max} - \sigma_{min})\epsilon^2.$$

This last expression will be positive for η sufficiently large, which implies that x^* is a strong local minimum of ϕ_F. Conversely, if x^* is a strong local minimum of ϕ_F then (4.6) must be positive for all y. This implies that HL^* must be positive definite on the null space of $\nabla h(x^*)^t$ and hence that x^* is a strong local minimum of (NLP). This establishes (i).

To show that the direction d_x is always a descent direction for ϕ_F, the inner product of both sides of (4.4) is taken with d_x to yield

$$
\begin{aligned}
d_x{}^t \nabla \phi_F(x^k; \eta) &= d_x{}^t \nabla f(x^k) + d_x{}^t \nabla h(x^k) \bar{u}(x^k) \\
&\quad + d_x{}^t \nabla \bar{u}(x^k) h(x^k) + \eta d_x{}^t \nabla h(x^k) h(x^k).
\end{aligned}
\qquad (4.7)
$$

Using (4.2) it follows that

$$
\begin{aligned}
d_x{}^t \nabla h(x^k) \bar{u}(x^k) &= -d_x{}^t \nabla h(x^k) \left[\nabla h(x)^t \nabla h(x) \right]^{-1} \nabla h(x)^t \nabla f(x) \\
&= -d_x{}^t Q^k \nabla f(x^k).
\end{aligned}
$$

(4.8)

Writing $d_x = Q^k d_x + P^k d_x$ and noting that $h(x^k) = -\nabla h(x^k)^t Q^k d_x$ results in

$$
\begin{aligned}
d_x{}^t \nabla \phi_F(x^k; \eta) &= P^k d_x{}^t \nabla f(x^k) - d_x{}^t \nabla \bar{u}(x^k) \nabla h(x^k)^t Q^k d_x \\
&\quad - \eta (Q^k d_x)^t \left[\nabla h(x^k) \nabla h(x^k)^t \right] (Q^k d_x).
\end{aligned}
$$

Finally, using the first order necessary conditions for (ECQP) and the fact that $P^k \nabla h(x^k)^t = 0$, the following is obtained:

$$
\begin{aligned}
d_x{}^t \nabla \phi_F(x^k; \eta) &= -(P^k d_x)^t B_k (P^k d_x) - (Q^k d_x)^t B_k d_x \\
&\quad - d_x{}^t \nabla \bar{u}(x^k) \nabla h(x^k)^t Q^k d_x \\
&\quad - \eta (Q^k d_x)^t \left[\nabla h(x^k) \nabla h(x^k)^t \right] (Q^k d_x).
\end{aligned}
$$

(4.9)

Now, from (4.3) and assumptions **C1** and **C2** it follows that

$$
d_x{}^t \nabla \bar{u}(x^k) \nabla h(x^k)^t Q^k d_x \le \gamma_1 \left\| Q^k d_x \right\| \| d_x \|
$$

for some constant γ_1. Dividing both sides of (4.9) by $\| d_x \|^2$, letting

$$
\epsilon = \frac{\left\| Q^k d_x \right\|}{\| d_x \|}
$$

and using **B1** yields

$$
\frac{\nabla \phi_F(x^k; \eta)^t d_x}{\| d_x \|^2} \le -\beta_1 + \gamma_2 \epsilon - \eta \gamma_3 \epsilon^2
$$

(4.10)

for constants γ_2 and γ_3. The quantity on the left of (4.10) is then negative and can be uniformly bounded away from zero provided η is sufficiently large, thus proving d_x is a descent direction for ϕ. $\square$

It is interesting to observe that the value of η necessary to obtain descent of d_x depends on the eigenvalue bounds on $\{B_k\}$, whereas the value of η necessary to ensure that x^* is a strong local minimizer of ϕ_F depends on the eigenvalues of $H\mathcal{L}^*$. Thus a strategy to adjust η to achieve a good descent direction may not be sufficient to prove that $H\phi_F(x^*)$ is positive definite. We comment further on this below.

This line of reasoning can be continued to obtain a more useful form of the descent result. From (4.4) and arguments similar to those above,

$$
\left\| \nabla \phi_F(x^k; \eta) \right\| \le \gamma_4 \| d_x \|
$$

for some constant γ_4. Thus from (4.10) it follows that there exists a constant γ_5 such that

$$\frac{\nabla\phi_F(x^k;\eta)^t d_x}{\left\|\nabla\phi_F(x^k;\eta)\right\|\left\|d_x\right\|} \leq -\gamma_5 < 0 \tag{4.11}$$

for all k. This inequality ensures that the cosine of the angle θ_k between the direction d_x and the negative gradient of ϕ_F is bounded away from zero, that is, for all k

$$\cos(\theta_k) = -\frac{\nabla\phi_F(x^k;\eta)^t d_x}{\left\|\nabla\phi_F(x^k;\eta)\right\|\left\|d_x\right\|} \geq \gamma_5 > 0. \tag{4.12}$$

This condition is sufficient to guarantee convergence of the sequence if it is coupled with suitable line-search criteria that impose restrictions on the steplength α. For example, the Wolfe conditions require a steplength α to satisfy

$$\phi_F(x^k + \alpha d_x;\eta) \leq \phi_F(x^k;\eta) + \sigma_1\alpha, \nabla\phi_F(x^k;\eta)^t d_x, \tag{4.13}$$
$$\nabla\phi_F(x^k + \alpha d_x;\eta)^t d_x \geq \sigma_2\nabla\phi_F(x^k;\eta)^t d_x, \tag{4.14}$$

where $0 < \sigma_1 < \sigma_2 < 1$. Inequality (4.13) ensures that there will be a sufficient reduction in ϕ_F while (4.14) guarantees that the steplength α will not be too small. An important property of these conditions is that if d_x is a descent direction for ϕ_F, then a steplength α satisfying (4.13)–(4.14) can always be found. Furthermore, the reduction in ϕ_F for such a step satisfies

$$\phi_F(x^{k+1};\eta) \leq \phi_F(x^k;\eta) - \gamma_6 \cos^2(\theta_k)\left\|\nabla\phi_F(x^k;\eta)\right\|^2,$$

for some positive constant γ_6. Therefore

$$\sum_{k=1}^{\infty} \cos^2(\theta_k)\left\|\nabla\phi_F(x^k;\eta)\right\|^2 < \infty \tag{4.15}$$

and, since $\cos(\theta_k)$ is uniformly bounded away from 0, it follows that

$$\lim_{k\to\infty} \nabla\phi_F(x^k;\eta) = 0.$$

This implies that $\{x^k\}$ converges to a critical point of ϕ_F. The following theorem results:

Theorem 10 *Assume that η is chosen such that (4.12) holds. Then the SQP algorithm with steplength α chosen to satisfy the Wolfe conditions (4.13)–(4.14) is globally convergent to a critical point of ϕ_F.*

If the critical points of ϕ_F all correspond to local minima of (NLP) then the algorithm will converge to a local solution. This, however, can rarely be guaranteed. Only convergence to a critical point, and not to a local minimum of ϕ_F, is guaranteed. It is easy to see from the following example that convergence can occur to a local maximum of (NLP). Consider

$$\begin{aligned}\underset{x}{\text{minimize}} \quad & x_1 \\ \text{subject to:} \quad & x_1^2 + x_2^2 - 1 = 0\end{aligned} \qquad (4.16)$$

with a starting approximation of $(2,0)$ and $B_k = I$ for all k. The iterates will converge from the right to the point $(1,0)$, which is a local maximum of the problem, but ϕ_F will decrease appropriately at each iteration. The point $(1,0)$, however, is a saddle point of ϕ_F. In practice convergence to a maximizer of (NLP) rarely occurs.

Theorems 9 and 10 require that η be sufficiently large. Since it is not known in advance how large η needs to be, it is necessary to employ an adaptive strategy for adjusting η when designing a practical code. Such adjustment strategies can have a dramatic effect on the performance of the implementation, as is discussed in Section 7.

Augmented Lagrangian merit functions have been extended to handle inequality constrained problems in several ways. Two successful approaches are described below.

In the first an *active set* strategy is employed, that is, at each iteration a set of the inequality constraints is selected and treated as if they were equality constraints; the remaining inequalities are handled differently. The active set is selected by using the multipliers from (QP). In particular,

$$\mathcal{I}_k = \left\{ i \ : \ g_i(x^k) \geq -\frac{(v_{\text{qp}})_i}{\eta} \right\}.$$

With this choice, $\mathcal{I}_k$ will contain all unsatisfied constraints and no 'safely satisfied' ones. The merit function at x^k is defined by

$$\begin{aligned}\phi_{FI}(x, u_{\text{qp}}, v_{\text{qp}}; \eta) \ = \ & f(x) + h(x)^t u_{\text{qp}}(x) + \tfrac{1}{2}\eta \, \|h(x)\|^2 \\ & + \sum_{i \in \mathcal{I}_k} \left(g_i(x)(v_{\text{qp}}(x))_i + \tfrac{1}{2}\eta g_i(x)^2 \right) \\ & + \frac{1}{2\eta} \sum_{i \notin \mathcal{I}_k} (v_{\text{qp}}(x))_i^2 .\end{aligned}$$

ϕ_{FI} is still differentiable and, as a result of **A3**, the correct active set is eventually identified in a neighbourhood of the solution. In this formulation, the multipliers are the multipliers from (QP), not the least squares multipliers. Therefore ϕ_{FI} is a function of both x and these multipliers and, consequently, the analysis for ϕ_{FI} is somewhat more complicated.

A second approach uses the idea of squared slack variables. One can consider the problem

$$\begin{aligned}
\underset{x,\,t}{\text{minimize}} \quad & f(x) \\
\text{subject to:} \quad h(x) &= 0, \\
g_i(x) + (t_i)^2 &= 0, \qquad i = 1, \dots, p,
\end{aligned} \qquad (4.17)$$

where t is the vector of slack variables. This problem is equivalent to (NLP) in the sense that both have a strong local solution at x^* where, in (4.17), $(t_i^*)^2 = -g_i(x^*)$. By writing out the function ϕ_F for the problem (4.17), it is observed that the merit function only involves the squares of t_i. Thus, by letting $z_i = t_i^2$ and setting

$$\overline{h}(x, z) = \left(\begin{array}{c} h(x) \\ g(x) + z \end{array} \right),$$

we can construct the merit function

$$\phi_{FZ}(x, z) = f(x) + \overline{h}(x, z)^{\text{t}} \overline{u}(x, z) + \frac{\eta}{2} \left\| \overline{h}(x, z) \right\|^2.$$

Here $\overline{u}(x,z)$ is the least squares estimate for all of the multipliers. Note that (4.17) is not used to create the quadratic subproblem, but rather (QP) is solved at each iteration to obtain the step d_x. The slack variables can then be updated at each iteration in a manner guaranteed to maintain the nonnegativity of z. For example, a step

$$d_z = -(\nabla g(x^k)^{\text{t}} d_x' + g(x^k) + z^k)$$

can be calculated. Then the constraints of (QP) imply that $z^{k+1} = z^k + \alpha d_z \geq 0$ if $z^k \geq 0$ and $\alpha \in (0, 1]$.

4.2. The ℓ_1 Merit Function

One of the first merit functions to be introduced was the ℓ_1 exact penalty function that, in the equality-constrained case, is given by

$$\phi_1(x; \rho) = f(x) + \rho \|h(x)\|_1, \qquad (4.18)$$

where ρ is a positive constant to be chosen. The properties of this function vis-à-vis the equality-constrained optimization problem have been well documented in the literature. For our purposes it is sufficient to note that ϕ_1, like the augmented Lagrangian of the previous section, is an 'exact' penalty function; that is, there exists a positive ρ^* such that for all $\rho \geq \rho^*$, an unconstrained minimum of ϕ_1 corresponds to a solution of (NLP). Note that ϕ_1 is not differentiable on the feasible set. If the penalty term were squared to achieve differentiability then the 'exact' property would be lost; minimizers of ϕ_1 would only converge to solutions of (NLP) as $\rho \to \infty$.

Although ϕ_1 is not differentiable, it does have a directional derivative along d_x. It can be shown that this directional derivative, denoted by the operator D, is given by

$$D(\phi_1(x^k; \rho); d_x) = \nabla f(x^k)^t d_x - \rho \left\| h(x^k) \right\|_1 . \tag{4.19}$$

Substituting the first order necessary conditions for (ECQP) into (4.19) yields

$$D(\phi_1(x^k; \rho); d_x) = -d_x{}^t B_k d_x - d_x{}^t \nabla h(x^k) u_{qp} - \rho \left\| h(x^k) \right\|_1 .$$

It follows from the linearized constraints of (ECQP) that

$$d_x{}^t \nabla h(x^k) u_{qp} = -h(x^k)^t u_{qp}$$

and, since

$$h(x^k)^t u_{qp} \leq \left\| u_{qp} \right\|_\infty \left\| h(x^k) \right\|_1 , \tag{4.20}$$

the inequality

$$D(\phi(x^k; \rho); d_x) \leq -d_x{}^t B_k d_x - (\rho - \left\| u_{qp} \right\|_\infty) \left\| h(x^k) \right\|_1 \tag{4.21}$$

is obtained. In order to have d_x be a descent direction for ϕ_1 and to obtain a convergence theorem it is sufficient to assume the uniform positive definiteness of $\{B_k\}$:

B4: For all $d \in \mathcal{R}^n$ there are positive constants β_1 and $\beta_2 > 0$ such that

$$\beta_1 \left\| d \right\|^2 \leq d^t B_k d \leq \beta_2 \left\| d \right\|^2$$

for all k.

The assumptions **B1** and **B2** that were made for the augmented Lagrangian are sufficient, but we make the stronger assumption to simplify the presentation. Using **B4**, the first term in (4.21) is always negative and thus d_x is a guaranteed descent direction for ϕ_1 if $\rho > \left\| u_{qp} \right\|_\infty$.

Global convergence of an algorithm that uses ϕ_1 as a merit function can be demonstrated using arguments similar to those for Theorem 9. First, at each iteration, the parameter ρ is chosen by

$$\rho = \left\| u_{qp} \right\|_\infty + \bar{\rho} \tag{4.22}$$

for some constant $\bar{\rho} > 0$. Next, the first Wolfe line-search condition, (4.13), is replaced by

$$\phi_1(x^k + \alpha d_x) \leq \phi_1(x^k) + \zeta \alpha D(\phi_1(x^k; \rho); d_x), \tag{4.23}$$

where $\zeta \in (0, \frac{1}{2})$. (The condition $\zeta < \frac{1}{2}$ is for technical reasons; in practice ζ is usually chosen to be much smaller.) Recall that the second of the Wolfe conditions, (4.14), is to prevent a steplength that is too small. The assumptions that have been made guarantee that the use of a backtracking line-search procedure produces steplengths that are uniformly bounded away from zero for all iterations. Denoting this lower bound by $\bar{\alpha}$, it now follows from (4.21) and (4.23) that the reduction in ϕ_1 at each step satisfies

$$\phi_1(x^k + \alpha d_x) - \phi_1(x^k) \leq -\zeta\bar{\alpha}\left[\mu_1\|d_x\|^2 + \bar{\rho}\|h(x^k)\|_1\right]. \qquad (4.24)$$

Assumption **C1** and (4.24) imply that

$$\sum_{k=1}^{\infty}\left[\|d_x(x^k)\|^2 + \|h(x^k)\|_1\right] \leq \infty$$

and therefore $\left[\|d_x(x^k)\|^2 + \|h(x^k)\|_1\right] \to 0$ as $k \to \infty$. Since $d_x = 0$ if and only if x^k is a feasible point satisfying **A1**, the following theorem results.

Theorem 11   *Assume that ρ is chosen such that (4.22) holds. Then the SQP algorithm started at any point x^0 with steplength $\alpha \geq \bar{\alpha} > 0$ chosen to satisfy (4.23) converges to a stationary point of ϕ_1.*

As in the case of ϕ_F convergence to a local minimum of (NLP) cannot be guaranteed. In fact, the same counterexample used in that case applies here.

An advantage of ϕ_1 over the augmented Lagrangian is that ϕ_1 is easily extended to handle inequalities. Since differentiability is not an issue in this case, the function ϕ_1 can simply be defined as

$$\phi_1(x) = f(x) + \rho\left[\|h(x)\|_1 + \|g^+(x)\|_1\right],$$

where

$$g_i^+(x) = \begin{cases} 0 & \text{if } g_i(x) \leq 0, \\ g_i(x) & \text{if } g_i(x) > 0. \end{cases}$$

All of the theoretical results continue to hold under this extension.

The merit function ϕ_1 has been popular because of its simplicity – it requires only the evaluation of f, h and g to check a prospective point. ϕ_F, on the other hand, is expensive to evaluate in that it requires the evaluation of f, h, g and their gradients just to check a prospective point. In addition, in the large scale case, the evaluation of $\bar{u}(x)$ involves nontrivial algebra. Implementations that use these merit functions partially circumvent this difficulty by using an approximation to ϕ_F, for example, by approximating $\bar{u}$ by a linear function or by keeping it fixed over the current step.

4.3. Notes and References

The augmented Lagrangian merit function was first proposed as an exact penalty function by Fletcher (1972). See also Bertsekas (1982) and Boggs and Tolle (1980). It was suggested as a merit function in Powell and Yuan (1986) and in a slightly different form in Boggs and Tolle (1984) and Boggs and Tolle (1989). See Byrd and Nocedal (1991) for its application to reduced Hessian methods. The Wolfe conditions have been studied by numerous authors; we recommend Nocedal (1992) or Dennis and Schnabel (1983).

Schittkowski (1981) and Schittkowski (1983) developed the form for inequalities given by ϕ_{FI}. This form has been further developed and incorporated into a highly successful algorithm by Gill et al. (1986). Boggs, Tolle and Kearsley (1991) proposed the form given by ϕ_{FZ}.

The ℓ_1 exact penalty function was originally suggested as a merit function by Han (1977) where he obtained the first global convergence results for SQP methods. See Fletcher (1981) for a good introduction to nondifferentiable optimization and the use of this function. See also Polak (1989) and Wolfe (1975). Byrd and Nocedal (1991) study the ℓ_1 merit function in the context of reduced Hessian methods. The analysis for the ℓ_1 merit function applies to the corresponding ℓ_p merit function with the only change ocurring in (4.20), where the ∞-norm and the 1-norm are replaced by the p-norm and the q-norm with $1/p + 1/q = 1$.

5. Global to Local Behaviour

In the previous two sections we have examined conditions that imply that the basic SQP algorithm will converge. Section 3 dealt with local convergence where the emphasis was on the rates of convergence, while Section 4 was concerned with obtaining convergence from remote starting points, the implicit hope being that the two theories would come together to produce a unified theory that would be applicable to a given algorithm. For the global convergence theories, where the process has to be controlled by a merit function, it was seen in Section 4 that convergence of $\{x^k\}$ to a critical point of ϕ is all that can be demonstrated. Assuming that this critical point is a solution of (NLP), the question that arises is whether or not the conditions for the local convergence theories are eventually satisfied by this sequence. If they are then the more rapid rates of local convergence can be achieved. In this section we discuss three questions concerning this possibility: will the correct active set be chosen by (QP) when x^k is close to x^*; will B_k eventually approximate HL^* in one of the ways stipulated in Section 3 that yields rapid local convergence; and will the merit function allow a steplength of one if the underlying process can achieve rapid local convergence?

Recall that the local convergence theory was proven for equality-constrained problems only. This was based on the assumption that the active inequality constraints at the solution of (NLP) would be identified by (SQP) when x^k gets close to the solution. The question of when the active constraints for (QP) will be the same as those for (NLP) is resolved by using a

perturbation theory result as follows. Consider (NLP) with only inequality
constraints

$$\underset{x}{\text{minimize}} \quad f(x)$$
$$\text{subject to: } g(x) \leq 0 \tag{5.1}$$

and the quadratic program with the same solution

$$\underset{x}{\text{minimize}} \quad \nabla f(x^*)^t(x - x^*) + \tfrac{1}{2}(x - x^*)^t B(x - x^*)$$
$$\text{subject to: } \nabla g(x^*)^t(x - x^*) + g(x^*) \leq 0, \tag{5.2}$$

where the only restriction on B is that

$$y^t B y > 0 \tag{5.3}$$

for all y such that $\nabla g(x^*)^t y = 0$. It is easily verified that (x^*, v^*) is a
strong local minimizer of both (5.1) and (5.2). This implies that the active
sets are the same and that strict complementary slackness (assumption **A3**)
holds for both. The quadratic programming approximation to (5.1) is

$$\underset{d_x}{\text{minimize}} \quad \nabla f(x^k)^t d_x + \tfrac{1}{2} d_x{}^t B_k d_x$$
$$\text{subject to: } \nabla g(x^k)^t d_x + g(x^k) \leq 0. \tag{5.4}$$

A standard perturbation argument now asserts that if x^k is close enough to
x^* and B_k is close enough to B, then the active set for (5.4) is the same as
the active set for (5.2). It follows that the active sets for (5.1) and (5.4) are
the same. As a result, if x^k is sufficiently close to x^* and B_k is sufficiently
close to any matrix B satisfying (5.3) then (5.4) will identify the correct
active set. The condition (5.3) will be satisfied if the B_k are always positive
definite or if they satisfy the condition

$$\lim_{k \to \infty} \mathcal{P}^k(B_k - H\mathcal{L}^*)\mathcal{P}^k = 0. \tag{5.5}$$

This latter condition implies two-step superlinear convergence by Theorem 5.

From Section 3 linear, superlinear or two-step superlinear convergence is
ensured if B_k approaches $H\mathcal{L}^*$ in one of the ways hypothesized in Theorems
2–5. Since the initial matrix is unlikely in practice to be a good approximation to $H\mathcal{L}^*$, it is reasonable to ask under what conditions the subsequent B_k
will be good enough approximations to $H\mathcal{L}^*$ for these results of Section 3 to
hold. If the B_k satisfy the secant equation (3.19) then it is reasonable to expect that the B_k will converge to $H\mathcal{L}^*$ provided that the directions, d_x, span
$\mathcal{R}^n$ repeatedly, for example, if each set of n directions is (uniformly) linearly
independent. Some recent research supports this expectation. However, if
positive definiteness is maintained, it is not possible for the B_k to converge to
$H\mathcal{L}^*$ unless the latter is at least positive semidefinite. So for the general case,

further analysis is necessary. As seen in Section 3, (3.16) is equivalent to

$$\lim_{k \to \infty} \left\{ \frac{\mathcal{P}^*(B_k - H\mathcal{L}^*)\mathcal{P}^* s_k}{\|s_k\|} + \frac{\mathcal{P}^*(B_k - H\mathcal{L}^*)\mathcal{Q}^* s_k}{\|s_k\|} \right\} = 0, \qquad (5.6)$$

where $s_k = (x^{k+1} - x^k)$. For superlinear convergence both of these terms must tend to zero if the directions $s_k/\|s_k\|$ repeatedly span $\mathcal{R}^n$. Thus for positive definite B_k superlinear convergence is not likely unless the second term goes to zero, which is the type of tangential convergence mentioned in Section 3.2. However, two-step superlinear convergence, in which the first term of (5.6) goes to zero, is possible for positive definite updates and for reduced Hessian updates as well, provided that convergence occurs and steplengths of one are acceptable. Thus, it is important from the point of view of obtaining a rapid local rate of convergence that a steplength of one be taken near x^*.

For the general algorithm, however, the use of the merit function with the line search continues throughout the course of the algorithm. Thus, the line search for the merit function should allow a steplength of one if the matrix approximations are such that rapid local convergence is possible. In the case of the augmented Lagrangian merit function it can be shown, using arguments similar to those in Section 4, that

$$\frac{\phi_F(x^{k+1}) - \phi_F(x^k)}{\|d_x\|^2} \leq -\tfrac{1}{2}\beta_1 + \gamma_2 \epsilon - \eta \gamma_3 \epsilon^2$$
$$+ \frac{\mathcal{P}^k[B_k - H\mathcal{L}(x^k, \overline{u}(x^k))]d_x}{\|d_x\|} \qquad (5.7)$$
$$+ \mathcal{O}(\|d_x\|) .$$

Then, if the process is converging superlinearly, the penultimate term in (5.7) tends to zero by Theorem 4 and the right-hand side of (5.7) is ultimately less than zero, thus showing that a steplength of one decreases ϕ_F. A slight extension to this argument shows that the Wolfe condition (4.13) also can be satisfied for a steplength of one.

A significant disadvantage of the merit function ϕ_1 is that it may not allow a steplength of $\alpha = 1$ near the solution, no matter how good the approximation of (QP) to (NLP). This phenomenon, which can prohibit superlinear convergence, is called the Maratos effect. Several procedures have been proposed to overcome this problem, including so-called nonmonotone line searches, that is, techniques that allow the merit function to increase over one or more iterations (see Section 7).

5.1. Notes and References

The perturbation argument showing conditions under which (QP) has the same active set as (NLP) is derived from the work of Robinson (1974) who proves a general perturbation result. For a discussion of the convergence of matrix updates see Boggs and Tolle (1994), Ge and Powell (1983) and Stoer (1984).

The Maratos effect, that is, the fact that the ℓ_1 merit function may not allow a steplength of one even when the iterates are close to the solution and B_k is a good approximation to HL^*, was discussed by Chamberlain et al. (1982). These authors suggested the 'watchdog' technique, which is essentially a nonmonotone line search method. (See also Bonnans et al. (1992)) The fact that the augmented Lagrangian merit function does not suffer from this problem has been shown by many authors.

6. SQP Trust Region Methods

Trust region algorithms have become a part of the arsenal for solving unconstrained optimization problems, so it is natural to attempt to extend the ideas to solving constrained optimization problems and, in particular, to SQP methods. In this section an outline of the basic ideas of this approach will be provided. As the subject is still in a state of flux, no attempt will be made to give a comprehensive account of the algorithms that have been proposed. The discussion will be limited to equality-constrained problems; the inclusion of inequality constraints into trust region algorithms has been the subject of little research.

A major difficulty associated with SQP methods arises from trying to ensure that (QP) has a solution. As was seen in the preceding, the requirement that the Hessian approximations be positive definite on the null space of the constraints is difficult to guarantee short of requiring the B_k to be positive definite. The trust region methods attempt to avoid this difficulty by the addition of a bounding constraint. Since this added constraint causes the feasible region to be bounded the subproblem is guaranteed to have a solution independently of the choice of B_k provided the feasible region is nonempty. To be precise, a straightforward trust region adaptation of the SQP method would generate the iterates by solving the problem

$$
\begin{aligned}
\underset{d_x}{\text{minimize}} \quad & \nabla f(x^k)^t d_x + \tfrac{1}{2} d_x{}^t B_k\, d_x \\
\text{subject to:} \quad & \nabla h(x^k)^t d_x + h(x^k) \;=\; 0, \\
& \|S d_x\|^2 \;\leq\; \triangle_k^2,
\end{aligned}
\tag{6.1}
$$

where S is a positive diagonal scaling matrix and $\triangle_k$ is the *trust region radius*. In this discussion we assume the norm is the Euclidean norm but other norms have been used in the trust region constraint. The trust region radius is updated at each iteration based on a comparison of the actual decrease in a merit function to the decrease predicted by the model. If there is good agreement the radius is maintained or increased; if not, it is decreased.

It is worth noting that the necessary condition for d_x to be a solution to (6.1) is that

$$(B_k + \mu S^2)d_x = -\nabla h(x^k)u - \nabla f(x^k) \qquad (6.2)$$

for some multiplier vector u and some nonnegative scalar μ. This equation is used to generate approximate solutions to (6.1).

The removal of the positive definite requirement on B_k does not come without cost. The additional constraint is a quadratic inequality constraint and hence it is not a trivial matter to find a good approximate solution to (6.1). Moreover, there is the bothersome possibility that the solution sets of the linear constraints and the trust region constraint may be disjoint. For this reason, research on trust region methods has centered on finding different types of subproblems that have feasible solutions but still capture the essence of the quadratic subproblem. Several avenues of investigation are summarized below.

One approach is to *relax* the linear constraints in such a way that the resulting problem is feasible. In this case the linear constraint in (6.1) is replaced by

$$\nabla h(x^k)^t d_x + \theta_k h(x^k) = 0, \qquad (6.3)$$

where $0 \le \theta_k \le 1$. A major difficulty with this approach is the problem of choosing θ_k so that (6.3) together with the trust region constraint has a solution.

A second approach is to replace the equality constraints by a least squares approximation. Then the equality constraints in (6.1) become the quadratic constraint

$$\left\| \nabla h(x^k)^t d_x + h(x^k) \right\|^2 \le (\rho_k)^2, \qquad (6.4)$$

where ρ_k is an appropriate value. One choice of ρ_k is the error in the linear constraints evaluated at the 'Cauchy' point. The Cauchy point is defined to be the optimal step in the steepest descent direction for the function

$$\left\| \nabla h(x^k)^t d_x + h(x^k) \right\|^2$$

that is, the step, s_{cp}, that minimizes this function in the steepest descent direction. This yields

$$(\rho_k)^2 = \left\| \nabla h(x^k)^t s_{cp} + h(x^k) \right\|^2$$

Another possibility is to take ρ_k to be any value of $\left\| \nabla h(x^k)^t s + h(x^k) \right\|$ for which

$$\sigma_1 \triangle_k \le \|s\| \le \sigma_2 \triangle_k,$$

where $0 < \sigma_1 \le \sigma_2 < 1$. This approach ensures that the quadratic subproblem is always feasible, but at the cost of some extra computation to obtain ρ_k. Moreover, while the resulting subproblem has only two quadratic constraints, finding a good approximate solution quickly is a matter that has not been completely resolved.

Finally, in a reduced Hessian approach to the trust region method the linear constraint in (6.1) is replaced by

$$\nabla h(x^k)^t d_x + \nabla h(x^k)^t s_k = 0, \tag{6.5}$$

where s_k is the solution of

$$\begin{aligned} \underset{s}{\text{minimize}} \quad & \left\| \nabla h(x^k)^t s + h(x^k) \right\|^2 \\ \text{subject to:} \quad & \|s\| \leq \tau \triangle_k \end{aligned}$$

for $\tau \in (0,1)$. Using the decomposition (3.27) from Section 3.4

$$d_x = Z_k p_Z + Y_k p_Y$$

where the columns of $Z(x^k)$ are a basis for the null space of $\nabla h(x^k)^t$ and the columns of Y_k are a basis for the range space of $\nabla h(x^k)$, it follows from (6.5) that

$$Y_k p_Y = -s_k.$$

As in Section 3.4, the null space component, p_Z, can now be seen to be the solution of the quadratic problem

$$\begin{aligned} \underset{p_Z}{\text{minimize}} \quad & \left[f(x^k) + B_k s_k \right]^t p_Z + \tfrac{1}{2} p_Z{}^t Z_k{}^t B_k Z_k p_Z \\ \text{subject to:} \quad & \left\| p_Z \right\|^2 \leq (\triangle_k)^2 - \|s_k\|^2 . \end{aligned}$$

A great deal of research has been done on the subject of minimizing a quadratic function subject to a trust region constraint so quick and accurate methods for approximating the solutions to these two problems are available.

Much of the work to be done in transforming these approaches into algorithms for which local and global convergence theorems can be proven is similar in nature to that which must be done for standard SQP methods. In particular, a method for updating the matrices B_k needs to be specified (a wider class of updating schemes is now available, at least in theory, because there is no necessity of requiring them to have the positive definiteness properties) and a merit function has to be specified. As was shown in Section 3, local superlinear convergence depends on the steps approaching the Newton-SQP steps as the solution is neared. In particular, this requires that the modified constraints reduce to the linearized constraint of (ECQP) and that the trust region constraint not be active so that steplengths of one can be accepted. Conditions under which these events will occur have not been established. Global convergence theorems have been proved for most of the above variations of the trust region method by utilizing either the ℓ_1 or the augmented Lagrangian merit function under the assumptions **C1** and **C2**.

6.1. Notes and References

An introductory exposition of trust region methods for unconstrained optimization can be found in the book by Dennis and Schnabel (1983). Methods for minimizing a quadratic function subject to a trust region constraint can be found in Gay (1981) and Moré and Sorensen (1983).

The first application of trust region methods to the constrained problem appears to be that of Vardi (1985), who uses the constraint (6.3) in place of the linearized equality constraints. This approach was also used by Byrd, Schnabel and Schultz (1985). The introduction of the quadratic constraint (6.4) as a substitute for the linear constraint is due to Celis, Dennis and Tapia (1985), who used the error at the 'Cauchy point' as the ρ_k. A global convergence theory for this strategy is given in El-Alem (1991). Powell and Yuan (1986) suggested the second version of this approach mentioned in the text. Yuan (1990) proposed solution techniques for the resulting subproblem. The reduced Hessian approach has been introduced by Omojokun (1989) who also considers the case when inequality constraints are present. This approach has also been used by Lalee, Nocedal and Plantega (1993) for large-scale problems.

7. Practical Considerations

Our goal throughout this survey has been to concentrate on those theoretical properties that bear on actual implementations, but we have mentioned some of the difficulties that arise in the implementation of an SQP algorithm when the assumptions that we have made are not satisfied. In this section we elaborate on a few of the more important of these difficulties and suggest some common computational techniques that can be used to work around them. No attempt is made to be complete; the object is to give an indication of the issues involved. We also discuss briefly the extension of the SQP algorithm to the large scale case, where special considerations are necessary to create an efficient algorithm.

7.1. The Solution of (QP)

The assumptions behind the theory in Sections 3 and 4 imply that (QP) always has a solution. In practice, of course, this may not be the case. (QP) may be infeasible or the objective function may be unbounded on the feasible set and have no local minima. As stated earlier, surveying methods for solving (QP) is beyond the scope of this paper, but we discuss some ways of continuing the computations when these difficulties arise.

One technique to avoid infeasibilities is to relax the constraints by using the trust region methods of Section 6. Another approach is to take d_x to be some convenient direction when (QP) is infeasible, for example, the steepest descent direction for the merit function. The infeasibility of (QP) is usually detected during a 'phase I' procedure that is used to obtain a feasible point with which to begin the quadratic programming algorithm. If the constraints are inconsistent, then there are no feasible points and the phase I procedure will fail. However, the direction, d_x, obtained in this case can sometimes be used to produce directions that reduce the constraint infeasibilities and can thus be used to improve x^k. For example, the phase I procedure known as the 'big M' method modifies (QP) by adding one new variable, say θ, in such a way that the new problem

$$\begin{aligned} \minimize_{d_x} \quad & \nabla f(x^k)^t d_x + \tfrac{1}{2} d_x{}^t B_k d_x + M\theta \\ \text{subject to:} \quad & \nabla h(x^k)^t d_x + h(x^k) \;=\; 0, \\ & \nabla g(x^k)^t d_x + g(x^k) - \theta e \;\leq\; 0 \end{aligned} \qquad (7.1)$$

is feasible. In this form M is a constant and e is the vector of ones. For $\theta^0 \equiv \max\{g_i(x^k) : g_i(x^k) > 0\}$ the initial point $(d_x, \theta) = (0, \theta^0)$ is feasible for (7.1). The constant M is chosen large enough that, as (7.1) is being solved, θ is forced to be reduced. Once $\theta \leq 0$, d_x is a feasible point for (QP). If θ cannot be reduced to zero, then the inequalities are inconsistent. Nevertheless, it can be shown under certain conditions that the resulting d_x is a descent direction for the merit function ϕ_F.

If the problem is unbounded, but feasible, then the difficulty is due to the structure of the Hessian approximation, B_k. For example, if B_k is not positive definite then a multiple of the identity (or non-negative diagonal matrix) can be added to B_k to ensure that it is positive definite. Note that adding a strictly positive diagonal matrix to B_k is equivalent to using a trust region constraint (see (6.2)).

Finally, in cases where (QP) does not yield approximate multipliers for nonlinear program such as when (QP) is infeasible or the matrix $G(x^k)$ has linearly dependent columns, any reasonable approximation of the multipliers will usually suffice. The most common approach in these cases is to use a least squares multiplier approximation. If (NLP) does not satisfy **A2** then at best the theoretical convergence rate will be slowed (if x^* is a simple degeneracy) and at worst even the Newton method may fail.

7.2. Adjusting the Merit Function Parameter

It was shown in Section 4 that setting the parameter ρ to ensure that the direction d_x is a descent direction for the merit function ϕ_1 given by (4.18) can be relatively straightforward. Setting the parameter for ϕ_F is only a little more difficult. In either case adjusting the parameter can cause both theoretical and computational difficulties. In theory, there is no problem with having the parameter large. Indeed, if only increases in the parameter are allowed, the assumptions of Section 4 coupled with a reasonable adjustment strategy will lead to a provably finite value of the parameter and convergence can be proved using the techniques of Section 4. Computationally, however, having a large value of the parameter implies that there will be too much emphasis on satisfying the constraints. The result will be that the iterates will be forced to follow the constraint boundary closely, which, in highly nonlinear problems, can cause the algorithm to be slow. A strategy for only allowing increases in the parameter can lead to an excessively large value due entirely to an early iterate's being far from the solution.

Ideally the parameter should be adjusted up or down at various stages of the iteration process to ensure both good theoretical and computational performance. Some strategies for this have been proposed. For example, it is possible to allow controlled decreases in the parameter that ensure that the predicted decrease in the merit function is not dominated by a decrease in the constraint infeasibilities. For such choices it is still possible to prove convergence.

7.3. Nonmonotone Decrease of the Merit Function

Global convergence results are usually proved by insisting that an appropriate merit function be sufficiently reduced at each iteration. Sometimes, however, it is more efficient computationally to be less conservative and to allow steps to be accepted even if the merit function is temporarily increased. If, for example, the merit function is forced to be reduced over any fixed number of iterations, then convergence follows. In practice such strategies have been quite successful, especially near the solution. As a particular example, note that the merit function ϕ_1 may not allow a steplength of one near the solution. One remedy for this is to accept the full step temporarily even if ϕ_1 increases. Then, if ϕ_1 is not sufficiently reduced after a small number of steps, the last good iterate is restored and reduction in ϕ_1 is required for the next iteration.

As a second example, it was noted in Section 4 that the use of ϕ_F requires the evaluation of f, h, g and their gradients just to test a prospective point for acceptance. This difficulty can be circumvented by using an 'approximate' or 'working' merit function that only requires evaluation of f, h and g, and no gradients, to test a point. For example, ϕ_F could be approximated (in the equality-constrained case) by

$$\phi_F^k(x) = f(x) + h(x)^{\mathrm{t}}\overline{u}(x^k) + \tfrac{1}{2}\eta \, \|h(x)\|^2 \,,$$

where $\overline{u}(x^k)$ stays fixed throughout the kth iteration. This is then coupled with a strategy to monitor the iterations to ensure that ϕ_F is sufficiently reduced after a certain number of iterations as discussed above.

7.4. Large Scale Problems

Efficient SQP algorithms in the large scale case depend on carefully addressing many factors. In this section we mention two of these, namely, problem structure and the solution of large scale quadratic programs.

Problems are considered large if, to solve them efficiently, either their structure must be exploited or the storage and manipulation of the matrices involved must be handled in a special manner. The most obvious structure, and the one most commonly considered, is sparsity of the matrices ∇h, ∇g and $H\mathcal{L}$. Typically in large scale problems most constraints depend on only a few of the variables and the objective function is 'partially separable' (a partially separable function is, for example, made up of a sum of functions, each of which depends on only a few of the variables). In such cases the matrices are sparse. In other cases, one or more of the matrices is of low rank, and this structure can also be exploited.

A key to using SQP in the large scale case is to be able to solve, or to solve approximately, the large quadratic programming subproblems. In the small scale case, it rarely matters how (QP) is solved and it usually makes sense to solve it completely. Depending on the algorithm used in the large scale case, it is often essential to realize that (QP) at iteration $k + 1$ differs only slightly from that at iteration k. In these cases, the active set may remain the same, or change only slightly, and it follows that solving (approximately) the next (QP) can be accomplished quickly. It must be shown, however, that approximate solutions of (QP) are descent directions for an appropriate merit function.

Recently, efficient interior point methods have been developed for solving large scale linear programs; these ideas are now being applied to large scale quadratic programs. One such method is the 'subspace' method where, at each iteration, a low-dimensional subspace is chosen and (QP) restricted to that subspace is solved. A step in the resulting direction is calculated and the procedure iterated. It has been shown that any number of iterations of this technique gives rise to a descent direction for an augmented Lagrangian merit function; an SQP algorithm based on this has shown promise.

7.5. Notes and References

See Fletcher (1981) for an introduction to the solution of quadratic programs using classical techniques, including the Big M method. Recently there have been numerous papers on the application of interior-point methods to quadratic programming problems. This is still the subject of intense research and there are no good survey papers on the subject. See Vanderbei and Carpenter (1993) and Boggs, Domich et al. (1991) or Boggs, Domich et al. (1994) for a discussion of two different approaches.

Adjusting the penalty parameter in the augmented Lagrangian merit function is discussed in Schittkowski (1981), Schittkowski (1983), Powell and Yuan (1986), Byrd and Nocedal (1991), Boggs, Tolle and Kearsley (1994), Byrd and Nocedal (1991), and El-Alem (1992). The idea of allowing the merit function to increase temporarily is an old one. It is sometimes referred to as a nonmonotone line search; for a general discussion see Grippo, Lampariello and Lucidi (1986). Its use with the ℓ_1 merit function is discussed in Chamberlain, Lemarechal, Pedersen and Powell (1982). Using approximate merit functions is suggested in Powell and Yuan (1986), Boggs and Tolle (1989) and Boggs, Tolle and Kearsley (1994).

Based on the work to date there is significant promise for SQP in the large scale case, but more work is required to compare SQP with other large scale techniques. For the work to date see Murray (1994), Murray and Prieto (1995) and Boggs, Tolle and Kearsley (1994).

Acknowledgments: The authors would like to thank Ubaldo Garcia-Palomares, Anthony Kearsley, Stephen Nash, Dianne O'Leary and G. W. Stewart for reading and commenting on early drafts of the paper.

REFERENCES

D. P. Bertsekas (1980), 'Variable metric methods for constrained optimization based on differentiable exact penalty functions', in *Proceedings of the Eighteenth Allerton Conference on Communications, Control and Computing*, Allerton Park, Illinois, pp. 584–593.

D. P. Bertsekas (1982), *Constrained Optimization and Lagrange Multipliers Methods*, Academic Press, London.

P. T. Boggs, P. D. Domich, J. E. Rogers and C. Witzgall (1991), 'An interior point method for linear and quadratic programming problems', *Mathematical Programming Society Committee on Algorithms Newsletter* **19**, 32–40.

P. T. Boggs, P. D. Domich, J. E. Rogers and C. Witzgall (1994), 'An interior-point method for general large scale quadratic programming problems', Internal report (in progress), National Institute of Standards and Technology.

P. T. Boggs and J. W. Tolle (1980), 'Augmented Lagrangians which are quadratic in the multiplier', *Journal of Optimization Theory and Applications* **31**, 17–26.

P. T. Boggs and J. W. Tolle (1984), 'A family of descent functions for constrained optimization', *SIAM Journal on Numerical Analysis* **21**, 1146–1161.

P. T. Boggs and J. W. Tolle (1989), 'A strategy for global convergence in a sequential quadratic programming algorithm', *SIAM Journal on Numerical Analysis* **21**, 600–623.

P. T. Boggs and J. W. Tolle (1994), 'Convergence properties of a class of rank-two updates', *SIAM Journal on Optimization* **4**, 262–287.

P. T. Boggs, J. W. Tolle and A. J. Kearsley (1991), 'A merit function for inequality constrained nonlinear programming problems', Internal Report 4702, National Institute of Standards and Technology.

P. T. Boggs, J. W. Tolle and A. J. Kearsley (1994), 'A practical algorithm for general large scale nonlinear optimization problems', Internal report, National Institute of Standards and Technology.

P. T. Boggs, J. W. Tolle and P. Wang (1982), 'On the local convergence of quasi-Newton methods for constrained optimization', *SIAM Journal on Control and Optimization* **20**, 161–171.

J. F. Bonnans, E. R. Panier, A. L. Tits and J. L. Zhou (1992), 'Avoiding the Maratos effect by means of a nonmonotone line search II. Inequality constrained problems—feasible iterates', *SIAM Journal on Numerical Analysis* **29**, 1187–1202.

C. G. Broyden, J. E. Dennis, Jr. and J. J. Moré (1973), 'On the local and superlinear convergence of quasi-Newton methods', *Journal of the Institute of Mathematical Applications* **12**, 223–246.

R. H. Byrd and J. Nocedal (1991), 'An analysis of reduced Hessian methods for constrained optimization', *Mathematical Programming* **49**, 285–323.

R. H. Byrd and R. B. Schnabel (1986), 'Continuity of the null space basis and constrained optimization', *Mathematical Programming* **35**, 32–41.

R. H. Byrd, R. B. Schnabel and G. A. Schultz (1985), 'A trust-region strategy for nonlinearly constrained optimization', *SIAM Journal on Numerical Analysis* **24**, 1152–1170.

R. H. Byrd, R. A. Tapia and Y. Zhang (1992), 'An SQP augmented Lagrangian BFGS algorithm for constrained optimization', *SIAM Journal on Optimization* **2**, 210–241.

M. R. Celis, J. E. Dennis, Jr. and R. A. Tapia (1985), 'A trust-region strategy for nonlinear equality constrained optimization', in *Numerical Optimization 1984* (P. Boggs, R. Byrd and R. Schnabel, eds.), SIAM, Philadelphia, pp. 71–82.

R. Chamberlain, C. Lemarechal, H. C. Pedersen and M. J. D. Powell (1982), 'The watchdog technique for forcing convergence in algorithms for constrained optimization', *Mathematical Programming Study* **16**, 1–17.

T. F. Coleman (1990), 'On characterizations of superlinear convergence for constrained optimization', in *Lectures in Applied Mathematics*, American Mathematical Society, Providence, Rhode Island, pp. 113–133.

T. F. Coleman and A. R. Conn (1984), 'On the local convergence of a quasi-Newton method for the nonlinear programming problem', *SIAM Journal on Numerical Analysis* **21**, 755–769.

T. F. Coleman and P. A. Feynes (1992), 'Partitioned quasi-Newton methods for nonlinear equality constrained optimization', *Mathematical Programming* **53**, 17–44.

T. F. Coleman and D. C. Sorensen (1984), 'A note on the computation of an orthonormal basis for the null space of a matrix', *Mathematical Programming* **29**, 234–242.

J. E. Dennis, Jr. and J. J. Moré (1977), 'Quasi-Newton methods, motivation and theory', *SIAM Review* **19**, 46–89.

J. E. Dennis, Jr. and R. B. Schnabel (1983), *Numerical Methods for Unconstrained Optimization and Nonlinear Equations*, Prentice-Hall, Englewood Cliffs, New Jersey.

M. M. El-Alem (1991), 'A global convergence theory for the Celis–Dennis–Tapia trust region algorithm for constrained optimization', *SIAM Journal on Numerical Analysis* **28**, 266–290.

M. M. El-Alem (1992), 'A robust trust region algorithm with nonmonotonic penalty parameter scheme for constrained optimization', Department of Mathematical Sciences 92-30, Rice University.

R. Fletcher (1972), 'A class of methods for nonlinear programming, iii: Rates of convergence', in *Numerical Methods for Nonlinear Optimization* (F. A. Lootsma, ed.), Academic Press, New York, pp. 371–382.

R. Fletcher (1981), *Practical Methods of Optimization*, volume 2, Wiley, New York.

R. Fontecilla (1988), 'Local convergence of secant methods for nonlinear constrained optimization', *SIAM Journal on Numerical Analysis* **25**, 692–712.

R. Fontecilla, T. Steihaug and R. A. Tapia (1987), 'A convergence theory for a class of quasi-Newton methods for constrained optimization', *SIAM Journal on Numerical Analysis* **24**, 1133–1151.

M. Fukushima (1986), 'A successive quadratic programming algorithm with global and superlinear convergence properties', *Mathematical Programming* **35**, 253–264.

D. Gabay (1982), 'Reduced quasi-Newton methods with feasibility improvement for nonlinearly constrained optimization', *Mathematical Programming Study* **16**, 18–44.

U. M. Garcia-Palomares and O. L. Mangasarian (1976), 'Superlinearly convergent quasi-Newton methods for nonlinearly constrained optimization problems', *Mathematical Programming* **11**, 1–13.

D. M. Gay (1981), 'Computing optimal locally constrained steps', *SIAM Journal on Scientific and Statistical Computing* **2**, 186–197.

R. Ge and M. J. D. Powell (1983), 'The convergence of variable metric matrices in unconstrained optimization', *Mathematical Programming* **27**, 123–143.

J. C. Gilbert (1993), 'Superlinear convergence of a reduced BFGS method with piecewise line-search and update criterion', Rapport de Recherche 2140, Institut National de Recherche en Informatique et en Automatique.

P. E. Gill, W. Murray, M. A. Saunders, G. W. Stewart and M. H. Wright (1985), 'Properties of a representation of a basis for the null space', *Mathematical Programming* **33**, 172–186.

P. E. Gill, W. Murray, M. A. Saunders and M. H. Wright (1986), 'User's guide for NPSOL (version 4.0): A Fortran package for nonlinear programming', Technical Report SOL 2, Department of Operations Research, Stanford University.

P. E. Gill, W. Murray and M. H. Wright (1981), *Practical Optimization*, Academic Press, New York.

S. T. Glad (1979), 'Properties of updating methods for the multipliers in augmented Lagrangians', *Journal of Optimization Theory and Applications* **28**, 135–156.

J. Goodman (1985), 'Newton's method for constrained optimization', *Mathematical Programming* **33**, 162–171.

L. Grippo, F. Lampariello and S. Lucidi (1986), 'A nonmonotone line search technique for Newton's method', *SIAM Journal on Numerical Analysis* **23**, 707–716.

S.-P. Han (1976), 'Superlinearly convergent variable metric algorithms for general nonlinear programming problems', *Mathematical Programming* **11**, 263–282.

S.-P. Han (1977), 'A globally convergent method for nonlinear programming', *Journal of Optimization Theory and Applications* **22**, 297–309.

M. Lalee, J. Nocedal and T. Plantega (1993), On the implementation of an algorithm for large-scale equality optimization, Department of Electrical Engineering and Computer Science NAM 09-93, Northwestern University.

D. G. Luenberger (1984), *Linear and Nonlinear Programming*, 2d edition, Addison-Wesley, Reading, Massachusetts.

J. Moré and D. C. Sorensen (1983), 'Computing a trust region step', *SIAM Journal of Scientific and Statistical Computing* **4**, 553–572.

W. Murray (1994), 'Algorithms for large nonlinear problems', in *Mathematical Programming: State of the Art 1994* (J. R. Birge and K. G. Murty, eds.), University of Michigan, Ann Arbor, MI, pp. 172–185.

W. Murray and F. J. Prieto (1995), 'A sequential quadratic programming algorithm using an incomplete solution of the subproblem', *SIAM Journal on Optimization, to appear.*

S. G. Nash and A. Sofer (1995), *Linear and Nonlinear Programming*, McGraw-Hill, New York.

J. Nocedal (1992), 'Theory of algorithms for unconstrained optimization', *Acta Numerica* **1991**, 199–242.

J. Nocedal and M. L. Overton (1985), 'Projected Hessian updating algorithms for nonlinearly constrained optimization problems', *SIAM Journal on Numerical Analysis* **22**, 821–850.

E. M. Omojokun (1989), 'Trust region algorithms for optimization with nonlinear equality and inequality constraints', PhD thesis, University of Colorado.

J. M. Ortega and W. C. Rheinboldt (1970), *Iterative Solution of Nonlinear Equations In Several Variables*, Academic Press, New York.

E. R. Panier and A. L. Tits (1993), 'On combining feasibility, descent and superlinear convergence in inequality constrained optimization', *Mathematical Programming* **59**, 261–276.

E. Polak (1989), 'Basics of minimax algorithms', in *Nonsmooth Optimization and Related Topics* (F. H. Clark, V. F. Dem'yanov and F. Giannessi, eds.), Plenum, New York, pp. 343–369.

M. J. D. Powell (1977), 'A fast algorithm for nonlinearly constrained optimization calculations', in *Numerical Analysis Dundee, 1977* (G. A. Watson, ed.), Springer-Verlag, Berlin, pp. 144–157.

M. J. D. Powell (1978a), 'Algorithms for nonlinear constraints that use Lagrangian functions', *Mathematical Programming* **14**, 224–248.

M. J. D. Powell (1978b), 'The convergence of variable metric methods for nonlinearly constrained optimization calculations', in *Nonlinear Programming 3* (O. Mangasarian, R. Meyer and S. Robinson, eds.), Academic Press, New York, pp. 27–64.

M. J. D. Powell and Y. Yuan (1986), 'A recursive quadratic programming algorithm that uses differentiable exact penalty functions', *Mathematical Programming* **35**, 265–278.

S. M. Robinson (1974), 'Perturbed Kuhn–Tucker points and rates of convergence for a class of nonlinear-programming algorithms', *Mathematical Programming* **7**, 1–16.

K. Schittkowski (1981), 'The nonlinear programming method of Wilson, Han, and Powell with an augmented Lagrangian type line search function, part 1: Convergence analysis', *Numerische Mathematik* **38**, 83–114.

K. Schittkowski (1983), 'On the convergence of a sequential quadratic programming method with an augmented Lagrangian line search function', *Math. Operationsforsch. U. Statist., Ser. Optimization* **14**, 197–216.

J. Stoer (1984), 'The convergence of matrices generated by rank-2 methods from the restricted β-class of Broyden', *Numerische Mathematik* **44**, 37–52.

R. A. Tapia (1974), 'A stable approach to Newton's method for optimization problems with equality constraints', *Journal of Optimization Theory and Applications* **14**, 453–476.

R. A. Tapia (1977), 'Diagonalized multiplier methods and quasi-Newton methods for constrained optimization', *Journal of Optimization Theory and Applications* **22**, 135–194.

R. A. Tapia (1978), 'Quasi-Newton methods for equality constrained optimization: Equivalence of existing methods and a new implementation', in *Nonlinear Programming 3* (O. Mangasarian, R. Meyer and S. Robinson, eds.), Academic Press, New York, pp. 125–164.

R. A. Tapia (1988), 'On secant updates for use in general constrained optimization', *Mathematics of Computation* **51**, 181–202.

R. J. Vanderbei and T. J. Carpenter (1993), 'Symmetric indefinite systems for interior point methods', *Mathematical Programming* **58**, 1–32.

A. Vardi (1985), 'A trust-region algorithm for equality constrained minimization and implementation', *SIAM Journal on Numerical Analysis* **22**, 575–591.

R. B. Wilson (1963), A simplicial algorithm for concave programming, PhD thesis, Harvard University.

P. Wolfe (1975), 'A method of conjugate subgradients for minimizing nondifferentiable functions', in *Mathematical Programming Study 3* (M. Balinski and P. Wolfe, eds.), North-Holland, Amsterdam, pp. 145–173.

Y. Yuan (1985), 'An only 2-step q-superlinear convergence example for some algorithms that use reduced Hessian approximations', *Mathematical Programming* **32**, 224–231.

Y. Yuan (1990), 'On a subproblem of trust region algorithms for constrained optimization', *Mathematical Programming* **47**, 53–63.

Acta Numerica (1995), *pp.* 53–103

A Taste of Padé Approximation

C. Brezinski

Laboratoire d'Analyse Numérique et d'Optimisation

UFR IEEA

Université des Sciences et Technologies de Lille

59655–Villeneuve d'Ascq cedex, France

E-mail: brezinsk@omega.univ-lille1.fr

J. Van Iseghem

UFR de Mathématiques Pures et Appliquées

Université des Sciences et Technologies de Lille

59655–Villeneuve d'Ascq cedex, France

E-mail: jvaniseg@ano.univ-lille1.fr

The aim of this paper is to provide an introduction to Padé approximation and related topics. The emphasis is put on questions relevant to numerical analysis and applications.

CONTENTS

Let f be a formal power-series. A Padé approximant of f is a rational function whose numerator and denominator are chosen so that its power series expansion (which is obtained by dividing the numerator by the denominator) agrees with f as far as possible, that is, at least up to the term whose degree equals the sum of the degrees of the numerator and the denominator of the rational function.

Such approximants have a long history and they have played an important rôle in the solution of many problems such as the transcendence of the numbers e and π and have given birth to some fundamental ideas in mathematics such as the spectral theory of operators. They are also closely connected to continued fractions; see Brezinski (1990) and Lorentzen and Waadeland (1992). Thirty years ago, Padé approximants were rediscovered by physicists and they proved to be a very efficient tool not only for improving existing methods but also for extracting important information from power series and thus leading to new possibilities which were not open before.

Let us give some examples and begin with a purely mathematical one. We shall consider the series

$$f(z) = z - z^2/2 + z^3/3 - z^4/4 + \cdots,$$

which is known to converge to $\ln(1+z)$ for $|z| \leq 1, z \neq -1$. Thus the simplest process for obtaining an approximate value of $\ln(1+z)$ is to sum the series up to a certain term. Let us call $f_k(z)$ the partial sum of f up to the term of degree k inclusive and let $[p/q]_f(z)$ be the Padé approximant of f whose numerator and denominator have the respective degrees p and q at most. As we shall see below, the computation of this Padé approximant requires the knowledge of the coefficients of f up to that of the power $p+q$.

For $z = 1$ we have $\ln 2 = 0.6931471805599453\ldots$. For $z = 2$, the series diverges and we have $\ln 3 = 1.098612288668110\ldots$. The Padé approximants give the results presented in the following table.

k	$f_{2k}(1)$	$[k/k]_f(1)$	k	$f_{2k}(2)$	$[k/k]_f(2)$
1	0.830	0.7	1	$0.260 \cdot 10^1$	1.14
2	0.783	0.6933	2	$0.506 \cdot 10^1$	1.101
3	0.759	0.693152	3	$0.126 \cdot 10^2$	1.0988
4	0.745	0.69314733	4	$0.375 \cdot 10^2$	1.098625
5	0.736	0.6931471849	5	$0.121 \cdot 10^3$	1.0986132
6	0.730	0.69314718068	6	$0.410 \cdot 10^3$	1.09861235
7	0.725	0.693147180563	7	$0.142 \cdot 10^4$	1.098612293
8	0.721	0.69314718056000	8	$0.504 \cdot 10^4$	1.0986122890
9	0.718	0.6931471805599485	9	$0.181 \cdot 10^5$	1.098612288692
10	0.716	0.6931471805599454	10	$0.655 \cdot 10^5$	1.0986122886698

Figures 1 and 2 below display respectively the partial sums $[3/0], [5/0]$ and $[7/0]$ of the series for $\arctan x$ for real values of x and its Padé approximants $[2/1], [3/2]$ and $[4/3]$. Each of them is easily recognizable. They clearly show that the domain of convergence of the series has been increased.

For other examples from physics, the interested reader is referred to the very complete book of Baker and Graves-Morris (1981), to the work of A. P. Magnus (1988) and to Guttmann (1989). Applications to numerical analysis, through the use of continued fractions, are described by Jones and Thron (1988). See Brezinski (1991a) for a bibliography.

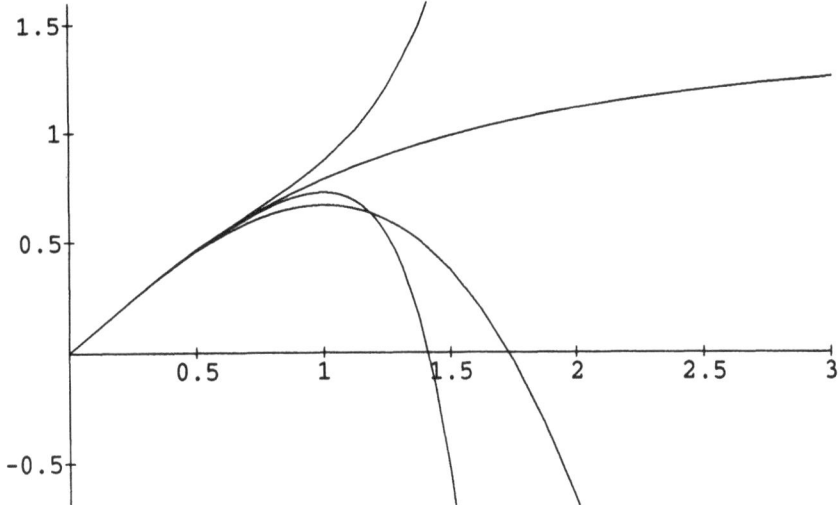

Fig. 1. Partial sums of arctan x

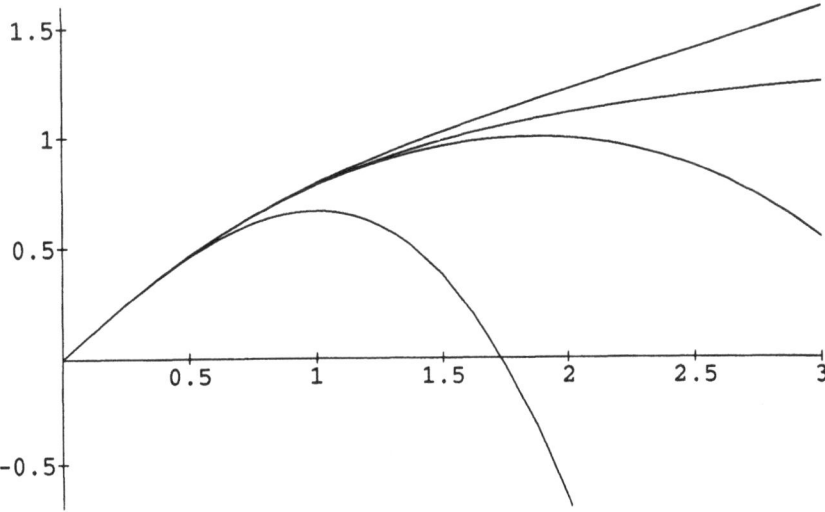

Fig. 2. Padé approximants of arctan x

1. Algebraic theory

Let us first begin with some definitions.

1.1. Definitions

Let us now give the exact definition of Padé approximants. We shall give two approaches to the subject: a direct one which is sufficient to understand it and a more complicated one which leads to a better grasp of its numerous relations with the theory of formal orthogonal polynomials and will serve as a basic tool for developing recurrence relations for the computation of Padé approximants and for other purposes that will be discussed later.

Let f be a formal power series with complex coefficients

$$f(z) = c_0 + c_1 z + c_2 z^2 + c_3 z^3 + \cdots.$$

Definition 1.1 The Padé approximant $[p/q]_f(z)$ is a rational function $N(z)/D(z)$ such that degree $N \leq p$, degree $D \leq q$ and

$$N(z) - f(z)D(z) = \mathcal{O}\left(z^{p+q+1}\right), \qquad z \to 0.$$

Let us write

$$N(z) = a_0 + a_1 z + \cdots + a_p z^p,$$

$$D(z) = b_0 + b_1 z + \cdots + b_q z^q.$$

Then the conditions of the definition lead to

$$
\begin{aligned}
a_0 &= c_0 b_0, \\
a_1 &= c_1 b_0 + c_0 b_1, \\
&\vdots \qquad \vdots \\
a_p &= c_p b_0 + c_{p-1} b_1 + \cdots + c_{p-q} b_q, \\
0 &= c_{p+1} b_0 + c_p b_1 + \cdots + c_{p-q+1} b_q, \\
&\vdots \qquad \vdots \\
0 &= c_{p+q} b_0 + c_{p+q-1} b_1 + \cdots + c_0 b_q
\end{aligned}
$$

with the convention that $c_i = 0$ for $i < 0$.

The last q equations contain $q + 1$ unknowns $b_0, \ldots, b_q$ and, thus, this system has a non-trivial solution. With knowledge of the b_i's, the first $p+1$ equations directly give the a_i's.

Two solutions of the problem lead to the same rational function since $N_1(z) - f(z)D_1(z) = \mathcal{O}(z^{p+q+1})$ and $N_2(z) - f(z)D_2(z) = \mathcal{O}(z^{p+q+1})$ implies $N_1(z)D_2(z) - D_1(z)N_2(z) = \mathcal{O}(z^{p+q+1})$. But the degree of $N_1 D_2 - D_1 N_2$ is at most $p + q$ and thus $N_1(z)D_2(z)$ is identical to $D_1(z)N_2(z)$. N and D can have a common factor. In particular if z^k is a factor of D, it is also a factor of N, as can be seen from the previous system, and thus $N(z)/D(z)$ cannot have a pole at the origin. Dividing by the highest power k in z contained in D gives a solution with $D(0) \neq 0$ and degree $N \leq p - k$, degree $D \leq q - k$, $N(z) - f(z)D(z) = \mathcal{O}\left(z^{p+q+1-k}\right)$, and we have

Theorem 1.1 Let $R(z) = N(z)/D(z)$ be an irreducible rational function with degree $N = p - k$, degree $D = q - k$, $k \geq 0$, and

$$N(z) - f(z)D(z) = \mathcal{O}\left(z^{p+q+1-k}\right), \qquad z \to 0.$$

Then, for $i, j = 0, \ldots, k$,

$$[p - k + i/q - k + j]_f(z) \equiv R(z)$$

and no other Padé approximant is identical to R if k is a maximum.

The result follows from the definition of Padé approximants and their uniqueness. This identity between Padé approximants can hold for all i and j (see Property 1.5) and, in that case, f is a rational function with a numerator of degree $p - k$ and a denominator of degree $q - k$ or there may exist a maximal value of k for which it holds and, in that case, no other Padé approximant is identical to R.

Usually the Padé approximants are arranged in a double-entry table known as the Padé table

$$
\begin{array}{cccc}
[0/0] & [0/1] & [0/2] & \cdots \\
[1/0] & [1/1] & [1/2] & \cdots \\
[2/0] & [2/1] & [2/2] & \cdots \\
\vdots & \vdots & \vdots & \ddots
\end{array}
$$

The theorem given above was proved by Henri Padé in 1892 (see Padé, 1984). It says that identical Padé approximants can only occur in square blocks of the table, a property known as the block structure of the Padé table. If the Padé table does not contain blocks, it is said to be normal; otherwise it is called non-normal. This block structure corresponds in fact to the block structure of the table of formal orthogonal polynomials (see Subsection 1.4) which itself mimics the block structure of the table of Hankel determinants (see Subsection 1.3). On these questions see Gragg (1972), de Bruin and Van Rossum (1975), Gilewicz (1978) and Draux (1983).

Other algebraic properties of Padé approximants will be given in Subsection 1.2.

Let us now come to the second approach to the subject.

Let c be the linear functional on the space of complex polynomials defined by

$$c(x^i) = c_i.$$

The functional c can be extended to the space of formal power series, thus leading to formal (that is, term-by-term) identities.

Our second approach is based on the following obvious formal identity which is given as a theorem since it is fundamental.

Theorem 1.2

$$f(z) = c\left(\frac{1}{1 - xz}\right).$$

The problem of approximating $f(z)$ is classical in numerical analysis. For example, if an approximation of

$$I = \int_a^b g(x)w(x)\,dx$$

is wanted, one can replace g by an interpolation polynomial and integrate it. This procedure leads to a so-called interpolatory quadrature formula which is exact on the space of polynomials of degree at most $k-1$ if k interpolation points are used. If these interpolation points are the zeros of the polynomial of degree k belonging to the family of orthogonal polynomials on $[a, b]$ with respect to w, the quadrature formula, called a Gaussian quadrature formula, becomes exact on the space of polynomials of degree at most $2k - 1$ (instead of $k - 1$). Thus, in order to obtain an approximation of $f(z)$, let us replace $1/(1 - xz)$ by its (Hermite) interpolation polynomial and then apply the functional c (which is analogous to integration). We have

Theorem 1.3 Let $v_k(x) = (x - x_1)^{k_1} \cdots (x - x_n)^{k_n}$, where $x_1, \ldots, x_n$ are distinct points in the complex plane and $k = k_1 + \cdots + k_n$. The polynomial

$$R_k(x) = \frac{1}{1 - xz}\left(1 - \frac{v_k(x)}{v_k(z^{-1})}\right)$$

is the Hermite interpolation polynomial of degree $k - 1$ of $(1 - xz)^{-1}$, that is, the polynomial such that

$$R_k^{(j)}(x_i) = \frac{d^j}{dx^j}(1 - xz)^{-1}\Big|_{x=x_i}$$

for $i = 1, \ldots, n$ and $j = 0, \ldots, k_i - 1$.

The proof of this result was given by Brezinski (1983a). Let us now apply the functional c to R_k in order to obtain an approximation of $f(z)$. We have

$$c(R_k(x)) = \frac{1}{v_k(z^{-1})}c\left(\frac{v_k(z^{-1}) - v_k(x)}{1 - xz}\right).$$

Setting

$$w_k(z) = c\left(\frac{v_k(z) - v_k(x)}{z - x}\right),$$

where c acts on x and z is a parameter, it is easy to see that w_k is a polynomial of degree $k - 1$ in z and that

$$c(R_k(x)) = \tilde{w}_k(z)/\tilde{v}_k(z),$$

where $\tilde{w}_k(z) = z^{k-1} w_k(z^{-1})$ and $\tilde{v}_k(z) = z^k v_k(z^{-1})$. Thus $c(R_k(x))$ is a rational function whose numerator has degree $k-1$ at most and whose denominator has degree k at most. Moreover

$$
\begin{aligned}
c(R_k(x)) &= c\left(\frac{1}{1-xz}\right) - \frac{z^k}{\tilde{v}_k(z)} c\left(\frac{v_k(x)}{1-xz}\right) \\
&= f(z) + \mathcal{O}\left(z^k\right).
\end{aligned}
$$

This property is quite similar to the property of interpolatory quadrature formulae to be exact on the space of polynomials of degree at most $k-1$. Thus $c(R_k(x))$ appears as a generalization of such formulae for the function $(1-xz)^{-1}$. Such rational functions, whose poles (the zeros of $\tilde{v}_k$) are arbitrarily chosen, are called Padé-type approximants of f. They will be denoted by $(p/q)_f(z)$. They generalize the Padé approximants. They have interesting properties and will be studied in Subsection 3.1.

From the above formula for the error, we have

$$
c(R_k(x)) = f(z) - \frac{z^k}{\tilde{v}_k(z)} c\left(v_k(x)\left(1 + xz + \cdots + x^{k-1}z^{k-1} + \frac{x^k z^k}{1-xz}\right)\right).
$$

The polynomial v_k, called the generating polynomial of the Padé-type approximant $(k-1/k)$, can be arbitrarily chosen and we have k degrees of freedom (its k zeros or k among its $k+1$ coefficients since the numerator and the denominator of a rational function are uniquely defined apart from a multiplying factor). Thus, let us take v_k such that

$$
c(x^i v_k(x)) = 0 \quad \text{for } i = 0, \ldots, k-1.
$$

In that case we shall have

$$
c(R_k(x)) = f(z) + \mathcal{O}\left(z^{2k}\right).
$$

But $2k = (k-1)+k+1$ which shows that $c(R_k(x))$ matches the original series f up to the degree of the numerator plus the degree of the denominator. Thus $c(R_k(x))$ is the Padé approximant $[k-1/k]$ of f. It can be understood as a generalization of Gaussian quadrature formulae for the function $(1-xz)^{-1}$ since it is exact on the space of polynomials of degree at most $2k-1$.

The relations $c(x^i v_k(x)) = 0$ for $i = 0, \ldots, k-1$ show that v_k is the polynomial of degree k belonging to the family of (formal) orthogonal polynomials with respect to the functional c. In that case v_k will be denoted by P_k. Thus formal orthogonal polynomials appear in a very natural way in the theory of Padé approximants. They form the basis of their algebraic study and lead to recurrence relationships for their computation. These questions will be studied in Subsection 1.5.

Moreover, by construction, we have the following error formula

$$f(z) - [k-1/k]_f(z) = \frac{z^{2k}}{\tilde{P}_k(z)} c \left(\frac{x^k P_k(x)}{1 - xz} \right).$$

But $(P_k(x) - P_k(z^{-1}))/(1 - xz)$ is a polynomial of degree $k - 1$ in x and, due to the orthogonality relations of P_k,

$$c \left(\frac{P_k(x) - P_k(z^{-1})}{1 - xz} P_k(x) \right) = 0 = c \left(\frac{P_k^2(x)}{1 - xz} \right) - P_k(z^{-1}) c \left(\frac{P_k(x)}{1 - xz} \right)$$

and thus

$$f(z) - [k-1/k]_f(z) = \frac{z^{2k}}{\tilde{P}_k^2(z)} c \left(\frac{P_k^2(x)}{1 - xz} \right).$$

These expressions are useful for estimating the error in Padé approximation. It is easy to see that

$$f(z) - [k-1/k]_f(z) = \frac{z^{2k}}{\tilde{P}_k(z)} \sum_{i=0}^{\infty} d_{i+k} z^i$$

with $d_i = c(x^i P_k(x)) = b_0 c_i + b_1 c_{i+1} + \cdots + b_k c_{i+k}$ and $P_k(x) = b_0 + \cdots + b_k x^k$.
Obviously, by the orthogonality property of P_k, $d_i = 0$ for $i = 0, \ldots, k - 1$.

1.2. Algebraic properties

In this subsection we shall give some algebraic properties of the Padé approximants. The first one is a determinantal formula which was obtained by Jacobi in 1846 using a determinantal formula due to Cauchy for interpolating rational functions.

We set

$$\begin{aligned} f_k(z) &= \sum_{i=0}^{k} c_i z^i \quad \text{for } k \geq 0, \\ &= 0 \qquad\quad \text{for } k < 0. \end{aligned}$$

Then, the following property holds:

Property 1.1

$$[p/q]_f(z) = \frac{\begin{vmatrix} z^q f_{p-q}(z) & z^{q-1} f_{p-q+1}(z) & \cdots & f_p(z) \\ c_{p-q+1} & c_{p-q+2} & \cdots & c_{p+1} \\ \vdots & \vdots & & \vdots \\ c_p & c_{p+1} & \cdots & c_{p+q} \end{vmatrix}}{\begin{vmatrix} z^q & z^{q-1} & \cdots & 1 \\ c_{p-q+1} & c_{p-q+2} & \cdots & c_{p+1} \\ \vdots & \vdots & & \vdots \\ c_p & c_{p+1} & \cdots & c_{p+q} \end{vmatrix}}.$$

Let us now assume that $f(0) = c_0 \neq 0$ and let g be the reciprocal series of f formally defined by

$$f(z)g(z) = 1.$$

Setting $g(z) = d_0 + d_1 z + d_2 z^2 + \cdots$ we have

$$c_0 d_0 = 1,$$
$$c_0 d_i + c_1 d_{i-1} + \cdots + c_i d_0 = 0, \qquad i \geq 1.$$

Then we have

Property 1.2
$\forall p, q, [p/q]_f(z)[q/p]_g(z) = 1.$

This property is very useful since it relates the two halves of the Padé table.

The other algebraic properties deal with transformations of the variable and of the series. They have been gathered in the two following properties:

Property 1.3

1. Let $g(z) = f(az), a \neq 0$. Then $[p/q]_g(z) = [p/q]_f(az)$.
2. Let $g(z) = f(z^k), k > 0$. Then, $\forall i, j$ such that $i + j \leq k - 1, [pk + i/qk + j]_g(z) = [p/q]_f(z^k)$.
3. Let $T(z) = Az^k/R(z), A \neq 0$, with R a polynomial of the degree $k > 0$ such that $R(0) \neq 0$. Let $g(z) = f(T(z))$. Then, $\forall i, j$ such that $i + j \leq k - 1, [pk + i/qk + j]_g(z) = [p/q]_f(T(z))$.

Property 1.4

1. Let $g(z) = z^k f(z)$. Then $[p + k/q]_g(z) = z^k [p/q]_f(z)$.
2. If $c_0 = \cdots = c_{k-1} = 0$ and $c_k \neq 0$ and if we set $g(z) = z^{-k} f(z)$ then $[p/q]_g(z) = z^{-k}[p + k/q]_f(z)$.
3. Let R be a polynomial of degree k. If $p \geq q + k$ then $[p/q]_{f+R}(z) = [p/q]_f(z) + R(z)$.

4. Let $g(z) = (A + Bf(z))/(C + Df(z))$ with $C + Dc_0 \neq 0$. Then

$$[p/p]_g(z) = \frac{A + B[p/p]_f(z)}{C + D[p/p]_f(z)}.$$

5. Let $g(z) = af(z), a \neq 0$. Then $[p/q]_g(z) = a[p/q]_f(z)$.

An important property is that of consistency.

Property 1.5 Let f be the power-series expansion of a rational function with a numerator of degree p and a denominator of degree q. Then $\forall i, j \geq 0, [p + i/q + j]_f(z) \equiv f(z)$.

A useful formula is the so-called Nuttall compact formula, obtained by Nuttall (1967). A generalization of it is

Property 1.6 Let $\{q_n\}$ be an arbitrary family of polynomials such that $\forall n, q_n$ has the exact degree n. Let V be the $k \times k$ matrix with elements $v_{ij} = c((1 - xz)q_{i-1}(x)q_{j-1}(x))$ for $i, j = 1, \ldots, k$, let v' be the vector with components $v'_i = c(q_{i-1}(x)(1 - v_k(x)/v_k(z^{-1})))$ for $i = 1, \ldots, k$ and let u be the vector with components $u_i = c(q_{i-1}(x))$ for $i = 1, \ldots, k$. Then

$$(k - 1/k)_f(z) = (u, V^{-1}v'),$$

where v_k is the generating polynomial of $(k - 1/k)$. If $v_k \equiv P_k$ then

$$[k - 1/k]_f(z) = (u, V^{-1}v').$$

If $q_n(x) = x^n$, then $v_{ij} = c_{i+j-2} - zc_{i+j-1}, u_i = c_{i-1}$ and the formula for $[k - 1/k]$ exactly reduces to Nuttall's. Since $(k - 1/k)$ only depends on $c_0, \ldots, c_{k-1}$ then, in the preceding formula, $c_k, \ldots, c_{2k-1}$ can be arbitrarily chosen. In particular they can be set to zero. If $q_n(x) = P_n(x)$, the preceding extension of Nuttall's formula is closely related to the matrix interpretation of Padé approximants; see Gragg (1972).

1.3. Formal orthogonal polynomials

As seen in Subsection 1.1, Padé approximants are based on formal orthogonal polynomials. Thus we shall now digress to treat this subject. The other approximants of the table are related to other families of orthogonal polynomials, called adjacent families of orthogonal polynomials, which will be studied in Subsection 1.4.

Let c be the linear functional on the space of complex polynomials defined by its moments c_i

$$c(x^i) = c_i, \qquad i \geq 0.$$

Let $\{P_k\}$ be a family of polynomials. $\{P_k\}$ is said to be the family of formal orthogonal polynomials with respect to c if, $\forall k \geq 0$,

1. P_k has the exact degree k;
2. $c(x^i P_k(x)) = 0$ for $i = 0, \ldots, k-1$.

Conditions (2) are the so-called orthogonality relations. They are equivalent to the condition $c(p(x)P_k(x)) = 0$ for any polynomial p of degree $k-1$ at most or to $c(P_n(x)P_k(x)) = 0, \forall n \neq k$. The usual orthogonal polynomials (that is, those orthogonal with respect to $c(\cdot) = \int_a^b (\cdot) \, d\alpha(x)$ with α bounded and nondecreasing in $[a, b]$) are known to satisfy a bunch of interesting properties such as a three-term recurrence relation, the Christoffel–Darboux identity, properties of their zeros, etc. Most of these properties still hold for formal orthogonal polynomials. However, in that case, the first question is that of existence. Let us write P_k as

$$P_k(x) = a_0 + a_1 x + \cdots + a_k x^k.$$

Then the orthogonality relations are equivalent to the system

$$a_0 c_i + a_1 c_{i+1} + \cdots + a_k c_{i+k} = 0, \qquad i = 0, \ldots, k-1.$$

Since P_k must have the exact degree k, a_k must be different from zero or, in other words, the Hankel determinant

$$H_k^{(0)} = \begin{vmatrix} c_0 & \cdots & c_{k-1} \\ c_1 & \cdots & c_k \\ \vdots & & \vdots \\ c_{k-1} & \cdots & c_{2k-2} \end{vmatrix}$$

must not vanish. Thus, in the sequel, we shall assume that $\forall k > 0, H_k^{(0)} \neq 0$. In that case, we shall say that the functional c is definite, a property clearly related to the normality of the Padé table. The case where the functional c is non-definite has been extensively studied by Draux (1983).

In the definite case, the polynomial P_k is uniquely determined apart from an arbitrary non-zero constant. Moreover we have the following determinantal formula

$$P_k(x) = D_k \begin{vmatrix} c_0 & c_1 & \cdots & c_k \\ c_1 & c_2 & \cdots & c_{k+1} \\ \vdots & & & \vdots \\ c_{k-1} & c_k & \cdots & c_{2k-1} \\ 1 & x & \cdots & x^k \end{vmatrix}$$

with $D_k \neq 0$ and $P_0(x) = D_0$. Let us set $P_k(x) = t_k x^k + s_k x^{k-1} + \cdots$. For a family of formal orthogonal polynomials we have

Theorem 1.4 $\forall k \geq 0$,

$$P_{k+1}(x) = (A_{k+1} x + B_{k+1}) P_k(x) - C_{k+1} P_{k-1}(x)$$

with $P_{-1}(x) = 0, P_0(x) = t_0$,

$$A_{k+1} = t_{k+1}/t_k, \quad B_{k+1} = -\frac{\alpha_k t_{k+1}}{h_k t_k},$$

$$C_{k+1} = \frac{t_{k-1}t_{k+1}}{t_k^2}\frac{h_k}{h_{k-1}},$$

$$\alpha_k = c(xP_k^2(x)), h_k = c(P_k^2(x)).$$

Since P_k is determined apart from a multiplying factor, the t_k's in the preceding recurrence relation can be arbitrarily chosen and thus this relation can be used for computing recursively the P_k's. In particular the choice $t_k = 1$ leads to monic orthogonal polynomials.

The reciprocal of this theorem was first proved by Favard (1935) for the usual orthogonal polynomials. It was extended by Shohat (1938) (see also Van Rossum (1953)) to the formal case.

Theorem 1.5 Let $\{P_k\}$ be a family of polynomials such that the relation of Theorem 1.4 holds with $t_0 \neq 0$ and $\forall k, A_k C_k \neq 0$. Then $\{P_k\}$ is a family of formal orthogonal polynomials with respect to a linear functional c whose moments c_i can be computed.

Let us now define the associated polynomials Q_k by

$$Q_k(z) = c\left(\frac{P_k(x) - P_k(z)}{x - z}\right),$$

where c acts on x and where z is a parameter. It is easy to see that Q_k is a polynomial of degree $k - 1$ in z, that $Q_0(z) = 0$ and that, for $k > 1$,

$$Q_k(z) = D_k \begin{vmatrix} c_0 & c_1 & c_2 & \cdots & c_k \\ \vdots & \vdots & \vdots & & \vdots \\ c_{k-1} & c_k & c_{k+1} & \cdots & c_{2k-1} \\ 0 & c_0 & c_0 z + c_1 & \cdots & (c_0 z^{k-1} + c_1 z^{k-2} + \cdots + c_{k-1}) \end{vmatrix}.$$

Theorem 1.6 The family $\{Q_k\}$ satisfies the three-term recurrence relation of Theorem 1.4 with $Q_{-1}(x) = -1, Q_0(x) = 0$ and $C_1 = A_1 c(P_0(x))$. Moreover, $\forall k \geq 0$

$$P_k(x)Q_{k+1}(x) - Q_k(x)P_{k+1}(x) = A_{k+1}h_k.$$

Some other relations satisfied by the P_k's and the Q_k's are given in the definite case by Brezinski (1980, chapter 2).

It follows from Theorem 1.6 that

$$[k/k+1]_f(z) = [k-1/k]_f(z) + \frac{A_{k+1}h_k}{\tilde{P}_k(z)\tilde{P}_{k+1}(z)}z^{2k},$$

a relation known as the Euler–Minding identity and which follows directly from the theory of continued fractions.

All these relations have been extended to the non-definite case by Draux (1983).

The zeros of the classical orthogonal polynomials are known to possess some properties. Not all of them extend to the formal case. In particular the zeros of formal orthogonal polynomials need not be simple or real. However, we have

Theorem 1.7   If c is definite, then $\forall k \geq 0$

1. P_k and P_{k+1} have no common zero,

2. Q_k and Q_{k+1} have no common zero,

3. P_k and Q_k have no common zero.

To end this subsection let us mention that a matrix formalism of orthogonality can be given via tridiagonal matrices. Orthogonal polynomials are known to play an important rôle in numerical analysis. In particular they are closely connected with projection methods used in the theory of linear operators, for example, with the method of moments, Lanczos's method and the conjugate gradient algorithms (see Section 4). All these connections were reviewed by Brezinski (1980, section 2.7, 1994); see also the works of Gutknecht (1990, 1992).

The notion of orthogonality studied in this subsection is a particular case of the more general notion of biorthogonality between a family of elements of a vector space and a family of elements of its dual. The notion of biorthogonality was extensively studied in Brezinski (1991b).

1.4. Adjacent families of orthogonal polynomials

Let us now define the linear functionals $c^{(n)}$ by

$$c^{(n)}(x^i) = c_{n+i}.$$

With the same convention as above, namely, that $c_i = 0$ for $i < 0$, these linear functionals $c^{(n)}$ can be defined even for negative values of the upper index n.

Let us denote by $\{P_k^{(n)}\}$ the family of formal orthogonal polynomials with respect to $c^{(n)}$. The family $\{P_k\}$ studied in Subsection 1.3 corresponds to $n = 0$. Such families are called adjacent families of orthogonal polynomials. They satisfy the same properties as above after replacing c by $c^{(n)}$ or, in other words, the sequence $c_0, c_1, \ldots$ by the sequence $c_n, c_{n+1}, \ldots$. In particular

$\{P_k^{(n)}\}$ exists only if $\forall k, n$

$$H_k^{(n)} = \begin{vmatrix} c_n & \cdots & c_{n+k-1} \\ \vdots & & \vdots \\ c_{n+k-1} & \cdots & c_{n+2k-2} \end{vmatrix} \neq 0.$$

In that case we shall say that the linear functional c is completely definite. For the non–completely definite case, we again refer the interested reader to Draux (1983).

The polynomials $P_k^{(n)}$ are usually placed in a double-entry table similar to the Padé table

$$
\begin{array}{ccccc}
P_{-1}^{(0)} & P_0^{(-1)} & P_1^{(-2)} & P_2^{(-3)} & \cdots \\
P_{-1}^{(1)} & P_0^{(0)} & P_1^{(-1)} & P_2^{(-2)} & \cdots \\
P_{-1}^{(2)} & P_0^{(1)} & P_1^{(0)} & P_2^{(-1)} & \cdots \\
P_{-1}^{(3)} & P_0^{(2)} & P_1^{(1)} & P_2^{(0)} & \cdots \\
\vdots & \vdots & \vdots & \vdots & \ddots
\end{array}
$$

Many relationships exist between adjacent polynomials of this table. First of all, each family of orthogonal polynomials satisfies a three-term recurrence relation similar to that of Theorem 1.4. Assuming all the polynomials to be monic we shall write this relation as

$$P_{-1}^{(n)}(x) = 0, \qquad P_0^{(n)}(x) = 1, \qquad \bullet$$

$$\bullet$$

$$P_{k+1}^{(n)}(x) = (x - q_{k+1}^{(n)} - e_k^{(n)})P_k^{(n)}(x) - q_k^{(n)}e_k^{(n)}P_{k-1}^{(n)}(x). \qquad *$$

A $\bullet$ indicates the position of a polynomial that is known in the table, while a $*$ indicates the position of the polynomial that is computed by the relation.

We also have

$$P_k^{(n+1)}(x) = P_k^{(n)}(x) - e_k^{(n)}P_{k-1}^{(n+1)}(x), \qquad \bullet \quad \bullet$$

$$*$$

$$P_{k+1}^{(n)}(x) = xP_k^{(n+1)}(x) - q_{k+1}^{(n)}P_k^{(n)}(x). \qquad \bullet$$

$$\bullet \quad *$$

Using alternatively these two relations allows us to compute recursively the two adjacent families $\{P_k^{(n)}\}$ and $\{P_k^{(n+1)}\}$.

It can be proved (see, for example, Brezinski (1980, section 2.8) where all these relations and the following ones are given) that the numbers $e_k^{(n)}$ and $q_k^{(n)}$ are related by

$$e_0^{(n)} = 0, \qquad q_1^{(n)} = c_{n+1}/c_n,$$

$$q_{k+1}^{(n)} + e_{k+1}^{(n)} = q_{k+1}^{(n+1)} + e_k^{(n+1)},$$
$$e_k^{(n)} q_{k+1}^{(n)} = e_k^{(n+1)} q_k^{(n+1)}$$

which is the so-called Qd algorithm. This algorithm can be used for their recursive computation. It is due to Rutishauser (1954) (see also Henrici (1974)) and was the basis for the development of the LR-algorithm for the computation of the eigenvalues of a matrix. Setting $x = 0$ in the preceding relations, it is easy to see that

$$q_k^{(n)} = H_k^{(n+1)} H_{k-1}^{(n)} / H_k^{(n)} H_{k-1}^{(n+1)},$$
$$e_k^{(n)} = H_{k-1}^{(n+1)} H_{k+1}^{(n)} / H_k^{(n)} H_k^{(n+1)}.$$

From these determinantal expressions and from the three preceding relations we can obtain the following ones

$$H_k^{(n+2)} H_k^{(n)} P_k^{(n)}(x) = [H_k^{(n+1)}]^2 P_k^{(n+1)}(x) + H_{k+1}^{(n)} H_{k-1}^{(n+2)} P_{k-1}^{(n+2)}(x),$$
$$H_k^{(n+1)} H_{k+1}^{(n-1)} P_{k+1}^{(n-1)}(x) = x H_{k+1}^{(n-1)} H_k^{(n+1)} P_k^{(n+1)}(x) - H_{k+1}^{(n)} H_k^{(n)} P_k^{(n)}(x),$$

$$x H_k^{(n)} H_{k+1}^{(n-1)} H_k^{(n+1)} P_k^{(n+1)}(x) = [x H_k^{(n+1)} H_{k+1}^{(n-1)} + H_{k+1}^{(n)} H_k^{(n)}] H_k^{(n)} P_k^{(n)}(x)$$
$$- H_{k+1}^{(n)} H_k^{(n+1)} H_k^{(n-1)} P_k^{(n-1)}(x),$$
$$H_k^{(n)} H_k^{(n+1)} H_{k+1}^{(n-1)} P_{k+1}^{(n-1)}(x) = [x H_{k+1}^{(n-1)} H_k^{(n+1)} - H_k^{(n)} H_{k+1}^{(n)}] H_k^{(n)} P_k^{(n)}(x)$$
$$- x H_{k+1}^{(n-1)} H_{k+1}^{(n)} H_{k-1}^{(n+1)} P_{k-1}^{(n+1)}(x).$$

Combining these relations leads to many other ones. However, the preceding eight relations are sufficient to follow any path in the table of the adjacent families of orthogonal polynomials. Of course similar relations hold among the associated polynomials

$$Q_k^{(n)}(z) = c^{(n)} \left(\frac{P_k^{(n)}(x) - P_k^{(n)}(z)}{x - z} \right).$$

1.5. Recursive computation of Padé approximants

Let us first relate all the approximants of the Padé table to the adjacent families of orthogonal polynomials defined in the Subsection 1.4. Thus the recurrence relations given there will provide recursive methods for computing any sequence of Padé approximants.

In Subsection 1.1, we saw that

$$[k - 1/k]_f(z) = \tilde{Q}_k^{(0)}(z) / \tilde{P}_k^{(0)}(z).$$

Making use of the convention that a sum with a negative upper index is equal to zero, it is easy to see that

Theorem 1.8 $\forall k \geq 0, \forall n \geq -k$

$$[n + k/k]_f(z) = \sum_{i=0}^{n} c_i z^i + z^{n+1} \tilde{Q}_k^{(n+1)}(z)/\tilde{P}_k^{(n+1)}(z)$$

with $\tilde{P}_k^{(n+1)}(z) = z^k P_k^{(n+1)}(z^{-1})$ and $\tilde{Q}_k^{(n+1)}(z) = z^{k-1} Q_k^{(n+1)}(z^{-1})$.

Let us set

$$[n + k/k]_f(z) = \tilde{N}_k^{(n+1)}(z)/\tilde{P}_k^{(n+1)}(z)$$

and

$$\tilde{P}_k^{(n)}(z) = \sum_{i=0}^{k} b_i^{(k,n)} z^i,$$

$$\tilde{N}_k^{(n)}(z) = \sum_{i=0}^{n+k-1} a_i^{(k,n)} z^i.$$

The relations of the preceding subsection give

$$\frac{\tilde{N}_{k+1}^{(n)}(z)}{\tilde{P}_{k+1}^{(n)}(z)} = \frac{a_{n+k}^{(k,n+1)} \tilde{N}_k^{(n+2)}(z) - z a_{n+k+1}^{(k,n+2)} \tilde{N}_k^{(n+1)}(z)}{a_{n+k}^{(k,n+1)} \tilde{P}_k^{(n+2)}(z) - z a_{n+k+1}^{(k,n+2)} \tilde{P}_k^{(n+1)}(z)},$$
●   *
●

$$\frac{\tilde{N}_k^{(n)}(z)}{\tilde{P}_k^{(n)}(z)} = \frac{a_{n+k}^{(k-1,n+2)} \tilde{N}_k^{(n+1)}(z) - a_{n+k}^{(k,n+1)} \tilde{N}_{k-1}^{(n+2)}(z)}{a_{n+k}^{(k-1,n+2)} \tilde{P}_k^{(n+1)}(z) - a_{n+k}^{(k,n+1)} \tilde{P}_{k-1}^{(n+2)}(z)}.$$
*
●   ●

These two relations are identical with a method due to Longman (1971) for computing recursively approximants located on an ascending staircase of the Padé table. They also cover an algorithm due to Baker (1970).

We have

$$\frac{\tilde{N}_k^{(n+1)}(z)}{\tilde{P}_k^{(n+1)}(z)} = \frac{\tilde{N}_k^{(n)}(z) - e_k^{(n)} z \tilde{N}_{k-1}^{(n+1)}(z)}{\tilde{P}_k^{(n)}(z) - e_k^{(n)} z \tilde{P}_{k-1}^{(n+1)}(z)},$$
●   ●
*

$$\frac{\tilde{N}_{k+1}^{(n)}(z)}{\tilde{P}_{k+1}^{(n)}(z)} = \frac{\tilde{N}_k^{(n+1)}(z) - q_{k+1}^{(n)} z \tilde{N}_k^{(n)}(z)}{\tilde{P}_k^{(n+1)}(z) - q_{k+1}^{(n)} z \tilde{P}_k^{(n)}(z)}$$
●
●   *

with $e_k^{(n)} = h_k^{(n)}/h_{k-1}^{(n+1)}, q_{k+1}^{(n)} = h_k^{(n+1)}/h_k^{(n)}$ and $h_k^{(n)} = c^{(n)}(P_k^{(n)^2}(x)) = \sum_{i=0}^{k} c_{n+k+i} b_{k-i}^{(k,n)}$. These two relations are identical to a method due to Watson (1973) to compute recursively approximants located on a descending staircase of the Padé table.

We also have

$$\frac{\tilde{N}_k^{(n+2)}(z)}{\tilde{P}_k^{(n+2)}(z)} = \frac{b_k^{(k,n+1)} \tilde{N}_{k+1}^{(n)}(z) - zb_{k+1}^{(k+1,n)} \tilde{N}_k^{(n+1)}(z)}{b_k^{(k,n+1)} \tilde{P}_{k+1}^{(n)}(z) - zb_{k+1}^{(k+1,n)} \tilde{P}_k^{(n+1)}(z)},$$

$$\frac{\tilde{N}_{k-1}^{(n+2)}(z)}{\tilde{P}_{k-1}^{(n+2)}(z)} = \frac{b_k^{(k,n+1)} \tilde{N}_k^{(n)}(z) - b_k^{(k,n)} \tilde{N}_k^{(n+1)}(z)}{b_k^{(k,n+1)} \tilde{P}_k^{(n)}(z) - b_k^{(k,n)} \tilde{P}_k^{(n+1)}(z)},$$

$$\frac{\tilde{N}_{k-1}^{(n+1)}(z)}{\tilde{P}_{k-1}^{(n+1)}(z)} = \frac{\tilde{N}_k^{(n)}(z) - \tilde{N}_k^{(n+1)}(z)}{\tilde{P}_k^{(n)}(z) - \tilde{P}_k^{(n+1)}(z)},$$

$$\frac{\tilde{N}_k^{(n)}(z)}{\tilde{P}_k^{(n)}(z)} = \frac{\tilde{N}_k^{(n+1)}(z) - \tilde{N}_{k+1}^{(n)}(z)}{\tilde{P}_k^{(n+1)}(z) - \tilde{P}_{k+1}^{(n)}(z)}.$$

Combining these relations together allows one to obtain all the other possible ones. Eight of them were used in a conversational program to compute recursively any sequence of Padé approximants in the normal case; see Brezinski (1980, appendix).

All these recurrence relations were extended by Draux (1983) to the non-normal case. A universal conversational program for computing any sequence of Padé approximants in the non-normal case was given by Draux and Van Ingelandt (1986). In order to avoid numerical instability and also for the detection of the block structure of the Padé table, it was necessary to program these relations in exact arithmetic that is in rational arithmetic coded on several words. All these programs are written in FORTRAN.

Setting for simplicity

$$[n + k - 1/k] = N,$$

$$[n + k/k - 1] = W, \qquad [n + k/k] = C, \qquad [n + k/k + 1] = E,$$

$$[n + k + 1/k] = S,$$

we obtain, after elimination among the preceding identities, the so-called cross rule of Wynn (1966)

$$(N - C)^{-1} + (S - C)^{-1} = (W - C)^{-1} + (E - C)^{-1}.$$

When a square block of size m occurs in the Padé table, this cross rule was extended by Cordellier (1979) who proved that

$$(N_i - C)^{-1} + (S_i - C)^{-1} = (W_i - C)^{-1} + (E_i - C)^{-1}$$

for $i = 0, \ldots, m$, where the N_i's and the W_i's are numbered from the upper left corner of the block and the S_i's and the E_i's from its lower right one. The initial values are

$$[-1/q]_f(z) = 0, \quad [p/-1]_f(z) = \infty,$$

$$[p/0]_f(z) = \sum_{i=0}^{p} c_i z^i, \quad [0/q]_f(z) = ([q/0]_g(z))^{-1} = \left(\sum_{i=0}^{q} d_i z^i\right)^{-1},$$

where the d_i's are the coefficients of the reciprocal series g of f.

Thus, the theory of formal orthogonal polynomials provides a basis for rediscovering known recursive methods for the computation of sequences of Padé approximants which were found more or less heuristically by their authors. This theory also gives us the possibility of computing any sequence of approximants by new recursive algorithms. It has been possible to extend the theory to the non-normal case, thus leading for the first time to all the possible recurrence relationships among the entries of a non-normal Padé table and to write the only existing complete subroutine.

1.6. The ε-algorithm

We shall now deal with a subject which, a priori, has nothing to do with Padé approximation but which is, in fact, closely related to it: convergence acceleration.

Let (S_n) be a sequence converging to S. If the convergence is slow, one can try to accelerate it. For that purpose, we shall transform the sequence (S_n) into another sequence (T_n) such that, if possible, (T_n) converges to S faster than (S_n), that is,

$$\lim_{n \to \infty} (T_n - S)/(S_n - S) = 0.$$

One of the most popular sequence transformations for that purpose is certainly Aitken's Δ^2 process which corresponds to

$$T_n = S_n - (\Delta S_n)^2/\Delta^2 S_n \quad \text{for } n = 0, 1, \ldots.$$

If the sequence (S_n) is such that $\exists a \neq 1, \lim_{n \to \infty} (S_{n+1} - S)/(S_n - S) = a$ then (T_n) obtained by Aitken's process converges to S faster than (S_n).

In 1955, Shanks (1955) gave a generalization of Aitken's process. He considered the various transformations $e_k : (S_n) \to (e_k(S_n))$ where

$$e_k(S_n) = \frac{\begin{vmatrix} S_n & \cdots & S_{n+k} \\ \vdots & & \vdots \\ S_{n+k} & \cdots & S_{n+2k} \end{vmatrix}}{\begin{vmatrix} \Delta^2 S_n & \cdots & \Delta^2 S_{n+k-1} \\ \vdots & & \vdots \\ \Delta^2 S_{n+k-1} & \cdots & \Delta^2 S_{n+2k-2} \end{vmatrix}}.$$

A recursive algorithm to compute the $e_k(S_n)$'s without computing the determinants involved in their definition was found one year later by Wynn (1956). It is the ε-algorithm whose rules are

$$\varepsilon_{-1}^{(n)} = 0, \qquad \varepsilon_0^{(n)} = S_n, \qquad n = 0, 1, \ldots,$$

$$\varepsilon_{k+1}^{(n)} = \varepsilon_{k-1}^{(n+1)} + \left[\varepsilon_k^{(n+1)} - \varepsilon_k^{(n)} \right]^{-1}, \qquad n, k = 0, 1, \ldots.$$

It is related to Shanks's transformation by

$$\varepsilon_{2k}^{(n)} = e_k(S_n).$$

The $\varepsilon_{2k+1}^{(n)}$'s are only intermediate quantities. The ε-algorithm is a quite powerful acceleration process which has been widely studied. For its theory one can consult, for example, Brezinski (1977). Subroutines and applications can be found in Brezinski and Redivo-Zaglia (1991).

The ε-algorithm is related to Padé approximants in the following way: if it is applied to the partial sums of the series f, that is, if

$$S_n = \sum_{i=0}^{n} c_i z^i, \qquad n, = 0, 1, \ldots,$$

then

$$\varepsilon_{2k}^{(n)} = [n + k/k]_f(z).$$

Thus the ε-algorithm can be used to compute recursively the lower half of the Padé table. The upper half of the Padé table can be computed by applying the ε-algorithm to the partial sums of the reciprocal series g of f as stated in Property 1.2. Let us mention that the elimination of the ε's with an odd lower index leads to Wynn's cross rule mentioned in the preceding subsection.

2. Convergence

2.1. Introduction

More complete results about convergence can be found in Brezinski and Van Iseghem (1994) or Baker and Graves-Morris (1981).

The problem of convergence of Padé approximants, which means the convergence of a sequence of Padé approximants when at least one of the degrees tends to infinity, is a difficult problem which can be studied from different points of view. The first one is to study all the abilities of convergence to one function and the first theorem (due to Padé) is an example of such a study for the exponential function. The history of numbers such as e or π tells us, through the link with continued fractions, that it is also possible to do so for functions such as $\tan x$, $\arctan x$ and some others. At the other end, it is possible to hope for convergence results for a whole class of functions. The most useful and well known examples are the meromorphic functions with a fixed number of poles in a disc (convergence of the columns) and the Stieltjes functions, for which the classical uniform convergence on compact subsets of $\mathbb{C}$ can be proved for the diagonal and paradiagonal sequences.

In each case, the problems are different if the sequences considered are in a column, a diagonal or a paradiagonal; close to these cases are, for example, sectorial sequences where m/n has lower and upper bounds as n goes to infinity. Other cases could also be considered.

We will first quote two results which show the most optimistic result that can be expected and a counterexample which limits our ambitions.

Theorem 2.1 For any sequence $(m_i, n_i), i \geq 1$, where $m_i + n_i$ tends to infinity, the poles of the Padé approximants of e^z tend to infinity and

$$\lim_{i \to \infty} [m_i/n_i](z) = e^z$$

uniformly on any compact set of $\mathbb{C}$.

The following result is due to Wallin (1974).

Theorem 2.2 There exists an entire function f such that the sequence of diagonal Padé approximants $([n/n]_f)$ is unbounded at every point of the complex plane except zero, and so no convergence result can be expected in any open set of the plane.

As we shall see below, the location of the poles of the approximants is of primary importance for studying convergence: for meromorphic functions, they are supposed to be known, and so Montessus de Ballore's theorem is obtained. For Stieltjes functions, the link with orthogonal polynomials is extensively used; they are, in that case, defined by a positive-definite functional, and so properties about the zeros are known.

As a consequence, functions with branch points, for example, are outside of our study, and convergence will be obtained on extremal subsets of $\mathbb{C}$ that localize the set of zeros and poles of the Padé approximants as a barrier to convergence.

Let us now give a very simple example showing the difficulties related with the convergence of Padé approximants (that is, the convergence of a sequence of approximants). We consider the series given by Bender and Orszag (1978)

$$f(z) = \frac{10 + z}{1 - z^2} = \sum_{i=0}^{\infty} c_i z^i$$

with $c_{2i} = 10$ and $c_{2i+1} = 1$. It converges for $|z| < 1$. We have

$$[k/1]_f(z) = \sum_{i=0}^{k-1} c_i z^i + \frac{c_k z^k}{1 - c_{k+1} z / c_k}.$$

When k is odd, $[k/1]$ has a simple pole at $z = 1/10$ while f has no pole. Thus the sequence $([k/1])$ cannot converge to f in $|z| < 1$.

This example shows that the poles of the Padé approximants can prevent convergence and that a sequence of approximants can be non-convergent in a domain where the series is. In order to prove the convergence of a sequence of Padé approximants in a domain D of the complex plane, it must be proved that the spurious poles of the approximants (that is, the poles that do not approximate poles of f) move out of D when the degree(s) of the approximants tends to infinity.

Another more paradoxical situation can arise: the zeros of the Padé approximants can also prevent convergence. Let us take the reciprocal series g of the series f of the preceding example

$$g(z) = \frac{1 - z^2}{10 + z}.$$

It converges in $|z| < 10$. We have

$$[1/k]_g(z) = 1/[k/1]_f(z).$$

Since $[1/2k + 1]_g(0.1) = 0$ and $g(0.1) \neq 0$ the sequence $([1/k]_g)$ cannot converge in $|z| < 10$ where the series g does.

Another counterexample is due to Perron (1957). Let an arbitrary sequence (z_n) of points of $\mathbb{C}$ be given, and let us define the following function

$$f(z) = \sum_{i=0}^{\infty} c_i z^i,$$

$$\text{if } |z_n| \leq 1, \quad c_{3n} = z_n/(3n + 2)!,$$
$$c_{3n+1} = c_{3n+2} = 1/(3n + 2)!,$$
$$\text{if } |z_n| > 1, \quad c_{3n} = c_{3n+1} = 1/(3n + 2)!,$$
$$c_{3n+2} = z_n^{-1}/(3n + 2)!.$$

We have $|c_i| < 1/i, \forall i \geq 0$. Thus f is an entire function and either $[3n/1]$ or $[3n + 1/1]$ has a pole at z_n. The sequence (z_n) is a subsequence of the poles of $([m/1])$ and if (z_n) is dense in $\mathbb{C}$, the sequence $([m/1])$ cannot converge in any open set of the complex plane.

2.2. Meromorphic functions

Let us first consider the convergence of the columns $([m/n])_m$ of the Padé table.

The most famous theorem is that of de Montessus de Ballore (1902). It is concerned with meromorphic functions with a fixed known number of poles in the disc of radius R centered at the origin.

An extension of this result has been given by Saff (1972) for the case of interpolating rational functions instead of Padé approximants. Montessus's theorem is then a particular case of it when all the interpolation points coincide at zero.

Theorem 2.3 (Montessus de Ballore's theorem). Let f be analytic at $z = 0$ and meromorphic with exactly n poles $\alpha_1, \ldots, \alpha_n$, counted with their multiplicities, in the disc $D_R = \{z, |z| < R\}$. Let D be the domain $D_R - \{\alpha_i\}_{i=1,\ldots,n}$.

The sequence $([m/n])_{m \geq 0}$ converges to f uniformly on every compact subset of D. The poles of $[m/n]$ approach the poles of f as m tends to infinity.

This result is optimal since, if there is a pole on the boundary of D_R, then divergence of the sequence $R_{m,n}$ occurs outside $\mathbb{C} - D$ as proved by Wallin (1987).

Let us now have a look at some simple examples to illustrate the different aspects of the result. The computations have been conducted with *Mathematica* and they are given up to the first inexact digit with at most 9 digits.

First of all, the convergence to the poles is obtained with a speed of convergence that is $\mathcal{O}(r/R)$, where r is the modulus of the pole to be computed and R is the radius of the largest disc of meromorphy of the function, or $\mathcal{O}(r)$ if the function is meromorphic in the whole plane.

For the function $\sin z/(z-1)(z-2)(z-3)$, we obtain the following results, where the various columns represent the three zeros of the Padé approximant $[n/3]$ (up to the first inexact digit).

$[n/3]$			
3	0.991	2.3	−0.5
5	1.0006	1.92	0.2
7	1.000001	1.996	−4.
9	0.99999991	2.001	2.6
11	1.000000000	1.99997	3.02
13	1.000000000	2.0000004	2.9989
15	1.000000000	1.999999994	3.00003
17	1.000000000	2.000000000	2.9999990

Now, let us consider the function $\log(a + z)/(1 - z^3)$. Here $R = |a|$ and we take $a = 1.1$ and $a = 3$. The two conjugate poles are in the same column. The convergence is, of course, much better for the last two columns which correspond to $a = 3$.

$\lceil n/3 \rceil$					
1	1.1	$-.59 \pm.71\,i$	1.01	$-.48$	$\pm.85\,i$
3	1.06	$-.52 \pm.91\,i$	1.0005	$-.501$	$\pm.865\,i$
5	1.04	$-.44 \pm.84\,i$	1.00004	$-.499992$	$\pm.8661\,i$
7	1.02	$-.52 \pm.84\,i$	1.000003	$-.499994$	$\pm.866021\,i$
9	1.01	$-.504 \pm.88\,i$	1.0000003	$-.5000005$	$\pm.8660251\,i$
11	1.01	$-.49 \pm.860\,i$	1.00000003	$-.499999993$	$\pm.86602546\,i$
13	1.007	$-.508 \pm.858\,i$	1.000000003	$-.499999995$	$\pm.866025400\,i$

Let us finally consider the function $1/\cos z$, which has an infinite number of simple poles. Although the theorem concerns the columns, this example shows that it is possible, in some cases, to obtain approximations of all the poles using the diagonals (only the positive poles are given).

$\lceil n/n \rceil$				
2	1.54			
4	1.57082	4.4		
6	1.570796320	4.72	6.9	
8	1.570796327	4.71231	8.0	9.0
exact	1.570796327	4.712388981	7.8539	10.9955

Montessus's theorem provides a result only if the exact number n of poles is known and only for the column sequence $([m/n])_{m\geq 0}$. The poles of f serve as *attractors* for the poles of $[m/n]$. But if f has less than n poles, then only some of the poles of $[m/n]$ are attracted and the other ones may go anywhere, destroying the convergence. If another column is considered, no result can be obtained as can be seen from counterexamples. The first one, due to Bender and Orszag (1978) and already quoted above, concerns the series $f(z) = (10+z)/(1-z^2)$ where the first column $([m/1])_{m\geq 0}$ cannot converge.

Taking into account the last counterexample (due to Perron (1957)), and coming back to a meromorphic function with n poles, it is now obvious that it is impossible to obtain a convergence result for all the sequences $([m/k])_m$ with k smaller or greater than n. It is a conjecture, made by Baker and Graves-Morris (1977), that at least a subsequence of $([m/k])_{m\geq 0}$ converges for $k \geq n$. Such a result was proved by Beardon (1968) for the column sequence $([m/1])_{m\geq 0}$:

Theorem 2.4 Let f be analytic in $|z| < R$. Then, for $r < R$, there exists a subsequence of $([m/1])_{m\geq 0}$ converging uniformly to f in the disc $|z| \leq r$.

The same result has been proved by Baker and Graves-Morris (1977) for the second and third columns. Buslaev, Gončar and Suetin (1984) established the conjecture for entire functions. For $R < \infty$ they showed that the conjecture is still true in a neighborhood of zero and they gave a counterexample for the whole disc.

2.3. Stieltjes series

The main references for this section are Baker (1975) and Baker and Graves-Morris (1981). The complete proofs are given in the last reference. A study of the subject can also be found in Karlson and Von Sydow (1976).

A Stieltjes series is a power series of the form

$$S(z) = \sum_{i=0}^{\infty} f_i(-z)^i \quad \text{with} \quad f_i = \int_0^{\infty} x^i \, d\varphi(x),$$

where φ is a positive, bounded, non-decreasing measure. The Stieltjes function associated to the Stieltjes series is

$$f(z) = \int_0^{\infty} \frac{d\varphi(x)}{1 + xz}.$$

So, the series S is the formal expansion of f into a power series, although this series may not converge except for $z = 0$ while the function is analytic in the cut plane $\mathbb{C} - (-\infty, 0)$. This is, for example, the case for the Euler series

$$f(z) = \int_0^{\infty} \frac{e^{-t} \, dt}{1 + tz}, \quad S(z) = \sum_{n=0}^{\infty} n!(-z)^n.$$

An important question is the *moment problem*, i.e. the existence and uniqueness of f corresponding to the moments f_i. If φ takes only a finite number of values, it is a step function: φ is constant on (u_i, u_{i+1}) for a finite number of u_i and so

$$f(z) = \sum_{1}^{p} \frac{\lambda_i}{1 + zu_i}.$$

To avoid this too-simple case, φ is assumed in the sequel to take an infinite number of different values.

The particularity of Stieltjes series is that the special form of the coefficients f_i allows us to study the corresponding Hankel determinants and thus to locate the zeros of the orthogonal polynomials $P_n^{(m)}$ where

$$D_{mn}(z) = z^n P_n^{(m)}(1/z)$$

is the denominator of $[m + n - 1/n]$.

So the most natural sequences to be considered are the paradiagonal sequences $([n + J/n])_n$, $J \geq -1$. From the first section we know that for each J the $P_n^{n+J} (\equiv P_n)$ satisfy the three-term recurrence relation

$$P_{n+1}(x) = (x - \beta_n^{n+J}) P_n(x) - \gamma_n^{n+J} P_{n-1}(x),$$

$$\gamma_n^{n+J} = h_n/h_{n-1}, h_n = c^{(J-1)}(P_n^2) = c^{(J-1)}(x^n P_n) = H_{n+1}^{(J-1)}/H_n^{(J-1)}.$$

It can be proved that all the Padé approximants exist for $m \geq n - 1$. So the Padé table is normal. Then, in the recurrence relation of $P_n^{(m)}$, all the γ_n^m are positive. From the theory of orthogonal polynomials, it means that each diagonal sequence $(P_n^{(m)})_n$, m fixed, is orthogonal with respect to a positive-definite functional. So $P_n^{(m)}$ has n real distinct zeros and the zeros of $P_n^{(m)}$ and $P_{n+1}^{(m)}$ interlace. We have the following theorem:

Theorem 2.5 All the zeros of $P_n^{(m)}$ are real, distinct and negative, for $m \geq n - 1 \geq 0$.

Let us now consider the convergence of paradiagonal sequences. As $P_n^{(m)}$ has n simple negative zeros α_i, the denominators D_{mn} have also n simple negative zeros $1/\alpha_i$. Thus, all the poles of the Padé approximants $([n + J/n])_n$ lie on the cut and there is no obstacle to the convergence in the cut plane $\mathbb{C} - (-\infty, 0]$.

We have the following theorem:

Theorem 2.6 Let $D(\Delta, r) = \{z \in \mathbb{C}, |z| \leq r$ and $\forall x \leq 0, d(z, x) \geq \Delta\}$. Then, for each $J \geq -1$, the sequence $([n + J/n])_n$ converges uniformly on $D(\Delta, r)$ to a function f^J analytic in the cut plane $\mathbb{C} - (-\infty, 0]$.

If the moment problem is determinate (i.e. if there exists a measure φ such that for all i the coefficients f_i of f are given by $f_i = \int_0^\infty x^i \, d\varphi(x)$), then all the f^J are identical to f. The problem is known to be determinate if the Stieltjes series has a nonzero radius of convergence R, or if $R = 0$ and the f_i's satisfy Carleman's condition: $\sum_{i \geq 1} (f_i)^{-1/2i}$ diverges.

For the Euler series, Carleman's condition is satisfied since

$$\sum_{n \geq 1} (1/n!)^{1/2n} \text{ is equivalent to } \sum_{n \geq 1} (1/n),$$

which diverges, and so the last theorem holds for the Euler function

$$f(z) = \int_0^\infty \frac{e^{-t} \, dt}{1 + tz}.$$

For such sequences, Padé approximants can be useful for reconstructing the function from its power series expansion.

In the case of convergent Stieltjes series of radius R, the last theorem can be put into a more precise form due to Markov (1948):

Theorem 2.7 $f(z) = \int_0^{1/R} \frac{d\varphi(u)}{1 + uz}$ is analytic in the cut plane $\mathbb{C} - (-\infty, -R]$. All the poles of $[n + J/n]$ lie in $(-\infty, -R]$. The convergence of the sequence $([n + J/n])_n$ is uniform in $D^+(\Delta, r)$, where

$$D^+(\Delta, r) = \{z, |z| \le r, \forall x \in (-\infty, -R], d(x, z) \ge \Delta\}.$$

A convergence result has also been proved by Prévost (1990) for the product of two Stieltjes functions:

Theorem 2.8　Let $f(z) = \displaystyle\int_0^a \frac{\mathrm{d}\alpha(x)}{1 - xz}$ and $g(z) = \displaystyle\int_{-b}^0 \frac{\mathrm{d}\beta(x)}{1 - xz}$, where a and b are finite and positive, and let α and β be positive, bounded and non-decreasing measures. Let also the integral $\int_{-b}^0 \int_0^a \frac{z}{z-x} \, \mathrm{d}\alpha(x) \, \mathrm{d}\beta(z)$ be assumed to exist. Then $f \cdot g(z) = \displaystyle\int_{-b}^a \frac{\mathrm{d}\gamma(x)}{1 - xz}$ is a Stieltjes function, and the sequence $([m_k + J, m_k]_{f \cdot g})_{m_k}$ $(J \ge -1$ and $\lim_{k \to \infty} m_k = +\infty)$ of Padé approximants of $f \cdot g$ converges uniformly on every compact subset of $\mathbb{C} - ((-\infty, -b^{-1}] \cup [a^{-1}, +\infty))$.

3. Generalizations

There exist many generalizations of Padé approximants. First of all, it is possible to define rational approximants to formal power series where the denominator is arbitrarily chosen and the numerator is then defined in order to achieve the maximum order of approximation. Such approximants are called *Padé-type approximants*. Their definition was given in the first section and we shall study below some of their convergence properties. In rational approximants, it is also possible to choose only a part of the denominator or a part of the numerator and the denominator. These are the *partial Padé approximants*. Such generalizations allow us to include into the construction of the approximant the information that is known about the zeros and the poles of the function being approximated, thus often leading to better convergence properties. *Multipoint Padé approximants* have expansions around several points which agree with the expansion of the function around the same points up to given orders. Padé approximants for series of functions have also received much attention. Other generalizations deal with Padé approximants for double series. Another important generalization is the vector case, which will be studied below. Series with coefficients in a non-commutative algebra have also received much interest, in particular the matrix case, due to their applications. Other types of approximants, such as the *Cauchy-type approximants* or the *Padé–Hermite approximants*, have been defined. It is, of course, possible to study combinations of these various generalizations such as the multipoint Padé–type approximants for multiple series of functions with matrix coefficients. Due to the space limitation of this article, we shall only present Padé–type approximants and the vector case and refer the interested reader to Brezinski and Van Iseghem

(1994) where more details about these generalizations can be found with the relevant references to the literature.

3.1. Padé–type approximants

As we saw in the first section, the function $f(z) = \sum_{n \geq 0} c_n z^n$ completely defines the linear functional c. It always has an integral representation in the complex field as stated by the following theorem:

Theorem 3.1 Let R be the radius of convergence of the series f and $\mathcal{H}_\alpha$ the space of holomorphic functions in the disc $D_{1/\alpha}$. Then c has the following representation

$$\forall g \in \mathcal{H}_\alpha, \quad c(g) = \frac{1}{2\pi i} \int_{|x|=r} f(x)g(1/x)\frac{\mathrm{d}x}{x}, \qquad \alpha < r < R.$$

In fact, in practical situations, the contour can be transformed continuously in such a way that $f(x)g(1/x)$ remains holomorphic in a neighborhood of it. An application of this result is the representation of the remainder term of Padé or Padé-type approximants $f(z) - P_{n-1}/Q_n(z)$.

$$f(z) = \sum_{n \geq 0} c_n z^n, \quad \tilde{v}_n(z) = z^n v_n(z^{-1}),$$

$$f(z) = c\left(\frac{1}{1-xz}\right), \quad c(g) = \frac{1}{2\pi i}\int_C f(x)g(1/x)\frac{\mathrm{d}x}{x},$$

$$f(z) - c(P) = \frac{z^n}{\tilde{v}_n(z)} c\left(\frac{v(x)}{1-xz}\right) = \frac{1}{2\pi i}\frac{z^n}{\tilde{v}_n(z)}\int_C f(x)\frac{v_n(x^{-1})}{x-z}\mathrm{d}x$$

$$= \frac{1}{2\pi i}\int_{-C^{-1}} \frac{v_n(x)}{v_n(z^{-1})}\frac{f(x^{-1})}{x(1-xz)}\mathrm{d}x$$

Finally, using the notation of Cala Rodriguez and Wallin (1992), conditions on $\Gamma = -C^{-1}$, f and v_n are now summarized, with $v_n(z) = \tilde{Q}_n(z) = \prod_{j=1}^n(1-\beta_{jn}z)$, which means that the β_{jn} are the poles of the approximant whose denominator is $Q_n(z) = \prod_{j=1}^n(z-\beta_{jn})$ and the following error formula is obtained.

Theorem 3.2 Let f be analytic in a domain D containing zero, let $\beta_{jn}, 1 \leq j \leq n, n \geq 1$, be given complex numbers and let $z \in D - \{\beta_{jn}^{-1}\}$. Let Γ be a contour in D^{-1} consisting of a finite number of piecewise continuously differentiable closed curves with index

$$\mathrm{ind}_\Gamma(a) = \begin{cases} 1 & \text{if } a \in \bar{\mathbb{C}} - D^{-1}, \\ 0 & \text{if } a = z^{-1}, z \neq 0. \end{cases}$$

Finally, let P_{n-1}/Q_n be the $(n-1/n)$ Padé-type approximant of f with preassigned poles at the zeros of Q_n. Then

$$f(z) - \frac{P_{n-1}(z)}{Q_n(z)} = \frac{1}{i} \int_\Gamma \frac{\tilde{Q}_n(t)\, f(t^{-1})}{\tilde{Q}_n(z^{-1})\, t(1-zt)} dt.$$

Let us now formulate the theorem of Eiermann (1984) in the form he proved it (for a complete and detailed proof, see Eiermann (1984) or Cala Rodriguez and Wallin (1992)).

Theorem 3.3 Let f be analytic in a domain $D \subset \mathbb{C}$ containing zero. Let β_{jn}, $j = 1, \ldots, n$, be the given zeros of Q_n, and let $\Lambda \subset \mathbb{C}^2$ containing $\mathbb{C} \times \{0\}$ and such that, uniformly for (x, z) in compact subsets of Λ,

$$\lim_{n\to\infty} \frac{\tilde{Q}_n(x)}{\tilde{Q}_n(z^{-1})} = 0.$$

Then, the sequence of Padé-type approximants P_{n-1}/Q_n converges to f uniformly on compact subsets of $A = \{z, \forall \xi \in \bar{\mathbb{C}}/D, (\xi^{-1}, z) \in \Lambda\}$.

If f is a Stieltjes function defined on $\mathbb{C} - (-\infty, -R]$, then $K = \mathbb{C} - D^{-1} = [-1/R, 0]$, and Γ is any contour containing K. The best choice is to take Γ as small as possible, and the assumption on Q_n becomes, for z in some compact subset F of D, $\lim_{n\to\infty} \sup_{z\in F} \sup_{x\in K} \left| \frac{\tilde{Q}_n(x)}{\tilde{Q}_n(z^{-1})} \right| = 0.$ This remark leads to the following alternative results. With the same notation as before, we get:

Theorem 3.4 Let $K = \bar{\mathbb{C}}/D^{-1}, O$ a neighborhood of K and F some compact subset of D. If

$$\lim_{n\to\infty} \left(\sup_{(t,z)\in \bar{O}\times F} \left| \frac{\tilde{Q}_n(t)}{\tilde{Q}_n(z^{-1})} \right| \right) = 0, \text{ then } \lim_{n\to\infty} \left(\max_{z\in F} \left| f(z) - \frac{P_{n-1}(z)}{Q_n(z)} \right| \right) = 0.$$

Similarly, we have the following theorem:

Theorem 3.5 Let $K = \bar{\mathbb{C}}/D^{-1}, O$ a neighborhood of K and F some compact subset of D. Then for the sup norm on F $(z \in F)$, if

$$\overline{\lim_{n\to\infty}} \left(\sup_{t\in \bar{O}} \left\| \frac{\tilde{Q}_n(t)}{\tilde{Q}_n(z^{-1})} \right\|_F^{1/n} \right) \le r < 1, \text{ then } \overline{\lim_{n\to\infty}} \left(\left\| f(z) - \frac{P_{n-1}(z)}{Q_n(z)} \right\|_F^{1/n} \right) \le r.$$

If the $\tilde{Q}_n$'s are some orthogonal polynomials for which asymptotic formulae are known, the preceding theorem leads to interesting results (Prévost, 1983).

Another idea is to take Q_n with one multiple zero, so that $|Q_n(x)|^{1/n} = |x - \beta_n|$. Two cases are to be considered: the first one with $\beta_n = \beta$ or $\lim_n \beta_n = \beta$ (which gives the same rate of convergence), the second one with

β_n depending on n (and which may tend to ∞) in the case of entire functions having one singularity at $+\infty$ and none at $-\infty$ such as the exponential function (Van Iseghem, 1992; Le Ferrand, 1992b).

3.2. The vector case

• *Introduction*

There are at least two ways for obtaining approximations for vector problems.

The first one consists in considering simultaneously d scalar functions defined by their power series expansions in a neighborhood of zero, organizing them as one power series with vector coefficients in $\mathbb{C}^d$ and then looking for a rational approximation following the idea of Padé approximation, that is, finding a *best* approximant. This approach has been developed through the simultaneous approximants by de Bruin (1984) and through the vector approximants by Van Iseghem (1985, 1987b). In each case, the result is a rational approximant $(P_1/Q, \ldots, P_d/Q)$. In the vector case, all the P_i's have the same degree while they have to satisfy some constraints in the simultaneous case. Although the ideas are rather similar, the approximants are not the same, except when $\deg Q = nd$ and $\deg P_i = n$, $i = 1, \ldots, d$. In both cases, the scalar Padé approximants are recovered if the dimension d of the vectors is one. Only the vector case, which seems to be simpler, will be explained here.

The second way for obtaining vector approximations is through extrapolation and acceleration of vector sequences. In the scalar case, the ε-algorithm provides a way for computing Padé approximants and such a link can also be developed through the vector ε-algorithm or the topological ε-algorithm, as we shall see below.

• *Vector Padé approximants*

For vector Padé approximants, giving m as the common degree of all the numerators P_i and n as the degree of the common denominator Q defines completely the approximant. The vector Padé approximant $\mathbf{R}(t) = (P_\alpha/Q)_{\alpha=1,\ldots,d}$ is the best in the sense that it is impossible to improve simultaneously the order of approximation of all the components.

Let $\mathbf{F} = (f_1, \ldots, f_d)$. If, for each $\alpha = 1, \ldots, d$, we write

$$f_\alpha(z) = \sum_{i \geq 0} c_i^\alpha z^i,$$

then we set $\Gamma_i = (c_i^1, \ldots, c_i^d)^T \in \mathbb{C}^d$ and we define the series $\mathbf{F}$ by

$$\mathbf{F}(z) = \sum_{i \geq 0} \Gamma_i z^i.$$

Let $\Gamma : \mathbb{C}[[x]] \to \mathbb{C}^d$ be the linear functional defined by

$$\Gamma(x^i) = \Gamma_i.$$

Taking the components of the vector approximant as the Padé-type approximants of the f_α's, we get the following theorem:

Theorem 3.6 Let P be the Hermite interpolation polynomial of $1/(1-xz)$ at $x_1, \ldots, x_n$. Setting

$$v(x) = \prod_{i=1}^{n}(x - x_i), \qquad \tilde{v}(t) = t^n v(t^{-1}), \qquad \mathbf{W}(z) = \Gamma\left(\frac{v(z) - v(x)}{z - x}\right),$$

where the functional Γ acts on x, we have $\Gamma(P) = \tilde{\mathbf{W}}(z)/\tilde{v}(z)$, and

$$\mathbf{F}(z) - \Gamma(P) = \frac{z^n}{\tilde{v}(z)}\Gamma\left(\frac{v(x)}{1 - xz}\right) = \frac{z^n}{\tilde{v}(z)}\sum_{i \geq 0}D_i z^i, \qquad D_i = \Gamma(x^i v(x)).$$

$\tilde{\mathbf{W}}(z)/\tilde{v}(z)$ is called the *Padé-type approximant* $(n - 1/n)$ of $\mathbf{F}$.

The proof of this result is similar to that of the scalar case because each component $(\tilde{W}(z)/\tilde{V}(z))_\alpha$ is the Padé-type approximant of f_α for $\alpha = 1, \ldots, d$.

In order to improve the order of approximation on all the components, we have to choose the polynomial v such that a maximum number of D_i are zero. D_i is a vector of $\mathbb{C}^d$ and so $D_i = 0$ represents d scalar equations with the coefficients of v as unknowns. Thus, n being the degree of the denominator, the best order of approximation by rational functions of type $(n - 1, n)$ is $n + [n/d]$, where $[n/d]$ is the integer part of n/d.

Padé-type approximants (s/r) for arbitrary degrees s and r can be defined as for the scalar case. For any integer h, positive or not, we get

$$\mathbf{F}(z) = \sum_{i=0}^{h-1}\Gamma_i z^i + \mathbf{F}_h(z), \qquad \Gamma_i = 0 \quad \text{if } i < 0,$$

$$(r + h - 1/r)\mathbf{F}(z) = \sum_{i=0}^{h-1}\Gamma_i z^i + (r - 1/r)\mathbf{F}_h(z).$$

The order of approximation is $r + h - 1$. It can be increased up to $r + h - 1 + [r/d]$ by choosing the generating polynomials v of the vector Padé-type approximants. Let r and h be arbitrary integers ($r = nd + k$, $0 \leq k < d$); let us denote by $\Gamma^{(h)}$ the linear functional defined by $\Gamma^{(h)}(x^i) = \Gamma_{i+h}$ and by $P_{nd+k}^{(h)}$ the polynomial defined by the following equations:

$$\left.\begin{array}{ll}\Gamma^{(h)}(x^i P_{nd+k}^{(h)}(x)) = 0, & i = 0, \ldots, n - 1, \\[2mm] c^\alpha(x^{n+h}P_{nd+k}^{(h)}(x)) = 0, & \alpha = 1, \ldots, k.\end{array}\right\} \tag{3.1}$$

The Padé-type approximant $(r+h-1/r)$ generated by $P_{nd+k}^{(h)}$ will have the maximal order of approximation: the order of approximation is $r+h+n-1$ at least, $r+h+n$ for the first k components. This approximant will be called the *vector Padé approximant* $[r+h-1/r]_{\mathbf{F}}$.

Writing the conditions (3.1) as a linear system, we get an expression of $P_r^{(h)}$ as a ratio of two determinants. As usual the determinantal expression for $P_r^{(h)}$ gives rise to a determinantal expression for the vector Padé approximants where only the last row of the numerator is a vector, all the other rows $(\Gamma_i, \ldots, \Gamma_{r+i})$ being put for d scalar rows and the last one $(\Gamma_{n+h}^{(k)}, \ldots)$ representing the first k components of $(\Gamma_{n+h}, \ldots)$. Setting $s = r+h$ and $\Sigma_k = \sum_{i=0}^{k} \Gamma_i z^i$, we have

$$P_r^{(h)}(x) = \frac{\begin{vmatrix} \Gamma_h & \cdots & \Gamma_s \\ \vdots & & \vdots \\ \Gamma_{n-1+h} & \cdots & \Gamma_{n+s-1} \\ \Gamma_{n+h}^{(k)} & \cdots & \Gamma_{n+s}^{(k)} \\ 1 & \cdots & x^r \end{vmatrix}}{\begin{vmatrix} \Gamma_h & \cdots & \Gamma_{s-1} \\ \vdots & & \vdots \\ \Gamma_{h+n-1} & \cdots & \Gamma_{n+s-1} \\ \Gamma_{n+h}^{(k)} & \cdots & \Gamma_{n+s-1}^{(k)} \end{vmatrix}}, \; [s-1/r]_{\mathbf{F}}(z) = \frac{\begin{vmatrix} \Gamma_h & \cdots & \Gamma_s \\ \vdots & & \vdots \\ \Gamma_{h+n-1} & \cdots & \Gamma_{n+s-1} \\ \Gamma_{h+n}^{(k)} & \cdots & \Gamma_{n+s}^{(k)} \\ z^r \Sigma_{h-1} & \cdots & \Sigma_{s-1} \end{vmatrix}}{\begin{vmatrix} \Gamma_h & \cdots & \Gamma_s \\ \vdots & & \vdots \\ \Gamma_{h+n}^{(k)} & \cdots & \Gamma_{n+s}^{(k)} \\ z^r & \cdots & 1 \end{vmatrix}}.$$

Similarly to the scalar case, the polynomials $(P_r^{(h)})_{r\geq 0}$ are the generating polynomials of $[h+r-1/r]$, and for each h they satisfy a recurrence formula, which, here, is of order $d+1$ (i.e. with $d+2$ terms):

$$P_{r+1}^s(x) = (x - \beta_r^s)P_r^s(x) - \sum_{\mu=1}^{d} \gamma_\mu^s P_{r-\mu}^s(x). \tag{3.2}$$

If all the $P_r^{(h)}$ exist, then the last coefficient γ_d^h is not zero. Since a theory analogous to the theory of orthogonal polynomials can be developed, these polynomials have been called *vector orthogonal polynomials* (or of dimension d) (Van Iseghem, 1985, 1987b, 1989). A Shohat–Favard theorem can be proved: given a family $(P_r)_{r\geq 0}$ satisfying the the relation (3.2), there exist d functionals $c^1, \ldots, c^d$ such that the P_r's satisfy the relation (3.1) with respect to $\Gamma = (c^1, \ldots, c^d)$. The space of all possible Γ's is a vector space of dimension $(d!)$. And so there is, as in the scalar case, an equivalence between the vector orthogonality defined by the relations (3.1) and the family of polynomials defined by the recurrence relations (3.2).

From an algorithmic point of view, a QD-type algorithm linking two diagonals $(P_r^{(h)})_r$ and $(P_r^{(h+1)})_r$ can be defined. It allows one to move in all directions either in the table of the polynomials $(P_r^{(h)})$ or in the table of the vector Padé approximants. The approximants can be also computed by algorithms such as the recursive projection algorithm (R.P.A.) or the compact recursive projection algorithm (C.R.P.A.) of Brezinski (1983b).

An example of an algorithm deduced from the recurrence relations in the table of the polynomials is the generalization of the cross rule of Wynn obtained by Van Iseghem (1986). In the scalar case, this cross rule involves five approximants in the following array:

$$
\begin{array}{ccc}
 & N & \\
W & C & E \\
 & S &
\end{array}
$$

and it can be written in two different forms

$$(E - N)(C - W)(S - C) = (C - N)(E - C)(S - W),$$

$$\frac{1}{C - N} + \frac{1}{C - S} = \frac{1}{C - E} + \frac{1}{C - W}.$$

The proof can be extended to the vector case. The approximants involved are the following: W_i and N_i lying on diagonals and

$$
\begin{array}{cc}
W_d & N_d \\
 & \ddots \\
\end{array}
\qquad
\begin{array}{ccc}
N_1 & N & \\
W_1 & C & E \\
 & S &
\end{array}
$$

The explicit form is given for all the components $(E - C)^\alpha, \alpha = 1, \ldots, d$, the D_i being $d \times d$ determinants and the vectors indicated being the columns of the determinants

$$D_1 = |C - W_1, W_d - W_{d-1}, \ldots, W_2 - W_1|, \quad D_2 = |S - C, N_{d-1} - N_{d-2}, \ldots, N_1 - C|,$$

$$D_3 = |S - W_1, W_d - W_{d-1}, \ldots, W_2 - W_1|, \quad D_4 = |C - N, N_{d-1} - N_{d-2}, \ldots, N_1 - C|,$$

$$\frac{1}{(E - C)^\alpha} + \frac{1}{(C - N)^\alpha} = \frac{1}{(C - N)^\alpha} \frac{D_3 D_4}{D_1 D_2}, \qquad \alpha = 1, \ldots, d.$$

As in the scalar case, symbolic negative columns are defined so that the algorithm can be used with the center term C in the column with subscript zero. So this algorithm can be used for computing E from the first column and it gives all the approximants $[p/q]$ with $p \geq q - 1$.

Vector Padé approximants can be used for accelerating the convergence of vector sequences. Let $\mathbf{F} = \sum_{i \geq 0} \Gamma_i z^i$. Then, a vector sequence (S_n) can be canonically associated to $\mathbf{F}$ by taking $\Gamma_i = \Delta S_i$. Thus S_n is the nth partial sum of the series $\mathbf{F}(1)$ and the vector Padé approximants are associated to a transformation of sequences. By combining the rows in the determinantal expression of the vector approximant of **F** it follows that

$$[r + h/r]_{\mathbf{F}}(1) = \psi_r(S_h) = \frac{\begin{vmatrix} S_h & \cdots & S_{h+r} \\ \Delta S_h & \cdots & \Delta S_{h+r} \\ \vdots & & \vdots \\ \Delta S_{h+n-1} & \cdots & \Delta S_{h+r+n-1} \\ \Delta S_{h+n}^{(k)} & \cdots & \Delta S_{h+r+n}^{(k)} \end{vmatrix}}{\begin{vmatrix} 1 & \cdots & 1 \\ \Delta S_h & \cdots & \Delta S_{h+r} \\ \vdots & & \vdots \\ \Delta S_{h+n-1} & \cdots & \Delta S_{h+r+n-1} \\ \Delta S_{h+n}^{(k)} & \cdots & \Delta S_{h+r+n}^{(k)} \end{vmatrix}}.$$

The first row is formed by vectors and the following rows stand for the d rows of their components except, as usual, the last one which contains only the first k components.

The basic result is the following theorem.

Theorem 3.7 A necessary and sufficient condition for $\psi_r(S_n) = S \, \forall n$ is that the sequence (S_n) satisfies a linear recurrence relationship, that is, $\forall n$

$$\sum_{i=0}^{r} a_i(S_{n+i} - S) = 0, \quad \text{with} \quad \sum_{i=0}^{r} a_i \neq 0, \quad a_i \in \mathbb{C}.$$

In the scalar case, Aitken's Δ^2 process is recovered for $r = 2$ and the Padé table, Shanks transformation and ε-algorithm for $r \geq 2$. If $d \neq 1$ and if $r < d$, we recover the MPE (Minimal Polynomial Extrapolation) algorithm studied by Sidi, Ford and Smith (1986). For $r = d$, a transformation due to Henrici (1964) is obtained. For $r > d$, the transformation does not seem to have been studied yet independently from the vector Padé approximants (Van Iseghem, 1994). Graves-Morris (1994) defined another kind of approximants, also called vector Padé approximants, different from those studied below, which are generated by the vector ε-algorithm.

From the relation of the theorem above, it is clear that the results of this extrapolation method are to be linked with those of other extrapolation methods such as the vector ε-algorithm or the topological ε-algorithm.

• *The vector ε-algorithm*
The rule of the ε-algorithm is

$$\varepsilon_{k+1}^{(n)} = \varepsilon_{k-1}^{(n+1)} + \left[\varepsilon_k^{(n+1)} - \varepsilon_k^{(n)} \right]^{-1}$$

with $\varepsilon_{-1}^{(n)} = 0$ and $\varepsilon_0^{(n)} = S_n$.

Now, if S_n is a vector of $\mathbb{C}^p$, the preceding rule can be applied if the inverse of a nonzero vector y is defined. Using the pseudo-inverse of the rectangular matrix y, Wynn (1962) took $y^{-1} = y/(y, y)$. For this algorithm, the following result holds.

Theorem 3.8 A sufficient condition for $\varepsilon_{2k}^{(n)} = S \,\forall n$ is that $\forall n$,

$$a_0(S_n - S) + \cdots + a_k(S_{n+k} - S) = 0$$

with $a_0 + \cdots + a_k \neq 0$ and $a_0 a_k \neq 0$.

This theorem was first given by McLeod (1971) in the case where the a_i's are real numbers. The full proof was obtained by Graves-Morris (1983) for complex coefficients. Contrary to the scalar case, only the sufficiency has been proved. The theory of this vector ε-algorithm and its applications were based only on this theorem and other results about it were quite difficult to obtain (see, for example, Cordellier (1977)). This was due to the nonexistence of determinantal formulae for the vectors $\varepsilon_k^{(n)}$. It has been proved by Salam (1994) that these vectors can be expressed as a ratio of two designants, a notion generalizing that of determinants in a non-commutative algebra. This approach should lead to new theoretical results about the vector ε-algorithm.

- *The topological ε-algorithm*

A drawback of the vector ε-algorithm was its lack of determinantal expressions for the $\varepsilon_k^{(n)}$, due to the fact that this algorithm was obtained directly from the rule of the scalar ε-algorithm, by defining the inverse of a vector. Thus a possible remedy was to construct a vector-sequence transformation following the ideas of Shanks (1955) for the scalar case and then to obtain a recursive algorithm for its implementation following Wynn (1956). We start by assuming that the sequence (S_n) satisfies $\forall n$,

$$a_0(S_n - S) + \cdots + a_k(S_{n+k} - S) = 0.$$

Since, as above, it is assumed that $a_0 + \cdots + a_k \neq 0$, it is not a restriction to set this sum to 1. Thus we have $\forall n$,

$$S = a_0 S_n + \cdots + a_k S_{n+k}.$$

For computing the a_i's, we subtract this relation from the next one and multiply it scalarly by an arbitrary vector y. Thus for $i = 0, \ldots, k - 1$, we have

$$a_0(y, \Delta S_{n+i}) + \cdots + a_k(y, \Delta S_{n+i+k}) = 0.$$

Solving this system provides the a_i's and S. If we set

$$
e_k(S_n) = \frac{\begin{vmatrix} S_n & \cdots & S_{n+k} \\ (y, \Delta S_n) & \cdots & (y, \Delta S_{n+k}) \\ \vdots & & \vdots \\ (y, \Delta S_{n+k-1}) & \cdots & (y, \Delta S_{n+2k-1}) \end{vmatrix}}{\begin{vmatrix} 1 & \cdots & 1 \\ (y, \Delta S_n) & \cdots & (y, \Delta S_{n+k}) \\ \vdots & & \vdots \\ (y, \Delta S_{n+k-1}) & \cdots & (y, \Delta S_{n+2k-1}) \end{vmatrix}},
$$

then, by construction, we have $\forall n, e_k(S_n) = S$ if the sequence (S_n) satisfies the relation given above. If the sequence does not satisfy such a relation, then the determinants appearing in the expression of $e_k(S_n)$ can, however, be computed (the determinant in the numerator denotes the vector obtained as the linear combination of the vectors in its first row given by using the classical rules for expanding a determinant), and $e_k(S_n)$ is a generalization of the Shanks transformation and Padé approximants. Now, from a practical point of view, it is necessary to find an algorithm for computing recursively the vectors $e_k(S_n)$ without computing explicitly the determinants involved in the formula. This algorithm was called the topological ε-algorithm. Its rules are the following (Brezinski, 1975), with $\varepsilon_{-1}^{(n)} = 0$ and $\varepsilon_0^{(n)} = S_n$:

$$
\varepsilon_{2k+1}^{(n)} = \varepsilon_{2k-1}^{(n+1)} + \frac{y}{(y, \varepsilon_{2k}^{(n+1)} - \varepsilon_{2k}^{(n)})},
$$

$$
\varepsilon_{2k+2}^{(n)} = \varepsilon_{2k}^{(n+1)} + \frac{\varepsilon_{2k}^{(n+1)} - \varepsilon_{2k}^{(n)}}{(\varepsilon_{2k+1}^{(n+1)} - \varepsilon_{2k+1}^{(n)}, \varepsilon_{2k}^{(n+1)} - \varepsilon_{2k}^{(n)})}.
$$

The topological ε-algorithm can be considered as the construction of Padé approximants in the direction of y. If d independent directions are chosen and if the rows $((y, \Delta S_{n+i}), \ldots), i = 0, \ldots, k - 1$, are replaced by the d rows $((y_i, \Delta S_n), \ldots), i = 1, \ldots, d$, then the vector Padé approximants (or more exactly $\psi_k(S_n)$) are recovered. For $k = d + 1$, Henrici's transformation is obtained.

All these algorithms are, in fact, constructed from the same idea which is that of Padé approximation of achieving the maximum degree of approximation at zero. So, as we shall see in the next section, they have similar properties for their applications, for example, in solving systems of linear and nonlinear equations.

4. Applications

Other applications can be found in Cuyt and Wuytack (1987).

4.1. A-acceptable approximations to the exponential function

Let us consider the differential equation $y'(x) = -\lambda y(x)$ where λ is a complex number whose real part is strictly positive. Thus the solution will satisfy $\lim_{x\to\infty} y(x) = 0$. This differential equation (with the initial condition $y(0) = y_0$) is integrated by a numerical method that computes approximations y_n of the exact solution $y(nh)$, where h is the step size. This numerical method is said to be A-stable if $\forall h\lambda$ such that $\mathrm{Re}(h\lambda) > 0, \lim_{n\to\infty} y_n = 0$, which means that both the exact and the approximate solutions tend to zero at infinity.

Of course, since the exact solution is $y(x) = y_0 e^{-\lambda x}$, we have $y(x_{n+1}) = e^{-h\lambda} y(x_n)$ with $x_n = nh$. When using either a one-step or a multistep method, it can be proved that the approximate solution satisfies

$$y_{n+1} = r(h\lambda) y_n,$$

where r is a rational function. Thus, if the numerical method has order p, we have

$$r(z) = e^{-z} + \mathcal{O}\left(z^{p+1}\right).$$

Moreover, if the method is A-stable we must have, $\forall z$ such that $\mathrm{Re}(z) > 0, |r(z)| < 1$ since $y_n = [r(h\lambda)]^n y_0$.

Such a rational approximation to the exponential function is called A-acceptable and, of course, Padé, Padé-type and partial Padé approximants are candidates for such an r.

Using the maximum-modulus principle it can be shown that r is A-acceptable if and only if $\forall t \in \mathbb{R}, |r(it)| \leq 1, \lim_{|z|\to\infty} |r(z)| \leq 1$ and r is analytic in the right half part of the complex plane; see Alt (1972).

The A-acceptability of Padé approximants to the exponential function was studied by Ehle (1973) who proved:

Theorem 4.1 The Padé approximants $[n/n], [n-1/n]$ and $[n-2/n]$ of e^{-z} are A-acceptable for all n.

Let us now turn to Padé-type approximants. The following result is an adaptation of that of Crouzeix and Ruamps (1977) for rational approximants to the exponential function.

Theorem 4.2 Let r be a Padé-type approximant of e^{-z} with real coefficients, whose numerator has degree k and whose denominator has degree $n + k$ $(n \geq 0)$. Let

$$|r(it)|^2 = \frac{1 + \beta_1 t^2 + \cdots + \beta_k t^{2k}}{1 + \alpha_1 t^2 + \cdots + \alpha_{n+k} t^{2(n+k)}}.$$

If the zeros of the denominator of r have negative reals parts, if $\beta_i \leq \alpha_i$

for $i = [k/2] + 1, \ldots, k$ and if $0 \leq \alpha_i$ for $i = k + 1, \ldots, k + n$, then r is A-acceptable. ($[x]$ denotes the integer part of the real number x.)

When solving a parabolic partial differential equation of the second order, one obtains, after discretization of the space variable, a differential system of the form

$$
\begin{aligned}
Cu'(t) &= -Au(t) + v(t), \\
u(0) &= u_0,
\end{aligned}
$$

where C and A are real square matrices whose elements are independent of the time t. Using a one-step method for integrating this differential equation leads to $Q_k(Bh)u_{n+1} = P_m(Bh)u_n + T_n$, where, $B = C^{-1}A\, T_n$, is a matrix depending on k and m, u_n is an approximation of the exact solution $u(t_n)$ at the point t_n and Q_k and P_m are matrix polynomials of the respective degrees k and m. As before, $[Q_k(Bh)]^{-1}P_m(Bh)$ must be an approximation of e^{-Bh} and the order of the method is determined by that of the approximation. This approximation must be A-acceptable if an A-stable one-step method is needed. The computation of u_{n+1} from u_n requires the computation of the inverse of the matrix $Q_k(Bh)$. This computation is greatly simplified if $Q_k(z) = (1 + \alpha_k z)^k$. Indeed, in that case, the computation of u_{n+1} reduces to the solution of k systems of linear equations with the same matrix

$$
(I + \alpha_k Bh)v_{p+1} = v_p, \ p = 0, \ldots, k - 1, \ v_0 = P_m(Bh)u_n + T_n, \ v_k = u_{n+1}.
$$

Of course, such a simplification is impossible with Padé approximants but it becomes possible with Padé-type approximants. For convergence reasons, since $\lim_{k \to \infty}(1 + z/k)^k = e^z$ we shall make the choice $\alpha_k = 1/k$ which corresponds to the generating polynomials $v_k(x) = (x + 1/k)^k$. The following result can be proved:

Theorem 4.3 The Padé-type approximants $(k - 1/k)$ of e^{-z} constructed with the generating polynomials $v_k(x) = (x + 1/k)^k$ are A-acceptable for $k = 1, 2, 3$. The Padé-type approximants (k/k) constructed with the same generating polynomials are A-acceptable for $k = 1, \ldots, 4$. $(6/6)$ is not A-acceptable.

The study of the convergence of these approximants is due to Van Iseghem (1984) who proved the following results:

Theorem 4.4 The sequences $((k - 1/k))$ and $((k/k))$ of Padé-type approximants to $\exp(-z)$ constructed with the generating polynomials $v_k(x) = (x + 1/k)^k$ converge to $\exp(-z)$ uniformly and geometrically on every compact subset of the complex plane.

Complementary results on the A-acceptability of Padé-type approximants with a single pole were given by González-Concepción (1987). See Wanner

(1987) for a review on the A-acceptability of Padé approximants to the exponential function.

4.2. Laplace and other transforms

In this section, we shall show how Padé approximants can be used in the numerical solution of problems related to the Laplace, the Borel and the z-transforms.

The Laplace transform of a function f is defined by

$$\bar{f}(p) = \int_0^\infty e^{-pt} f(t) \mathrm{d}t.$$

The Borel transform is a Laplace transform, $B(z) = \bar{f}(1/z)$, and the z-transform is a discrete equivalent of the Laplace transform, as will be shown at the end of this section.

The Laplace transform is used in several cases: if f satisfies a functional equation, $\bar{f}$ satisfies a simpler one which can be solved more easily. For example, a linear ordinary differential equation is replaced by an algebraic equation, a linear partial differential equation is replaced by an ordinary differential equation and so forth. For example, the partial differential equation

$$\left(\frac{\partial^2}{\partial r^2} - 2\frac{\partial^2}{\partial r \partial x} + \frac{1}{r}\left(\frac{\partial}{\partial r} - \frac{\partial}{\partial x} \right) - \frac{m^2}{r^2} \right) R(x, r) = 0$$

is transformed, with respect to x, into

$$\left(\frac{\partial}{\partial r^2} + \left(\frac{1}{r} - 2p \right) \frac{\partial}{\partial r} - \left(\frac{p}{r} + \frac{m^2}{r^2} \right) \right) \bar{R}(p, r) = 0.$$

The problem is to find $f(t)$ from $\bar{f}(p)$. It requires the construction of approximate methods for computing inverse Laplace transforms that permit us to find the original function in a broad class of cases.

It must be noticed at once that the problem is always unstable: if $\bar{f}(p) = 1/(p - \alpha)$, then $f(t) = e^{\alpha t}$. So if there is an error in α, of say ε, then there is an amplification of the error: $f_\varepsilon = e^{\varepsilon t} f$. Conversely, if f is modified on a small interval, then $\bar{f}$ will have a very smooth change.

An extensive literature exists on the subject and on the different methods for inverting the Laplace transform. A review can be found, for example, in Luke (1969). We will only discuss here the problem of inversion by use of rational approximants.

In many applications, one knows explicitly $\bar{f}(p)$ and one wants to find $f(t)$ numerically. One way of obtaining approximations to the inverse $f(t)$ of $\bar{f}(p)$ is by approximating $\bar{f}(p)$ by a sequence of rational functions $\bar{f}_n(p)$, $n \geq 1$, and then inverting the $\bar{f}_n(p)$ exactly to obtain the sequence $f_n(t)$, $n \geq 1$. The hope is that, if the sequence $(\bar{f}(p))_n$ converges to $\bar{f}(p)$ quickly, then the sequence $(f_n(t))_n$ will converge to $f(t)$ also, more or less quickly. There are several ways of obtaining rational approximations to a given function, one of them being by expanding it into a Taylor series, and then forming the Padé table associated with the Taylor series. Detailed discussions and references to various applications can be found in the paper by Longman (1973).

The link with Prony's method (interpolation by a sum of exponential functions) has been generalized by Sidi (1981).

If $\bar{f}_n(p)$ has n poles, not necessarily distinct but of multiplicity n_i, with $\sum_1^m n_i \leq n$, then

$$\bar{f}_n(p) = \sum_{i=1}^m \sum_{j=1}^{n_i} \frac{A_{ij}}{(p-\alpha_i)^j} \Leftrightarrow f_n(t) = \sum_{i=1}^m \sum_{j=1}^{n_i} A_{ij} t^{j-1} e^{\alpha_i t}.$$

So the problem is the approximation of $f(t)$ by functions of the same type as $f_n(t)$. The result proved by Sidi is as follows:

Theorem 4.5 Define the set G_n as

$$G_n = \left\{ g(t) = \sum_{i=1}^m \sum_{j=1}^{n_i} B_{ij} t^{j-1} e^{\alpha_i t}, \quad \alpha_i \neq \alpha_j, \ \sum_1^m n_i \leq n, B_{ij} \in \mathbb{C} \right\}.$$

Now, let $g_n(t)$ be the function, if it exists, belonging to G_n that approximates $f(t)$ in $[0, \infty)$ in the following weak sense:

$$\int_0^\infty t^N e^{-wt} (f(t) - g_n(t)) t^i dt = 0, \quad i = 0, \ldots, 2n-1.$$

Then $\bar{g}_n(p)$, the Laplace transform of $g_n(t)$, is the Padé approximant $[n-1/n]$ of $\bar{f}(p-w)$. Furthermore, $g_n(t)$ is a real function of t if $\bar{f}(p)$ is real for real p.

As seen before, Padé-type approximants can give better numerical results than Padé approximants if some information about the poles is known. Various investigations have been conducted in the past ten years concerning the convergence of methods using Padé or Padé-type approximants.

One research path, followed by Van Iseghem (1987a), is through orthogonal polynomials and Padé-type approximants with one multiple pole. The basic remark is the following one, due to Tricomi (see Sneddon (1972)), about the Laguerre polynomials of order zero:

$$f(t) = e^{\lambda t} L_k(2\lambda t) \Leftrightarrow \bar{f}(p) = \frac{(p-\lambda)^k}{(p+\lambda)^{k+1}}.$$

So, the Laplace transform formally achieves a correspondence between a series in powers of $(p-\lambda)/(p+\lambda)$ and an expansion in Laguerre polynomials. Convergence in the least-squares sense is to be expected for (f_n), but better results are obtained. We have

$$\bar{f}(p) = \frac{1}{p+\lambda} \sum_{k\geq 0} a_k \left(\frac{p-\lambda}{p+\lambda}\right)^k.$$

The partial sum $\bar{f}_n(p)$ of this series is the $(n/n+1)$ Padé-type approximant of $\bar{f}(p)$ with denominator $(p+\lambda)^{n+1}$. If $\bar{f}(p) = \sum_{i\geq 0} c_i(p-\lambda)^i$, then the a_k's are easily computed:

$$a_k = \sum_{i=0}^k \binom{k}{i} c_i(2\lambda)^i.$$

The following results are obtained:

Theorem 4.6   Let us assume that $\bar{f}$ is analytic in the half plane $\mathrm{Re}(p) > 0$, $f(t)$ exists and $\int_0^\infty f^2(t)dt < \infty$. Then the sequence $(\bar{f}(p))_n$ converges to $\bar{f}$, uniformly on every compact set of the half plane $\mathrm{Re}(p) > 0$. The sequence $(f_n(t))_n$ converges to f in the least-squares sense. Furthermore if $(p+\lambda)\bar{f}(p)$ is analytic at infinity and if $\bar{f}(p)$ is analytic in $\mathrm{Re}(p) \geq 0$, then the sequence $(f_n(t))_n$ converges to $f(t)$ uniformly on compact sets of $\mathbb{R}^+$.

This theorem can be improved in two directions:
- $\bar{f}$ is analytic in $\mathrm{Re}(p) > w$ (instead of $w = 0$);
- $(p+\lambda)\bar{f}^{(k)}(p)$ is analytic at infinity (instead of $k = 0$).

With the first assumption, we get the following sequence $(h_n(t))$:

$$h_n(t) = \frac{1}{t^k} e^{(w-\lambda)t} \sum_0^n a_i L_i(2\lambda t), \quad \lim_{n\to\infty} \int_0^\infty t^{2k} e^{-2wt}(f(t) - h_n(t))^2 dt = 0.$$

With the second one, the same sequence converges uniformly to f on compact sets of $\mathbb{R}^+$.

The idea is to obtain a quickly decreasing sequence $a_i(\lambda)$, and the computations are very sensitive to that choice, even if the theoretical results are not. The choice of λ must be made in order to speed up the convergence of the power series $\sum_{i\geq 0} a_i u^i$ (i.e. the convergence of $\bar{f}_n(p)$ to $\bar{f}(p)$). Let

$$\varphi(u) = \sum_i a_i u^i, \quad (p+\lambda)\bar{f}(p) = \varphi(u), \quad u = \frac{p-\lambda}{p+\lambda}.$$

The singularities of φ are $(\alpha - \lambda)/\alpha + \lambda)$ with α a singularity of $\bar{f}$. So R_λ, the radius of convergence of φ, is $R_\lambda = 1/(\max_i |LA_i|)$ with L being the point represented by the complex number 1, and A_i being represented by $2\lambda/(\lambda - \alpha_i)$. So, the best λ is obtained by minimizing $\max_i |LA_i|$. Computations have been made and compared with those of Longman (1973) using Padé approximants. For the examples studied (with poles, branch points or isolated essential singularities) they give better results.

Let us now summarize two examples: the first one shows the improvement due to the choice of λ for the approximant $F_7(t)$ (the case $\lambda = 1$ has been obtained by Ward in Sneddon (1972) and the exact inverse $F(t) = e^{-t} - e^{-2t}$ is known); for the second example the results are obtained with the weight function, depending on k, $t^{2k}e^{-2kt}$, the choice of λ being optimal. The last line of each array contains the exact results when known.

$$\bar{f}(p) = 1/(p+1)(p+2)$$

λ	$t = 4$	$t = 1$	$t = 0.5$
1	.01797	.23263	.23866
1.2	.0179797	.2325430	.2386533
1.4	.0179802	.2325440	.2386513
	.0179801	.2325441	.2386512

$$\bar{f}(p) = (1/p)\ln(1+p)$$

k	$t = 4$	$t = 2$	$t = 0.5$
1	.013179	.049133	.558368
3	.013044	.048899	.559835
5	.013048	.048900	.559787
7	.013054	.048897	.559662
		.048900	.559773

It is obvious that for such unstable problems, no single method will give optimal results for all purposes and all occasions.

As the Borel transform is also a Laplace transform, Padé and Padé-type approximants can be used for its inversion, as explained by Marziani (1987). Let us first recall the basis of the Borel method, namely, the Watson–Nevanlinna theorem:

Theorem 4.7 Let $\alpha > 0$, $R > 0$ and $A > 0$ be given. We set $D_{\alpha,R} = \{z \in \mathbb{C}, 0 < |z| < R, |\arg z| \le \alpha + \pi/2\}$, $T_{\alpha,A} = \{z \in \mathbb{C}, |z| < 1/A\} \cup \{z \in \mathbb{C}, |\arg z| < \alpha\}$. Let f be analytic in $D_{\alpha,R}$ and continuous on $\bar{D}_{\alpha,R}$ and have there the asymptotic expansion $f(z) = \sum_{n=0}^{\infty} c_n z^n$ $(z \to 0)$. We assume that there exists $C > 0$ such that $\forall z \in D_{\alpha,R}$ and $\forall N$

$$\left| f(z) - \sum_{n=0}^{N} c_n z^n \right| \le C(N+1)! A^{N+1} |z|^{N+1}.$$

Then the Borel transform series $B(z) = \sum_{n=0}^{\infty} \frac{c_n}{n!} z^n$ converges in $\{z \in \mathbb{C}, |z| <$
$1/A\}$, $B(z)$ has an analytic continuation $g(z)$ in $T_{\alpha,A}$, the integral $F(z) = \frac{1}{z} \int_0^{\infty} e^{-t/z} g(t) dt$ is absolutely convergent $\forall z \in \{z \in \mathbb{C}, |z| < R, |\arg z| < \alpha\}$ and $F(z) = f(z)$.

Thus the integral $F(z)$ provides a formal sum for the asymptotic series $f(z)$ and the Taylor expansion of F around the origin coincides with the series f.

The main drawback of this method is that usually g is not known since, in practice, only a finite set of numerically computed coefficients c_n is available. The series B cannot be used either since it converges only for $|z| < 1/A$. Thus, the idea was to replace $g(t)$ in the definition of F by $[n + k/k]_B(t)$

with $n \geq -1$, giving rise to the Borel–Padé approximants

$$F_B^{[n+k/k]}(z) = \frac{1}{z} \int_0^\infty e^{-t/z}[n+k/k]_B(t)dt.$$

To prove that these Borel–Padé approximants tend to $f(z)$ when k tends to infinity, one has first to prove that $[n+k/k]_B$ tends to B uniformly when $k \to \infty$. Usually this is not possible and this was the reason why Marziani (1987) replaced the Padé approximant $[n+k/k]_B$ by the Padé-type approximant $(n+k/k)$ thus obtaining the so-called Borel–Padé-type approximant denoted by $F_B^{(n+k/k)}$. Using the convergence results for Padé-type approximants, he was able to prove the following theorem:

Theorem 4.8   Let f be an analytic function satisfying the assumptions of the preceding theorem with α arbitrarily close to π. If the generating polynomials v_k are the Chebyshev polynomials of the first kind $v_k(x) = T_k(2x/A+1)$, then $\forall n \geq -1$

$$f(z) = \lim_{k \to \infty} F_B^{(n+k/k)}(z)$$

for every z in the half plane $\{z, \mathrm{Re}(z) > 0\}$.

The numerical results given by Marziani (1987) show that the Borel–Padé-type approximants converge with almost the same rate as the Borel-Padé approximants. The main advantage is that one has complete control of the poles when using the Padé-type approximants and thus a proof of the convergence of the method can be obtained.

Let us end this section with the z-transform. It is a functional transformation of sequences that can be considered as equivalent to the Laplace transform for functions. While the Laplace transform is useful in solving differential equations, the z-transform plays a central rôle in the solution of difference equations. If one changes z into z^{-1}, it is identical to the method of generating functions introduced by the French mathematician François Nicole (1683–1758) and developed by Joseph Louis Lagrange (1736–1813). It has many applications in digital filtering and in signal processing as exemplified by Vich (1987). By signal processing we mean the transformation of a function f of the time t, called the input signal, into an output signal h. This transformation is realized via a system G called a digital filter. f can be known for all values of t, and in that case we speak of a continuous signal and a continuous filter, or it can only be known at equally spaced values of t, $t_n = nT$ for $n = 0, 1, \ldots$, where T is the period, and, in that case, we speak of a discrete signal and a discrete filter. The z-transform of a discrete signal is given by $F(z) = \sum_{n=0}^\infty f_n z^{-n}$, where $f_n = f(nT)$. Corresponding to the input sequence (f_n) is the output sequence $(h_n = h(nT))$. If we set $H(z) = \sum_{n=0}^\infty h_n z^{-n}$ then the system G can be represented by its so-called transfer function $G(z)$ such that $H(z) = G(z)F(z)$. In other words, if we write $G(z) = \sum_{n=0}^{\infty} g_n z^{-n}$ then $h_n = \sum_{k=0}^{n} f_k g_{n-k}$, $n = 0, 1, \ldots$. Thus if (f_n) and (h_n) are known, then (g_n) can be computed. An important problem in the analysis of digital filters is the identification of the transfer function when (f_n) and (h_n) are known. If the filter is linear then G is a rational function of z and, if not, its transfer function can be approximated by a rational function $R(z) = P(z)/Q(z)$ which is in fact the Padé approximant $[s/s]_G(z)$.

4.3. Systems of equations

As explained in Section 1.6, the scalar ε-algorithm (Wynn, 1956) is a recursive method for computing Padé approximants. It is also a powerful convergence-acceleration process (see, for example, Brezinski and Redivo-Zaglia (1991)). Since, in numerical analysis, one often has to deal with vector sequences, the ε-algorithm was generalized to the vector case. Also as usual, when generalizing from one dimension to several, several possible generalizations exist. However, in our case, they all have some properties in common since they were all built in order to compute exactly the vector S for sequences of vectors (S_n) such that, $\forall n$,

$$a_0(S_n - S) + \cdots + a_k(S_{n+k} - S) = 0,$$

where $a_0, \ldots, a_k$ are scalars such that $a_0 + \cdots + a_k \neq 0$.

Due to this property, the various generalizations of the ε-algorithm (which obviously give rise to the corresponding generalizations of Padé approximants for series with vector coefficients) have applications in linear algebra. Indeed, let us consider the sequence of vectors (x_n) generated by

$$x_{n+1} = Bx_n + b,$$

where B is a square matrix such that $A = I - B$ is regular and b is a vector. Let x be the unique solution of the system $Ax = b$. Then $x_n - x = B^n(x_0 - x)$, where x_0 is the initial vector of the sequence (x_n). Let $P_k(t) = a_0 + a_1 t + \cdots + a_k t^k$ be the minimal polynomial of the matrix B for the vector $x_0 - x$, that is, the polynomial of the minimal degree such that $P_k(A)(x_0 - x) = 0$. Then, $\forall n$, $A^n P_k(A)(x_0 - x) = a_0(x_n - x) + \cdots + a_k(x_{n+k} - x) = 0$. Moreover, since A is assumed to be regular, 1 is not an eigenvalue of B, that is, $P_k(1) = a_0 + \cdots + a_k \neq 0$. It follows, from the aforementioned property shared by the various generalizations of the ε-algorithm, that, when applied to the sequence (x_n), they will all lead to $\varepsilon_{2k}^{(n)} = x \forall n$. Thus, all these generalizations (and also the scalar ε-algorithm applied componentwise) are direct methods for solving systems of linear equations.

As we shall see below, the connection between extrapolation methods and linear algebra is still deeper. Moreover, such algorithms can also be used for solving systems of nonlinear equations. They are derivative-free and exhibit a quadratic convergence under the usual assumptions.

- *The vector ε-algorithm*

Let us give an application of Theorem 3.8 to the solution of systems of linear equations. We have the following theorem (Brezinski, 1974):

Theorem 4.9   Let us apply the vector ε-algorithm to the sequence

$$x_{n+1} = Bx_n + b, \qquad n = 0, 1, \ldots,$$

where x_0 is a given arbitrary vector.

If $A = I - B$ is regular, if m is the degree of the minimal polynomial of B for the vector $x_0 - x$ and if 0 is a zero of multiplicity r (possibly $= 0$) of this polynomial, then $\forall n \geq 0$

$$\varepsilon_{2(m-r)}^{(n+r)} = x.$$

If A is singular, if b belongs to its range (which means that the system has infinitely many solutions), if m and r are defined as above and if q denotes the multiplicity of the zero equal to 1 of this polynomial, then, if $q = 1$ we have $\forall n \geq 0$

$$\varepsilon_{2(m-r)-2)}^{(n+r)} = x,$$

where x is one of the solutions of the system. If $q = 2$, then $\forall n \geq 0$

$$\varepsilon_{2(m-r)-3}^{(n)} = y,$$

where y is a constant vector independent of n.

If A is singular, if b does not belong to its range (which means that the system has no solution), if m is the degree of the minimal polynomial of B for the vector $x_1 - x_0$, if 0 is a zero of multiplicity r (possibly $= 0$) of this polynomial and if q denotes the multiplicity of its zero equal to 1, then, if $q = 1$, we have $\forall n \geq 0$

$$\varepsilon_{2(m-r)-1}^{(n+r)} = z,$$

where z is a constant vector independent of n.

This theorem was generalized to the case where the sequence (x_n) is generated by

$$x_{n+1} = \sum_{i=0}^{k} B_i x_{n-i} + b,$$

where the B_i's are square matrices (Brezinski, 1974).

• *The topological ε-algorithm*

We saw above that the coefficients a_i appearing in the recurrence relation assumed to be satisfied by the S_n's are obtained by writing

$$a_0(y, \Delta S_{n+i}) + \cdots + a_k(y, \Delta S_{n+k+i}) = 0$$

for $i = 0, \ldots, k-1$, where y is an arbitrary vector. Another possibility consists in taking $i = 0$ in this relation and choosing several linearly independent vectors y_i instead of y. Again, solving the system of equations (together with $a_0 + \cdots + a_k = 1$) provides the a_i's and thus S. A recursive algorithm for implementing this procedure (Brezinski, 1975) was obtained by Jbilou (1988). It is called the $S\beta$-algorithm. When $y_i = e_i$, the method proposed by Henrici (1964) is recovered. Its theory was developed in detail by Sadok (1990) and a recursive algorithm for its implementation, called the H-algorithm, was given by Brezinski and Sadok (1987).

Thus, from Theorem 3.8, the scalar (applied component-wise), the vector and the topological ε-algorithms and the $S\beta$-algorithm are direct methods for solving systems of linear equations. However, it must be noticed that, due to their storage requirements, the ε-algorithms are not competitive with other direct methods from the practical point of view.

The topological ε-algorithm is also related to the method due to Lanczos (1952) for solving a system of linear equations $Ax = b$. This method consists in constructing a sequence of vectors (x_k) such that

- $\quad x_k - x_0 \in \text{span}(r_0, Ar_0, \ldots, A^{k-1}r_0)$,
- $\quad r_k = b - Ax_k \perp \text{span}(y, A^*y, \ldots, A^{*k-1}y)$,

where x_0 and y are arbitrary vectors and $r_0 = b - Ax_0$. These relations completely define the vectors x_k if they exist. An important property of the Lanczos method is its finite termination, namely, that $\exists k \leq p$ (the dimension of the system) such that $x_k = x$.

The vectors x_k and r_k can be recursively computed by several algorithms, the most well known being the biconjugate-gradient method due to Fletcher (1976) which becomes the conjugate gradient of Hestenes and Stiefel (1952) when the matrix A is symmetric and positive definite. The other algorithms for implementing the Lanczos method can be deduced from the theory of formal orthogonal polynomials (Brezinski and Sadok, 1993) thus showing the link with Padé approximants as studied by Gutknecht (1990). Thanks to the theory of formal orthogonal polynomials, the vectors r_k of the Lanczos method can be expressed as the ratios of two determinants. After some manipulations on the rows and the columns of these determinants and using the relation $B = I - A$, it can be proved (Brezinski, 1980) that the vectors x_k generated by the Lanczos method are identical to the vectors $\varepsilon_{2k}^{(0)}$ obtained by applying the topological ε-algorithm to the sequence $y_{n+1} = By_n + b$ with $y_0 = x_0$. With the determinantal formula of the topological ε-algorithm, a determinantal expression for the iterates of the CGS algorithm of Sonneveld (1989), which consists in squaring the formal orthogonal polynomials involved in the Lanczos method, was obtained by Brezinski and Sadok (1993).

There are many more connections between methods of numerical linear algebra and extrapolation algorithms but it is not our purpose here to emphasize this point. We shall refer the interested reader to Brezinski (1980), Sidi (1988), Brezinski and Redivo-Zaglia (1991), Brezinski and Sadok (1992) and Brezinski (1993).

- *Systems of nonlinear equations*

Let us consider the nonlinear fixed-point problem $x = F(x)$, where F is an application of $\mathbb{R}^p$ into itself. This problem can be solved by Newton's method which constructs a sequence of vectors converging quadratically to x under some assumptions. The main drawback of Newton's method is that it needs the knowledge of the Jacobian matrix of F, which is not always easily available. Quasi-Newton methods provide an alternative but with a slower rate of convergence. When $p = 1$, a well-known method is Steffensen's, which has a quadratic convergence under the same assumptions as Newton's method. Steffensen's method is based on Aitken's Δ^2 process and it does not need the knowledge of the derivative of F. Since the ε-algorithm generalizes Aitken's process, the problem arises of finding a generalization of Steffensen's method for solving a system of p nonlinear equations written in the form $x = F(x)$. This algorithm (Brezinski, 1970; Gekeler, 1972) is as follows:

- choose x_0,
- for $n = 0, 1, \ldots$ until convergence
- set $u_0 = x_n$,
- compute $u_{i+1} = F(u_i)$ for $i = 0, \ldots, 2p_n - 1$, where p_n is the degree of the minimal polynomial of $F'(x)$ for the vector $x_n - x$,
- apply the ε-algorithm to the vectors $u_0, \ldots, u_{2p_n}$,
- set $x_{n+1} = \varepsilon_{2p_n}^{(0)}$.

In this method, either the scalar (component-wise), the vector or the topological ε-algorithm could be used. However, the following theorem (Le Ferrand, 1992a) was only proved in the case of the topological ε-algorithm although all the numerical experiments show that it might also be true for the two other ε-algorithms. The proof is based on the determinantal expression of the vectors computed by the topological ε-algorithm. A similar result was also proved to hold for the vector Padé approximants (Van Iseghem, 1994). Let H_n be the matrix

$$H_n = \frac{1}{\|\Delta u_0\|} \begin{pmatrix} 1 & \cdots & 1 \\ (y, \Delta u_0) & \cdots & (y, \Delta u_{p_n}) \\ \vdots & & \vdots \\ (y, \Delta u_{p_n - 1}) & \cdots & (y, \Delta u_{2p_n - 1}) \end{pmatrix}.$$

Theorem 4.10 If the matrix $I - F'(x)$ is regular, if F' satisfies a Lipschitz condition and if $\exists N, \exists \alpha > 0$ such that $\forall n > N, |\det H_n| > \alpha$, then the sequence (x_n) generated by the previous algorithm converges quadratically to x for any x_0 in a neighborhood of x.

Henrici's method is also a method with a quadratic convergence for systems of nonlinear equations (Ortega and Rheinboldt, 1970). It is equivalent to the vector Padé approximants when $k = d + 1$. It has the same general structure as the algorithm given above after replacing $2p_n$ by $p_n + 1$ and using either the $S\beta$-algorithm or the H-algorithm instead of one of the ε-algorithms.

REFERENCES

R. Alt (1972),'Deux théorèmes sur la A-stabilité des schémas de Runge–Kutta simplement implicites', *RAIRO* **R3**, 99–104.

G. A. Baker, Jr. (1970), 'The Padé approximant method and some related generalizations', in *The Padé Approximant in Theoretical Physics* (G. A. Baker, Jr., and J. L. Gammel, eds.), Academic Press (New York), 1–39.

G. A. Baker, Jr. (1975), *Essentials of Padé Approximants*, Academic Press (New York).

G. A. Baker, Jr. and P. R. Graves-Morris (1977),'Convergence of rows of the Padé table', *J. Math. Anal. Appl.* **57**, 323–339.

G. A. Baker, Jr. and P. R. Graves-Morris (1981), *Padé Approximants*, Addison-Wesley (Reading).

A. F. Beardon (1968), 'The convergence of Padé approximants', *J. Math. Anal Appl.* **21**, 344–346.

C. M. Bender and S. A. Orszag (1978), *Advanced Mathematical Methods for Scientists and Engineers*, McGraw–Hill (New York).

C. Brezinski (1970), 'Application de l'ε-algorithme à la résolution des systèmes non linéaires', *C.R. Acad. Sci. Paris* **271 A**, 1174–1177.

C. Brezinski (1974), 'Some results in the theory of the vector ε-algorithm', *Linear Algebra Appl.* **8**, 77–86.

C. Brezinski (1975), 'Généralisation de la transformation de Shanks, de la table de Padé et de l'ε-algorithme', *Calcolo* **12**, 317–360.

C. Brezinski (1977), *Accélération de la Convergence en Analyse Numérique*, LNM 584, Springer-Verlag (Berlin).

C. Brezinski (1980), *Padé-Type Approximation and General Orthogonal Polynomials*, ISNM vol. 50, Birkhäuser (Basel).

C. Brezinski (1983a), 'Outlines of Padé approximation', in *Computational Aspects of Complex Analysis* (H. Werner et al., eds.), Reidel (Dordrecht), 1–50.

C. Brezinski (1983b), 'Recursive interpolation, extrapolation and projection', *J. Comput. Appl. Math.* **9**, 369–376.

C. Brezinski (1991), *History of Continued Fractions and Padé Approximants*, Springer-Verlag (Berlin).

C. Brezinski (1991a), *A Bibliography on Continued Fractions, Padé Approxima-tion, Extrapolation and Related Subjects*, Prensas Universitarias de Zaragoza (Zaragoza).

C. Brezinski (1991b), *Biorthogonality and Its Applications to Numerical Analysis*, Dekker (New York).

C. Brezinski (1993), 'Biorthogonality and conjugate gradient-type algorithms', in *Contributions in Numerical Mathematics* (R. P. Agarwal ed.), World Scientific (Singapore), 55–70.

C. Brezinski (1994), 'The methods of Vorobyev and Lanczos', *Linear Algebra Appl.*, to appear.

C. Brezinski and M. Redivo-Zaglia (1991), *Extrapolation Methods: Theory and Practice*, North-Holland (Amsterdam).

C. Brezinski and H. Sadok (1987), 'Vector sequence transformations and fixed point methods', in *Numerical Methods in Laminar and Turbulent Flows*, vol. 1 (C. Taylor et al., eds.), Pineridge Press (Swansea), 3–11.

C. Brezinski and H. Sadok (1992), 'Some vector sequence transformations with applications to systems of equations', *Numerical Algor.* **3**, 75–80.

C. Brezinski and H. Sadok (1993), 'Lanczos-type algorithms for solving systems of linear equations', *Appl. Numer. Math.* **11**, 443–473.

C. Brezinski and J. Van Iseghem (1994), 'Padé approximations', in *Handbook of Numerical Analysis*, vol. 3 (P. G. Ciarlet and J. L. Lions, eds.), North-Holland (Amsterdam).

M. G. de Bruin (1984), 'Simultaneous Padé approximation and orthogonality', in *Polynômes Orthogonaux et Applications* (C. Brezinski et al., eds.), LNM 1171, Springer-Verlag (Berlin), 74–83.

M. G. de Bruin and H. Van Rossum (1975), 'Formal Padé approximation', *Nieuw Arch. Wiskd. Ser. 23*, **3** 115–130.

V. I. Buslaev, A. A. Gončar and S. P. Suetin (1984), 'On the convergence of sub-sequences of the mth row of the Padé Table', *Math. USSR Sb.* **48**, 535–540.

F. Cala Rodriguez and H. Wallin (1992), 'Padé-type approximants and a summa-bility theorem by Eiermann', *J. Comput. Appl. Math.* **39**, 15–21.

F. Cordellier (1977), 'Particular rules for the vector ε-algorithm', *Numer. Math.* **27**, 203–207.

F. Cordellier (1979), 'Démonstration algébrique de l'extension de l'identité de Wynn aux tables de Padé non normales', in *Padé Approximation and Its Applications* (L. Wuytack ed.), LNM 765, Springer-Verlag (Berlin), 36–60.

M. Crouzeix and F. Ruamps (1977), 'On rational approximations to the exponen-tial', *RAIRO Anal. Numer.* **R11**, 241–243.

A. A. M. Cuyt and L. Wuytack (1987), *Nonlinear Methods in Numerical Analysis*, North-Holland (Amsterdam).

A. Draux (1983), *Polynômes Orthogonaux Formels: Applications*, LNM 974, Springer-Verlag (Berlin).

A. Draux and P. Van Ingelandt (1986), *Polynômes Orthogonaux et Approximants de Padé: Logiciels*, Technip (Paris).

B. L. Ehle (1973), 'A-stable methods and Padé approximation to the exponential', *SIAM J. Math. Anal.* **4**, 671–680.

M. Eiermann (1984), 'On the convergence of Padé-type approximants to analytic functions', *J. Comput. Appl. Math.* **10**, 219–227.

J. Favard (1935), 'Sur les polynômes de Tchebicheff', *C.R. Acad. Sci. Paris* **200**, 2052–2053.

R. Fletcher (1976), 'Conjugate gradient methods for indefinite systems', in *Numerical Analysis, Dundee 1975*, LNM 506, Springer-Verlag (Berlin), 73–89.

E. Gekeler (1972), 'On the solution of systems of equations by the epsilon algorithm of Wynn', *Math. Comput.* **26**, 427–436.

J. Gilewicz (1978), *Approximants de Padé*, LNM 667, Springer-Verlag (Berlin).

C. González-Concepción (1987), 'On the A-acceptability of Padé-type approximants to the exponential with a single pole', *J. Comput. Appl. Math.* **19**, 133–140.

W. B. Gragg (1972), 'The Padé table and its relation to certain algorithms of numerical analysis', *SIAM Rev.* **14**, 1–62.

P. R. Graves-Morris (1983), 'Vector-valued rational interpolants I', *Numer. Math.* **42**, 331–348.

P. R. Graves-Morris (1994), 'A review of Padé methods for the acceleration of convergence of a sequence of vectors', *Appl. Numer. Math.*, to appear.

M. H. Gutknecht (1990), 'The unsymmetric Lanczos algorithms and their relations to Padé approximation, continued fractions, and the QD algorithm', in *Proceedings of the Copper Mountain Conference on Iterative Methods*.

M. H. Gutknecht (1992), 'A completed theory of the unsymmetric Lanczos process and related algorithms, Part I', *SIAM J. Matrix Anal. Appl.* **13**, 594–639.

A. J. Guttmann (1989), 'Asymptotic analysis of power-series expansions', in *Phase Transitions and Critical Phenomena*, vol. 13 (C. Domb and J. L. Lebowitz, eds.), Academic Press (New York), 1–234.

P. Henrici (1964), *Elements of Numerical Analysis*, Wiley (New York).

P. Henrici (1974), *Applied and Computational Complex Analysis*, vol. 1, Wiley (New York).

M. R. Hestenes and E. Stiefel (1952), 'Methods of conjugate gradients for solving linear sytems', *J. Res. Natl. Bur. Stand.* **49**, 409–436.

K. Jbilou (1988), 'Méthodes d'Extrapolation et de Projection: Applications aux Suites de Vecteurs', Thesis, Université des Sciences et Technologies de Lille.

W. B. Jones and W. J. Thron (1988), 'Continued fractions in numerical analysis', *Appl. Numer. Math.* **4**, 143–230.

J. Karlson and B. Von Sydow (1976), 'The convergence of Padé approximants to series of Stieltjes', *Arkiv for Math.* **14**, 44–53.

C. Lanczos (1952), 'Solution of systems of linear equations by minimized iterations', *J. Res. Natl. Bur. Stand.* **49**, 33–53.

H. Le Ferrand (1992a), 'The quadratic convergence of the topological epsilon algorithm for systems of nonlinear equations', *Numerical Algor.* **3**, 273–284.

H. Le Ferrand (1992b), 'Convergence et Applications d'Approximations Rationnelles Vectorielles', Thesis, Université des Sciences et Technologies de Lille.

I. Longman (1971), 'Computation of the Padé table', *Int. J. Comput. Math.* **3B**, 53–64.

I. Longman (1973), 'On the generation of rational function approximations for Laplace transform inversion with an application to viscoelasticity', *SIAM J. Appl. Math.* **24**, 429–440.

L. Lorentzen and H. Waadeland (1992), *Continued Fractions with Applications*, North-Holland (Amsterdam).

Y. L. Luke (1969), *The Special Functions and Their Approximation, vol. 2*, Academic Press (New York).

A. P. Magnus (1988), *Approximations Complexes Quantitatives*, Cours de Questions Spéciales en Mathématiques, Facultés Universitaires Notre Dame de la Paix, Namur.

A. A. Markov (1948), *Selected Papers on Continued Fractions and the Theory of Functions Deviating Least from Zero* (in Russian) (N. I. Achyeser, ed.), Ogiz (Moscow).

M. F. Marziani (1987), 'Convergence of a class of Borel-Padé type approximants', *Nuovo Cimento* **99 B**, 145–154.

J. B. McLeod (1971), 'A note on the ε-algorithm', *Computing* **7**, 17–24.

R. de Montessus de Ballore (1902), 'Sur les fractions continues algébriques', *Bull. Soc. Math. France* **30**, 28–36.

J. M. Ortega and W. C. Rheinboldt (1970), *Iterative Solution of Nonlinear Equations in Several Variables*, Academic Press (New York).

J. Nuttall (1967), 'Convergence of Padé approximants for the Bethe–Salpeter amplitude', *Phys. Rev.* 157, 1312–1316

H. Padé (1984), *Oeuvres* (C. Brezinski, ed.), Librairie Scientifique et Technique Albert Blanchard (Paris).

O. Perron (1957), *Die Lehre von den Kettenbrüchen, vol. 2*, Teubner (Stuttgart).

M. Prévost (1983), 'Padé-type approximants with orthogonal generating polynomials', *J. Comput. Appl. Math.* **9**, 333–346.

M. Prévost (1990), 'Cauchy product of distributions with applications to Padé approximants', Note ANO–229, Université des Sciences et Technologies de Lille.

H. Rutishauser (1954), 'Der Quotienten–Differenzen–Algorithms', *Z. Angew. Math. Physik* **5**, 233–251.

H. Sadok (1990), 'About Henrici's transformation for accelerating vector sequences', *J. Comput. Appl. Math.* **29**, 101–110.

E. B. Saff (1972), 'An extension of Montessus de Ballore's theorem on the convergence of interpolating rational functions', *J. Approx. Theory* **6**, 63–67.

A. Salam (1994), 'Non-commutative extrapolation algorithms', *Numerical Algor.* **7**, 225–251.

D. Shanks (1955), 'Nonlinear transformations of divergent and slowly convergent sequences', *J. Math. Phys.* **34**, 1–42 .

J. Shohat (1938), 'Sur les polynômes orthogonaux généralisés', *C.R. Acad. Sci. Paris* **207**, 556–558.

A. Sidi (1981), 'The Padé table and its connection with some weak exponential function approximations to Laplace transform inversion', in *Padé Approximation and Its Applications, Amsterdam 1980* (M. G. De Bruin and H. Van Rossum, eds.), LNM 888, Springer-Verlag (Berlin), 352–362.

A. Sidi (1988), 'Extrapolation vs. projection methods for linear systems of equations', *J. Comput. Appl. Math.* **22**, 71–88.

A. Sidi, W. F. Ford and D. A. Smith (1986), 'Acceleration of convergence of vector sequences', *SIAM J. Numer. Anal.* **23**, 178–196. ff

I. N. Sneddon (1972), *The Use of Integral Transforms*, McGraw–Hill (New York).

P. Sonneveld (1989), 'CGS, a fast Lanczos-type solver for nonsymmetric linear systems', *SIAM J. Sci. Stat. Comput.* **10**, 36–52.

J. Van Iseghem (1984), 'Padé-type approximants of $\exp(-z)$ whose denominators are $(1 + z/n)^n$', *Numer. Math.* **43**, 282–292.

J. Van Iseghem (1985), 'Vector Padé approximants', in *Numerical Mathematics and Applications* (R. Vichnevetsky and J. Vignes, eds.), North-Holland (Amsterdam), 73–77.

J. Van Iseghem (1986), 'An extended cross rule for vector Padé approximants', *Appl. Numer. Math.* **2**, 143–155.

J. Van Iseghem (1987a), 'Laplace transform inversion and Padé-type approximants', *Appl. Numer. Math.* **3** , 529–538.

J. Van Iseghem (1987b), 'Approximants de Padé Vectoriels', Thesis, Université des Sciences et Technologies de Lille.

J. Van Iseghem (1989), 'Convergence of the vector QD algorithm. Zeros of vector orthogonal polynomials', *J. Comput. Appl. Math.* **25**, 33–46. ff

J. Van Iseghem (1992), 'Best choice of the poles for the Padé-type approximant of a Stieltjes function', *Numerical Algor.* **3**, 463–476.

J. Van Iseghem (1994), 'Convergence of vectorial sequences: applications', *Numer. Math.*, to appear.

H. Van Rossum (1953), 'A Theory of Orthogonal Polynomials Based on the Padé Table', Thesis, University of Utrecht, Van Gorcum (Assen).

R. Vich (1987), *Z-Transform: Theory and Applications*, Reidel (Dordrecht).

H. Wallin (1974), 'The convergence of Padé approximants and the size of the power series coefficients', *Applicable Analysis* **4**, 235–251.

H. Wallin (1987), 'The divergence problem for multipoint Padé approximants of meromorphic functions', *J. Comput. Appl. Math.* **19**, 61–68.

G. Wanner (1987), 'Order stars and stability', in *The State of the Art in Numerical Analysis* (A. Iserles and M. J. D. Powell, eds.), Clarendon Press (Oxford), 451–471.

P. J. S. Watson (1973), 'Algorithms for differentiation and integration', in *Padé Approximants and Their Applications* (P. R. Graves-Morris, ed.), Academic Press (New York), 93–97.

P. Wynn (1956), 'On a device for computing the $e_m(S_n)$ transformation', *MTAC* **10**, 91–96.

P. Wynn (1962), 'Acceleration techniques for iterated vector and matrix problems', *Math. Comput.* **16**, 301–322.

P. Wynn (1966), 'Upon systems of recursions which obtain among the quotients of the Padé table', *Numer. Math.* **8**, 264–269.

Acta Numerica (1995), pp. 105–158

Introduction to Adaptive Methods for Differential Equations

Kenneth Eriksson

Mathematics Department,
Chalmers University of Technology,
412 96 Göteborg
kenneth@math.chalmers.se

Don Estep

School of Mathematics,
Georgia Institute of Technology,
Atlanta GA 30332
estep@pmath.gatech.edu

Peter Hansbo

Mathematics Department,
Chalmers University of Technology,
412 96 Göteborg
hansbo@math.chalmers.se

Claes Johnson

Mathematics Department,
Chalmers University of Technology,
412 96 Göteborg
claes@math.chalmers.se

Knowing thus the Algorithm of this calculus, which I call Differential Calculus, all differential equations can be solved by a common method (Gottfried Wilhelm von Leibniz, 1646–1719).

When, several years ago, I saw for the first time an instrument which, when carried, automatically records the number of steps taken by a pedestrian, it occurred to me at once that the entire arithmetic could be subjected to a similar kind of machinery so that not only addition and subtraction, but also multiplication and division, could be accomplished by a suitably arranged machine easily, promptly and with sure results.... For it is unworthy of excellent men to lose hours like slaves in the labour of calculations, which could safely be left to anyone else if the machine was used.... And now that we may give final praise to the machine, we may say that it will be desirable to all who are engaged in computations which, as is well known, are the managers of financial affairs, the administrators of others estates, merchants, surveyors, navigators, astronomers, and those connected with any of the crafts that use mathematics (Leibniz).

CONTENTS

1. Leibniz's vision

Newton and Leibniz invented calculus in the late 17th century and laid the foundation for the revolutionary development of science and technology up to the present day. Already 300 years ago, Leibniz sought to create a 'marriage' between calculus and computation, but failed because the calculator he invented was not sufficiently powerful. However, the invention of the modern computer in the 1940s started a second revolution and today, we are experiencing the realization of the original Leibniz vision. A concrete piece of evidence of the 'marriage' is the rapid development and spread of mathematical software such as Mathematica, Matlab and Maple and the large number of finite-element codes.

The basic mathematical models of science and engineering take the form of differential equations, typically expressing laws of physics such as conservation of mass or momentum. By determining the solution of a differential equation for given data, one may gain information concerning the physical process being modelled. Exact solutions may sometimes be determined through symbolic computation by hand or using software, but in most cases this is not possible, and the alternative is to approximate solutions with numerical computations using a computer. Often massive computational effort is needed, but the cost of computation is rapidly decreasing and new possibilities are quickly being opened. Today, differential equations modelling complex phenomena in three space dimensions may be solved using desktop workstations.

As a familiar example of mathematical modelling and numerical solution, consider weather prediction. Weather forecasting is sometimes based on solving numerically a system of partial differential equations related to the Navier–Stokes equations that model the evolution of the atmosphere beginning from initial data obtained from measuring the physical conditions – temperature, wind speed, etc. – at certain locations. Such forecasts sometimes give reasonably correct predictions but are also often incorrect. The sources of errors affecting the reliability are data, modelling and computation. The initial conditions at the start of the computer simulation are measured only approximately, the set of differential equations in the model only approximately describe the evolution of the atmosphere, and finally the differential equations can be solved only approximately. All these contribute to the total error, which may be large. It is essential to be able to estimate the total error by estimating individually the contributions from the three sources and to improve the precision where most needed. This example contains the issues in mathematical modelling that are common to all applications.

In these notes, we present a framework for the design and analysis of computational methods for differential equations. The general objective is to achieve reliable control of the total error in mathematical modelling including data, modelling and computation errors, while making efficient use of computational resources. This goal may be achieved using adaptive methods with feedback from computations. The framework we describe is both simple enough to be introduced early in the mathematical curriculum and general enough to be applied to problems on the frontiers of research. We see a great advantage in using this framework in a mathematics education program. Namely, its simplicity suggests that numerical methods for differential equations could be introduced even in the calculus curriculum, in line with the Leibniz idea of combining calculus and computation. In these notes, we hope to reach a middle ground between mathematical detail and ease of understanding. In Eriksson, *et al.* (1994), we give an even more simplified version aimed at early incorporation in a general mathematical curriculum. These notes, together with the software Femlab implementing the adaptive methods for a variety of problems, are publicly available through the Internet; see below. In the textbook Eriksson, *et al.* (in preparation), we develop the framework in detail and give applications not only to model problems, but also to a variety of basic problems in science including, for example, the Navier–Stokes equations for fluid flow.

We begin by discussing the basic concepts of predictability and computability, which are quantitative measures of the accuracy of prediction from the computational solution of a mathematical model consisting of differential equations.

In the next part, we present an abstract framework for discretization, error estimation and adaptive error control. We introduce the fundamental concepts of the framework: reliability and efficiency, a priori and a posteriori error estimates, accuracy and stability. We then recall the basic principles underlying the Galerkin finite-element method (Fem), which we use as a general method of discretization for all differential equations. We then describe the fundamental ingredients of error estimates for Galerkin discretization, including stability, duality, Galerkin orthogonality and interpolation. We also discuss data, modelling, quadrature and discrete-solution errors briefly.

In the last part, we apply this framework to a variety of model problems. We begin by recalling some essential facts from interpolation theory. We next consider a collection of model problems including stationary as well as time-dependent, linear and non-linear, ordinary and partial differential equations. The model problems represent a spectrum of differential equations including problems of elliptic, parabolic and hyperbolic type, as well as general systems of ordinary differential equations. In each case, we derive a posteriori and a priori error bounds and then construct an adaptive algorithm based on feedback from the computation. We present a sample of computations to illustrate the results. We conclude with references to the literature and some reflections on future developments and open problems.

2. Computability and predictability

Was man mit Fehlerkontrolle nicht berechnen kann, darüber muss mann schweigen (Wittgenstein).

The ability to make predictions from a mathematical model is determined by the concepts of *computability* and *predictability*. We consider a mathematical model of the form

$$A(u) = f, \qquad (2.1)$$

where A represents a differential operator with specified coefficients (including boundary and initial conditions) on some domain, f is given data and u is the unknown solution. Together, A and f define the mathematical model. We assume that by numerical and/or symbolic computation an approximation U of the exact solution u is computed, and we define the *computational error* $e_c \equiv u - U$. The solution u, and hence U, is subject to perturbations from the data f and the operator A. Letting the unperturbed form of (2.1) be

$$\hat{A}(\hat{u}) = \hat{f}, \qquad (2.2)$$

with unperturbed operator $\hat{A}$, data $\hat{f}$ and corresponding solution $\hat{u}$, we define the *data-modelling error* $e_{\mathrm{dm}} \equiv \hat{u} - u$. In a typical situation, the unperturbed problem (2.2) represents a complete model that is computationally too complex to allow direct computation, and (2.1) a simplified model that is actually used in the computation. For example, (2.2) may represent the full Navier–Stokes equations, and (2.1) a modified Navier–Stokes equations that is determined by a turbulence model that eliminates scales too small to be resolved computationally.

We define the total error e as the sum of the data-modelling and computational errors,

$$e \equiv \hat{u} - U = \hat{u} - u + u - U \equiv e_{\mathrm{dm}} + e_{\mathrm{c}}. \tag{2.3}$$

Basic problems in computational mathematical modelling are: (i) estimate quantitatively the total error by estimating both the data-modelling error e_{dm} and the computational error e_{c}, and (ii) control any components of the data-modelling and computational errors that can be controlled. Without some quantitative estimation and even control of the total error, mathematical modelling loses its meaning.

We define the solution $\hat{u}$ of the unperturbed model (2.2) to be *predictable* with respect to a given norm $\| \cdot \|$ and tolerance $TOL > 0$ if $\|e_{\mathrm{dm}}\| \leq TOL$. We define the solution u of the perturbed model (2.1) to be *computable* with respect to a given norm $\| \cdot \|$, tolerance TOL and computational work, if $\|e_{\mathrm{c}}\| \leq TOL$ with the given computational work. Note that the choice of norm depends on how the error is to be measured. For example, the L^2 and L^∞ norms are appropriate for the standard goal of approximating the values of a solution. Other norms are appropriate if some qualitative feature of the solution is the goal of approximation. We note that the predictability of a solution is quantified in terms of the norm $\| \cdot \|$ and the tolerance TOL, whereas the computability of a solution is quantified in terms of the available computational power, the norm $\| \cdot \|$ and the tolerance TOL. There is a natural scale for computability for all models, namely, the level of computing power. The scale for predictablity depends on the physical situation underlying the model. The relevant level of the tolerance TOL and the choice of norm depend on the particular application and the nature of the solution $\hat{u}$.

A mathematical model with predictable and computable solutions may be useful since computation with the given model and data may yield relevant information concerning the phenomenon being modelled.

If the uncertainty in data and/or modelling is too large, individual solutions may effectively be non-predictable, but may still be computable in the sense that the computational error for each choice of data is below the chosen tolerance. In such cases, accurate computations on a set of data may give useful information of a statistical nature. Thus, models with non-predictable but computable solutions may be considered partially deterministic, that is, deterministic from the computational point of view but non-deterministic from the data-modelling point of view. One may think of weather prediction again, in which it is not possible to describe the initial state as accurately as one could enter the data into a computation. Finally, models for which solutions are non-computable do not seem to be useful from a practical point of view.

The computational error e_c is connected to perturbations arising from discretization through a certain stability factor S_c measuring the sensitivity of u to perturbations from discretization. The computability of a problem may be estimated quantitatively in terms of the stability factor S_c and a quantity Q related to the nature of the solution u being computed and the tolerance level. The basic test for computability reads: if

$$S_c \times Q \leq P \tag{2.4}$$

where P is the available computational power, then the problem is numerically computable, whereas if $S_c \times Q > P$, then the problem is not computable. In this way we may give the question of numerical computability a precise quantitative form for a given exact solution, norm, tolerance and amount of computational power. Note that $S_c \times Q$ is related to the complexity of computing the solution u, and an uncomputable solution has a very large stability factor S_c. This occurs with pointwise error control of direct simulation of turbulent flow without turbulence modelling over a long time, for example.

Similarly, the sensitivity of $\hat{u}$ to data errors may be measured in terms of a stability factor S_d. If $S_d \times \delta$ is sufficiently small, where δ measures the error in the data, then the problem is predictable from the point of view of data error. In addition, some kind of modelling errors can be associated to a stability factor S_m and if $S_m \times \mu$ is sufficiently small, where μ measures the error in the model, then the problem is predictable from the point of view of modelling.

Different perturbations propagate and accumulate differently, and this is reflected in the different stability factors. The various stability factors are approximated by numerically solving auxiliary linear problems. In the adaptive algorithms to be given, these auxiliary computations are routinely carried out as a part of the adaptive algorithm and give critically important information on perturbation sensitivities.

3. The finite-element method

$(Fe)^m$: Finite elements, For everything, For everyone, 'For ever' $(m = 4)$.

Fem is based on

- Galerkin's method for discretization,
- piecewise-polynomial approximation in space, time or space/time.

In a Galerkin method, the approximate solution is determined as the member of a specified finite-dimensional space of trial functions for which the residual error is orthogonal to a specified set of test functions. The residual error, or simply the residual, is obtained by inserting the approximate solution into the given differential equation. The residual of the exact solution is zero, whereas the residual of approximate solutions deviates from zero. Applying this method for a given set of data leads to a system of equations that is solved using a computer to produce the approximate solution. In Fem, the trial and test functions are piecewise polynomials. A piecewise polynomial is a function that is equal to a polynomial, for example, a linear function, on each element of a partition of a given domain in space, time or space-time into subdomains. The subdivision is referred to as a mesh and the subdomains as elements. In the simplest case, the trial and test space are the same.

If the trial functions are continuous piecewise polynomials of degree q and the test functions are continuous or discontinuous, we refer to the resulting methods as continuous Galerkin or $cG(q)$ methods. With discontinuous piecewise polynomials of degree q in both trial and test space, we obtain discontinuous Galerkin methods or $dG(q)$ methods.

4. Adaptive computational methods

The goal of the design of any numerical computational method is

- *reliability,*
- *efficiency.*

Reliability means that the computational error is controlled on a given tolerance level; for instance, the numerical solution is guaranteed to be within 1 per cent of the exact solution at every point. Efficiency means that the computational work to compute a solution within the given tolerance is essentially as small as possible.

The computational error of a Fem has three sources:

- Galerkin discretization,
- quadrature,
- solution of the discrete problem.

The Galerkin discretization error arises because the solution is approximated by piecewise polynomials. The quadrature error comes from evaluating the integrals arising in the Galerkin formulation using numerical quadrature, and the discrete-solution error results from solving the resulting discrete systems only approximately, using Newton's method or multigrid methods, for example. It is natural to seek to balance the contribution to the total computational error from the three sources.

To achieve the goals of reliability and efficiency, a computational method must be adaptive with feedback from the computational process. An *adaptive method* consists of a discretization method together with an *adaptive algorithm*. An adaptive algorithm consists of

- a *stopping criterion* guaranteeing error control to a given tolerance level,
- a *modification strategy* in case the stopping criterion is not satisfied.

The adaptive algorithm is used to optimize the computational resources to achieve both reliability and efficiency. In practice, optimization is performed by an iterative process, where in each step an approximate solution is computed on a given mesh with piecewise polynomials of a certain degree, a certain quadrature and a discrete-solution procedure. If the stopping criterion is satisfied, then the approximate solution is accepted. If the stopping criterion is not satisfied, then a new mesh, polynomial approximation, set of quadrature points and discrete-solution process are determined through the modification strategy and the process is continued. To start the procedure, a coarse mesh, low-order piecewise-polynomial approximation, set of quadrature points and discrete-solution procedure are needed.

Feedback is centrally important to the optimization process. The feedback is provided by the computational information used in the stopping criteria and the modification strategy.

We now consider the Galerkin discretization error. Adaptive control of this error is built on *error estimates*. The control of quadrature and discrete-solution errors is largely parallel, but each error has its own special features to be taken into account.

5. General framework for analysis of Fem

5.1. Error estimates

Error estimates for Galerkin discretizations come in two forms:

- *a priori error estimates,*
- *a posteriori error estimates.*

An a priori estimate relates the error between the exact and the approximate solution to the regularity properties of the exact (unknown) solution. In an a posteriori estimate, the error is related to the regularity of the approximation.

The basic concepts underlying error estimates are

- *accuracy,*
- *stability.*

Accuracy is a measure of the level of the discretization at each point of the domain, while stability is a measure of the degree to which discretization errors throughout the domain interact and accumulate to form the total error. These properties enter in different forms in the a posteriori and a priori error estimates. The a posteriori version may be expressed conceptually as follows:

$$\text{small residual} + \text{stability of the continuous problem} \Longrightarrow \text{small error,} \quad (5.1)$$

where the continuous problem refers to the given differential equation. The a priori version takes the conceptual form

$$\text{small interpolation error} + \text{stability of the discrete problem} \Longrightarrow \text{small error,} \quad (5.2)$$

where the interpolation error is the difference between the exact solution and a piecewise polynomial in the Fem space that interpolates the exact solution in some fashion. Note that the a posteriori error estimate involves the stability of the continuous problem and the a priori estimate the stability of the discrete problem. We see that accuracy is connected to the size of the residual in the a posteriori case, and to the interpolation error in the a priori case.

Both the residual and the interpolation error contribute to the total error in the Galerkin solution. The concept of stability measures the accumulation of the contributions and is therefore fundamental. The stability is measured by a multiplicative *stability factor*. The size of this factor reflects the computational difficulty of the problem. If the stability factor is large, then the problem is sensitive to perturbations from the Galerkin discretization and more computational work is needed to reach a certain error tolerance.

In general, there is a trade-off between the norms used to measure stability and accuracy, that is, using a stronger norm to measure stability allows a weaker norm to be used to measure accuracy. The goal is to balance the measurements of stability and accuracy to obtain the smallest possible bound on the error. The appropriate stability concept for Galerkin discretization methods on many problems is referred to as *strong stability* because the norms used to measure stability involve derivatives. Strong stability is possible because of the orthogonality built into the Galerkin discretization. For some problems, Galerkin's method needs modification to enhance stability.

The stopping criterion is based solely on the a posteriori error estimate. The modification strategy, in addition, may build on a priori error estimates. The adaptive feature comes from the information gained through computed solutions.

5.2. A posteriori error estimates

The ingredients of the proofs of *a posteriori* error estimates for Galerkin discretization are:

1 Representation of the error in terms of the residual of the finite-element solution and the solution of a continuous (linearized) dual problem.
2 Use of Galerkin orthogonality.
3 Local interpolation estimates for the dual solution.
4 Strong-stability estimates for the continuous dual problem.

We describe (1)–(4) in an abstract situation for a symbolic linear problem of the form

$$Au = f, \tag{5.3}$$

where $A: V \to V$ is a given linear operator on V, a Hilbert space with inner product $(\cdot, \cdot)$ and corresponding norm $\| \cdot \|$, and $f \in V$ is given data. The corresponding Galerkin problem is: find $U \in V_h$ such that

$$(AU, v) = (f, v), \quad \forall v \in V_h,$$

where $V_h \subset V$ is a finite-element space. In many cases, $V = L^2(\Omega)$, where Ω is a domain in $\mathbb{R}^n$. We let $e \equiv u - U$.

1 Error representation via a dual problem:

$$\|e\|^2 = (e, e) = (e, A^*\varphi) = (Ae, \varphi) = (f - AU, \varphi) = -(R(U), \varphi),$$

where φ solves the dual problem

$$A^*\varphi = e,$$

with A^* denoting the adjoint of A, and $R(U)$ is the residual defined by

$$R(U) \equiv AU - f.$$

2 Galerkin orthogonality: Since $(Ae, v) = -(R(U), v) = 0, \quad \forall v \in V_h,$

$$\|e\|^2 = -(R(U), \varphi - \pi_h\varphi),$$

where $\pi_h\varphi \in V_h$ is an interpolant of φ.
3 Interpolation estimate:

$$\|h^{-\beta}(\varphi - \pi_h\varphi)\| \le C_i\|D^\beta\varphi\|,$$

where C_i is an interpolation constant, h is a measure of the size of the discretization and $D^\beta\varphi$ denotes derivatives of order β of the dual solution φ. Such estimates follow from classical interpolation theory when the solution is smooth.
4 Strong-stability estimate for the dual continuous problem:

$$\|D^\beta\varphi\| \le S_c\|e\|,$$

where S_c is a strong-stability factor.

Combining (1)–(4), we obtain

$$\|e\|^2 = (R(U), \pi_h \varphi - \varphi) \le S_c C_i \|h^\beta R(U)\| \, \|e\|,$$

which gives the following a posteriori error estimate

$$\|u - U\| \le S_c C_i \|h^\beta R(U)\|. \tag{5.4}$$

The indicated framework for deriving a posteriori error estimates for Galerkin methods is very general. In particular, it extends to problems $A(u) = f$, where A is a nonlinear operator. In such a case, the operator A^* in the dual problem is the adjoint of the Fréchet derivative of A (linearized form of A) evaluated between u and U. Details are given below in the context of systems of nonlinear ordinary differential equations.

5.3. A priori error estimates

Proofs of a priori error estimates have similar ingredients:

1.   Representation of the error in terms of the exact solution and a discrete linearized dual problem.
2.   Use of Galerkin orthogonality to introduce the interpolation error in the error representation.
3.   Local estimates for the interpolation error.
4.   Strong-stability estimates for the discrete dual problem.

We give more details for the above abstract case.

1.   Error representation via a discrete dual problem:

$$\|e_h\|^2 = (e_h, e_h) = (e_h, A^* \varphi_h) = (A e_h, \varphi_h),$$

where $e_h \equiv \pi_h u - U$, for $\pi_h u$ an interpolant of u in V_h, and the discrete dual problem with solution $\varphi_h \in V_h$ is defined by

$$(v, A^* \varphi_h) = (v, e_h), \quad \forall v \in V_h.$$

2.   Galerkin orthogonality: Using $(Ae, v) = 0$, $\forall v \in V_h$, gives

$$\|e_h\|^2 = (A(\pi_h u - U), \varphi_h) = (\pi_h u - u, A^* \varphi_h).$$

3.   Interpolation estimate:

$$\|u - \pi_h u\| \le C_i \|h^\alpha D^\alpha u\|,$$

where C_i is an interpolation constant.

4.   Strong-stability estimate for the discrete dual problem:

$$\|A^* \varphi_h\| \le S_{c,h} \|e_h\|,$$

where $S_{c,h}$ is discrete strong-stability factor.

Combining (1)–(4), we get an a priori error estimate:

$$\|u - U\| \le C_i(S_{c,h} + 1)\|h^\alpha D^\alpha u\|. \tag{5.5}$$

The interplay between the 'strong' norm, used in the strong stability involving A^*, and the corresponding 'weak' norm, used to estimate the interpolation error, is crucial.

Note that the stability of a continuous dual problem is used in the a posteriori error analysis whereas the stability of a discrete dual problem is used to prove the a priori error estimate. In both cases, the stability of the dual problems reflect the error accumulation and propagation properties of the discretization procedure.

5.4. Adaptive algorithms

Suppose the computational goal is to determine an approximate solution U such that $\|u - U\| \le TOL$, where TOL is a given tolerance. The corresponding stopping criterion reads:

$$S_c C_i \|h^\beta R(U)\| \le TOL, \tag{5.6}$$

which guarantees the desired error control via the a posteriori error estimate (5.4). The strategy of adaptive error control can be posed as a constrained nonlinear optimization problem: compute an approximation U satisfying (5.6) with minimal computational effort. In the case of Galerkin discretization, the control parameters are the local mesh size h and the local degree of the piecewise polynomials q and we seek h and q that minimize computational effort. We solve this problem iteratively, where the modification strategy indicates how to compute an improved iterate from the current iterate. The modification strategy is based on both the a posteriori error estimate (5.4) and the a priori error estimate (5.5). The mesh modification itself requires a mesh generator capable of generating a mesh with given mesh sizes. Mesh modification may also involve stretching and orientating the mesh. Such mesh generators in two and three dimensions are available today.

Adaptive algorithms using (5.6) as stopping criterion are reliable in the sense that by (5.4) the error control $\|u - U\| \le TOL$ is guaranteed. The efficiency of the adaptive algorithm depends on the quality of the mesh-modification strategy.

5.5. The stability factors and interpolation constants

The stability factor S_c and the interpolation constant C_i in the a posteriori error estimate defining the stopping criterion have to be computed to give the error control a quantitative meaning.

The stability factor S_c depends in general on the particular solution being approximated, since it is defined in terms of the linearized continuous dual problem. In some cases, all solutions have the same stability factors. For example, for typical elliptic problems with analysis in the energy norm, $S_c = S_{c,h} = 1$ with a suitable definition of norms. In general, this is not true. Analytic upper bounds on S_c often are too crude to be useful for quantitative error control. Hence, in the adaptive algorithms the stability factors S_c are approximated by solving the linearized continuous dual problem numerically. The amount of work required to compute S_c with sufficient accuracy is problem and solution dependent and depends on the degree of reliability desired. For complex problems, complete reliability cannot be reached, but the degree of reliability may be increased by spending more on the computation of the S_c.

The interpolation constants C_i depend on the shape of the elements, the local order of the polynomial approximation and the choice of norms, but not on the particular solution being approximated or the mesh size. Bounds for the interpolation constants C_i may be determined analytically or numerically from interpolation theory. Alternatively, once stability factors have been computed, the interpolation constants may be determined through calibration by numerically solving problems with known exact solutions.

5.6. Error estimates for quadrature, discrete-solution, data and modelling errors

A posteriori estimates of quadrature, discrete-solution and data-modelling errors are performed in a similar fashion to the analysis of the Galerkin-discretization error. The key difference is due to the fact that different perturbations accumulate at different rates. In particular, perturbations satisfying an orthogonality relation are connected to strong stability. For example, orthogonality is the basis of the Galerkin discretization and multigrid methods for discrete solutions. In general, weak stability must be used. A typical weak-stability estimate for the dual continuous problem takes the form

$$\|\varphi\| \leq \tilde{S}_c \|e\|,$$

where the dual solution φ is estimated in terms of the data e and the weak-stability factor. The corresponding a posteriori error estimate takes the form

$$\|u - U\| \leq \tilde{S}_c C_i \|R(U)\|. \tag{5.7}$$

Note that the factor h^β that resulted from the use of strong stability is missing.

6. Piecewise-polynomial approximation

In this section, we review results on piecewise-polynomial interpolation that are used below. For the sake of simplicity, we limit ourselves to discontinuous piecewise-constant approximation and continuous piecewise-linear approximation on an interval $I \equiv (a, b)$ in space or time. Higher-order results are similar. Two-dimensional analogues are given below.

Assuming first that $I \equiv (0, 1)$ is a space interval, let $0 \equiv x_0 < x_1 < x_2 < \cdots < x_{M+1} \equiv 1$ be a subdivision of I into subintervals $I_j \equiv (x_{j-1}, x_j)$ of length $h_j \equiv x_j - x_{j-1}$. We use the notation $T_h \equiv \{I_j\}$ for the corresponding mesh and define its mesh function $h(x)$ by $h(x) = h_j$ for x in I_j.

We also consider the situation in which I is a time interval, for example, $I \equiv (0, \infty)$. In this case, the mesh is given by a sequence of discrete time levels $0 \equiv t_0 < t_1 < \cdots < t_n < \cdots$, with corresponding time intervals $I_n \equiv (t_{n-1}, t_n)$, time steps $k_n \equiv t_n - t_{n-1}$ and mesh function $k(t)$ defined by $k(t) = k_n$ for t in I_n. We denote the corresponding mesh by T_k.

We let W_h denote the space of discontinuous piecewise-constant functions $v = v(x)$ on I, that is, v is constant on each subinterval I_j. We define the interpolant $\pi_h v \in W_h$ of an integrable function v by

$$\int_{I_j} (v - \pi_h v)\, dx = 0, \tag{6.1}$$

that is,

$$\pi_h v = \frac{1}{h_j} \int_{I_j} v(x)\, dx \text{ on } I_j \tag{6.2}$$

is the average of v on each element. It is easy to show that for $1 \le p \le \infty$,

$$\|v - \pi_h v\|_p \le \|h v'\|_p, \tag{6.3}$$

where $\| \cdot \|_p$ denotes the usual $L^p(I)$ norm.

We let V_h denote the space of functions that are continuous on I and linear on each subinterval I_j, $j = 1, ..., M + 1$. We denote by V_h^0 the subspace of functions $v \in V_h$ satisfying $v(0) = v(1) = 0$. For a continuous function v on I, we define the nodal interpolant $\pi_h v$ in V_h by

$$\pi_h v(x_j) = v(x_j), \quad j = 0, \ldots, M + 1. \tag{6.4}$$

Then, there are constants $C_{i,k}$ such that for $1 \le p \le \infty$,

$$\|h^{-2}(v - \pi_h v)\|_p \le C_{i,1}\|v''\|_p, \tag{6.5}$$

$$\|h^{-1}(v - \pi_h v)\|_p \le C_{i,2}\|v'\|_p, \tag{6.6}$$

$$\|h^{-1}(v - \pi_h v)'\|_p \le C_{i,3}\|v''\|_p. \tag{6.7}$$

Remark 1 The interpolation constants $C_{i,k}$ have the values $1/8$, 1 and $1/2$ for $k = 1, 2, 3$ respectively.

Remark 2 Below, we use weighted L^2 norms. Given a continuous positive function $a(x)$, a weighted L^2 norm is defined by

$$\| \cdot \|_a \equiv \|\sqrt{a}(\cdot)\|_2.$$

The interpolation results (6.3), (6.5)–(6.7) hold in the weighted norm with C_i depending on $\max_j (\max_{I_j} a / \min_{I_j} a)$.

7. An elliptic model problem in one dimension

As a model case, we consider a two-point boundary-value problem: find $u(x)$ such that

$$
\begin{aligned}
-(a(x)u')' + b(x)u' + c(x)u &= f(x), \quad x \in I = (0,1), \\
u(0) &= 0, \quad u(1) = 0,
\end{aligned}
\tag{7.1}
$$

where $a(x)$, $b(x)$ and $c(x)$ are given coefficients with $a(x) > 0$, and $f = f(x)$ is a given source term. This is a model for a stationary diffusion–convection–absorption process in one dimension. If $|b|/a$ is not large, then this problem has elliptic character, while if $|b|/a$ is large then the character is hyperbolic. We first consider the elliptic case with $b = 0$ for simplicity, and comment on the hyperbolic case with $|b|/a$ large in Remark 3 below.

In the elliptic case, we first assume $c = 0$. The variational formulation of (7.1) with $b = c = 0$, resulting from integration by parts, takes the form

$$\int_I au'v'\, dx = \int_I fv\, dx, \quad \forall v \in V_h^0. \tag{7.2}$$

The cG(1) method for (7.1) reads: find $U \in V_h^0$ such that

$$\int_I aU'v'\, dx = \int_I fv\, dx, \quad \forall v \in V_h^0. \tag{7.3}$$

This expresses the Galerkin orthogonality condition on the residual error which is apparent upon subtracting (7.3) from (7.2) to obtain

$$\int_I a(u - U)'v'\, dx = 0, \quad \forall v \in V_h^0. \tag{7.4}$$

Representing the finite-element solution as

$$U(x) = \sum_{j=1}^{M} \xi_j \varphi_j(x), \tag{7.5}$$

where $\{\varphi_j\}_{j=1}^{M}$ is the set of basis functions associated with the interior nodes, we find that (7.3) is equivalent to a matrix equation for the vector $\xi = (\xi_j)$:

$$A\xi = b, \tag{7.6}$$

where $A = (a_{ij})$ is the $M \times M$ stiffness matrix with coefficients

$$a_{ij} \equiv \int_I a\varphi_j'\varphi_i'\, dx, \tag{7.7}$$

and $b = (b_i)$ is the load vector with elements

$$b_i \equiv \int_I f\varphi_i \, dx. \tag{7.8}$$

The system matrix A is positive definite and tridiagonal and is easily solved by Gaussian elimination to give the approximate solution U.

The basic issue is the size of the error $u - U$. We first prove an a posteriori error estimate and then an a priori error estimate.

7.1. A posteriori error estimate in the energy norm

We prove an a posteriori estimate of the error $e \equiv u - U$ in the *energy norm* $\| \cdot \|_E$ defined for functions v with $v(0) = v(1) = 0$ by

$$\|v\|_E = \|v'\|_a \equiv \left(\int_I a(v')^2 \, dx \right)^{\frac{1}{2}}.$$

Using Galerkin orthogonality (7.3) by choosing $v = \pi_h e \in V_h^0$, we obtain the error representation:

$$
\begin{aligned}
\|e'\|_a^2 &= \int_I ae'e' \, dx = \int_I au'e' \, dx - \int_I aU'e' \, dx = \int_I fe \, dx - \int_I aU'e' \, dx \\
&= \int_I f(e - \pi_h e) \, dx - \int_I aU'(e - \pi_h e)' \, dx \\
&= \int_I f(e - \pi_h e) \, dx - \sum_{i=1}^{M+1} \int_{I_j} aU'(e - \pi_h e)' \, dx.
\end{aligned}
\tag{7.9}
$$

In this case, the solution of the dual problem is the error itself. We integrate by parts over each subinterval I_j in the last term, and use the fact that all the boundary terms disappear, to get

$$\|e'\|_a^2 = \int_I R(U)(e - \pi_h e) dx \le \|hR(U)\|_{\frac{1}{a}} \|h^{-1}(e - \pi_h e)\|_a,$$

where $R(U)$ is the residual defined on each subinterval I_j by

$$R(U) \equiv f + (aU')'.$$

Recalling (6.6) for the weighted L^2 norm, one proves:

Theorem 1   The finite-element solution U satisfies

$$\|u - U\|_E \le C_i \|hR(U)\|_{\frac{1}{a}}. \tag{7.10}$$

7.2. An adaptive algorithm

We design an adaptive algorithm for automatic control of the energy-norm error $\|u - U\|_E$ using the a posteriori error estimate (7.10) as follows:

1   Choose an initial mesh $T_{h(0)}$ of mesh size $h^{(0)}$.
2   Compute the corresponding cG(1) finite-element solution $U^{(0)}$ in $V_{h(0)}$.

3   Given a computed solution $U^{(m-1)}$ in $V_{h^{(m-1)}}$ on a mesh with mesh size $h^{(m-1)}$, stop if

$$C_i \| h^{(m-1)} R(U^{(m-1)}) \|_{\frac{1}{a}} \leq TOL. \tag{7.11}$$

4   If not, determine a new mesh $T_{h^{(m-1)}}$ with mesh function $h^{(m-1)}$ of maximal size such that

$$C_i \| h^{(m-1)} R(U^{(m-1)}) \|_{\frac{1}{a}} = TOL \tag{7.12}$$

and continue.

We note that (7.11) is the stopping criterion and (7.12) defines the mesh-modification strategy. By Theorem 1, it follows that the error $\|u - U\|_E$ is controlled to the tolerance TOL if the stopping criterion (7.11) is reached with $U = U^{(m-1)}$. The relation (7.12) defines the new mesh size $h^{(m-1)}$ by maximality, that is, we seek a mesh function $h^{(m-1)}$ as large as possible (to maintain efficiency) such that (7.12) holds. In general, maximality in $\| \cdot \|$ is obtained by the 'equidistribution' of error such that the error contributions from the individual intervals I_j are kept equal.

Equidistribution of the error results in the equation

$$a(x_j)^{-1}(h_j^{(m)} \| R(U^{(m-1)}) \|_{L^\infty I_j^{(m)}})^2 h_j^{(m)} = \frac{TOL^2}{N^{(m)}}, \quad j = 1, \ldots, N^{(m)}, \tag{7.13}$$

where $N^{(m)}$ is the number of intervals in $T_{h^{(m)}}$. The equation reflects the fact that the total error is given by the sum of the errors from each interval, and so the error on each interval must be a fraction of the total error. In practice, this nonlinear equation is simplified by replacing $N^{(m)}$ by $N^{(m-1)}$.

Example 1. Consider problem (7.1) with $a(x) = x + \varepsilon$, $\varepsilon = 0.01$, $b = c = 0$ and $f(x) \equiv 1$. Because the diffusion coefficient a is small near $x = 0$, the solution u and its derivatives u' and u'' there change rapidly with x; see Figure 1a. To compute u, we use the code Femlab which contains an implementation of the adaptive algorithm just described. Figure 1b shows the residual $R(U)$ of the computed solution U, and Figure 1c shows, the local mesh size $h(x)$ when the stopping criterion with $TOL = 0.05$ was reached. Note that the mesh size is small near $x = 0$ where the residual is large.

7.3. A priori error estimate in the energy norm

We now prove an a priori energy-norm error estimate for (7.3) by comparing the Galerkin-discretization error to the interpolation error. Using Galerkin

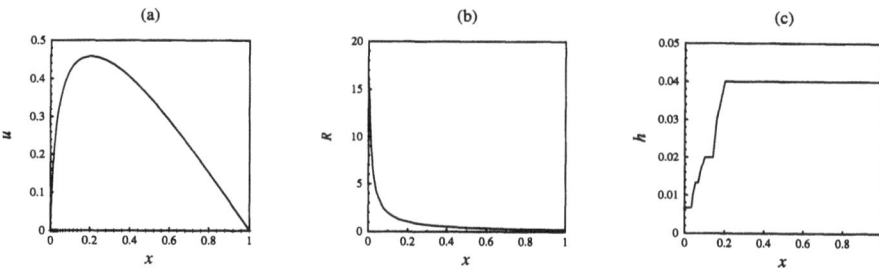

Fig. 1. Solution, residual and mesh size for Example 1

orthogonality (7.4) with $v = U - \pi_h u$, we obtain

$$\int_I a(u-U)'(u-U)' \, dx = \int_I a(u-U)'(u-\pi_h u)' \, dx \leq \|u' - U'\|_a \|(u-\pi_h u)'\|_a,$$

so

$$\|u' - U'\|_a \leq \|(u - \pi_h u)'\|_a \tag{7.14}$$

This shows that the Galerkin approximation is optimal because its error in the energy norm is less than the error of the interpolant. Together with (6.7), this proves:

Theorem 2 The finite-element solution U satisfies

$$\|u' - U'\|_a \leq C_i \|h u''\|_a. \tag{7.15}$$

Remark 3 It is easy to show that

$$C_i \|h R(U)\|_{\frac{1}{a}} \leq C C_i \|h u''\|_a,$$

with C a constant depending on a, indicating that the a posteriori energy error estimate is optimal in the same sense as the a priori estimate.

7.4. A posteriori error estimate in the L^2 norm

We prove an a posteriori error estimate in the L^2-norm, allowing the absorption coefficient c in (7.1) to be nonzero. The extension of (7.3) to this case is direct by including $\int_I c U v \, dx$ on the left-hand side. We introduce the dual problem

$$
\begin{aligned}
-(a\varphi')' + c\varphi &= e, & x \in I, \\
\varphi(0) &= 0, & \varphi(1) = 0,
\end{aligned}
\tag{7.16}
$$

which takes the same form as the original problem (7.1). We use Galerkin orthogonality (7.3), by choosing $v = \pi_h e \in V_h^0$, to get

$$
\begin{aligned}
\|e\|_2^2 &= \int_I e(-(a\varphi')' + c\varphi)\,dx = \int_I (ae'\varphi' + ce\varphi)\,dx \\
&= \int_I (au'\varphi' + cu\varphi)\,dx - \int_I (aU'\varphi' + cU\varphi)\,dx \\
&= \int_I f\varphi\,dx - \int_I (aU'\varphi' + cU\varphi)\,dx \\
&= \int_I f(\varphi - \pi_h\varphi)\,dx - \sum_{i=1}^{M+1} \int_{I_j} (aU'(\varphi - \pi_h\varphi)' + cU(\varphi - \pi_h\varphi))\,dx.
\end{aligned}
$$

We now integrate by parts over each subinterval I_j, using the fact that all the boundary terms disappear, to get

$$
\|e\|_2^2 \leq \|h^2 R(U)\|_2 \|h^{-2}(\varphi - \pi_h\varphi)\|_2,
$$

where $R(U) \equiv f + (aU')' + cU$ on each subinterval. Using (6.3) and defining the strong-stability factor S_c by

$$
S_c \equiv \max_{g \in L^2(I)} \frac{\|\psi_g''\|_2}{\|g\|_2}, \tag{7.17}
$$

where ψ_g satisfies

$$
\begin{aligned}
-(a\psi_g')' + c\psi_g &= g, \quad x \in I, \\
\psi_g(0) &= 0, \quad \psi_g(1) = 0,
\end{aligned} \tag{7.18}
$$

we obtain:

Theorem 3 The finite-element solution U satisfies

$$
\|u - U\|_2 \leq S_c C_i \|h^2 R(U)\|_2. \tag{7.19}
$$

Example 2. In Figure 2a, we plot the computed solution in the case $a = 0.01$, $c = 1$ and $f(x) = 1/x$ with L^2 error control based on (7.19) with $TOL = .01$. The residual and mesh size are plotted in Figures 2b and 2c. In this example, there are two sources of singularities in the solution. First, because the diffusion coefficient a is small, the solution may have boundary layers; second, the source term f is large near $x = 0$. The singularity in the data f enters only through the residual, while the smallness of a enters both through the residual and through the stability factor S_c. The adaptive algorithm computes the stability factor S_c by solving the dual problem (7.16) with e replaced by an approximation obtained by subtracting approximate solutions on two different grids. In this example, $S_c \sim 37$.

7.5. A priori error estimate in the L^2 norm

We now prove an a priori error estimate in the L^2 norm, assuming for simplicity that the mesh size h is constant and that $c = 0$.

Theorem 4 The finite-element solution U satisfies

$$
\|u - U\|_2 \leq C_i S_c \|h(u - U)'\|_2 \leq C_i S_c \|h^2 u''\|_2, \tag{7.20}
$$

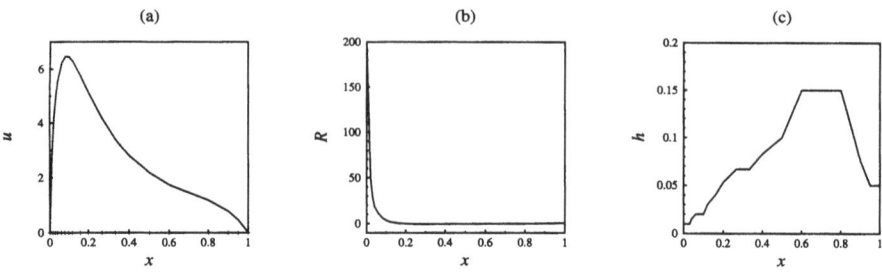

Fig. 2. Approximation, residual and mesh size for Example 2

where $S_c \equiv \max_{g \in L^2(I)} \|\psi_g''\|_a / \|g\|_2$, with ψ_g satisfying (7.18).

Proof. By (7.4) and (6.7), for φ satisfying (7.16) with $c = 0$, we have

$$
\begin{aligned}
\|e\|_2^2 &= \int_I ae'\varphi' \, dx = \int_I ae'(\varphi - \pi_h\varphi)' \, dx \\
&\leq \|he'\|_a \|h^{-1}(\varphi - \pi_h\varphi)'\|_a \leq C_i \|he'\|_a \|\varphi''\|_a.
\end{aligned}
$$

The proof is finished by noting that multiplying the energy-norm error estimate by h gives

$$\|he'\|_a \leq C_i \|h^2 u''\|_a. \tag{7.21}$$

$\square$

This estimate generalizes to the case of variable h assuming that the mesh size h does not change too rapidly from one element to the next, (cf. Eriksson (1994)).

7.6. Data and modelling errors

We make an a posteriori estimate of data and modelling errors. Suppose that $a(x)$ and $f(x)$ in (7.3) are approximations of the correct coefficient $\hat{a}(x)$ and data $\hat{f}(x)$ and let $\hat{u}$ be the corresponding correct solution. We seek an a posteriori error estimate of the total error $\hat{e} \equiv \hat{u} - U$ including Galerkin-discretization, data and modelling errors. We start from a modified form of the error representation (7.9),

$$
\begin{aligned}
\|(\hat{u} - U)'\|_{\hat{a}}^2 &= \int_I f(\hat{e} - \pi_h\hat{e}) \, dx - \sum_{j=1}^{M+1} \int_{I_j} aU'(\hat{e} - \pi_h\hat{e})' \, dx \\
&\quad + \int_I (\hat{f} - f)\hat{e} \, dx - \sum_{j=1}^{M+1} \int_{I_j} (\hat{a} - a)U'\hat{e}' \, dx \\
&\equiv I + II - III,
\end{aligned}
$$

with the obvious definition of I, II and III. The first term I is estimated as above. For the new term III, we have

$$III \leq C_i \|(\hat{a} - a)\hat{U}'\|_{\frac{1}{\hat{a}}} \|e'\|_{\hat{a}}.$$

Similarly, integration by parts gives

$$II \leq \|F - \hat{F}\|_{\frac{1}{\hat{a}}} \|e'\|_{\hat{a}},$$

where $F' = f$, $\hat{F}' = \hat{f}$ and $F(0) = \hat{F}(0) = 0$. Altogether, we obtain:

Theorem 5 The finite-element solution U satisfies

$$\|\hat{u}' - U'\|_{\hat{a}} \leq C_i(\|hR(U)\|_{\frac{1}{\hat{a}}} + \|F - \hat{F}\|_{\frac{1}{\hat{a}}} + \|(\hat{a} - a)U'\|_{\frac{1}{\hat{a}}}). \qquad (7.22)$$

An adaptive algorithm for control of both Galerkin and data-modelling errors can be based on (7.22). It is natural to assume that $\|\hat{a} - a\|_\infty \leq \mu$ or $\|(\hat{a} - a)\hat{a}^{-1}\|_\infty \leq \mu$, corresponding to an absolute or relative error in $\hat{a}$ on the level μ. In the first case, we obtain $\|(\hat{a} - a)U'\|_{\frac{1}{\hat{a}}} \leq \mu \|U'\|_{\frac{1}{\hat{a}}}$, and in the second case, $\|(\hat{a} - a)U'\|_{\frac{1}{\hat{a}}} \leq \mu \|U'\|_{\hat{a}}$. μ is supplied by the user while the relevant norm on U is computed by the program. For example, for the problem in Example 2 we find that $\|U'\|_{\frac{1}{\hat{a}}} = 13.3531$ while $\|U'\|_{\hat{a}} = 0.5406$.

Remark 4 If $|b|/a$ is large then the problem (7.1) has a hyperbolic character. If $a < h$ then a modified Galerkin method with improved stability properties is used which is called the streamline diffusion method. The modifications consist of a weighted least-squares stabilization that gives extra control of the residual $R(U)$ and a modification of the viscosity coefficient a. L^2 error estimates for this method are derived similarly to the case with $b = 0$. The resulting L^2 a posteriori error estimate has essentially the form (7.19), where the stability constant S_c contains a dependence on the viscosity a. In the generic case with a constant, we have $S_c \sim \frac{1}{a}$. The result of using strong stability and Galerkin orthogonality is a factor $\frac{h^2}{a}$ coupled with the residual $R(U)$. In a direct approach that uses weak stability, the result does not contain the factor $\frac{h^2}{a}$. Thus, an improvement results if $a > h^2$. In particular, if $a > h$ then the standard unmodified Galerkin method may be used and the above analysis applies. The condition $a > h$ may be satisfied on the last mesh in the sequence of meshes used in the adaptive process. In this case, the streamline diffusion modification is used only on the initial coarse meshes. Details of this extension to hyperbolic convection–diffusion problems are given in Eriksson and Johnson (1993), (to appear).

Example 3. Consider problem (7.1) with $a(x) = 0.02$, $b(x) = 1$, $c(x) = 0$ and $f(x) = 1$. In Figure 3, we plot the computed solution together with the residual and mesh size obtained using an adaptive algorithm based on an

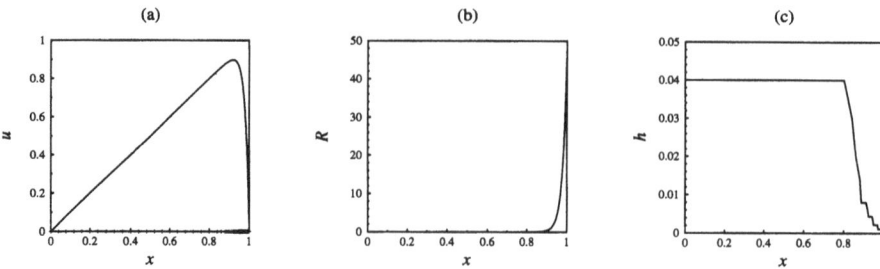

Fig. 3. Solution, residual and mesh size for Example 3

a posteriori error estimate of the form (7.19) with $TOL = .02$. Notice the singularity in u in the boundary layer near $x = 1$.

8. Basic time-dependent model problems

As a first example, we consider the scalar linear initial-value problem: find $u = u(t)$ such that

$$
\begin{aligned}
u' + a(t)u &= f(t), \quad t > 0, \\
u(0) &= u_0,
\end{aligned}
\tag{8.1}
$$

where $a(t)$ is a given coefficient, $f(t)$ is a given source term and $v' = \frac{dv}{dt}$ now denotes the time derivative of v. The exact solution $u(t)$ is given by the formula

$$
u(t) = e^{-A(t)}u_0 + \int_0^t e^{-(A(t)-A(s))}f(s)\,ds,
\tag{8.2}
$$

where $A' = a$ and $A(0) = 0$, from which we can draw some conclusions about the dependence of u on a. In general, the exponential factors may become large with time. However, if $a(t) \geq 0$ for all t, then $A(t) \geq 0$ and $A(t) - A(s) \geq 0$ for all $t \geq s$, and both u_0 and f are multiplied by quantities that are less than or equal to one. We shall see that if $a(t) \geq 0$, then Galerkin-discretization errors accumulate in such a way that accurate long-time computation is possible. The problem (8.1) with $a(t) \geq 0$ is a model for a class of parabolic problems that includes generalizations of (8.1) with the coefficient a replaced by $-\nabla \cdot (a\nabla)$ with $a \geq 0$. The analysis for the case $a \geq 0$ allowing long-time integration without error accumulation extends directly to this more complex case.

For the sake of simplicity, we consider the dG(0) method, which reads: find U in W_k such that for all polynomials v of degree 0 on I_n,

$$
\int_{I_n} (U' + a(t)U)v\,dt + [U_{n-1}]v_{n-1}^+ = \int_{I_n} fv\,dt,
\tag{8.3}
$$

where $[v_n] = (v_n^+ - v_n^-)$, $v_n^\pm = \lim_{s\to\pm 0} v(t_n + s)$ and $U_0^- = u_0$. We note that (8.3) says that the 'sum' of the residual $U_t' + a(t)U - f$ in I_n, and the 'jump'

$[U_{n-1}]$ is orthogonal to all discrete test functions. Since U is a constant on I_n, $U' \equiv 0$ on I_n.

If U_n denotes the constant value of U on the time interval I_n, then the dG(0) method (8.3) satisfies

$$U_n - U_{n-1} + U_n \int_{I_n} a \, dt = \int_{I_n} f \, dt, \qquad n = 1, 2, \ldots, \qquad (8.4)$$

where $U_0 = u_0$. The classical backward Euler method is thus the dG(0) method with the rectangle rule applied to the integrals. We assume that if $a(t)$ is negative, then the time step is small enough that $|\int_{I_n} a \, dt| < 1$, in which case (8.4) defines U_n uniquely. We use the notation $\|v\|_I \equiv \max_{t \in [0,T]} |v(t)|$, where $I \equiv (0, T)$ is a given time interval.

8.1. An a posteriori error estimate

To derive an a posteriori error estimate for the error $e_N \equiv u(t_N) - U_N$, $N \geq 1$, we introduce the continuous dual 'backward' problem,

$$\begin{aligned} -\varphi' + a(t)\varphi &= 0, \quad t \in (0, t_N), \\ \varphi(t_N) &= e_N, \end{aligned} \qquad (8.5)$$

with solution given by $\varphi(t) = e^{A(t) - A(t_N)} e_N$. Integration by parts over each subinterval I_n gives

$$\begin{aligned} e_N^2 &= e_N^2 + \sum_{n=1}^{N} \int_{I_n} e(-\varphi' + a\varphi) \, dt \\ &= \sum_{n=1}^{N} \int_{I_n} (e' + ae)\varphi \, dt + \sum_{n=1}^{N-1} [e_n]\varphi_n^+ + (u_0 - U_0^+)\varphi_0^+ \\ &= \sum_{n=1}^{N} \left(\int_{I_n} (f - aU)\varphi \, dt - [U_{n-1}]\varphi_{n-1}^+ \right), \end{aligned} \qquad (8.6)$$

where in the last step we use the facts that $U' = 0$ on each I_n and $U_0^- = u_0$. Now we use Galerkin orthogonality (8.3) by taking $v = \pi_k \varphi$ with

$$\int_{I_n} (\varphi - \pi_k \varphi) \, dt = 0, \qquad n = 1, \ldots, N,$$

to obtain the error representation formula:

$$e_N^2 = \sum_{n=1}^{N} \left(\int_{I_n} (f - aU)(\varphi - \pi_k \varphi) \, dt - [U_{n-1}](\varphi - \pi_k \varphi)_{n-1}^+ \right).$$

Using (6.3), we obtain

$$\begin{aligned} e_N^2 &\leq \int_0^{t_N} |\varphi'| \, dt \, \max_{n=1,\ldots,N} (|[U_{n-1}]| + \|k(f - aU)\|_{I_n}) \\ &\leq S_c(t_N)(|[U_{n-1}]| + \|k(f - aU)\|_{I_n})|e_N|, \end{aligned} \qquad (8.7)$$

where $S_c(t_N)$ is the stability factor defined by

$$S_c(t_N) \equiv \max_{\varphi(t_N)} \frac{\int_0^{t_N} |\varphi_t| \, dt}{|\varphi(t_N)|}. \qquad (8.8)$$

To complete the proof of the a posteriori error estimate, we need to estimate $S_c(t_N)$. The following lemma presents such a stability estimate in both the general case and the dissipative case when $a(t) \geq 0$ for all t. We also state an estimate for φ itself.

Lemma 1   If $|a(t)| \leq A$ for $t \in (0, t_N)$, then φ satisfies for all $t \in (0, t_N)$:

$$|\varphi(t)| \leq \exp(At_N)|e_N|, \tag{8.9}$$

and

$$S_c(t_N) \leq At_N \exp(At_N). \tag{8.10}$$

If $a(t) \geq 0$ for all t, then φ satisfies for all $t \in (0, t_N)$:

$$|\varphi(t)| \leq |e_N|, \tag{8.11}$$

and

$$S_c(t_N) \leq 1. \tag{8.12}$$

Proof. The first and second estimates follow from the boundedness assumption on a. The third estimate follows from the fact that $A(t_N) - A(t)$ is non-negative for $t \leq t_N$. Further, since a is non-negative,

$$\begin{aligned} \int_I |\varphi'| dt &= |e_N| \int_I a(t) \exp(A(t_N) - A(t)) dt \\ &= |e_N|(1 - \exp(A(0) - A(t_N))) \leq |e_N|, \end{aligned}$$

which completes the proof. $\square$

We insert the strong-stability estimates (8.10) or (8.12) into (8.7) and obtain the a posteriori error estimate:

Theorem 6   The finite-element solution U satisfies for $N = 1, 2, \ldots$

$$|u(t_N) - U_N| \leq S_c(t_N)|kR(U)|_{(0,t_N)},$$

where

$$R(U) \equiv \frac{|U_n - U_{n-1}|}{k_n} + |f - aU|_{I_n}, \quad t \in I_n.$$

8.2. An a priori error estimate

The a priori error estimate for (8.3) reads as follows:

Theorem 7   If $|a(t)| \leq A$ for all t, then there is a constant $C > 0$ such that U satisfies for $N = 1, 2, \ldots$.

$$|u(t_N) - U_N| \leq CAt_N \exp(CAt_N)|ku'|_I,$$

and if $a(t) \geq 0$ for all t, then for $N = 1, 2, \ldots$,

$$|u(t_N) - U_N| \leq |ku'|_{(0,t_N)}. \tag{8.13}$$

We note the optimal nature of the estimate compared to interpolation in the case $a(t) \geq 0$.

Proof. We introduce the discrete dual backward problem: find $\Phi \in W_k$ such that for $n = N, N-1, \ldots, 1$,

$$\int_{I_n} (-\Phi' + a(t)\Phi)v \, dt - [\Phi_n]v_n^- = 0, \quad \forall v \in W_k, \tag{8.14}$$

where $\Phi_N^+ = (\pi_k u - U)_N^-$. It suffices to estimate the 'discrete' error $\bar{e} \equiv \pi_k u - U$ in W_k since $u - \pi_k u$ is already known. With the choice $v = \bar{e}$, the Galerkin orthogonality allows U to be replaced by u and we obtain the following representation:

$$\begin{aligned}
|\bar{e}_N^-|^2 &= \sum_{n=1}^{N} \int_{I_n} (-\Phi' + a(t)\Phi)\bar{e} \, dt - \sum_{n=1}^{N-1} [\Phi_n]\bar{e}_n^- + \Phi_N^- \bar{e}_N^- \\
&= \sum_{n=1}^{N} \int_{I_n} (-\Phi' + a(t)\Phi)(\pi_k u - u) \, dt \\
&\quad - \sum_{n=1}^{N-1} [\Phi_n](\pi_k u - u)_n^- + \Phi_N^-(\pi_k u - u)_N^- \\
&= -\int_I (a\Phi(u - \pi_k u)) \, dt + \sum_{n=1}^{N-1} [\Phi_n](u - \pi_k u)_n^- - \Phi_N^-(u - \pi_k u)_N^-,
\end{aligned}$$

where we use $\Phi' = 0$ on each time interval. Recalling (6.3), we get the desired result follows from a lemma expressing the weak and strong stability of the discrete dual problem (8.14). $\square$

Lemma 2 When $|a(t)| \leq A$ for all t, then there is a constant $C > 0$ such that the solution of the discrete dual problem (8.14) satisfies

$$|\Phi_n^-| \leq \exp(CAt_N)|\bar{e}_N^-|, \tag{8.15}$$

$$\sum_{n=1}^{N-1} |[\Phi_n]| \leq CAt_N \exp(CAt_N)|\bar{e}_N^-|, \tag{8.16}$$

$$\sum_{n=1}^{N} \left| \int_{I_n} a|\Phi_n| \, dt \right| \leq CAt_N \exp(CAt_N)|\bar{e}_N^-|. \tag{8.17}$$

If $a(t) \geq 0$ for all t, then

$$|\Phi_n^-| \leq |\bar{e}_N^-|, \tag{8.18}$$

$$\sum_{n=1}^{N-1} |[\Phi_n]| \leq |\bar{e}_N^-|, \tag{8.19}$$

$$\sum_{n=1}^{N} \left| \int_{I_n} a|\Phi_n|dt \right| \le |\bar{e}_N^-|. \tag{8.20}$$

Proof. The discrete dual problem (8.14) takes the form

$$-\Phi_{n+1} + \Phi_n + \Phi_n \int_{I_n} a(t)dt = 0, \quad n = N, N-1, \ldots, 1,$$

$$\Phi_{N+1} = \bar{e}_N^-,$$

where Φ_n denotes the value of Φ on I_n, so

$$\Phi_n = \prod_{j=n}^{N} \left(1 + \int_{I_j} a\,dt\right)^{-1} \Phi_{N+1}.$$

In the case where a is bounded, the results follow from standard estimates. When a is nonnegative, this proves (8.18) immediately. To prove (8.19), we assume without loss of generality that Φ_{N+1} is positive, so the sequence Φ_n decreases when n decreases, and

$$\sum_{n=1}^{N} |[\Phi_n]| = \sum_{n=1}^{N} [\Phi_n] = \Phi_{N+1} - \Phi_1 \le |\Phi_{N+1}|.$$

Finally, (8.20) follows from the discrete equation. $\square$

We note that the a priori error estimate (8.13) is optimal compared to interpolation in the case $a \ge 0$.

Remark 5   It is important to compare the general results of Theorem 6 and Theorem 8.13, when a is only known to be bounded, to the result for dissipative problems with $a \ge 0$. In the first case, the errors can accumulate at an 'exponential' rate, and, depending on $\mathbf{A}$, $S_c(t_N)$ can become so large that controlling the error is no longer possible. In the case $a \ge 0$, there is no accumulation of error so accurate computation is possible over arbitrarily long times. Note that we do not require $a(t)$ to be positive and bounded away from zero; it is enough to assume that a is non-negative.

Example 4.   Consider the dissipative problem $u' + u = \sin t$, $u(0) = 1$ with solution $u(t) = 1.5e^{-t} + .5(\sin t - \cos t)$. We compute with dG(0) and plot the solution and the approximation in Figure 4a. The approximation is computed with an error tolerance of .001. In Figure 4b, we plot $S_c(t)$ versus time. Note that $S_c(t)$ tends to 1 as t increases, indicating that the numerical error does not grow significantly with time, and accurate computations can be made over arbitrarily long time intervals.

Example 5.   We now consider the problem $u' - u = 0$, $u(0) = 1$ with solution $u(t) = e^t$. We compute with dG(0) keeping the error below .025. Since the

(a)   (b)

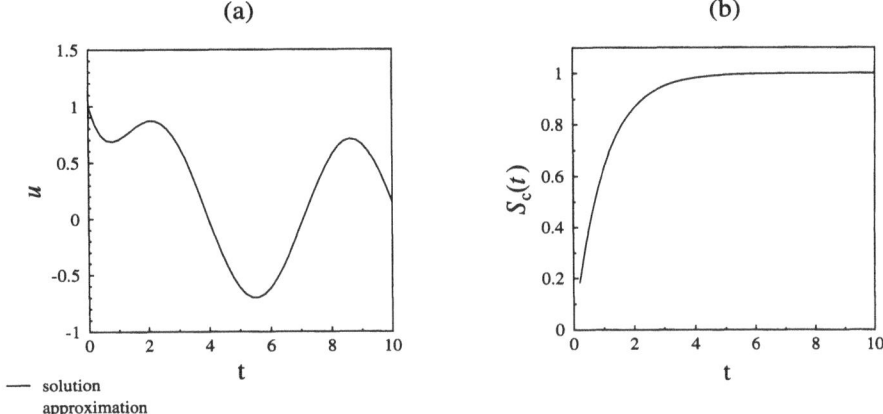

— solution
approximation

Fig. 4. Solution, approximation and stability factor for Example 4

(a)   (b)

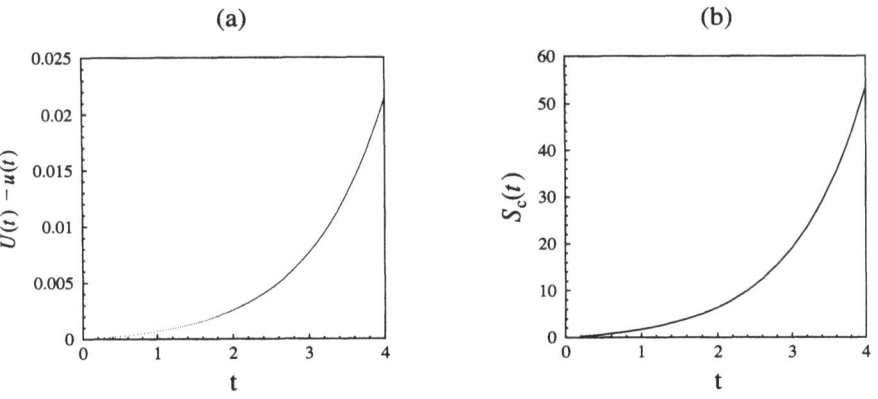

Fig. 5. Error and stability factor for Example 5

problem is not dissipative, we expect to see the error grow. The difference $U(t) - u(t)$ is plotted in Figure 5a and the exponential growth rate is clearly visible. Given a certain amount of computational power, for example, a fixed precision or a fixed amount of computing time, there is some point in time at which accurate computation is no longer possible. $S_c(t)$ is plotted in Figure 5b, and we note that it reflects the rate of instability precisely.

8.3. Adaptive error control

An adaptive algorithm based on the a posteriori error estimate takes the form: determine the time steps k_n so that

$$\hat{S}_c(t_N)(|U_n - U_{n-1}| + k_n|f - aU|_{I_n}) = TOL, \quad n = 1, ..., N,$$

where $\hat{S}_c(t_N) \equiv \max_{1\leq n\leq N} S_c(t_n)$. This guarantees that

$$|u(t_n) - U_n| \leq TOL, \quad n = 1, \cdots N.$$

As mentioned above, $\hat{S}_c(t_N)$ is approximated in an auxiliary computation solving the backward problem with chosen initial data; see below and [25] and [8] for more details.

Example 6. We consider a more complicated problem,

$$u' + (.25 + 2\pi \sin(2\pi t))u = 0, \quad t > 0,$$
$$u(0) = 1,$$

with solution

$$u(t) = \exp(-.25t + \cos(2\pi t) - 1).$$

The unstable solution oscillates as time passes, but the oscillations dampen. In Figure 6a, we plot the solution together with the dG(0) approximation computed with error below .12. In Figure 6b, we plot the time steps used for the computation. We see that the steps are adjusted for each oscillation and in addition that there is an overall trend to increasing the steps as the size of the solution decreases.

In addition, the solution has changing stability characteristics. In Figure 7a, we plot the stability factor versus time, and it is evident that the numerical error decreases and increases in alternating periods of time. If a crude 'exponential' bound on the stability factor is used instead of a computational estimate, then the error is greatly overestimated with the consequence that the computation can only be done over a much smaller interval. To demonstrate the effectiveness of the a posteriori estimate for error control, we plot the ratio of the true error to the computed bound versus time in Figure 7b. The ratio quickly settles down to a constant, which means that the bound is predicting the behaviour of the error in spite of the fact that the error oscillates a good deal.

8.4. Quadrature errors

We now consider the error arising from computing the integrals in the dG(0) method (8.3) approximately using quadrature. We focus on the error from computing $\int_{I_n} f \, dt$ using quadrature. To illustrate essential aspects, we consider the midpoint rule,

$$\int_{I_n} f \, dt \approx k_n f(t_{n-\frac{1}{2}}), \quad t_{n-\frac{1}{2}} = \frac{1}{2}(t_{n-1} + t_n), \tag{8.21}$$

and also the rectangle rule,

$$\int_{I_n} f \, dt \approx k_n f(t_n). \tag{8.22}$$

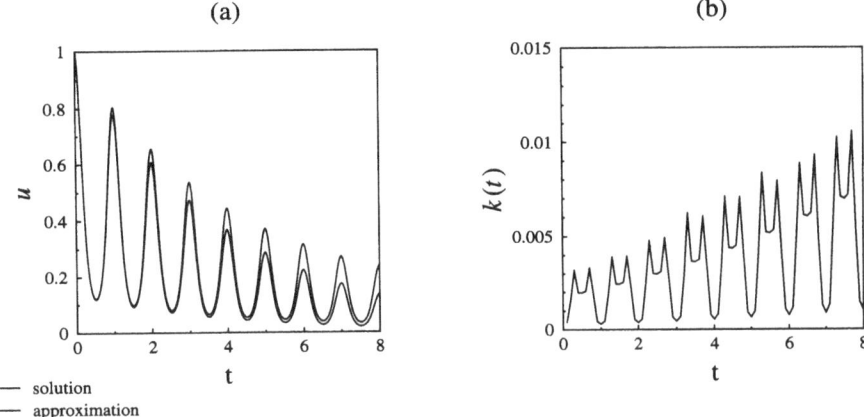

Fig. 6. Solution, approximation and step sizes for Example 6

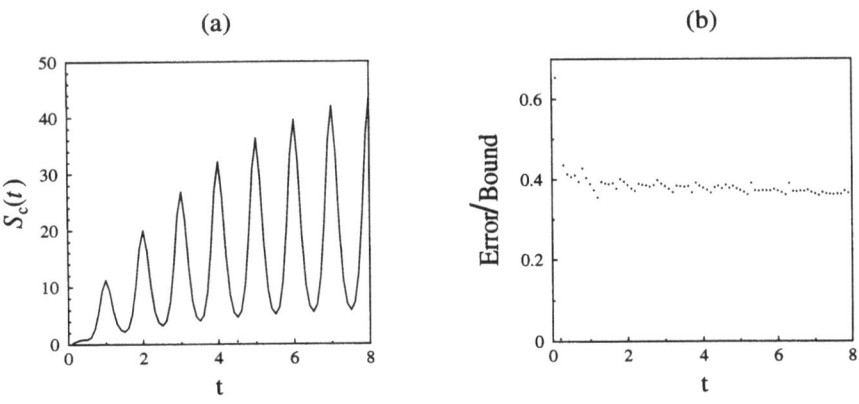

Fig. 7. Stability factor and error/bound ratio for Example 6

We recall that the backward Euler scheme is generated by using the rectangle rule. We compare dG(0) approximations computed with the two quadratures (8.21) and (8.22) and conclude that the classical choice (8.22) is less accurate for many problems. The analysis shows the advantage of separating the Galerkin and quadrature errors since they accumulate differently.

For the midpoint rule (8.21), the quadrature error on a single interval is bounded by

$$\left| \int_{I_n} f \, dt - k_n f(t_{n-\frac{1}{2}}) \right| \le \min \left\{ \int_{I_n} |k f'| \, dt, \frac{1}{2} \int_{I_n} |k^2 f''| \, dt \right\}. \qquad (8.23)$$

The corresponding error estimate for the rectangle rule reads

$$\left| \int_{I_n} f \, dt - k_n f(t_n) \right| \leq \int_{I_n} |k f'| \, dt. \tag{8.24}$$

We notice that the midpoint rule is more accurate unless $|f''| \gg |f'|$, while the cost of the two rules is the same.

We now determine the effect of quadrature on the final error $u(t_N) - U_N$ after N steps. We start with the modified form of the the error representation

$$
\begin{aligned}
e_N^2 \;=\; & \sum_{n=1}^{N} \left(\int_{I_n} (\hat{f} - aU)(\varphi - \pi_k \varphi) \, dt - [U_{n-1}](\varphi - \pi_k \varphi)_{n-1} \right. \\
& \left. + \int_{I_n} (f - \hat{f}) \varphi \, dt \right).
\end{aligned}
\tag{8.25}
$$

where for $t \in I_n$ we define $\hat{f}(t) \equiv f(t_{n-\frac{1}{2}})$ for the midpoint rule and $\hat{f}(t) \equiv f(t_n)$ for the rectangle rule. Introducing the weak-stability factor

$$\tilde{S}_c(t_N) \equiv \frac{\int_0^{t_N} |\varphi| \, dt}{|\varphi(t_N)|},$$

we obtain a modified a posteriori error estimate that includes the quadrature errors.

Theorem 8 U satisfies for $N = 1, 2, \ldots,$

$$|u(t_N) - U_N| \leq S_c(t_N)|k\hat{R}(U)|_{(0,t_N)} + \tilde{S}_c(t_N) C_{qj} \int_0^{t_N} k^j |f^{(j)}| \, dt,$$

where

$$\hat{R}(U) = \frac{|U_n - U_{n-1}|}{k_n} + |\hat{f} - aU|_{I_n}, \quad t \in I_n,$$

and $j = 1$ for the rectangle rule, $j = 2$ for the midpoint rule, $C_{q1} = 1$, $C_{q2} = 1/2$, $f^{(1)} = f'$ and $f^{(2)} = f''$.

We note that this estimate includes the factor $\int_0^{t_N} k^j |f^{(j)}| \, dt$ that grows linearly with t_N if the integrand is bounded. This linear growth in time, representing the accumulation of quadrature errors, is also present in the case $a \geq 0$ when $\tilde{S}(t_N) \leq 1$. For long-time integration in the case $a \geq 0$, it is thus natural to use the midpoint rule, since the accumulation of quadrature error can be compensated by the second-order accuracy.

In general, since the computational cost of the quadrature is usually small compared to the Galerkin computational work (which requires the solution of a system of equations), the precision of the quadrature may be increased if needed without significantly increasing the overall work. This illustrates the

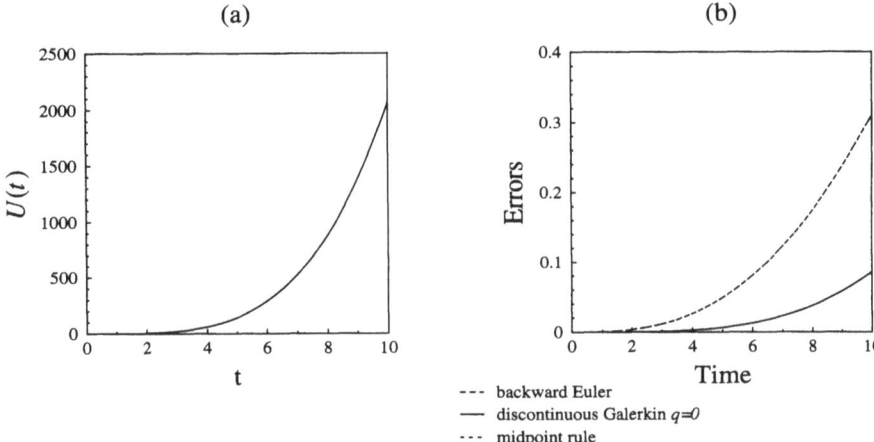

Fig. 8. Approximation and errors for Example 7

importance of separating Galerkin-discretization and quadrature errors since they accumulate at different rates. These errors should not be combined as happens in the classic analysis of difference schemes, leading to non-optimal performance.

Example 7. We consider the approximation of $u' - .1u = t^3$, $u(0) = 1$. We compute using the dG(0) method, the backward Euler scheme (rectangle rule quadrature), and the midpoint rule, with accuracies plotted below. The approximation is plotted in Figure 8a; the problem is not dissipative, so we expect error accumulation. We plot the errors of the three computations in Figure 8b. The dG(0) and the dG(0) with midpoint rule approximations are very close in accuracy, while the backward-Euler computation errors accumulate at a much faster rate.

8.5. A 'hyperbolic' model problem

We consider the ordinary differential equation model for a 'hyperbolic' problem: find $u = (u_1, u_2)$ such that

$$
\begin{aligned}
u_1' + au_2 &= f_1, \quad t > 0, \\
u_2' - au_1 &= f_2, \quad t > 0, \\
u_1(0) &= u_{10}, \quad u_2(0) = u_{20},
\end{aligned}
\tag{8.26}
$$

where the $a = a(t)$ is a given bounded coefficient with $|a| \leq A$, and the f_i and u_{i0} are given data. This is a simple model for wave propagation.

We study the application of the cG(1) method to (8.26), where V_k is the set of continuous piecewise-linear functions $v = (v_1, v_2)$ on a partition T_k. This method takes the form: find $U = (U_1, U_2)$ in V_k such that for $n = 1, 2, \ldots,$

$$
\begin{aligned}
\int_{I_n} (U_1' + aU_2)\, dt &= \int_{I_n} f_1\, dt, \\
\int_{I_n} (U_2' - aU_1)\, dt &= \int_{I_n} f_2\, dt, \\
U_1(0) = u_{10}, \quad U_2(0) &= u_{20},
\end{aligned} \tag{8.27}
$$

corresponding to using piecewise-constant test functions on each interval I_n. We use piecewise-constant test functions because there are only first-order derivatives in (8.26), in contrast to the elliptic problem discussed above. In the case where a is constant, with the notation $U_{i,n} = U_i(t_n)$, the method (8.27) reduces to:

$$
\begin{aligned}
U_{1,n} - U_{1,n-1} + k_n a(U_{2,n} + U_{2,n-1})/2 &= \int_{I_n} f_1\, dt, \\
U_{2,n} - U_{2,n-1} - k_n a(U_{1,n} + U_{1,n-1})/2 &= \int_{I_n} f_2\, dt, \\
U_1(0) = u_{10}, \quad U_2(0) &= u_{20},
\end{aligned} \tag{8.28}
$$

from which the classical Crank–Nicolson method can be obtained by an appropriate choice of quadrature. The method cG(1) has less dissipation and better accuracy than dG(0), and it is advantageous to use it in this problem since the solution is smooth.

8.6. An a posteriori error estimate

To derive an a posteriori error estimate for the error $e_N = u(t_N) - U_N$, $U_N \equiv U(t_N)$, we introduce the dual problem: find $\varphi = (\varphi_1, \varphi_2)$ such that

$$
\begin{aligned}
-\varphi_1' + a\varphi_2 &= 0, + \in (0_1 + N), \\
-\varphi_2' - a\varphi_1 &= 0, + \in (0_1 + N), \\
\varphi(t_N) &= e_N.
\end{aligned} \tag{8.29}
$$

Again using Galerkin orthogonality, we obtain an error representation formula:

$$
\|e_N\|^2 = -\int_0^{t_N} R \cdot (\varphi - \pi_k \varphi)\, dt,
$$

where

$$
R_1 = U_1' + aU_2 - f_1, \quad R_2 = U_2' - aU_1 - f_2
$$

and π_k is the nodal interpolation operator into V_k. Multiplying the equations by φ_1 and φ_2 respectively and using the cancellation of the terms $\pm a\varphi_1\varphi_2$, we obtain the following stability estimates:

$$
\max_{0 \le t \le t_N} \|\varphi(t)\| \le \|e_N\|
$$

and

$$
\max_{0 \le t \le t_N} \|\varphi'(t)\| \le \mathbb{A}\|e_N\|.
$$

Combining the error representation with the strong-stability estimate and using the interpolation estimate (6.3) we have proved:

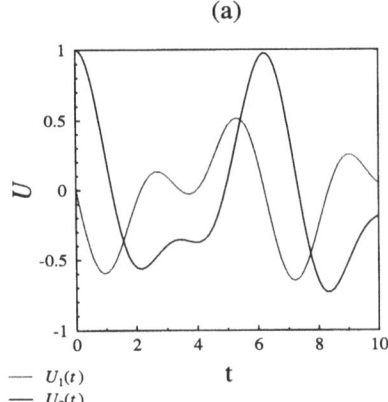
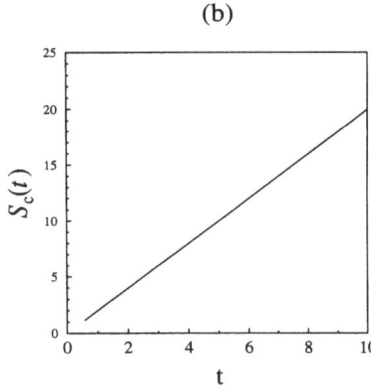

Fig. 9. Approximation and stability factors for Example 8

Theorem 9 U satisfies for $N = 1, 2, \ldots$,

$$\|u(t_N) - U_N\| \leq \mathbb{A} \int_0^{t_N} k\|R\| \, dt.$$

8.7. An a priori error estimate

The corresponding a priori error estimate takes the form:

Theorem 10

$$\|u(t_N) - U_N\| \leq \mathbb{A} \int_0^{t_N} \|k^2 u''\| \, dt \leq \mathbb{A} t_N \|k^2 u''\|_{(0,t_N)}.$$

We note the linear growth of error with time that is characteristic of a hyperbolic problem.

Example 8. We compute for the problem:

$$\begin{aligned}
u_1' + 2u_2 &= \cos(\pi t/3), \quad t > 0, \\
u_2' - 2u_1 &= 0, \quad t > 0, \\
u_1(0) &= 0, \quad u_2(0) = 1,
\end{aligned}$$

using the cG(1) method with error below .07. The two components of the approximation are plotted in Figure 9a. This demonstrates that different components of a system of equations may behave differently at different times. The error control discussed here must choose the time steps to maintain accuracy in all components simultaneously. In Figure 9b, we plot the stability factor, and the linear growth of error is evident.

9. Nonlinear systems of ordinary differential equations

The framework for a posteriori error analysis described above directly extends to initial-value problems for nonlinear systems of differential equations in $\mathbb{R}^d, d \geq 1$ (or more generally a Hilbert space). The ease of the extension depends on the definition of the stability factors occurring in the a posteriori analysis. In the adaptive algorithms built on the a posteriori error estimates, the stability factors are estimated by computation and not by analysis. Thus the essential computational difficulty is the approximation of the stability factors and the essential mathematical difficulty is the justification of this process. We return to this issue after presenting the extension.

We consider the computation of solutions $u = u(t)$ of the following initial-value problem:

$$\begin{aligned} u' + f(t, u) &= 0, \quad t > 0, \\ u(0) &= u_0, \end{aligned} \tag{9.1}$$

where $f(t, \cdot) : \mathbb{R}^d \to \mathbb{R}^d$ is a given function and u_0 given initial data. We assume that f and u_0 are perturbations of correct $\hat{f}$ and $\hat{u}_0$, and denote by $\hat{u}$ the corresponding exact solution satisfying

$$\begin{aligned} \hat{u}' + \hat{f}(t, \hat{u}) &= 0, \quad t > 0, \\ \hat{u}(0) &= \hat{u}_0. \end{aligned} \tag{9.2}$$

We seek an a posteriori error bound for the complete error $\hat{e} \equiv \hat{u} - U$, where U is the dG(0) approximate solution of (9.1) defined by: find U in W_k such that for all constant vectors v,

$$\int_{I_n} (U' + f(t, U)) \cdot v \, dt + [U_{n-1}] \cdot v_{n-1}^+ = 0, \tag{9.3}$$

where $[v_n] \equiv v_n^+ - v_n^-$, $v_n^\pm \equiv \lim_{s \to \pm 0} v(t_n + s)$ and $U_0^- = \hat{u}_0$. With the notation $U_n \equiv U|_{I_n}$, the dG(0) method (9.3) takes the form

$$U_n - U_{n-1} + \int_{I_n} f(t, U_n) \, dt = 0, \quad n = 1, 2, \ldots, \tag{9.4}$$

where $U_0 = \hat{u}_0$. Again, this is an improved variation of the classical backward Euler method with exact evaluation of the integral with integrand $f(t, U)$.

9.1. An a posteriori error estimate

To derive an a posteriori error estimate for e_N for $N \geq 1$ including data, modelling and Galerkin-discretization errors, we introduce the continuous dual 'backward' problem

$$\begin{aligned} -\varphi' + \hat{A}(t)^* \varphi &= 0, \quad t \in (0, t_N), \\ \varphi(t_N) &= \hat{e}_N, \end{aligned} \tag{9.5}$$

where

$$\hat{A}(t) \equiv \int_0^1 \hat{f}_u(t, su + (1 - s)U) \, ds,$$

where $\hat{f}_u(t, \cdot)$ denotes the Jacobian of $\hat{f}(t, \cdot)$ and $*$ denotes the transpose. Note that

$$\hat{A}(t)e = \int_0^1 \hat{f}_u(t, s\hat{u} + (1-s)U)\hat{e}\, ds$$

and

$$\int_0^1 \frac{d}{ds}\hat{f}(t, s\hat{u} + (1-s)U)\, ds = \hat{f}(t, \hat{u}) - \hat{f}(t, U).$$

Integrating by parts, with $\| \cdot \|$ denoting the Euclidean norm, we get

$$
\begin{aligned}
\|e_N\|^2 &= \|e_N\|^2 + \sum_{n=1}^{N} \int_{I_n} e \cdot (-\varphi' + \hat{A}^*\varphi)\, dt \\
&= \sum_{n=1}^{N} \int_{I_n} (e' + \hat{A}(t)e) \cdot \varphi\, dt + \sum_{n=0}^{N-1} [e_n] \cdot \varphi_n^+ + e_0^- \cdot \varphi(0) \\
&= -\sum_{n=1}^{N} \left(\int_{I_n} (U' + f(t, U)) \cdot \varphi\, dt + [U_{n-1}] \cdot \varphi_{n-1}^+ \right) \\
&\quad + e_0^- \cdot \varphi(0) + \sum_{n=1}^{N} \int_{I_n} (f(t, U) - \hat{f}(t, U)) \cdot \varphi dt.
\end{aligned}
$$

Now we use Galerkin orthogonality (9.3) to insert $\pi_k\varphi$ in the first term on the right, and we obtain, since $U' = 0$ on I_n, the following error representation formula:

$$
\begin{aligned}
\|e_N\|^2 &= -\sum_{n=1}^{N} \left(\int_{I_n} (U' + f(t, U)) \cdot (\varphi - \pi_k\varphi)\, dt + [U_{n-1}] \cdot (\varphi - \pi_k\varphi)_{n-1}^+ \right) \\
&\quad + e_0^- \cdot \varphi(0) + \sum_{n=1}^{N} \int_{I_n} (f(t, U) - \hat{f}(t, U)) \cdot \varphi dt \\
&= -\int_0^{t_N} f(t, U) \cdot (\varphi - \pi_k\varphi)\, dt - \sum_{n=0}^{N-1} [U_n] \cdot (\varphi - \pi_k\varphi)_n^+ + e_0^- \cdot \varphi(0) \\
&\quad + \int_0^{t_N} (f(t, U) - \hat{f}(t, U)) \cdot \varphi dt.
\end{aligned}
$$

Recalling the interpolation estimate (6.3), we see that

$$\|e_N\|^2 \leq \int_0^{t_N} \|\varphi'\| dt \max_{1 \leq n \leq N} (\|[U_{n-1}]\| + k_n \|f(\cdot, U_n)\|_{I_n})$$

$$+ \|e_0^-\|\|\varphi(0)\| + \int_0^{t_N} \|\varphi\| dt \max_{1 \leq n \leq N} \|f(\cdot, U_n) - \hat{f}(\cdot, U_n)\|_{I_n}.$$

Finally, we define the strong-stability factor $S_c(t_N)$ by

$$S_c(t_N) \equiv \frac{\int_0^{t_N} \|\varphi'\| dt}{\|\varphi(t_N)\|}, \tag{9.6}$$

and the data and modelling stability factors by

$$S_d(t_N) \equiv \frac{\|\varphi(0)\|}{\|\varphi(t_N)\|}, \qquad S_m(t_N) \equiv \frac{\int_0^{t_N} \|\varphi(s)\| ds}{\|\hat{\varphi}(t_N)\|},$$

and arrive at an a posteriori error estimate for data, modelling and Galerkin-discretization errors.

Theorem 11 U satisfies for $N = 1, 2, \ldots,$

$$\|u(t_N) - U_N\| \leq S_c(t_N) \max_{1 \leq n \leq N} k_n R(n, U)$$
$$+ S_d(t_N) \|\hat{u}_0 - u_0\|$$
$$+ S_m(t_N) \max_{1 \leq n \leq N} \|\hat{f}(\cdot, U_n) - f(\cdot, U_n)\|_{I_n},$$

where

$$R(n, U) \equiv \|U_n - U_{n-1}\|/k_n + \|f(\cdot, U_n)\|_{I_n}.$$

Remark 6   There is a corresponding a priori result with stability factors related to discrete dual problems.

9.2. Computational evaluation of stability factors

To give the a posteriori estimate concrete meaning, the stability factors have to be determined. Accurate analytic estimates are possible only in a few special cases, and in general we resort to numerical integration of the dual linearized problems. The critical mathematical issue is the reliability of this evaluation, since this directly translates into the reliability of the adaptive algorithm. The basic sources of error in the computational evaluation of stability factors are (i) the choice of linearization and (ii) the choice of data, and (iii) the numerical solution of the dual linearized problem. In practice, the problem is linearized around an approximation rather than the mean value that involves the unknown exact solution used in the definition of the dual problems. Moreover, the current error is unknown, and hence the true initial data for the dual problems cannot be used. Finally, the resulting problem must be approximated numerically. The reliability of the computation of stability factors related to (i) and (ii) may be guaranteed for certain classes of problems but in the general case, little is known. Reliability with respect to (iii) seems to be an issue of smaller magnitude.

In many experiments, (see Estep (to appear), Estep and French, (to appear), Eslep and Johnson (1994)), we have seen that the choice of initial data in the dual problem is often immaterial provided the time interval is

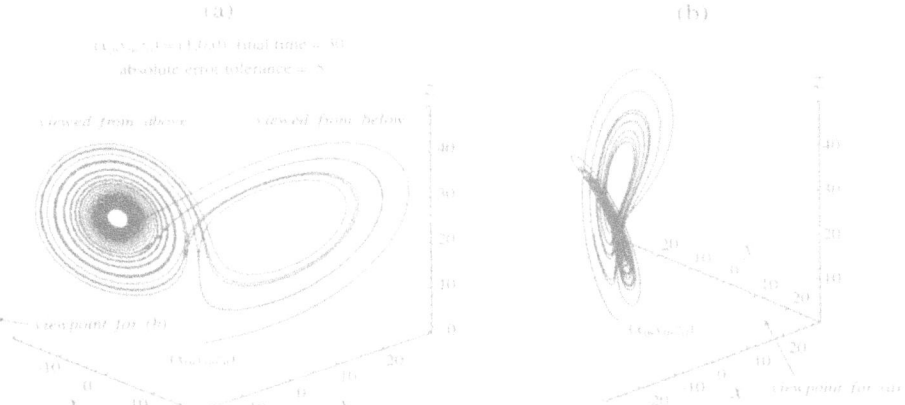

Fig. 10. Two views of a solution of the Lorenz system

sufficiently long. Otherwise, computing dual problems using several different initial values improves reliability. Moreover, unless grossly inaccurate, approximate trajectories seem to provide reasonably accurate stability factors.

Example 9. In the early 1960s, the meteorologist E. Lorenz presented a simple model in order to explain why weather forecasts over more than a couple of days are unreliable. The model is derived by taking a three-element Fem space discretization of the Navier–Stokes equations for fluid flow (the 'fluid' being the atmosphere in this case) and simply ignoring the discretization error. This gives a three-dimensional system of ODE's in time:

$$
\begin{aligned}
x' &= -\sigma x + \sigma y, && t > 0, \\
y' &= -rx - y - xz, && t > 0, \\
z' &= -bz + xy, && t > 0, \\
x(0) &= x_0, y(0) = y_0, z(0) = z_0,
\end{aligned}
\tag{9.7}
$$

where σ, r and b are positive constants. These were determined originally as part of the physical problem, but the interest among mathematicians quickly shifted to studying (9.7) for values of the parameters that make the problem *chaotic*.

A precise definition of chaotic behaviour seems difficult to give, but we point out two distinguishing features: while confined to a fixed region in space, the solutions do not 'settle down' into a steady state or periodic state; and the solutions are *data sensitive*, which means that perturbations of the initial data of a given solution eventually cause large changes in the solution. This corresponds to the 'butterfly effect' in meteorology in which small causes may sometimes have large effects on the evolution of the weather. In such situations, numerical approximations always become inaccurate after

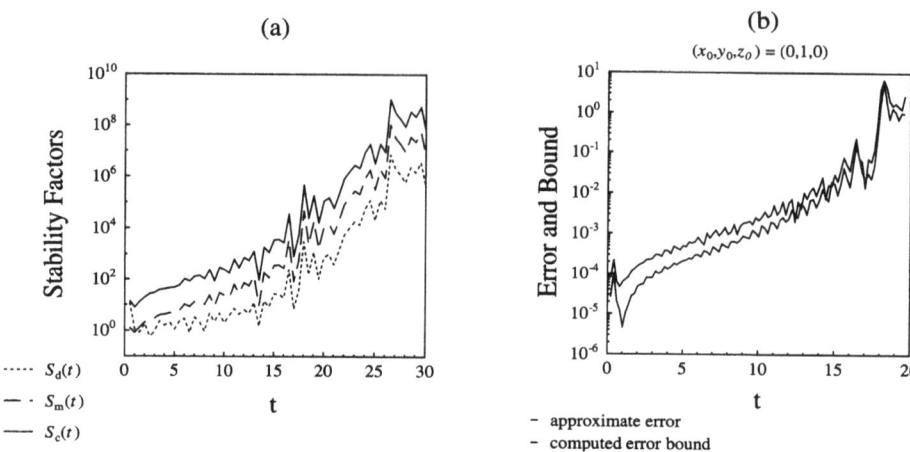

Fig. 11. Stability factors and error bound for the Lorentz system

some time. An important issue is to determine this time, since, for example, it is related to the maximal length of a weather forecast.

We choose standard values $\sigma = 10$, $b = 8/3$ and $r = 28$, and we compute with the dG(1) method. In Figure 10, we plot two views of the solution corresponding to initial data $(1, 0, 0)$ computed with an error of .5 up to time 30. The solutions always behave similarly: after some short initial time, they begin to 'orbit' around one of two points, with an occasional 'flip' back and forth between the points. The chaotic nature of the solutions is this flipping that occurs at apparently random times. In fact, accurate computation can reveal much detail about the behaviour of the solutions; see Eriksson *et al.* (1994*b*).

Here, we settle for demonstrating the quality of the error control explained in these notes. In Figure 11a, we plot the approximate stability factors on a logarithmic scale. The data sensitivity of this problem is reflected in the overall exponential growth of the factors, and it is clear that any computation becomes inaccurate at some point. The error control allows this time to be determined. Note, however, that the factors do not grow uniformly rapidly and there are periods of time with different data sensitivity. It is important for the error control to detect these to avoid gross overestimation of the error. To test this, we do an experiment. We compute using two error tolerances, one 10^{-5} smaller than the other, and then we subtract the less accurate computation from the more accurate computation. This should be a good approximation to the true error (which is unknown of course). In Figure 11b, we plot this approximate error together with the error bound predicted by the error control based on a posteriori estimates as we have described. There is remarkable agreement.

Finally, in Figure 12a, we plot the S_c for the various trajectories computed

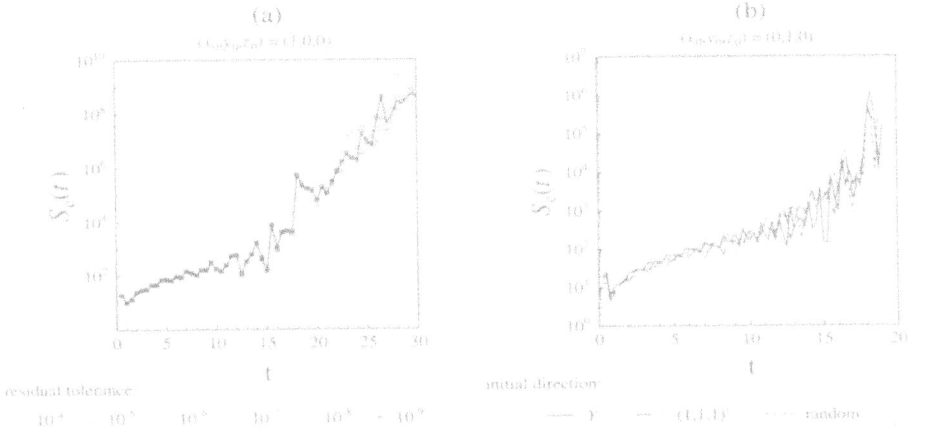

Fig. 12. Stability factors for the Lorentz system

with different tolerances. Overall, the stability factors are roughly the same order of magnitude for all trajectories. The stability factors agree as long as the trajectories are near each other, but variations occur as some trajectories enter more data-sensitive areas than others at the same time. In Figure 12b, we plot the approximation to S_c computed for three different choices of initial data for the dual problem (9.5).

10. An elliptic model problem in two dimensions

In this section, we consider Fem for Poisson's equation in two dimensions. We discuss a priori and a posteriori error estimates for the Galerkin-discretization error and also the discrete-solution error for a multigrid method, and design corresponding adaptive methods. The analysis is largely parallel to that of the one-dimensional model problem, though the analysis of the multigrid method is more technical.

Consider the Poisson equation with homogeneous Dirichlet boundary conditions: find $u = u(x)$ such that

$$\begin{aligned} -\Delta u &= f, & x \in \Omega, \\ u &= 0, & x \in \partial\Omega, \end{aligned} \tag{10.1}$$

where Ω is a bounded domain in $\mathbb{R}^2$ with boundary $\partial\Omega$, $x \equiv (x_1, x_2)$, Δ is the Laplace operator and $f = f(x)$ is given data. The variational form of (10.1) reads: find $u \in H_0^1(\Omega)$ such that

$$(\nabla u, \nabla v) = (f, v), \quad \forall v \in H_0^1(\Omega), \tag{10.2}$$

where $(w, v) \equiv \int_\Omega wv \, dx$, $(\nabla w, \nabla v) \equiv \int_\Omega \nabla w \cdot \nabla v \, dx$ and $H_0^1(\Omega)$ is the Sobolev space of square-integrable functions with square integrable derivatives on Ω that vanish on $\partial\Omega$. We recall that $\|\nabla \cdot\|_2$ is a norm in $H_0^1(\Omega)$

that is equivalent to the $H^1(\Omega)$ norm. The existence of a unique solution of (10.2) follows by the Riesz representation theorem if $f \in H^{-1}(\Omega)$, where $H^{-1}(\Omega)$ is the dual space of $H_0^1(\Omega)$ with norm

$$\|f\|_{H^{-1}(\Omega)} \equiv \sup_{v \in H_0^1(\Omega), \|\nabla v\|_2 = 1} (f, v).$$

We recall strong-stability (or elliptic regularity) estimates for the solution of (10.2) to be used below. The estimates are termed 'strong' because derivatives of the solution u are estimated. We use the notation $D^0 v \equiv v$, $D^1 v \equiv |\nabla v|$ and $D^2 v \equiv (\sum_{i=1}^2 (\frac{\partial^2 v}{\partial x_i \partial x_j})^2)^{\frac{1}{2}}$. Further, we use $\|\cdot\| = \|\cdot\|_\Omega$ to denote the $L^2(\Omega)$ norm.

Lemma 3   The solution u of (10.2) satisfies

$$\|\nabla u\| \leq \|f\|_{H^{-1}(\Omega)}. \tag{10.3}$$

Furthermore, if Ω is convex with polygonal boundary, or if $\partial\Omega$ is smooth, then there is a constant S_c independent of f, such that

$$\|D^2 u\| \leq S_c \|f\|. \tag{10.4}$$

If Ω is convex, then $S_c = 1$.

10.1. Fem for Poisson's equation

The simplest Fem for (10.1) results from applying Galerkin's method to the variational formulation (10.2) using a finite-dimensional subspace V_h $H_0^1(\Omega)$ based on piecewise-linear approximation on triangles. For simplicity, we consider the case of a convex polygonal domain. Let $T_h = \{K\}$ be a finite-element triangulation of Ω into triangles K of diameter h_K with associated set of nodes $N_h = \{N\}$ such that each node N is the corner of at least one triangle. We require that the intersection of any two triangles K' and K'' in T_h be either empty, a common triangle side or a common node.

To the mesh T_h we associate a mesh function $h(x)$ satisfying, for some positive constant c_1,

$$c_1 h_K \leq h(x) \leq h_K, \quad \forall x \in K, \quad \forall K \in T_h. \tag{10.5}$$

We further assume that there is a positive constant c_2 such that

$$c_2 h_K^2 \leq 2|K|, \quad \forall K \in T_h. \tag{10.6}$$

This is a 'minimum angle' condition stating that angles of triangles in T_h are bounded from below by the constant c_2. As usual, C_i denotes an interpolation constant related to piecewise-linear interpolation on the mesh T_h. In this case, C_i depends on c_1 and c_2, but not on h otherwise.

With $V_h \subset H_0^1(\Omega)$ denoting the standard finite-element space of piecewise-linear functions on T_h, the Fem for (10.1) reads: find $u_h \in V_h$ such that

$$(\nabla u_h, \nabla v) = (f, v), \quad \forall v \in V_h. \tag{10.7}$$

Galerkin orthogonality for (10.7), resulting from (10.2) and (10.7), takes the form:

$$(\nabla(u - u_h), \nabla v) = 0, \quad \forall v \in V_h. \tag{10.8}$$

We write $u_h = \sum_{i=1}^{M} \xi_i \varphi_i$, where $\{\varphi_i\}$ is the usual basis for V_h associated to the set of nodes $N_h^0 = \{N_i\}_{i=1}^{M}$ in the interior of Ω and $\xi_i = u_h(N_i)$. Then, (10.7) is equivalent to the linear system of equations

$$A\xi = b, \tag{10.9}$$

where $\xi = (\xi_i)_{i=1}^{M}$, $A = (a_{ij})_{j,i=1}^{M}$ is the $M \times M$ stiffness matrix with elements $a_{ij} \equiv (\nabla \varphi_i, \nabla \varphi_j)$ and $b = (b_j)_{j=1}^{M}$ is the load vector with $b_j \equiv (f, \varphi_j)$. We use a multigrid method to solve the discrete system (10.9), producing an approximation $\tilde{u}_h \in V_h$ of the exact discrete solution u_h.

We require an error estimate for interpolation by piecewise-linear functions, where the piecewise-linear nodal interpolant $\pi_h w \in V_h$ of a given function $w \in H_0^1(\Omega) \cap H^2(\Omega)$ is defined by $\pi_h w(N) = w(N)$, $\forall N \in N_h^0$. We also need an analogous estimate for a 'quasi-interpolant' of $w \in H_0^1(\Omega)$ that requires less regularity where the 'quasi-interpolant' interpolates local mean values of w over neighbouring elements. We use the same notation for the nodal interpolant and the quasi-interpolant. The basic interpolation estimate is:

Lemma 4 For $s \le m + 1$, $m \in \{0, 1\}$, there are constants C_i depending only on c_1 and c_2 such that for $w \in H_0^1(\Omega) \cap H^{m+1}(\Omega)$

$$\|h^{-m-1+s}D^s(w - \pi_h w)\| + \left(\sum_{K \in T_h} h_K^{-2m-1} \|w - \pi_h w\|_{\partial K}^2 \right)^{1/2} \le C_i \|D^{m+1} w\|.$$

10.2. The discrete and continuous residuals

We shall prove an a posteriori error estimate for the total error $e \equiv u - \tilde{u}_h = u - u_h + u_h - \tilde{u}_h$ including the Galerkin-discretization error $u - u_h$ and the discrete-solution error $u_h - \tilde{u}_h$. The a posteriori error estimate involves both a discrete residual $R_h(\tilde{u}_h)$ related to solving the discrete system (10.7) approximately and an estimate $R(\tilde{u}_h)$ of the residual related to the continuous problem (10.1).

To define $R_h(\tilde{u}_h)$, we introduce the L^2-projection $P_h \colon L^2(\Omega) \to V_h$ defined by $(u - P_h u, v) = 0$, $\forall v \in V_h$, and the 'discrete Laplacian' $\Delta_h \colon V_h \to V_h$ on V_h defined by $(\Delta_h w, v) = -(\nabla w, \nabla v)$, $\forall v, w \in V_h$. We may then write

(10.7) equivalently as $R_h(u_h) = 0$, where for $w \in V_h$ the discrete residual $R_h(w)$ is defined as

$$R_h(w) \equiv \Delta_h w + P_h f. \tag{10.10}$$

For the approximate solution $\tilde{u}_h$, we have $R_h(\tilde{u}_h) \neq 0$.

Remark 7 Letting $\tilde{U}_h$ denote the nodal-valued vector of the approximate solution $\tilde{u}_h$, we define the 'algebraic' residual $r_h(\tilde{U}_h)$ by $r_h(\tilde{U}_h) \equiv b - A\tilde{U}_h$. By the definition of $R_h(\tilde{u}_h)$, it follows that $r_h(\tilde{U}_h) = M_h \hat{R}_h(\tilde{u}_h)$, where $M_h = (m_{ij})_{i,j=1}^M$ is the mass matrix with elements $m_{ij} = (\varphi_i, \varphi_j)$ and $\hat{R}_h(\tilde{u}_h)$ is the nodal-valued vector of $R(\tilde{u}_h)$. Thus, $R_h(\tilde{u}_h)$ is computable from the algebraic residual $r_h(\tilde{U}_h)$ by applying M_h^{-1}. In practice, M_h may be replaced by a diagonal matrix corresponding to 'mass lumping'.

The estimate $R(\tilde{u}_h)$ for the continuous residual is defined on each element $K \in T_h$ by

$$R(\tilde{u}_h) \equiv |f + \Delta \tilde{u}_h| + D_h^2 \tilde{u}_h, \quad x \in K, \tag{10.11}$$

where for $v \in V_h$

$$D_h^2 v|_K \equiv \frac{1}{2\sqrt{h_K}} \|h_K^{-1}[\nabla v]\|_{\partial K}, \tag{10.12}$$

where $[\nabla v]$ denotes the jump in ∇v across ∂K. Note that D_h^2 resembles a second derivative in the case of piecewise-linear approximation when ∇v is constant on each element K. The factor $1/2$ arises naturally because the jump is associated to two neighbouring elements. We note that $D_h^2 v$ is a piecewise-constant function and thus in particular belongs to $L^2(\Omega)$.

The residual function $R(\tilde{u}_h)$ also belongs to $L^2(\Omega)$ and has two parts: the 'interior' part $|f + \Delta \tilde{u}_h|$ and the 'boundary' part $D_h^2 \tilde{u}_h$. The boundary part can be made to vanish in the one-dimensional problems considered above, because the interpolation error vanishes at inter-element boundaries. In the present case with piecewise-linear approximation, $R(u_h) = |f| + D_h^2 u_h$, $x \in K$, while in the case of higher-order polynomials, Δu_h no longer vanishes on each triangle and has to be taken into account.

In the proofs below, we use the following crucial estimate:

Lemma 5 For $m \in \{0, 1\}$, there is a constant C_i such that $\forall v \in H_0^1(\Omega) \cap H^{m+1}(\Omega)$,

$$|(f, v - \pi_h v) - (\nabla \tilde{u}_h, \nabla(v - \pi_h v))| \leq C_i \|h^{m+1} R(\tilde{u}_h)\| \|D^{m+1} v\|. \tag{10.13}$$

Appropriate values of the constant C_i in (10.13) can be calculated analytically or numerically. If $c_1 \sim 1$ and $c_2 \sim 1$, then $C_i \sim 0.2$ for $m = 0, 1$ (cf. Johnson and Hansbo (1992b), Becker et al. (1994)).

Proof. By integration by parts, observing that $v - \pi_h v$ is continuous, we have

$$
\begin{aligned}
(f, v - \pi_h v) \ - \ &(\nabla \tilde{u}_h, \nabla(v - \pi_h v)) \\
= \ &\sum_{K \in T_h} \{(f + \Delta \tilde{u}_h, v - \pi_h v)_K - (\partial_n \tilde{u}_h, v - \pi_h v)_{\partial K}\} \\
= \ &\sum_{K \in T_h} \left\{(f + \Delta \tilde{u}_h, v - \pi_h v)_K - \frac{1}{2}([\nabla \tilde{u}_h], v - \pi_h v)_{\partial K}\right\},
\end{aligned}
$$

with $[\nabla \tilde{u}_h]$ denoting the jump of $\nabla \tilde{u}_h$ across the element edges, from which the result follows by Lemma 4. $\square$

10.3. A priori estimates of the Galerkin-discretization error

We first give an a priori error estimate of the Galerkin-discretization error $u - u_h$ in the energy norm.

Theorem 12   There exists a constant C_i depending only on c_1 and c_2 such that

$$\|\nabla(u - U)\| \leq \|\nabla(u - \pi_h u)\| \leq C_i \|hD^2 u\|. \tag{10.14}$$

Proof. In (10.8), we choose $v = U - \pi_h u$ and use Cauchy's inequality to get

$$
\begin{aligned}
\|\nabla e\|^2 &= (\nabla e, \nabla(u - U)) = (\nabla e, \nabla(u - U)) + (\nabla e, \nabla(U - \pi_h u)) \\
&= (\nabla e, \nabla(u - \pi_h u)) \leq \|\nabla e\| \, \|\nabla(u - \pi_h u)\|,
\end{aligned}
\tag{10.15}
$$

from which the desired result follows from Lemma 4. $\square$

We next give an a priori error estimate in the L^2 norm.

Theorem 13   There exists a constant C_i only depending on c_1 and c_2 such that

$$\|u - U\| \leq S_c C_i \|h\nabla(u - U)\|, \tag{10.16}$$

where

$$S_c \equiv \max_{g \in L^2(\Omega)} \frac{\|D^2 \varphi\|}{\|g\|},$$

with $\varphi \in H_0^1(\Omega)$ satisfying $-\Delta \varphi = g$ in Ω. Further, if $|\nabla h(x)| \leq \mu$, $x \in \Omega$, with μ a sufficiently small positive constant, then

$$\|h\nabla(u - U)\| \leq C_i \|h^2 D^2 u\|, \tag{10.17}$$

where C_i now depends also on μ.

Proof. The proof of (10.16) uses duality in a manner similar to that of the proof of Theorem 4. Note that (10.17) follows directly from the energy-norm error estimate if h is constant. $\square$

10.4. A posteriori error estimates of Galerkin and discrete-solution errors

We now turn to a posteriori error estimates, including the discrete-solution error in the case of multigrid methods. Let T_j, $j = 0, 1, 2, \ldots, k$, be a hierarchy of successively refined meshes with corresponding nested sequence of finite-element spaces V_j and mesh functions h_j, where the final mesh T_k corresponds to the mesh T_h in the above presentation. We seek to compute an approximation $\tilde{u}_k \in V_k$ of the finite-element solution $u_k \in V_k$ on the final mesh T_k, using a multigrid algorithm based on the hierarchy of meshes. For $j \in \{0, \ldots, k\}$, define the residual $R_j(\tilde{u}_k) \in V_j$ related to the mesh T_j by the relation

$$R_j(\tilde{u}_k) \equiv P_j(f + \Delta_k \tilde{u}_k),   \tag{10.18}$$

where P_j is the L^2-projection onto V_j and $\Delta_k : V_k \to V_k$ is the discrete Laplacian on V_k.

The multigrid algorithm consists of a sequence of smoothing operations $V_j \to V_j$ (e.g. Jacobi, Gauss–Seidel or ILU iterations) on the different meshes T_j, which are together with grid transfer operations (prolongations and restrictions). The objective of the multigrid algorithm is to make the residual $R_k(\tilde{u}_k)$ on the final mesh T_k small, which is realized in a hierarchical process that also makes the residuals $R_j(\tilde{u}_k)$ small for $j = 0, 1, \ldots, k - 1$. We assume that $R_0(\tilde{u}_k) = 0$, which corresponds to solving the discrete equations exactly on the coarsest mesh. The details of the multigrid method are immaterial for the a posteriori error estimate to be given.

We now state and prove the a posteriori error estimate and then briefly discuss a corresponding adaptive algorithm.

Theorem 14   For $m \in \{0, 1\}$, there are constants C_i and S_c such that, if u is the solution of (10.1) and $\tilde{u}_k \in V_k$ is an approximate finite-element solution with $R_0(\tilde{u}_k) = 0$, then

$$\|D^m(u - \tilde{u}_k)\| \leq S_c C_i \left\{ \|h_k^{2-m} R(\tilde{u}_k)\| + \sum_{j=1}^{k} \|h_{j-1}^{2-m} R_j(\tilde{u}_k)\| \right\}.   \tag{10.19}$$

If $m = 1$, or if $m = 0$ and Ω is convex, then $S_c = 1$.

Proof.   For $m = 0$ or $m = 1$, let $\varphi \in H_0^1(\Omega)$ be the solution to the dual continuous problem

$$(\nabla v, \nabla \varphi) = (D^m v, D^m e), \quad \forall v \in H_0^1(\Omega).$$

Taking $v \equiv e$, we obtain the error representation

$$\|D^m e\|^2 = (\nabla e, \nabla \varphi) = (f, \varphi) - (\nabla \tilde{u}_k, \nabla \varphi) \equiv \langle r(\tilde{u}_k), \varphi \rangle.$$

For $j \leq k$, we have the telescoping identity

$$\langle r(\tilde{u}_k), \varphi \rangle = \langle r(\tilde{u}_k), \varphi - \pi_k \varphi \rangle + \sum_{j=1}^{k} \langle r(\tilde{u}_k), \pi_j \varphi - \pi_{j-1} \varphi \rangle + \langle r(\tilde{u}_k), \pi_0 \varphi \rangle,$$

where π_j denotes the interpolation operator into V_j related to the mesh T_j. Observing that for $v \in V_j$, since $V_j \subset V_k$,

$$\langle r(\tilde{u}_k), v \rangle = (P_j(f + \Delta_k \tilde{u}_k), v) = (R_j(\tilde{u}_k), v),$$

and that $R_0(\tilde{u}_k) = 0$ by assumption, we reduce this to

$$\langle r(\tilde{u}_k), \varphi \rangle = \langle r(\tilde{u}_k), \varphi - \pi_k \varphi \rangle + \sum_{j=1}^{k} (R_j(\tilde{u}_k), \pi_j \varphi - \pi_{j-1} \varphi).$$

Hence, we obtain using Lemma 4 and Lemma 5

$$\|D^m e\|^2 \leq C_i \left\{ \|h_k^{2-m} R(\tilde{u}_k)\| \|D^{2-m} \varphi\| + \sum_{j=1}^{k} \|h_{j-1}^{2-m} R_j(\tilde{u}_k)\| \|D^{2-m} \varphi\| \right\},$$

from which the assertion follows using Lemma 3. □

Remark 8 For the exact solution u_h of the finite-element equation (10.7), the a posteriori error estimate has the familiar form

$$\|D^m(u - \tilde{u}_h)\| \leq S_c C_i \|h_k^{2-m} R(\tilde{u}_h)\|. \tag{10.20}$$

Remark 9 The effect of round-off in the computation of the discrete solution $\tilde{u}_k$ may be taken into account as follows: Suppose the multigrid computation is carried out in single precision. The a posteriori error estimate is valid if the residuals $R(\tilde{u}_k)$ and $R_j(\tilde{u}_k)$, $j = 0, 1, \ldots, k$, are evaluated exactly. In practice, this means in double precision. We can also add the term $\|R_0(\tilde{u}_k)\|$ to take into account that $R_0(\tilde{u}_k) = 0$ in single precision only. If the a posteriori error estimators evaluated in single and double precision differ by more than the chosen tolerance, then the entire computation should be redone in double precision.

11. Adaptive algorithms

The stopping criterion of an adaptive algorithm based on Theorem 14 takes the form

$$S_c C_i \max \left(\sum_{j=1}^{k} \|h_{j-1}^{2-m} R_j(\tilde{u}_k)\|, \|h_k^{2-m} R(\tilde{u}_k)\| \right) \leq \frac{1}{2} TOL, \tag{11.1}$$

which ensures that the Galerkin-discretization and the discrete-solution errors are equilibriated. The form of the stopping criterion for the discrete-solution error suggests how to monitor the smoothing process on the different levels to make the different residuals $R_j(\tilde{u}_k)$ appropriately small.

11.1. A posteriori error estimates in the $L^\infty(\Omega)$ norm

We now give an a posteriori error estimate for the Galerkin-discretization error $u - u_h$ in the $L^\infty(\Omega)$ norm $\|\cdot\|_\infty$.

Theorem 15 There is a constant C_i such that

$$\|u - u_h\|_\infty \leq S_c L_h C_i \|h^2 R(u_h)\|_\infty, \tag{11.2}$$

with $S_c \equiv \max_{y\in\bar\Omega} \|\log(|\,-y|)^{-1} D^2 \psi_y\|_1$, where ψ_y is the Green's function for Δ on Ω with pole at $y \in \Omega$, and $L_h \equiv (1 + \log(1/h_{\min}))$, where $h_{\min}$ is the minimal mesh size of T_h. There is a constant C such that $S_c \leq C$ for all polygonal domains Ω of diameter at most one.

Proof. The proof is based on the following error representation:

$$(u - u_h)(y) = \int_\Omega f(\psi_y - \pi_h \psi_y)\,dx - \int_\Omega \nabla u_h \cdot \nabla(\psi_y - \pi_h \psi_y)\,dx,$$

from which the desired estimate follows by arguments analogous to those used above. $\square$

Example 10. We now present results obtained using adaptive algorithms based on Theorem 15 for L^∞ control and Theorem 14 for energy-norm control with $m = 1$ and $S_c = 1$, where Ω is the L-shaped domain $(-1, 1) \times (-1, 1) \setminus (0, 1) \times (-1, 0)$. We consider a case with an exact solution u with a singularity at the nonconvex corner, given by $u(r, \theta) = r^{\frac{2}{3}} \sin(\frac{2}{3}\theta)$ in polar coordinates.

In the case of maximum-norm control, the stability factor S_c is determined by computing approximately ψ_y for some sample points y. In this case apparently, a few choices of y are sufficient. The interpolation constant is set to $C_i = 1/8$. In Figure 13, we present the initial mesh (112 nodes and 182 elements) and the level curves of the exact solution. In Figure 14, we show the final mesh (217 nodes and 382 triangles) produced by the adaptive algorithm with $TOL = 0.005$. Figure 15 shows the variation of the efficiency index and the stability constant S_c as functions of the number of refinement levels. The efficiency index, defined as the ratio of the error to the computed bound, increases slightly from the coarsest to the finest grid. In Figure 16, we show the final mesh (295/538) using energy-norm error control with $TOL = 0.005$. Note that the final meshes in the two cases are considerably different.

12. The heat equation

We briefly consider the extension of the results for the scalar equation $u' + au = f$ with $a \geq 0$ given above, to the heat equation, the standard model

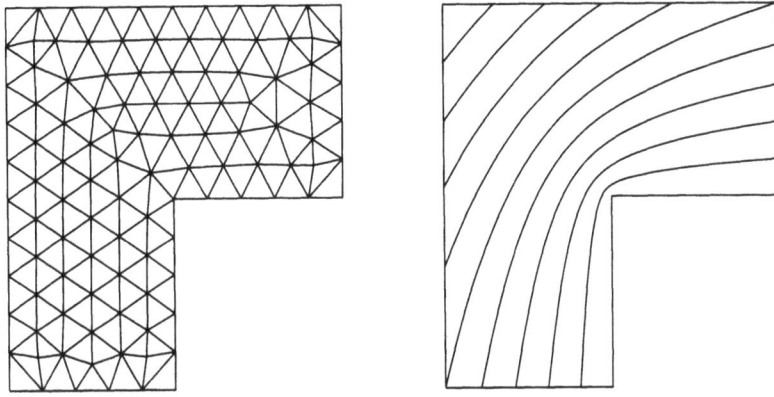

Fig. 13. Original mesh and isolines of the solution on a fine mesh

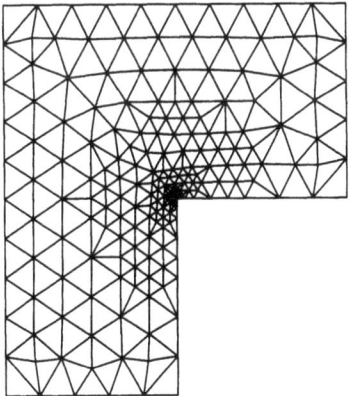

Fig. 14. Maximum-norm control of the error

problem of parabolic type: find $u = u(x, t)$ such that

$$
\begin{aligned}
u_t - \Delta u &= f, & (x, t) &\in \Omega \times I, \\
u &= 0, & (x, t) &\in \partial\Omega \times I, \\
u &= u_0, & x &\in \Omega,
\end{aligned}
\tag{12.1}
$$

where Ω is a bounded polygonal domain in $\mathbb{R}^2$, $I = (0, T)$ is a time interval, $u_t \equiv \partial u / \partial t$ and the functions f and u_0 are given data.

For discretization of (12.1) in time and space we use the dG(r) method based on a partition $0 \equiv t_0 < t_1 < \cdots < t_n < \cdots < t_N \equiv T$ of I and associate with each time interval $I_n \equiv (t_{n-1}, t_n]$ of length $k_n \equiv t_n - t_{n-1}$ a triangulation $\mathcal{T}_n$ of Ω with mesh function h_n and the corresponding space

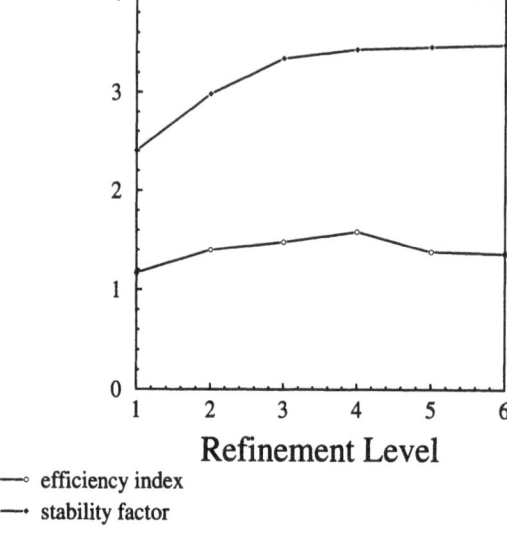

Refinement Level

∘—∘ efficiency index
•—• stability factor

Fig. 15. Stability constant and efficiency index on the different refinement levels

$V_n \in H_0^1(\Omega)$ of piecewise-linear functions as in Section 5. Note that we allow the space discretizations to change with time. We define

$$V_{rn} \equiv \left\{ v : v = \sum_{j=0}^r t^j \varphi_j, \ \varphi_j \text{ in } V_n \right\},$$

and discretize (12.1) as follows: find U such that for $n = 1, 2, \ldots, U|_{\Omega \times I_n} \in V_{rn}$ and

$$\int_{I_n} \{(U_t, v) + (\nabla U, \nabla v)\} dt + ([U]_{n-1}, v_{n-1}^+) = \int_{I_n} (f, v) dt, \quad \forall v \in V_{rn}, \quad (12.2)$$

where $[w]_n \equiv w_n^+ - w_n^-$, $w_n^{+(-)} \equiv \lim_{s \to 0+(-)} w(t_n + s)$ and $U_0^- = u_0$.

As above, if $r = 0$, then (12.2) reduces to a variant of the Euler backward method, and for $r = 1$ it reduces to a variant of the subdiagonal Padé scheme of order $(2,1)$, that is third-order accurate in U_n^- at the nodal points t_n.

The a posteriori error estimate in the case $r = 0$ has the form

$$\|u(t_N) - U_N\|_2 \le C_i L_N \max_{n=1,\ldots,N} (\|h_n^2 R(U_n)\| + \|[U_{n-1}]\| + \|h_n^2 k_n^{-1}[U_{n-1}]\|^*),$$

$$(12.3)$$

where $R(U)$ is defined by (10.11), $L_N \equiv \max_{n=1,\ldots,N}(1 + \log(\frac{t_n}{k_n}))^{\frac{1}{2}}$ is a logarithmic factor and the starred term is present only if the space mesh changes at time t_{n-1}. The analogous a priori error estimate assuming $h_n^2 \le k_n$

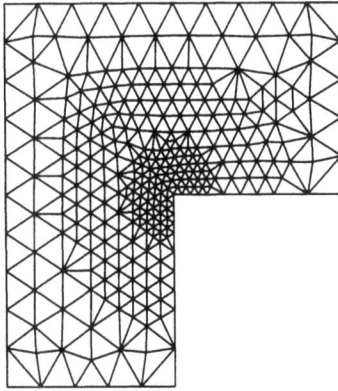

Fig. 16. Energy-norm control of the error

takes the form

$$\|u(t_N) - U_N\| \leq C_i L_N \max_{n=1,\ldots,N} (\|h_n^2 D^2 u\|_{I_n} + \|k_n u_t\|_{I_n}), \qquad (12.4)$$

An adaptive algorithm may be based on (12.3). We note the optimal character of (12.4) and (12.3), that in particular allows long-time integration without error accumulation.

13. References

We give here a brief account of the current status of the development of the framework for adaptive approximation of differential equations that we have described. We also give some references to the extensive literature on adaptive methods that are of particular relevance for our work.

Adaptive methods for linear elliptic problems with energy-norm control were first developed by Babuška *et al.* (see Babuška (1986) and references therein) and Bank *et al.* (see Bank (1986)). In both cases, a posteriori error estimates were obtained by solving local problems with the residual acting as data. Residual-based a posteriori energy-norm error estimates were also derived for Stokes's equations in Verfürth (1989).

The basic approach we use for adaptive methods for linear elliptic problems, including a priori and a posteriori error estimates in H^1, L^2 and L^∞ norms, is presented in Eriksson and Johnson (1991) and Eriksson (to appear). Extensions to adaptive control of the discrete-solution error using multigrid methods is developed in Becker *et al.* (1994). Nonlinear elliptic problems including obstacle and plasticity problems are considered in Johnson (1992a), Johnson and Hansbo (1992a) and Johnson and Hansbo

(1992*b*). Recently, applications to eigenvalue problems have been given in Nystedt (in preparation).

Early a posteriori error analysis for ordinary differential equations was used in Cooper (1971) and Zadunaisky (1976). These approaches are quite different from ours. We develop adaptive global error control for systems of ordinary differential equations in Johnson (1988), Estep (to appear), Estep and French (to appear), Estep and Johnson (1994), and Estep and Williams (in preparation). Lippold (1988) had an influence on our early work.

The series Eriksson and Johnson (1991), Eriksson and Johnson (1994*a, b, c, d*), (to appear), Eriksson, Johnson and Larsson (1994) develops adaptive Fem for a class of parabolic problems in considerable generality including space-time discretization that is variable in space-time, and applications to nonlinear problems.

Adaptive Fem for linear convection–diffusion problems is considered in Johnson (1990), Eriksson and Johnson (1993) and Eriksson and Johnson (to appear). Extensions to the compressible Euler equations are given in Hansbo and Johnson (1991) and Johnson and Szepessy (to appear). Extensions to the Navier–Stokes equations for incompressible flow are given in Johnson, et al. (to appear), Johnson and Rannacher (1993) and Johnson, *et al.* (1994). Second order wave equations are considered in Johnson (1993).

The presented framework also applies to Galerkin methods for integral equations. An application to integral equations is given in Asadzadeh and Eriksson (to appear). The potential of the framework is explored in Carstensen and Stephan (1993).

14. Conclusion, open problems

The framework for deriving a posteriori error estimates and designing adaptive algorithms for quantitative error control may be applied to virtually any differential equation. The essential difficulties are (i) the computational estimates of stability factors and (ii) the design of the modification strategy. The reliability depends on the accuracy of the computed stability factors and may be increased by increasing the fraction of the total work spent on stability factors. Optimization of computations of stability factors is an important open problem. Optimal design of the modification criterion is also largely an open matter for complex problems. Thus, the contours of a general methodology for adaptive error control seem to be visible, but essential concrete algorithmic problems connected mainly with (i) and (ii) remain to be solved. The degree of difficulty involved depends on the features of the underlying problem related to, for example, stability and nonlinearities.

The concept of computability as a measure of computational complexity is central. A basic problem in mathematical modelling is to develop mathematical models for which solutions are computable. A basic problem of this form is turbulence modelling. Isolating computational errors from modelling errors gives the possibility of evaluating and improving the quality of mathematical models.

To sum up, it appears to be possible to develop reliable and efficient adaptive computational software for a large class of differential and integral equations arising in applications, which could be made available to a large group of users from calculus students to engineers and scientists. If such a program can be successfully realised, it will open up entirely new possibilites in mathematical modelling.

15. How to obtain Femlab and 'Introduction to Numerical Methods for Differential Equations'

Femlab contains software for solving: (i) one dimensional, two point boundary value problems (Femlab-1d); (ii) initial value problems for general systems of ordinary differential equations (Femlab-ode); and two dimensional boundary valve problems (Femlab-2d). Femlab, together with the educational material Eriksson et al. (1994), can be obtained over the Internet. Femlab-ode can be obtained by anonymous ftp to

ftp.math.gatech.edu.

Change to directory */pub/users/estep* and get *femlabode.tar.* This tar file contains the codes and a brief user's manual. In that same directory is *intro.ps.Z*, a compressed postscript version of *Eriksson et al.* [1994].

To obtain Femlab-1d and Femlab-2d open to the WWW (World Wide Web) address

http://www.math.chalmers.se/ kenneth

using (for instance) the Mosaic program. There is a README file located there that gives further instructions.

Femlab-1d consists of a number of Matlab script files and M-files. You can import these files using the 'save as ...' command under the 'file' menu. To run the code, you then just start your local Matlab program and give the command adfem, calling the script file adfem.m. For more details, see the README file.

REFERENCES

M. Asadzadeh and K. Eriksson (1994), 'An adaptive finite element method for a potential problem', *M3AS*, to appear.

I. Babuška (1986), 'Feedback, adaptivity and *a posteriori* estimates in finite elements: Aims, theory, and experience', in *Accuracy Estimates and Adaptive Refinements in Finite Element Computations*, (I. Babuška, O. C. Zienkiewicz, J. Gago and E. R. de A. Oliveira, eds.), Wiley, New York, 3–23.

R. Bank (1986), 'Analysis of local a posteriori error estimate for elliptic equations', in *Accuracy Estimates and Adaptive Refinements in Finite Element Computations* (I. Babuška, O. C. Zienkiewicz, J. Gago and E. R. de A. Oliveira, eds.), Wiley, New York.

R. Becker, C. Johnson and R. Rannacher (1994), 'An error control for multigrid finite element methods', Preprint #1994-36, Department of Mathematics, Chalmers University of Technology, Göteborg.

C. Carstensen and E. Stephan (1993), 'Adaptive boundary element methods', Institute für Angewandt Mathematik, Universität Hannover.

G. Cooper (1971), 'Error bounds for numerical solutions of ordinary differential equations', *Num. Math.* **18**, 162–170.

K. Eriksson (1994), 'An adaptive finite element method with efficient maximum norm error control for elliptic problems', *M3AS*, to appear.

K. Eriksson, D. Estep, P. Hansbo and C. Johnson (1994a), *Adaptive Finite Element Methods*, North Holland, Amsterdam, in preparation.

Eriksson, *et al* (1994) K. Eriksson, D. Estep, P. Hansbo and C. Johnson (1994b), 'Introduction to Numerical Methods for Differential Equations', Department of Mathematics, Chalmers University of Technology, Göteborg.

K. Eriksson and C. Johnson (1988), 'An adaptive finite element method for linear elliptic problems', *Math. Comput.* **50**, 361–383.

K. Eriksson and C. Johnson (1991), 'Adaptive finite element methods for parabolic problems I: A linear model problem', *SIAM J. Numer. Anal.* **28**, 43–77.

K. Eriksson and C. Johnson (1993), 'Adaptive streamline diffusion finite element methods for stationary convection–diffusion problems', *Math. Comp.* **60**, 167–188.

K. Eriksson and C. Johnson (1994a), 'Adaptive finite element methods for parabolic problems II: Optimal error estimates in $L_\infty L_2$ and $L_\infty L_\infty$', *SIAM J. Numer Anal.* to appear.

K. Eriksson and C. Johnson (1994b), 'Adaptive finite element methods for parabolic problems III: Time steps variable in space'. in preparation.

K. Eriksson and C. Johnson (1994c), 'Adaptive finite element methods for parabolic problems IV: Non-linear problems', *SIAM J. Numer Anal.* to appear.

K. Eriksson and C. Johnson (1994d), 'Adaptive finite element methods for parabolic problems V: Long-time integration', *SIAM J. Numer Anal.* to appear.

K. Eriksson and C. Johnson (1994e), 'Adaptive streamline diffusion finite element methods for time-dependent convection–diffusion problems', *Math. Comp.* to appear.

K. Eriksson, C. Johnson and S. Larsson (1994), 'Adaptive finite element methods for parabolic problems VI: Analytic semigroups', Preprint, Department of Mathematics, Chalmers University of Technology, Göteborg.

D. Estep (1994), 'A posteriori error bounds and global error control for approximations of ordinary differential equations', *SIAM J. Numer. Anal.* to appear.

D. Estep and D. French (1994), 'Global error control for the continuous Galerkin finite element method for ordinary differential equations', *RAIRO M.M.A.N.* to appear.

D. Estep and C. Johnson (1994a), 'The computability of the Lorenz system', Preprint #1994-33, Department of Mathematics, Chalmers University of Technology, Göteborg.

D. Estep and C. Johnson (1994b), 'An analysis of quadrature in Galerkin finite element methods for ordinary differential equations', in preparation.

D. Estep and S. Larsson (1993), 'The discontinuous Galerkin method for semilinear parabolic problems', *RAIRO M.M.A.N.* **27**, 611–643.

D. Estep and A. Stuart (1994), 'The dynamical behavior of Galerkin methods for ordinary differential equations and related quadrature schemes', in preparation.

D. Estep and R. Williams (1994), 'The structure of an adaptive differential equation solver', in preparation.

P. Hansbo and C. Johnson (1991), 'Adaptive streamline diffusion finite element methods for compressible flow using conservation variables', *Comput. Methods Appl. Mech. Engrg.* **87**, 267–280.

C. Johnson (1988), 'Error estimates and adaptive time step control for a class of one step methods for stiff ordinary differential equations', *SIAM J. Numer. Anal.* **25**, 908–926.

C. Johnson (1990), 'Adaptive finite element methods for diffusion and convection problems', *Comput. Methods Appl. Mech. Engrg.* **82**, 301–322.

C. Johnson (1992a), 'Adaptive finite element methods for the obstacle problem, *Math. Models Methods Appl. Sci.* **2**, 483–487.

C. Johnson (1992b), 'A new approach to algorithms for convection problems based on exact transport + projection', *Comput. Methods Appl. Mech. Engrg.* **100**, 45–62.

C. Johnson (1993a), 'Discontinuous Galerkin finite element methods for second order hyperbolic problems', *Comput. Methods Appl. Mech. Engrg.* **107**, 117–129.

C. Johnson (1993b), 'A new paradigm for adaptive finite element methods', in *Proc. Mafelap 93, Brunel Univ.*, Wiley Comput. Methods Appl. Mech. Engrg. vol. 107, Wiley, New York, 117–129.

C. Johnson and P. Hansbo (1992a), 'Adaptive finite element methods for small strain elasto-plasticity', in *Finite Inelastic Deformations – Theory and Applications* (D. Besdo and E. Stein, eds.), Springer, Berlin, 273–288.

C. Johnson and P. Hansbo (1992b), 'Adaptive finite element methods in computational mechanics', *Comput. Methods Appl. Mech. Engrg.* **101**, 143–181.

C. Johnson and R. Rannacher (1994), 'On error control in CFD', in *Proc from Conf. on Navier–Stokes Equations Oct 93, Vieweg*, to appear.

C. Johnson, R. Rannacher, and M. Boman (1994a), 'Numerics and hydrodynamic stability: Towards error control in CFD', *SIAM J. Numer. Anal.* to appear.

C. Johnson, R. Rannacher, and M. Boman (1994b), 'On transition to turbulence and error control in CFD', Preprint #1994-26, Department of Mathematics, Chalmers University of Technology, Göteborg.

C. Johnson and A. Szepessy (1994), 'Adaptive finite element methods for conservation laws based on a posteriori error estimates', *Comm. Pure Appl. Math.* to appear.

G. Lippold (1988), 'Error estimates and time step control for the approximate solution of a first order evolution equation', preprint, Akademie der Wissenschaften der Karl Weierstrass Institut für Mathematik, Berlin.

C. Nystedt (1994), 'Adaptive finite element methods for eigenvalue problems', Licentiate Thesis, Department of Mathematics, Chalmers University of Technology, Göteborg in preparation.

R. Verfürth (1989), 'A posteriori error estimators for the Stokes equations', *Numer. Math.* **55**, 309–325.

P. Zadunaisky (1976), 'On the estimation of errors propagated in the numerical integration of ordinary differential equations', *Numer. Math.* **27**, 21–39.

Acta Numerica (1996), *pp.* 159–333

Exact and approximate controllability for distributed parameter systems

R. Glowinski

University of Houston, Houston, Texas, USA
Université Pierre et Marie Curie, Paris, France
C.E.R.F.A.C.S., Toulouse, France

J.L. Lions

Collège de France, Rue d'Ulm, 75005 Paris, France

This is the second part of an article which was started in the previous volume of *Acta Numerica*. References in the text to Section 1 refer to the preceding article.

CONTENTS: PART I

CONTENTS: PART II

2. BOUNDARY CONTROL

2.1. Dirichlet control (I): Formulation of the control problem

We consider again the *state equation*

$$\frac{\partial y}{\partial t} + Ay = 0 \text{ in } Q, \tag{2.1}$$

where the *second-order elliptic operator* A is as in Section 1.1, and where the control v is now a *boundary control of Dirichlet* type, namely

$$y = \begin{cases} v \text{ on } \Sigma_0 = \Gamma_0 \times (0, T), \\ 0 \text{ on } \Sigma \backslash \Sigma_0, \end{cases} \tag{2.2}$$

where Σ_0 is a (regular) subset of Γ.

The *initial condition* is (for simplicity)

$$y(0) = 0. \tag{2.3}$$

In (2.2) we assume that

$$v \in L^2(\Sigma_0). \tag{2.4}$$

Then, assuming that the coefficients of operator A are smooth enough (cf. Lions and Magenes (1968) for precise statements), the parabolic problem (2.1)–(2.3) has a *unique solution* such that

$$y \in L^2(0, T; L^2(\Omega)) \ (= L^2(Q)), \quad \frac{\partial y}{\partial t} \in L^2(0, T; H^{-2}(\Omega)), \tag{2.5}$$

so that

$$y \in C^0([0, T]; H^{-1}(\Omega)). \tag{2.6}$$

Remark 2.1 The solution y to (2.1)–(2.3) is defined, as usual, by *transposition*. Properties (2.5) and (2.6) still hold true if $v \in L^2(0, T; H^{-1/2}(\Gamma_0))$ (the notation is that used in Lions and Magenes (1968)).

Concerning *controllability*, the key result is given by the following:

Proposition 2.1 *When v spans $L^2(\Sigma_0)$, the function $y(T; v)$ spans a dense subspace of $H^{-1}(\Omega)$.*

Proof. We shall give a (nonconstructive) proof based on the *Hahn–Banach theorem*. Consider, thus, $f \in H_0^1(\Omega)$ such that

$$\langle y(T; v), f \rangle = 0, \quad \forall v \in L^2(\Sigma_0), \tag{2.7}$$

where, in (2.7), $\langle \cdot, \cdot \rangle$ denotes the duality pairing between $H^{-1}(\Omega)$ and $H_0^1(\Omega)$; next, define ψ by

$$-\frac{\partial \psi}{\partial t} + A^* \psi = 0 \text{ in } Q, \quad \psi(T) = f, \ \psi = 0 \text{ on } \Sigma. \tag{2.8}$$

Multiplying both sides of the first equation in (2.8) by the solution $\{x, t\} \rightarrow y(x, t; v)$ of problem (2.1)–(2.3) we obtain after integration by parts

$$\langle y(T; v), f \rangle = -\int_{\Sigma_0} \frac{\partial \psi}{\partial n_{A^*}} v \, d\Gamma \, dt, \tag{2.9}$$

where $\partial/\partial n_{A^*}$ denotes the *conormal derivative* operator associated with A^* (if $A = A^* = -\Delta$, then $\partial/\partial n_A = \partial/\partial n_{A^*} = \partial/\partial n$ where $\partial/\partial n$ is the usual outward normal derivative operator at Γ). Then (2.7) is equivalent to

$$\partial\psi/\partial n_{A^*} = 0 \text{ on } \Sigma_0. \tag{2.10}$$

It follows from (2.8), (2.10) that the *Cauchy data* of ψ are zero on Σ_0; using again the *Mizohata's uniqueness theorem*, we obtain that $\psi = 0$ in Q, so that $f = 0$, which completes the proof of the proposition. □

We can now formulate the following *approximate controllability problems* (where $d\Sigma = d\Gamma \, dt$):

Problem 1. It is defined by

$$\inf_v \frac{1}{2} \int_{\Sigma_0} v^2 \, d\Sigma, \ v \in L^2(\Sigma_0), \ y(T; v) \in y_T + \beta B_{-1}, \tag{2.11}$$

where, in (2.11), y_T is given in $H^{-1}(\Omega)$, $\beta > 0$, B_{-1} denotes the unit ball of $H^{-1}(\Omega)$ and $t \rightarrow y(t; v)$ is the solution of (2.1)–(2.3) associated with the control v.

Problem 2. It is the variant of problem (2.11) defined by

$$\inf_{v \in L^2(\Sigma_0)} \left[\frac{1}{2} \int_{\Sigma_0} v^2 \, d\Sigma + \tfrac{1}{2} k \| y(T; v) - y_T \|^2_{-1} \right], \tag{2.12}$$

where, in (2.12), $k > 0$, y_T and $y(T; v)$ are as in (2.11), and where

$$\begin{cases} \forall y \in H^{-1}(\Omega), \|y\|_{-1} = \left(\int_{\Omega} |\nabla\varphi|^2 \, dx \right)^{1/2} \\ \text{with } \varphi \text{ the unique solution in } H_0^1(\Omega) \text{ of} \\ \int_{\Omega} \nabla\varphi \cdot \nabla\theta \, dx = \langle y, \theta \rangle, \ \forall\theta \in H_0^1(\Omega). \end{cases}$$

Both problems (2.11) and (2.12) have a unique solution.

2.2. Dirichlet control (II): Optimality conditions and dual formulations

We discuss first problem (2.12) which is simpler than problem (2.11). Let us denote by $J_k(\cdot)$ the cost function in (2.12); using the relation

$$(J_k'(v), w)_{L^2(\Sigma_0)} = \lim_{\substack{\theta \to 0 \\ \theta \neq 0}} \frac{J_k(v + \theta w) - J_k(v)}{\theta}, \ \forall v, w \in L^2(\Sigma_0), \tag{2.13}$$

we can show that

$$(J_k'(v), w)_{L^2(\Sigma_0)} = \int_{\Sigma_0} \left(v - \frac{\partial p}{\partial n_{A^*}} \right) w \, d\Sigma, \quad \forall v, w \in L^2(\Sigma_0), \qquad (2.14)$$

where, in (2.14), the *adjoint state function* p is obtained from v via the solution of (2.1)–(2.3) and of the *adjoint state equation*

$$-\frac{\partial p}{\partial t} + A^* p = 0 \text{ in } Q, \quad p = 0 \text{ on } \Sigma,$$

$$p(T) \in H_0^1(\Omega) \text{ and } -\Delta p(T) = k(y(T) - y_T) \text{ in } \Omega. \qquad (2.15)$$

Suppose now that u is *the* solution of the control problem (2.12); since $J_k'(u) = 0$, we have then the following optimality system satisfied by u and the corresponding state and adjoint state functions:

$$u = \frac{\partial p}{\partial n_{A^*}}\Big|_{\Sigma_0},$$

$$\frac{\partial y}{\partial t} + Ay = 0 \text{ in } Q, \quad y(0) = 0, \ y = 0 \text{ on } \Sigma \backslash \Sigma_0 \text{ and } y = \frac{\partial p}{\partial n_{A^*}} \text{ on } \Sigma_0,$$

$$-\frac{\partial p}{\partial t} + A^* p = 0 \text{ in } Q, \quad p = 0 \text{ on } \Sigma, \ p(T) = f,$$

where f is the *unique* solution in $H_0^1(\Omega)$ of the Dirichlet problem

$$-\Delta f = k(y(T) - y_T) \text{ in } \Omega, f = 0 \text{ on } \Gamma. \qquad (2.16)$$

In order to identify the dual problem of (2.12), we proceed as in the above sections by introducing (in the spirit of the *Hilbert Uniqueness Method*) the operator $\Lambda \in \mathcal{L}(H_0^1(\Omega); H^{-1}(\Omega))$ defined by

$$\Lambda \hat{f} = -\hat{\varphi}(T), \quad \forall \hat{f} \in H_0^1(\Omega), \qquad (2.17)$$

where the function $\hat{\varphi}$ is obtained from $\hat{f}$ as follows:
Solve first

$$-\frac{\partial \hat{\psi}}{\partial t} + A^* \hat{\psi} = 0 \text{ in } Q, \quad \hat{\psi} = 0 \text{ on } \Sigma, \ \hat{\psi}(T) = \hat{f} \qquad (2.18)$$

and then,

$$\frac{\partial \hat{\varphi}}{\partial t} + A\hat{\varphi} = 0 \text{ in } Q, \quad \hat{\varphi}(0) = 0, \ \hat{\varphi} = 0 \text{ on } \Sigma \backslash \Sigma_0, \ \hat{\varphi} = \frac{\partial \hat{\psi}}{\partial n_{A^*}} \text{ on } \Sigma_0. \ (2.19)$$

We can easily show that (with obvious notation)

$$\langle \Lambda f_1, f_2 \rangle = \int_{\Sigma_0} \frac{\partial \psi_1}{\partial n_{A^*}} \frac{\partial \psi_2}{\partial n_{A^*}} \, d\Gamma \, dt, \quad \forall f_1, \ f_2 \in H_0^1(\Omega). \qquad (2.20)$$

It follows from (2.20) that the operator Λ is *self-adjoint* and *positive semi-*

definite; indeed, it follows from *Mizohata's uniqueness theorem* that the operator Λ is *positive definite*. However, the operator Λ *is not an isomorphism from $H_0^1(\Omega)$ onto $H^{-1}(\Omega)$* (implying that, in general, *we do not have exact boundary controllability here*).

Back to (2.16) we observe that from the definition of Λ we have $y(T) = -\Lambda f$, which implies in turn that f is the unique solution in $H_0^1(\Omega)$ of

$$- k^{-1}\Delta f + \Lambda f = -y_T. \tag{2.21}$$

Problem (2.21) is precisely the dual problem we are looking for. From the properties of operator $-k^{-1}\Delta + \Lambda$, problem (2.21) can be solved by a *conjugate gradient algorithm* operating in the space $H_0^1(\Omega)$; we shall return to this issue in Section 2.3.

Let us consider the control problem (2.11); using the *Fenchel–Rockafellar convex duality theory* as in the above sections, we can show that *the* solution u of problem (2.11) is characterized by the following optimality system

$$u = \frac{\partial p}{\partial n_{A^*}}\Big|_{\Sigma_0}, \tag{2.22}$$

$$\frac{\partial y}{\partial t} + Ay = 0 \text{ in } Q, \quad y(0) = 0, \ y = 0 \text{ on } \Sigma\backslash\Sigma_0 \text{ and } y = \frac{\partial p}{\partial n_{A^*}} \text{ on } \Sigma_0, \tag{2.23}$$

$$-\frac{\partial p}{\partial t} + A^*p = 0 \text{ in } Q, \quad p = 0 \text{ on } \Sigma, \ p(T) = f, \tag{2.24}$$

where f is the *unique* solution of the following *variational inequality* (with $\|\hat{f}\|_{H_0^1(\Omega)} = (\int_\Omega |\nabla\hat{f}|^2\,dx)^{1/2}, \ \forall\hat{f} \in H_0^1(\Omega))$:

$$\begin{cases} f \in H_0^1(\Omega), \\ \langle\Lambda f, \hat{f} - f\rangle + \beta\|\hat{f}\|_{H_0^1(\Omega)} - \beta\|f\|_{H_0^1(\Omega)} + \langle y_T, \hat{f} - f\rangle \geq 0, \ \forall\hat{f} \text{ in } H_0^1(\Omega). \end{cases} \tag{2.25}$$

Problem (2.25) is precisely the dual problem to (2.11). The solution of problem (2.25) will be discussed in Section 2.3.

2.3. Dirichlet control (III): Iterative solution of the control problems

2.3.1. Conjugate gradient solution of problem (2.12)

It follows from Section 2.2 that solving the control problem (2.12) is equivalent to solving the linear equation

$$J_k'(u) = 0, \tag{2.26}$$

where operator J_k' is defined by (2.1)–(2.3), (2.14), (2.15). It is fairly easy to show that the *linear* part of operator J_k', namely

$$v \rightarrow J_k'(v) - J_k'(0),$$

is *symmetric* and *strongly elliptic* over $L^2(\Sigma_0)$. From these properties, problem (2.26) can be solved by a *conjugate gradient algorithm* operating in the space $L^2(\Sigma_0)$. It follows from Section 1.8.2 that this algorithm is as follows.

Description of the conjugate gradient algorithm:

$$u^0 \text{ is given in } L^2(\Sigma_0); \tag{2.27}$$

solve

$$\frac{\partial y^0}{\partial t} + Ay^0 = 0 \text{ in } Q, \quad y^0(0) = 0, \; y^0 = u^0 \text{ on } \Sigma_0, \; y^0 = 0 \text{ on } \Sigma\backslash\Sigma_0, \tag{2.28}$$

and then

$$\begin{cases} f^0 \in H_0^1(\Omega), \\ -\Delta f^0 = k(y^0(T) - y_T) \text{ in } \Omega, \end{cases} \tag{2.29}$$

and finally

$$-\frac{\partial p^0}{\partial t} + A^* p^0 = 0 \text{ in } Q, \quad p^0 = 0 \text{ on } \Sigma, \; p^0(T) = f^0. \tag{2.30}$$

Set

$$g^0 = u^0 - \frac{\partial p^0}{\partial n_{A^*}}\Big|_{\Sigma_0}, \tag{2.31}$$

and then

$$w^0 = g^0. \tag{2.32}$$

For $n \geq 0$, assuming that u^n, g^n, w^n are known, compute u^{n+1}, g^{n+1}, w^{n+1} as follows:
Solve

$$\frac{\partial \bar{y}^n}{\partial t} + A\bar{y}^n = 0 \text{ in } Q, \quad \bar{y}^n(0) = 0, \; \bar{y}^n = w^n \text{ on } \Sigma_0, \; \bar{y}^n = 0 \text{ on } \Sigma\backslash\Sigma_0, \tag{2.33}$$

and then

$$\begin{cases} \bar{f}^n \in H_0^1(\Omega), \\ -\Delta \bar{f}^n = k\bar{y}^n(T) \text{ in } \Omega, \end{cases} \tag{2.34}$$

and finally

$$-\frac{\partial \bar{p}^n}{\partial t} + A^* \bar{p}^n = 0 \text{ in } Q, \quad \bar{p}^n = 0 \text{ on } \Sigma, \; \bar{p}^n(T) = \bar{f}^n. \tag{2.35}$$

Compute

$$\bar{g}^n = w^n - \frac{\partial \bar{p}^n}{\partial n_{A^*}}\Big|_{\Sigma_0}, \tag{2.36}$$

and then

$$\rho_n = \int_{\Sigma_0} |g^n|^2 \, d\Gamma \, dt \Big/ \int_{\Sigma_0} \bar{g}^n w^n \, d\Gamma \, dt, \tag{2.37}$$

$$u^{n+1} = u^n - \rho_n w^n, \tag{2.38}$$

$$g^{n+1} = g^n - \rho_n \bar{g}^n. \tag{2.39}$$

If $\|g^{n+1}\|_{L^2(\Sigma_0)}/\|g^0\|_{L^2(\Sigma_0)} \le \epsilon$ take $u = u^{n+1}$, else compute

$$\gamma_n = \|g^{n+1}\|^2_{L^2(\Sigma_0)}/\|g^n\|^2_{L^2(\Sigma_0)} \tag{2.40}$$

and update w^n by

$$w^{n+1} = g^{n+1} + \gamma_n w^n. \tag{2.41}$$

Do $n = n + 1$ and go to (2.33).

Remark 2.2 The number of iterations necessary to obtain the convergence varies here too, as $k^{1/2} \ln \epsilon^{-1/2}$.

2.3.2. Conjugate gradient solution of the dual problem (2.21)
   We mentioned in Section 2.2 that the *dual problem* (2.21), namely

$$-k^{-1}\Delta f + \Lambda f = -y_T,$$

can be solved by a *conjugate gradient algorithm* operating in the space $H_0^1(\Omega)$; from the definition of operator Λ (see (2.17)–(2.19)), and from Section 1.8.2, this algorithm takes the following form:

$$f^0 \text{ is given in } H_0^1(\Omega); \tag{2.42}$$

solve

$$-\frac{\partial p^0}{\partial t} + A^* p^0 = 0 \text{ in } Q, \quad p^0 = 0 \text{ on } \Sigma, \quad p^0(T) = f^0, \tag{2.43}$$

and

$$\frac{\partial y^0}{\partial t} + Ay^0 = 0 \text{ in } Q, \quad y^0(0) = 0, \; y^0 = 0 \text{ on } \Sigma\backslash\Sigma_0, \; y^0 = \frac{\partial p^0}{\partial n_{A^*}} \text{ on } \Sigma_0. \tag{2.44}$$

Solve now

$$\begin{cases} g^0 \in H_0^1(\Omega), \\ \displaystyle\int_\Omega \nabla g^0 \cdot \nabla z \, dx = k^{-1} \int_\Omega \nabla f^0 \cdot \nabla z \, dx + \langle y_T - y^0(T), z \rangle, \; \forall z \in H_0^1(\Omega), \end{cases} \tag{2.45}$$

and set

$$w^0 = g^0. \tag{2.46}$$

Then, for $n \ge 0$, assuming that f^n, g^n, w^n are known, compute f^{n+1}, g^{n+1}, w^{n+1} as follows:
Solve

$$-\frac{\partial \bar{p}^n}{\partial t} + A^* \bar{p}^n = 0 \text{ in } Q, \quad \bar{p}^n = 0 \text{ on } \Sigma, \quad \bar{p}^n(T) = w^n, \tag{2.47}$$

and

$$\frac{\partial \bar{y}^n}{\partial t} + A\bar{y}^n = 0 \text{ in } Q, \quad \bar{y}^n(0) = 0, \ \bar{y}^n = 0 \text{ on } \Sigma \backslash \Sigma_0, \ \bar{y}^n = \frac{\partial \bar{p}^n}{\partial n_{A^*}} \text{ on } \Sigma_0.$$

$$(2.48)$$

Solve now

$$\begin{cases} \bar{g}^n \in H_0^1(\Omega), \\ \displaystyle\int_\Omega \boldsymbol{\nabla}\bar{g}^n \cdot \boldsymbol{\nabla} z \, \mathrm{d}x = k^{-1} \int_\Omega \boldsymbol{\nabla} w^n \cdot \boldsymbol{\nabla} z \, \mathrm{d}x - \langle \bar{y}^n(T), z \rangle, \ \forall z \in H_0^1(\Omega). \end{cases}$$

$$(2.49)$$

Compute

$$\rho_n = \int_\Omega |\boldsymbol{\nabla} g^n|^2 \, \mathrm{d}x \Big/ \int_\Omega \boldsymbol{\nabla}\bar{g}^n \cdot \boldsymbol{\nabla} w^n \, \mathrm{d}x \qquad (2.50)$$

and then

$$f^{n+1} = f^n - \rho_n w^n, \qquad (2.51)$$

$$g^{n+1} = g^n - \rho_n \bar{g}^n. \qquad (2.52)$$

If $\|g^{n+1}\|_{H_0^1(\Omega)}/\|g^0\|_{H_0^1(\Omega)} \le \epsilon$ *take* $f = f^{n+1}$ *and solve* (2.24) *to obtain* $u = \partial p/\partial n_{A^*}|_{\Sigma_0}$; *if the above stopping test is not satisfied, compute*

$$\gamma_n = \int_\Omega |\boldsymbol{\nabla} g^{n+1}|^2 \, \mathrm{d}x \Big/ \int_\Omega |\boldsymbol{\nabla} g^n|^2 \, \mathrm{d}x \qquad (2.53)$$

and then

$$w^{n+1} = g^{n+1} + \gamma_n w^n. \qquad (2.54)$$

Do $n = n+1$ *and go to* (2.47).

Remark 2.3 Remark 2.2 still holds for algorithm (2.42)–(2.54).

The *finite element* implementation of the above algorithm will be discussed in Section 2.5, while the results of numerical experiments will be presented in Section 2.6.

2.3.3. Iterative solution of problem (2.25)

Problem (2.25) can also be written as

$$- y_T \in \Lambda f + \beta \partial j(f), \qquad (2.55)$$

which is a *multivalued* equation in $H^{-1}(\Omega)$, the unknown function f belonging to $H_0^1(\Omega)$; in (2.55), $\partial j(f)$ denotes the *subgradient* at f of the convex function $j: H_0^1(\Omega) \to \mathbb{R}$ defined by

$$j(\hat{f}) = \left(\int_\Omega |\boldsymbol{\nabla}\hat{f}|^2 \, \mathrm{d}x \right)^{1/2}, \ \forall \hat{f} \in H_0^1(\Omega).$$

Problem (2.25), (2.55) is clearly a variant of problem (1.237) (see Section 1.8.8) and as such can be solved by those *operator splitting methods* advocated in Section 1.8.8. To derive these methods we associate with the *'elliptic problem'* (2.55) the following *initial value problem*

$$\begin{cases} \dfrac{\partial}{\partial \tau}(-\Delta f) + \Lambda f + \beta \partial j(f) = -y_T, \\ f(0) = f_0 (\in H_0^1(\Omega)), \end{cases} \quad (2.56)$$

where, in (2.56), τ is a *pseudo-time*.

To capture the steady-state solution of (2.56) (i.e. the solution of problem (2.25), (2.55)) we can approximately integrate (2.56) from $\tau = 0$ to $\tau = +\infty$ by a *Peaceman–Rachford scheme*, like the one described just below:

$$f^0 = f_0 \text{ given in } H_0^1(\Omega); \quad (2.57)$$

then, for $m \geq 0$, *compute* $f^{m+1/2}$ *and* f^{m+1}, *from* f^m, *by solving in* $H_0^1(\Omega)$ the following problems:

$$\frac{(-\Delta f^{m+1/2}) - (-\Delta f^m)}{\Delta \tau / 2} + \beta \partial j(f^{m+1/2}) + \Lambda f^m = -y_T, \quad (2.58)$$

and

$$\frac{(-\Delta f^{m+1}) - (-\Delta f^{m+1/2})}{\Delta \tau / 2} + \beta \partial j(f^{m+1/2}) + \Lambda f^{m+1} = -y_T, \quad (2.59)$$

where $\Delta \tau (> 0)$ is a (pseudo) time discretization step.

As in Section 1.8.8, for problem (1.237), the *convergence* of $\{f^m\}_{m \geq 0}$ to the solution of (2.25), (2.55) is a direct consequence of P.L. Lions and B. Mercier (1979), Gabay (1982; 1983) and Glowinski and Le Tallec (1989); the convergence results proved in the above references apply to the present problem since operator Λ (respectively functional $j(\cdot)$) is *linear, continuous and positive definite* (respectively *convex* and *continuous*) over $H_0^1(\Omega)$. As in Section 1.8.8, we can also use a θ-scheme to solve problem (2.25), (2.55); we shall not describe this scheme here since it is a straightforward variant of algorithm (1.242)–(1.245) (actually such an algorithm is described in Carthel *et al.* (1994), where it has been applied to the solution of the boundary control problem (2.11), (2.25) in the particular case where $\Gamma_0 = \Gamma$).

Back to algorithm (2.57)–(2.59) we observe that problem (2.59) can also be written as

$$\frac{-\Delta f^{m+1} + 2\Delta f^{m+1/2} - \Delta f^m}{\Delta \tau / 2} + \Lambda f^{m+1} = \Lambda f^m. \quad (2.60)$$

Problem (2.60) is a particular case of problem (2.21); it can be solved therefore by the conjugate gradient algorithm described in Section 2.3.2. Concerning the solution of problem (2.58), we observe that the solution of a closely related problem (namely problem (1.343) in Section 1.10.4) has already been discussed; since the solution methods for problem (1.343) and

(2.58) are essentially the same we shall not discuss the solution of (2.58) further.

2.4. Dirichlet control (IV): Approximation of the control problems

2.4.1. Generalities and synopsis

It follows from Section 2.3.3 that the solution of the *state constrained* control problem (2.11) (in fact of its dual problem (2.25)) can be reduced to a sequence of problems similar to (2.21), which is itself the dual problem of the control problem (2.12) (where the closeness of $y(T)$ to the target y_T is forced via *penalty*); we shall therefore concentrate our discussion on the approximation of the control problem (2.12), only.

We shall address both the 'direct' solution of problem (2.12) and the solution of the dual problem (2.21).

The notation will be essentially as in Sections 1.8 and 1.10.6.

2.4.2. Time discretization of problems (2.12) and (2.21)

The *time discretization* of problems (2.12) and (2.21) can be achieved using either first-order or second-order accurate time discretization schemes, very close to those already discussed in Sections 1.8 and 1.10.6 (see also Carthel *et al.* (1994, Sections 5 and 6)). Instead of essentially repeating the discussion which took place in the above sections and reference, we shall describe another *second-order accurate* time discretization scheme, recently introduced by Carthel (1994); actually, the numerical results shown in Section 2.6 have been obtained using this new scheme.

The time discretization of the control problem (2.12) is defined as follows (where $\Delta t = T/N$, N being a positive integer):

$$\underset{\mathbf{v} \in (L^2(\Gamma_0))^{N-1}}{\text{Min}} \quad J_k^{\Delta t}(\mathbf{v}), \tag{2.61}$$

where $\mathbf{v} = \{v^n\}_{n=1}^{N-1}$ and

$$J_k^{\Delta t}(\mathbf{v}) = \frac{\Delta t}{2} \sum_{n=1}^{N-1} a_n \|v^n\|_{L^2(\Gamma_0)}^2 + \frac{k}{2} \|y^N - y_T\|_{-1}^2; \tag{2.62}$$

in (2.62) we have $a_n = 1$ for $n = 1, 2, \ldots, N - 2$, $a_{N-1} = 3/2$ and y^N obtained from $\mathbf{v}$ as follows:

$$y^0 = 0; \tag{2.63}$$

to obtain y^1 (respectively y^n, $n = 2, \ldots, N-1$) we solve the following *elliptic* problem

$$\frac{y^1 - y^0}{\Delta t} + A(\tfrac{2}{3}y^1 + \tfrac{1}{3}y^0) = 0 \text{ in } \Omega, \quad y^1 = v^1 \text{ on } \Gamma_0, \quad y^1 = 0 \text{ on } \Gamma \backslash \Gamma_0 \tag{2.64}$$

(respectively

$$\frac{\frac{3}{2}y^n - 2y^{n-1} + \frac{1}{2}y^{n-2}}{\Delta t} + Ay^n = 0 \text{ in } \Omega, \quad y^n = v^n \text{ on } \Gamma_0, \ y^n = 0 \text{ on } \Gamma \backslash \Gamma_0);$$

(2.65)

finally y^N is defined via

$$\frac{2y^N - 3y^{N-1} + y^{N-2}}{\Delta t} + Ay^{N-1} = 0.$$

(2.66)

Problem (2.61) has a *unique* solution.

In order to discretize the *dual problem* (2.21) we look for the dual problem of the discrete control problem (2.61). The simplest way to derive the dual of problem (2.61) is to start from the *optimality condition*

$$\nabla J_k^{\Delta t}(\mathbf{u}^{\Delta t}) = 0,$$

(2.67)

where, in (2.67), $\mathbf{u}^{\Delta t} = \{u^n\}_{n=1}^{N-1}$ is *the* solution of the discrete control problem (2.61), and where $\nabla J_k^{\Delta t}$ denotes the gradient of the discrete cost function $J_k^{\Delta t}$. Suppose that the discrete control space $\mathcal{U}^{\Delta t} = (L^2(\Gamma_0))^{N-1}$ is equipped with the scalar product

$$(\mathbf{v}, \mathbf{w})_{\Delta t} = \Delta t \sum_{n=1}^{N-1} a_n \int_{\Gamma_0} v^n w^n \, d\Gamma, \quad \forall \mathbf{v}, \mathbf{w} \in \mathcal{U}^{\Delta t};$$

(2.68)

then a tedious calculation will show that $\forall \mathbf{v}, \mathbf{w} \in \mathcal{U}^{\Delta t}$

$$(\nabla J_k^{\Delta t}(\mathbf{v}), \mathbf{w})_{\Delta t}$$

$$= \Delta t \sum_{n=1}^{N-2} \int_{\Gamma_0} \left(v^n - \frac{\partial p^n}{\partial n_{A^*}} \right) w^n \, d\Gamma$$

$$+ \frac{3}{2} \Delta t \int_{\Gamma_0} \left[v^{N-1} - \left(\frac{2}{3} \frac{\partial p^{N-1}}{\partial n_{A^*}} + \frac{1}{3} \frac{\partial p^N}{\partial n_{A^*}} \right) \right] w^{N-1} \, d\Gamma, \quad (2.69)$$

where, in (2.69), the *adjoint state vector* $\{p^n\}_{n=1}^N$ belongs to $(H_0^1(\Omega))^N$ and is obtained as follows.

First, compute p^N as the solution in $H_0^1(\Omega)$ of the elliptic problem

$$-\Delta p^N = k(y^N - y_T) \text{ in } \Omega, \quad p^N = 0 \text{ on } \Gamma,$$

(2.70)

then p^{N-1} (respectively p^n, $n = N-2, \ldots, 2, 1$) as the solution in $H_0^1(\Omega)$ of the elliptic problem

$$\frac{p^{N-1} - p^N}{\Delta t} + A^*(\tfrac{2}{3}p^{N-1} + \tfrac{1}{3}p^N) = 0 \text{ in } \Omega, \quad p^{N-1} = 0 \text{ on } \Gamma$$

(2.71)

(respectively

$$\frac{\frac{3}{2}p^n - 2p^{n+1} + \frac{1}{2}p^{n+2}}{\Delta t} + A^*p^n = 0 \text{ in } \Omega, \quad p^n = 0 \text{ on } \Gamma).$$

(2.72)

Combining (2.67) and (2.69) shows that the *optimal triple*

$$\{u^{\Delta t}, \{y^n\}_{n=1}^N, \{p^n\}_{n=1}^N\}$$

is *characterized* by

$$u^n = \frac{\partial p^n}{\partial n_{A^*}}\Big|_{\Gamma_0} \text{ if } n = 1, \ldots, N-2, \quad u^{N-1} = \left(\frac{2}{3}\frac{\partial p^{N-1}}{\partial n_{A^*}} + \frac{1}{3}\frac{\partial p^N}{\partial n_{A^*}}\right)\Big|_{\Gamma_0},$$
$$(2.73)$$

to be completed by (2.70)–(2.72) and by

$$y^0 = 0, \tag{2.74}$$

$$\frac{y^1 - y^0}{\Delta t} + A(\tfrac{2}{3}y^1 + \tfrac{1}{3}y^0) = 0 \text{ in } \Omega, \quad y^1 = u^1 \text{ on } \Gamma_0, \quad y^1 = 0 \text{ on } \Gamma\backslash\Gamma_0, \tag{2.75}$$

$$\frac{\tfrac{3}{2}y^n - 2y^{n-1} + \tfrac{1}{2}y^{n-2}}{\Delta t} + Ay^n = 0 \text{ in } \Omega, \quad y^n = u^n \text{ on } \Gamma_0, \quad y^n = 0 \text{ on } \Gamma\backslash\Gamma_0$$
$$(2.76)$$

if $n = 2, \ldots, N-1$,

$$\frac{2y^N - 3y^{N-1} + y^{N-2}}{\Delta t} + Ay^{N-1} = 0. \tag{2.77}$$

Following Section 2.2 we define $\Lambda^{\Delta t} \in \mathcal{L}(H_0^1(\Omega), H^{-1}(\Omega))$ by

$$\Lambda^{\Delta t}\hat{f} = -\hat{\varphi}^N, \quad \forall \hat{f} \in H_0^1(\Omega), \tag{2.78}$$

where $\hat{\varphi}^N$ is obtained from $\hat{f}$ via the solution of the *discrete backward parabolic problem*

$$\hat{\psi}^N = \hat{f}, \tag{2.79}$$

$$\frac{\hat{\psi}^{N-1} - \hat{\psi}^N}{\Delta t} + A^*(\tfrac{2}{3}\hat{\psi}^{N-1} + \tfrac{1}{3}\hat{\psi}^N) = 0 \text{ in } \Omega, \quad \hat{\psi}^{N-1} = 0 \text{ on } \Gamma, \tag{2.80}$$

$$\frac{\tfrac{3}{2}\hat{\psi}^n - 2\hat{\psi}^{n+1} + \tfrac{1}{2}\hat{\psi}^{n+2}}{\Delta t} + A^*\hat{\psi}^n = 0 \text{ in } \Omega, \quad \hat{\psi}^n = 0 \text{ on } \Gamma \tag{2.81}$$

for $n = N-2, \ldots, 1$, and then of the *discrete forward parabolic problem*

$$\hat{\varphi}^0 = 0, \tag{2.82}$$

$$\frac{\hat{\varphi}^1 - \hat{\varphi}^0}{\Delta t} + A(\tfrac{2}{3}\hat{\varphi}^1 + \tfrac{1}{3}\hat{\varphi}^0) = 0 \text{ in } \Omega, \quad \hat{\varphi}^1 = \frac{\partial \hat{\psi}^1}{\partial n_{A^*}} \text{ on } \Gamma_0, \quad \hat{\varphi}^1 = 0 \text{ on } \Gamma\backslash\Gamma_0,$$
$$(2.83)$$

$$\frac{\tfrac{3}{2}\hat{\varphi}^n - 2\hat{\varphi}^{n-1} + \tfrac{1}{2}\hat{\varphi}^{n-2}}{\Delta t} + A\hat{\varphi}^n = 0 \text{ in } \Omega, \quad \hat{\varphi}^n = \frac{\partial \hat{\psi}^n}{\partial n_{A^*}} \text{ on } \Gamma_0, \quad \hat{\varphi}^n = 0 \text{ on } \Gamma\backslash\Gamma_0$$
$$(2.84)$$

if $n = 2, \ldots, N - 2$,

$$\frac{\frac{3}{2}\hat{\varphi}^{N-1} - 2\hat{\varphi}^{N-2} + \frac{1}{2}\hat{\varphi}^{N-3}}{\Delta t} + A\hat{\varphi}^{N-1} = 0 \text{ in } \Omega,$$

$$\hat{\varphi}^{N-1} = \frac{2}{3}\frac{\partial \hat{\psi}^{N-1}}{\partial n_{A*}} + \frac{1}{3}\frac{\partial \hat{\psi}^N}{\partial n_{A*}} \text{ on } \Gamma_0, \ \hat{\varphi}^{N-1} = 0 \text{ on } \Gamma\backslash\Gamma_0, \qquad (2.85)$$

$$\frac{2\hat{\varphi}^N - 3\hat{\varphi}^{N-1} + \hat{\varphi}^{N-2}}{\Delta t} + A\hat{\varphi}^{N-1} = 0. \qquad (2.86)$$

We can show that (with obvious notation) we have, $\forall f_1, f_2 \in H_0^1(\Omega)$,

$$\langle \Lambda^{\Delta t} f_1, f_2 \rangle = \Delta t \Bigg[\sum_{n=1}^{N-2} \int_{\Gamma_0} \frac{\partial \psi_1^n}{\partial n_{A*}} \frac{\partial \psi_2^n}{\partial n_{A*}} \, d\Gamma + \frac{3}{2} \int_{\Gamma_0} \left(\frac{2}{3}\frac{\partial \psi_1^{N-1}}{\partial n_{A*}} + \frac{1}{3}\frac{\partial \psi_1^N}{\partial n_{A*}} \right)$$

$$\times \left(\frac{2}{3}\frac{\partial \psi_2^{N-1}}{\partial n_{A*}} + \frac{1}{3}\frac{\partial \psi_2^N}{\partial n_{A*}} \right) d\Gamma \Bigg], \qquad (2.87)$$

where, in (2.87), $\langle \cdot, \cdot \rangle$ denotes the duality pairing between $H^{-1}(\Omega)$ and $H_0^1(\Omega)$.

It follows from (2.87) that operator $\Lambda^{\Delta t}$ is self-adjoint and positive semi-definite over $H_0^1(\Omega)$.

Back to the optimality system (2.70)–(2.77), let us denote by $f^{\Delta t}$ the function p^N; it follows then from the definition of $\Lambda^{\Delta t}$ that (2.70) can be reformulated as

$$- k^{-1}\Delta f^{\Delta t} + \Lambda^{\Delta t} f^{\Delta t} = -y_T, \qquad (2.88)$$

which is precisely the dual problem we have been looking for. The full space/time discretization of problems (2.12) and (2.21) will be discussed in the following.

2.4.3. Full space/time discretization of problems (2.12) and (2.21)

The full discretization of control problems, related to (2.12) and (2.21), has been already discussed in Sections 1.8.4 and 1.10.6. Despite many similarities, the boundary control problems discussed here are substantially more complicated to fully discretize than the above distributed and pointwise control problems. The main reason for this increased complexity arises from the fact that we still intend to employ low-order finite element approximations – as in Sections 1.8 and 1.10 – to space discretize the parabolic state problem (2.1)–(2.3) and the corresponding adjoint system (2.15). With such a choice the 'obvious' approximations of $\partial/\partial n_{A*}|_{\Gamma_0}$ will be fairly inaccurate. In order to obtain second-order accurate approximations of $\partial/\partial n_{A*}|_{\Gamma_0}$, we shall rely on a discrete Green's formula, following a strategy which has been successfully used in, e.g., Glowinski et al. (1990), Glowinski (1992a) (for the boundary control of the wave equation) and Carthel et al. (1994) (for the boundary control of the heat equation).

We suppose for simplicity that Ω is a *bounded* polygonal domain of $\mathbb{R}^2$. We introduce then, as in Sections 1.8.4. and 1.10.6, a *triangulation* $\mathcal{T}_h$ of Ω (h: largest length of the edges of the triangles of $\mathcal{T}_h$). Next, we approximate $H^1(\Omega), L^2(\Omega)$ and $H_0^1(\Omega)$ by

$$H_h^1 = \{z_h | z_h \in C^0(\bar{\Omega}), z_h|_T \in P_1, \forall T \in \mathcal{T}_h\}, \tag{2.89}$$

$$H_{0h}^1 = \{z_h | z_h \in H_h^1, z_h = 0 \text{ on } \Gamma\}(= H_0^1(\Omega) \cap H_h^1), \tag{2.90}$$

respectively (with, as usual, P_1 the space of polynomials in x_1, x_2 of degree ≤ 1). Another important finite element space is

$$V_{0h} = \{z_h \mid z_h \in H_h^1, z_h = 0 \text{ on } \Gamma \backslash \Gamma_0\}; \tag{2.91}$$

if $\int_{\Gamma \backslash \Gamma_0} d\Gamma > 0$ we shall assume that those boundary points at the interface of Γ_0 and $\Gamma \backslash \Gamma_0$ are vertices of $\mathcal{T}_h$. Finally, the role of $L^2(\Gamma_0)$ will be played by the space $M_h(\subset V_{0h})$ defined as follows:

$$M_h \oplus H_{0h}^1 = V_{0h}, \ \mu_h \in M_h \Rightarrow \mu_h |_T = 0, \ \forall T \in \mathcal{T}_h, \text{ such as } \partial T \cap \Gamma = \emptyset. \tag{2.92}$$

Space M_h is clearly *isomorphic* to the boundary space consisting of the traces on Γ of those functions belonging to V_{0h}; also, $\dim(M_h)$ is equal to the number of $\mathcal{T}_h$ boundary vertices interior to Γ_0 and the following bilinear form

$$\{\lambda_h, \mu_h\} \to \int_{\Gamma_0} \lambda_h \mu_h \, d\Gamma$$

defines a scalar product on M_h.

Since the full space/time discretization of problems (2.12) and (2.21) will rely on *variational* techniques, it is convenient to introduce the bilinear form $a: H^1(\Omega) \times H_0^1(\Omega) \to \mathbb{R}$ defined by

$$a(y, z) = \langle Ay, z \rangle, \ \forall y \in H^1(\Omega), \ \forall z \in H_0^1(\Omega), \tag{2.93}$$

where $\langle \cdot, \cdot \rangle$ denotes the duality pairing between $H^{-1}(\Omega)$ and $H_0^1(\Omega)$. Assuming that the coefficients of the second-order elliptic operator A are sufficiently smooth we also have

$$a(y, z) = \int_\Omega (A^* z) y \, dx + \int_\Gamma \frac{\partial z}{\partial n_{A^*}} y \, d\Gamma, \ \forall y \in H^1(\Omega), \ \forall z \in H_0^1(\Omega) \cap H^2(\Omega), \tag{2.94}$$

which is definitely a generalization of the well-known *Green's formula*

$$\int_\Omega \nabla y \cdot \nabla z \, dx = -\int_\Omega \Delta z y \, dx + \int_\Gamma \frac{\partial z}{\partial n} y \, d\Gamma, \ \forall y \in H^1(\Omega), \ \forall z \in H_0^1(\Omega) \cap H^2(\Omega).$$

Following Section 2.4.2 we approximate the control problem (2.12) by

$$\min_{\mathbf{v} \in \mathcal{U}_h^{\Delta t}} J_h^{\Delta t}(\mathbf{v}), \tag{2.95}$$

where, in (2.95), we have $\mathcal{U}_h^{\Delta t} = (M_h)^{N-1}$, $\mathbf{v} = \{v^n\}_{n=1}^{N-1}$ and

$$J_h^{\Delta t}(\mathbf{v}) = \frac{\Delta t}{2} \sum_{n=1}^{N-1} a_n \int_\Omega |v^n|^2 \, d\Gamma + \frac{k}{2} \int_\Gamma |\nabla \phi^N|^2 \, dx, \qquad (2.96)$$

with, in (2.96), ϕ^N obtained from $\mathbf{v}$ via the solution of the following well-posed *discrete parabolic* and *elliptic* problems

Parabolic problem.

$$y^0 = 0; \qquad (2.97)$$

compute y^1 from

$$\begin{cases} y^1 \in V_{0h}, \ y^1 = v^1 \text{ on } \Gamma_0, \\ \int_\Omega \dfrac{y^1 - y^0}{\Delta t} z \, dx + a(\tfrac{2}{3} y^1 + \tfrac{1}{3} y^0, z) = 0, \ \forall z \in H_{0h}^1, \end{cases} \qquad (2.98)$$

then y^n from

$$\begin{cases} y^n \in V_{0h}, \ y^n = v^n \text{ on } \Gamma_0, \\ \int_\Omega \dfrac{\tfrac{3}{2} y^n - 2y^{n-1} + \tfrac{1}{2} y^{n-2}}{\Delta t} z \, dx + a(y^n, z) = 0, \ \forall z \in H_{0h}^1 \end{cases} \qquad (2.99)$$

for $n = 2, \ldots, N-1$, and y^N from

$$\begin{cases} y^N \in H_{0h}^1, \\ \int_\Omega \dfrac{2y^N - 3y^{N-1} + y^{N-2}}{\Delta t} z \, dx + a(y^{N-1}, z) = 0, \ \forall z \in H_{0h}^1. \end{cases} \qquad (2.100)$$

Elliptic problem.

$$\begin{cases} \phi^N \in H_{0h}^1, \\ \int_\Omega \nabla \phi^N \cdot \nabla z \, dx = \int_\Omega (y^N - y_T) z \, dx, \ \forall z \in H_{0h}^1. \end{cases} \qquad (2.101)$$

We then have the following

Proposition 2.2   *The discrete control problem (2.95) has a unique solution* $\mathbf{u}_h^{\Delta t} = \{u^n\}_{n=1}^{N-1}$. *If we denote by* $\mathbf{y}_h^{\Delta t} = \{y^n\}_{n=0}^{N}$ *the solution of (2.97)–(2.100) associated with* $\mathbf{v} = \mathbf{u}_h^{\Delta t}$, *the optimal pair* $\{\mathbf{u}_h^{\Delta t}, \mathbf{y}_h^{\Delta t}\}$ *is characterized by the existence of* $\mathbf{p}_h^{\Delta t} = \{p^n\}_{n=1}^{N} \in (H_{0h}^1)^N$ *such that*

$$\begin{cases} p^N \in H_{0h}^1, \\ \int_\Omega \nabla p^N \cdot \nabla z \, dx = k \left[\int_\Omega y^N z \, dx - \langle y_T, z \rangle \right], \ \forall z \in H_{0h}^1, \end{cases} \qquad (2.102)$$

($\langle \cdot, \cdot \rangle$: *duality pairing between* $H^{-1}(\Omega)$ *and* $H_0^1(\Omega)$),

$$\begin{cases} p^{N-1} \in H_{0h}^1, \\ \int_\Omega \dfrac{p^{N-1} - p^N}{\Delta t} z \, dx + a(z, \tfrac{2}{3} p^{N-1} + \tfrac{1}{3} p^N) = 0, \ \forall z \in H_{0h}^1, \end{cases} \qquad (2.103)$$

$$\begin{cases} p^n \in H_{0h}^1, \\ \int_\Omega \dfrac{\frac{3}{2}p^n - 2p^{n+1} + \frac{1}{2}p^{n+2}}{\Delta t} z \, dx + a(z, p^n) = 0, \ \forall z \in H_{0h}^1 \end{cases} \quad (2.104)$$

for $n = N - 2, \ldots, 1$, and also

$$\begin{cases} u^n \in M_h, \\ \int_{\Gamma_0} u^n \mu \, d\Gamma = \int_\Omega \dfrac{\frac{3}{2}p^n - 2p^{n+1} + \frac{1}{2}p^{n+2}}{\Delta t} \mu \, dx + a(\mu, p^n), \ \forall \mu \in M_h \end{cases} \quad (2.105)$$

if $n = 1, 2, \ldots, N - 2$, and finally

$$\begin{cases} u^{N-1} \in M_h, \\ \int_\Gamma u^{N-1} \mu \, d\Gamma = \int_\Omega \dfrac{p^{N-1} - p^N}{\Delta t} \mu \, dx + a(\mu, \frac{2}{3}p^{N-1} + \frac{1}{3}p^N), \ \forall \mu \in M_h \end{cases} \quad (2.106)$$

Proof. The *existence* and *uniqueness* properties are obvious. Concerning now the relations characterizing the optimal pair $\{u_h^{\Delta t}, y_h^{\Delta t}\}$ they follow from the *optimality condition*

$$\nabla J_h^{\Delta t}(u_h^{\Delta t}) = 0, \quad (2.107)$$

where $\nabla J_h^{\Delta t}$ is the gradient of the functional $J_h^{\Delta t}$. Indeed, if we use

$$(\mathbf{v}, \mathbf{w})_{\Delta t} = \Delta t \sum_{n=1}^{N-1} a_n \int_{\Gamma_0} v^n w^n \, d\Gamma$$

as the scalar product over $\mathcal{U}_h^{\Delta t}$, it can be shown that, $\forall \mathbf{v}, \mathbf{w} \in \mathcal{U}_h^{\Delta t}$, we have

$$(\nabla J_h^{\Delta t}(\mathbf{v}), \mathbf{w})_{\Delta t}$$

$$= \Delta t \sum_{n=1}^{N-2} \left[\int_{\Gamma_0} v^n w^n \, d\Gamma - \int_\Omega \frac{\frac{3}{2}p^n - 2p^{n+1} + \frac{1}{2}p^{n+2}}{\Delta t} w^n \, dx - a(w^n, p^n) \right]$$

$$+ \frac{3}{2}\Delta t \left[\int_{\Gamma_0} v^{N-1} w^{N-1} \, d\Gamma - \int_\Omega \frac{p^{N-1} - p^N}{\Delta t} w^{N-1} \, dx \right.$$

$$\left. - a(w^{N-1}, \frac{2}{3}p^{N-1} + \frac{1}{3}p^N) \right], \quad (2.108)$$

where, in (2.108), $\{p^n\}_{n=1}^{N-1}$ is obtained from $\mathbf{v} = \{v^n\}_{n=1}^{N-1}$ via the solution of the discrete parabolic and elliptic problems (2.97)–(2.100), (2.102) and (2.103), (2.104). Relations (2.107) and (2.108) clearly imply (2.102)–(2.106).

Remark 2.4 Relations (2.105), (2.106) are not that mysterious. For the continuous problem (2.12), we know (see Section 2.2) that the optimal control u satisfies

$$u = \frac{\partial p}{\partial n_{A^*}} \text{ on } \Sigma_0, \quad (2.109)$$

where p is the solution of the corresponding adjoint system (2.15). We have thus $\partial p/\partial t = A^*p$, which combined with Green's formula (2.94), implies that a.e. on $(0, T)$ we have

$$\int_{\Gamma_0} u\mu \, d\Gamma = -\int_\Omega \frac{\partial p}{\partial t}\mu \, dx + a(\mu, p), \quad \forall \mu \in H^1(\Omega), \quad \mu = 0 \text{ on } \Gamma\backslash\Gamma_0. \quad (2.110)$$

Relations (2.105), (2.106) are clearly discrete analogues of (2.110).

To obtain the *fully discrete* analogue of the *dual problems* (2.21) and (2.88) we introduce $\Lambda_h^{\Delta t} \in \mathcal{L}(H_{0h}^1, H_{0h}^1)$ defined as follows

$$\Lambda_h^{\Delta t}\hat{f} = -\hat{\varphi}^N, \quad \forall \hat{f} \in H_{0h}^1, \quad (2.111)$$

where $\hat{\varphi}^N$ is obtained from $\hat{f}$ via the solution of the *fully discrete backward parabolic problem*

$$\hat{\psi}^N = \hat{f}, \quad (2.112)$$

$$\begin{cases} \hat{\psi}^{N-1} \in H_{0h}^1, \\ \int_\Omega \dfrac{\hat{\psi}^{N-1} - \hat{\psi}^N}{\Delta t} z \, dx + a(z, \tfrac{2}{3}\hat{\psi}^{N-1} + \tfrac{1}{3}\hat{\psi}^N) = 0, \quad \forall z \in H_{0h}^1, \end{cases} \quad (2.113)$$

$$\begin{cases} \hat{\psi}^n \in H_{0h}^1, \\ \int_\Omega \dfrac{\tfrac{3}{2}\hat{\psi}^n - 2\hat{\psi}^{n+1} + \tfrac{1}{2}\hat{\psi}^{n+2}}{\Delta t} z \, dx + a(z, \hat{\psi}^n) = 0, \quad \forall z \in H_{0h}^1 \end{cases} \quad (2.114)$$

for $n = N-2, \ldots, 1$, and then of the *fully discrete forward parabolic problem*

$$\hat{\varphi}^0 = 0, \quad (2.115)$$

$$\begin{cases} \hat{\varphi}^1 \in V_{0h}, \ \hat{\varphi}^1 = \hat{u}^1 \text{ on } \Gamma_0, \\ \int_\Omega \dfrac{\hat{\varphi}^1 - \hat{\varphi}^0}{\Delta t} z \, dx + a(\tfrac{2}{3}\hat{\varphi}^1 + \tfrac{1}{3}\hat{\varphi}^0, z) = 0, \quad \forall z \in H_{0h}^1, \end{cases} \quad (2.116)$$

$$\begin{cases} \hat{\varphi}^n \in V_{0h}, \ \hat{\varphi}^n = \hat{u}^n \text{ on } \Gamma_0, \\ \int_\Omega \dfrac{\tfrac{3}{2}\hat{\varphi}^n - 2\hat{\varphi}^{n-1} + \tfrac{1}{2}\hat{\varphi}^{n-2}}{\Delta t} z \, dx + a(\hat{\varphi}^n, z) = 0, \quad \forall z \in H_{0h}^1 \end{cases} \quad (2.117)$$

for $n = 2, \ldots, N-1$, and finally

$$\begin{cases} \hat{\varphi}^N \in H_{0h}^1, \\ \int_\Omega \dfrac{2\hat{\varphi}^N - 3\hat{\varphi}^{N-1} + \hat{\varphi}^{N-2}}{\Delta t} z \, dx + a(\hat{\varphi}^{N-1}, z) = 0, \quad \forall z \in H_{0h}^1; \end{cases} \quad (2.118)$$

in (2.116), (2.117) the vector $\{\hat{u}^n\}_{n=1}^{N-1}$ is defined from $\{\hat{\psi}^n\}_{n=1}^N$ as follows

$$\begin{cases} \hat{u}^n \in M_h, \\ \int_{\Gamma_0} \hat{u}^n \mu \, d\Gamma = \int_\Omega \dfrac{\tfrac{3}{2}\hat{\psi}^n - 2\hat{\psi}^{n+1} + \tfrac{1}{2}\hat{\psi}^{n+2}}{\Delta t} \mu \, dx + a(\mu, \hat{\psi}^n), \quad \forall \mu \in M_h \end{cases}$$
$$(2.119)$$

if $n = 1, 2, \ldots, N - 2$, and

$$\begin{cases} \hat{u}^{N-1} \in M_h, \\ \int_{\Gamma_0} \hat{u}^{N-1} \mu \, d\Gamma = \int_{\Omega} \frac{\hat{\psi}^{N-1} - \hat{\psi}^N}{\Delta t} \mu \, dx \\ \qquad + a(\mu, \tfrac{2}{3}\hat{\psi}^{N-1} + \tfrac{1}{3}\hat{\psi}^N), \ \forall \mu \in M_h. \end{cases} \tag{2.120}$$

We can show that

$$\int_{\Omega} (\Lambda_h^{\Delta t} f_1) f_2 \, dx = \Delta t \sum_{n=1}^{N-1} a_n \int_{\Gamma_0} u_1^n u_2^n \, d\Gamma, \ \forall f_1, f_2 \in H_{0h}^1, \tag{2.121}$$

where, in (2.121), $\{u_i^n\}_{n=1}^{N-1}$, $i = 1, 2$ is obtained from f_i via (2.112)–(2.114), (2.119), (2.120).

It follows from (2.121) that operator $\Lambda_h^{\Delta t}$ is *symmetric* and *positive semi-definite* over H_{0h}^1. $\square$

Let us consider now the *optimal triple* $\{\mathbf{u}_h^{\Delta t}, \mathbf{y}_h^{\Delta t}, \mathbf{p}_h^{\Delta t}\}$ and define $f_h^{\Delta t} \in H_h^1$ by

$$f_h^{\Delta t} = p^N. \tag{2.122}$$

It follows then from Proposition 2.2 and from the definition of $\Lambda_h^{\Delta t}$ that

$$\Lambda_h^{\Delta t} f_h^{\Delta t} = -y^N. \tag{2.123}$$

Combining (2.123) with (2.102) we obtain

$$\begin{cases} f_h^{\Delta t} \in H_{0h}^1, \\ k^{-1} \int_{\Omega} \nabla f_h^{\Delta t} \cdot \nabla z \, dx + \int_{\Omega} \Lambda_h^{\Delta t} f_h^{\Delta t} z \, dx = -\langle y_T, z \rangle, \ \forall z \in H_{0h}^1. \end{cases} \tag{2.124}$$

Problem (2.124) is precisely the *fully discrete dual problem* we were looking for. From the properties of $\Lambda_h^{\Delta t}$ (*symmetry* and *semi-positiveness*), problem (2.124) can be solved by a *conjugate gradient* algorithm operating in H_{0h}^1 (a fully discrete analogue of algorithm (2.42)–(2.54)); we shall describe this algorithm in Section 2.5.

Remark 2.5 From a practical point of view, it makes sense to use the *trapezoidal rule* to (approximately) compute the various $L^2(\Omega)$ and $L^2(\Gamma_0)$ scalar products occurring in the definition of the approximate control problem (2.95), and of its dual problem (2.124). If this approach is retained, the corresponding operator $\Lambda_h^{\Delta t}$ has the same basic properties as that defined by (2.111), namely *symmetry* and *semi-positiveness*, implying that the corresponding variant of problem (2.121) can also be solved by a *conjugate gradient algorithm* operating in the space H_{0h}^1.

2.5. Dirichlet control (V): Iterative solution of the fully discrete dual problem (2.124)

We have described in Section 2.3 a *conjugate gradient algorithm* for solving the control problem (2.12), either *directly* (by algorithm (2.27)–(2.41); see Section 2.3.1) or via the solution of the *dual problem* (2.21) (by algorithm (2.42)–(2.54); see Section 2.3.2). Since the numerical results presented in the following section were obtained via the solution of the dual problem we shall focus our discussion on the *iterative solution* of the *fully discrete* approximation of problem (2.21) (i.e. problem (2.124)). From the properties of $\Lambda_h^{\Delta t}$ problem (2.124) can be solved by a *conjugate gradient algorithm* operating in the finite dimensional space H_{0h}^1. From Sections 1.8.2 and 2.3.2 this algorithm takes the following form:

$$f_0 \text{ is given in } H_{0h}^1; \qquad (2.125)$$

set

$$p_0^N = f_0 \qquad (2.126)$$

and solve first

$$\begin{cases} p_0^{N-1} \in H_{0h}^1, \\ \int_\Omega \dfrac{p_0^{N-1} - p_0^N}{\Delta t} z \, dx + a(z, \tfrac{2}{3}p_0^{N-1} + \tfrac{1}{3}p_0^N) = 0, \ \forall z \in H_{0h}^1 \end{cases} \qquad (2.127)$$

and

$$\begin{cases} u_0^{N-1} \in M_h, \\ \int_{\Gamma_0} u_0^{N-1}\mu \, d\Gamma = \int_\Omega \dfrac{p_0^{N-1} - p_0^N}{\Delta t}\mu \, dx + a(\mu, \tfrac{2}{3}p_0^{N-1} + \tfrac{1}{3}p_0^N), \ \forall \mu \in M_h, \end{cases} \qquad (2.128)$$

and then for $n = N - 2, \ldots, 1$

$$\begin{cases} p_0^n \in H_{0h}^1, \\ \int_\Omega \dfrac{\tfrac{3}{2}p_0^n - 2p_0^{n+1} + \tfrac{1}{2}p_0^{n+2}}{\Delta t} z \, dx + a(z, p_0^n) = 0, \ \forall z \in H_{0h}^1, \end{cases} \qquad (2.129)$$

$$\begin{cases} u_0^n \in M_h, \\ \int_{\Gamma_0} u_0^n \mu \, d\Gamma = \int_\Omega \dfrac{\tfrac{3}{2}p_0^n - 2p_0^{n+1} + \tfrac{1}{2}p_0^{n+2}}{\Delta t}\mu \, dx + a(\mu, p_0^n), \ \forall \mu \in M_h. \end{cases} \qquad (2.130)$$

Solve next the following fully discrete forward parabolic problem

$$y_0^0 = 0, \qquad (2.131)$$

$$\begin{cases} y_0^1 \in V_{0h}, \ y_0^1 = u_0^1 \text{ on } \Gamma_0, \\ \int_\Omega \dfrac{y_0^1 - y_0^0}{\Delta t} z \, dx + a(\tfrac{2}{3}y_0^1 + \tfrac{1}{3}y_0^0, z) = 0, \ \forall z \in H_{0h}^1, \end{cases} \qquad (2.132)$$

$$\begin{cases} y_0^n \in V_{0h}, \ y_0^n = u_0^n \text{ on } \Gamma_0, \\ \displaystyle\int_\Omega \frac{\frac{3}{2}y_0^n - 2y_0^{n-1} + \frac{1}{2}y_0^{n-2}}{\Delta t} z \, dx + a(y_0^n, z) = 0, \ \forall z \in H_{0h}^1 \end{cases} \quad (2.133)$$

for $n = 2, \dots, N - 1$ and finally

$$\begin{cases} y_0^N \in H_{0h}^1, \\ \displaystyle\int_\Omega \frac{2y_0^N - 3y_0^{N-1} + y_0^{N-2}}{\Delta t} z \, dx + a(y_0^{N-1}, z) = 0, \ \forall z \in H_{0h}^1. \end{cases} \quad (2.134)$$

Solve now

$$\begin{cases} g_0 \in H_{0h}^1, \\ \displaystyle\int_\Omega \nabla g_0 \cdot \nabla z \, dx = k^{-1} \int_\Omega \nabla f_0 \cdot \nabla z \, dx + \langle y_T, z \rangle - \int_\Omega y_0^N z \, dx, \ \forall z \in H_{0h}^1 \end{cases} \quad (2.135)$$

and set

$$w_0 = g_0. \quad (2.136)$$

Then for $m \geq 0$, assuming that f_m, g_m, w_m are known, compute f_{m+1}, g_{m+1}, w_{m+1} as follows:
Take

$$\bar{p}_m^N = w_m \quad (2.137)$$

and solve

$$\begin{cases} \bar{p}_m^{N-1} \in H_{0h}^1, \\ \displaystyle\int_\Omega \frac{\bar{p}_m^{N-1} - \bar{p}_m^N}{\Delta t} z \, dx + a(z, \tfrac{2}{3}\bar{p}_m^{N-1} + \tfrac{1}{3}\bar{p}_m^N) = 0, \ \forall z \in H_{0h}^1 \end{cases} \quad (2.138)$$

and

$$\begin{cases} \bar{u}_m^{N-1} \in M_h, \\ \displaystyle\int_{\Gamma_0} \bar{u}_m^{N-1} \mu \, d\Gamma = \int_\Omega \frac{\bar{p}_m^{N-1} - \bar{p}_m^N}{\Delta t} \mu \, dx + a(\mu, \tfrac{2}{3}\bar{p}_m^{N-1} + \tfrac{1}{3}\bar{p}_m^N) = 0, \ \forall \mu \in M_h, \end{cases} \quad (2.139)$$

and then for $n = N - 2, \dots, 1$

$$\begin{cases} \bar{p}_m^n \in H_{0h}^1, \\ \displaystyle\int_\Omega \frac{\frac{3}{2}\bar{p}_m^n - 2\bar{p}_m^{n+1} + \frac{1}{2}\bar{p}_m^{n+2}}{\Delta t} z \, dx + a(z, \bar{p}_m^n) = 0, \ \forall z \in H_{0h}^1, \end{cases} \quad (2.140)$$

$$\begin{cases} \bar{u}_m^n \in M_h, \\ \displaystyle\int_{\Gamma_0} \bar{u}_m^n \mu \, d\Gamma = \int_\Omega \frac{\frac{3}{2}\bar{p}_m^n - 2\bar{p}_m^{n+1} + \frac{1}{2}\bar{p}_m^{n+2}}{\Delta t} \mu \, dx + a(\mu, \bar{p}_m^n), \ \forall \mu \in M_h. \end{cases} \quad (2.141)$$

Solve next the following discrete forward parabolic problem

$$\bar{y}_m^0 = 0, \quad (2.142)$$

$$\begin{cases} \bar{y}_m^1 \in V_{0h}, \; \bar{y}_m^1 = \bar{u}_m^1 \text{ on } \Gamma_0, \\ \displaystyle\int_\Omega \frac{\bar{y}_m^1 - \bar{y}_m^0}{\Delta t} z \, dx + a(\tfrac{2}{3}\bar{y}_m^1 + \tfrac{1}{3}\bar{y}_m^0, z) = 0, \; \forall z \in H_{0h}^1, \end{cases} \tag{2.143}$$

$$\begin{cases} \bar{y}_m^n \in V_{0h}, \; \bar{y}_m^n = \bar{u}_m^n \text{ on } \Gamma_0, \\ \displaystyle\int_\Omega \frac{\tfrac{3}{2}\bar{y}_m^n - 2\bar{y}_m^{n-1} + \tfrac{1}{2}\bar{y}_m^{n-2}}{\Delta t} z \, dx + a(\bar{y}_m^n, z) = 0, \; \forall z \in H_{0h}^1 \end{cases} \tag{2.144}$$

for $n = 2, \ldots, N-1$, and finally

$$\begin{cases} \bar{y}_m^N \in H_{0h}^1, \\ \displaystyle\int_\Omega (2\bar{y}_m^N - 3\bar{y}_m^{N-1} + \bar{y}_m^{N-2}) z \, dx + \Delta t a(\bar{y}_m^{N-1}, z) = 0, \; \forall z \in H_{0h}^1. \end{cases} \tag{2.145}$$

Solve now

$$\begin{cases} \bar{g}_m \in H_{0h}^1, \\ \displaystyle\int_\Omega \boldsymbol{\nabla}\bar{g}_m \cdot \boldsymbol{\nabla}z \, dx = k^{-1} \int_\Omega \boldsymbol{\nabla}w_m \cdot \boldsymbol{\nabla}z \, dx - \int_\Omega \bar{y}_m^N z \, dx, \; \forall z \in H_{0h}^1, \end{cases} \tag{2.146}$$

and compute

$$\rho_m = \int_\Omega |\boldsymbol{\nabla}g_m|^2 \, dx \Big/ \int_\Omega \boldsymbol{\nabla}\bar{g}_m \cdot \boldsymbol{\nabla}w_m \, dx, \tag{2.147}$$

$$f_{m+1} = f_m - \rho_m w_m, \tag{2.148}$$

$$g_{m+1} = g_m - \rho_m \bar{g}_m. \tag{2.149}$$

If $\|g_{m+1}\|_{H_0^1(\Omega)}/\|g_0\|_{H_0^1(\Omega)} \le \epsilon$ take $f_h^{\Delta t} = f_{m+1}$ and solve (2.112)–(2.120) with $\hat{f} = f_h^{\Delta t}$ to obtain $\mathbf{u}_h^{\Delta t} = \{u^n\}_{n=1}^{N-1}$; if the above stopping test is not satisfied, compute

$$\gamma_m = \int_\Omega |\boldsymbol{\nabla}g_{m+1}|^2 \, dx \Big/ \int_\Omega |\boldsymbol{\nabla}g_m|^2 \, dx, \tag{2.150}$$

and then

$$w_{m+1} = g_{m+1} + \gamma_m w_m. \tag{2.151}$$

Do $m = m + 1$ and go to (2.137).

Remark 2.6 Algorithm (2.125)–(2.151) may seem complicated (27 instructions); in fact it is quite easy to implement since it essentially requires a *fast elliptic solver*; for the calculations presented in Section 2.6 we have been using a *multigrid* based elliptic solver (see, e.g., Hackbush (1985), Yserentant (1993) and the references therein for a thorough discussion of the solution of discrete elliptic problems by multigrid methods).

Remark 2.7 If h and Δt are sufficiently small Remarks 2.2 and 2.3 still hold for algorithm (2.125)–(2.151).

2.6. Dirichlet control (VI): Numerical experiments

2.6.1. First test problem

The first test problem is one for which the *exact controllability property holds*; indeed, to construct more easily a test problem whose exact solution is known we have taken a *nonzero source term* in the right-hand side of the state equation (2.1), obtaining thus

$$\frac{\partial y}{\partial t} + Ay = s \text{ in } Q, \qquad (2.1)'$$

and also replaced the initial condition (2.3) by

$$y(0) = y_0, \qquad (2.3)'$$

with $y_0 \neq 0$. For these numerical experiments we have taken $\Omega = (0,1) \times (0,1)$, $\Gamma_0 = \Gamma$, $T = 1$ and $A = -\nu\Delta$, with $\nu > 0$ ((2.1)$'$ is therefore a *heat equation*); the *source term* s, the *initial value* y_0 and the *target function* y_T are defined by

$$s(x_1, x_2, t) = 3\pi^3 \nu e^{2\pi^2 \nu t}(\sin \pi x_1 + \sin \pi x_2), \qquad (2.152)$$

$$y_0(x_1, x_2) = \pi(\sin \pi x_1 + \sin \pi x_2), \qquad (2.153)$$

$$y_T(x_1, x_2) = \pi e^{2\pi^2 \nu}(\sin \pi x_1 + \sin \pi x_2), \qquad (2.154)$$

respectively.

With these data the (unique) solution u of the optimal control problem

$$\operatorname*{Min}_{v \in \mathcal{U}_f} J(v) \qquad (2.155)$$

(with

$$
\begin{aligned}
J(v) &= \frac{1}{2} \int_\Sigma |v|^2 \, d\Gamma \, dt, \\
\mathcal{U}_f &= \{v \mid v \in L^2(\Sigma), \text{ the pair } \{v, y\} \\
&\qquad \text{satisfies } (2.1)', (2.2), (2.3)' \text{ and } y(T) = y_T\})
\end{aligned}
$$

is given by

$$
\begin{cases}
u(x_1, x_2, t) = \pi e^{2\pi^2 \nu t} \sin \pi x_1 \text{ if } 0 < x_1 < 1 \text{ and } x_2 = 0 \text{ or } 1, \\
u(x_1, x_2, t) = \pi e^{2\pi^2 \nu t} \sin \pi x_2 \text{ if } 0 < x_2 < 1 \text{ and } x_1 = 0 \text{ or } 1,
\end{cases} \qquad (2.156)
$$

the corresponding function y being defined by

$$y(x_1, x_2, t) = \pi e^{2\pi^2 \nu t}(\sin \pi x_1 + \sin \pi x_2). \qquad (2.157)$$

Concerning now the *dual problem* of (2.155) we can easily show that it is defined by

$$\Lambda f = Y_0(T) - y_T, \qquad (2.158)$$

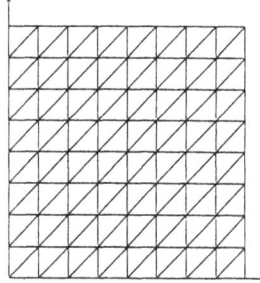

Fig. 1. A regular triangulation of $(0,1) \times (0,1)$.

where operator Λ is still defined by (2.17)–(2.19) and where the function Y_0 is the solution of

$$\frac{\partial Y_0}{\partial t} + AY_0 = s \text{ on } Q, \quad Y_0(0) = y_0, \ Y_0 = 0 \text{ on } \Sigma. \tag{2.159}$$

Since the data have been chosen so that we have exact controllability, the dual problem (2.158) has a unique solution which, in the particular case discussed here, is given by

$$f(x_1, x_2) = \pi e^{2\pi^2 \nu} \sin \pi x_1 \sin \pi x_2. \tag{2.160}$$

To approximate problem (2.158) (and therefore problem (2.155)) we have used the method described in Sections 2.3 to 2.5, namely: *time discretization* by a *second-order accurate scheme*, *space discretization* by *finite element methods* (using *regular* triangulations $\mathcal{T}_h$ like the one in Figure 1) and *iterative solution* by a trivial variant of algorithm (2.125)–(2.151) (with $k = +\infty$) with $\epsilon = 10^{-4}$ for the stopping criterium.

The above solution methodology has been tested for various values of h and Δt; for all of them, we have taken $\nu = 1/2\pi^2 (= 5.066059 \ldots \times 10^{-2})$. On Table 1 we have summarized the results which have been obtained (we have used a $*$ to indicate a *computed* quantity). All the calculations have been done with $f_0 = 0$ as initializer for the conjugate gradient algorithms.

The results presented in Table 1 deserve some comments:

1   The convergence of the conjugate gradient algorithm is fairly fast if we keep in mind that the solution $f_h^{\Delta t}$ of the discrete problem which has been solved can be viewed as a vector with $(31)^2 = 961$ components if $h = \Delta t = 1/32$ (respectively $(63)^2 = 3969$ components if $h = \Delta t = 1/64$).

2   The target function y_T has been reached within a good accuracy, similar comments holding for the approximation of the optimal control u and of the solution f to the dual problem (2.158).

3   For information, we have $\|u\|_{L^2(\Sigma)} = \pi \sqrt{e^2 - 1} = 7.94087251 \ldots$ and $\|f\|_{H_0^1(\Omega)} = \pi e / \sqrt{2} = 6.03850398 \ldots$.

Table 1. *Summary of numerical results.*

	$h = \Delta t = \frac{1}{32}$	$h = \Delta t = \frac{1}{64}$
Number of iterations	10	11
$\dfrac{\|y_T^* - y_T\|_{-1}}{\|y_T\|_{-1}}$	2.24×10^{-5}	1.78×10^{-5}
$\|u^*\|_{L^2(\Sigma)}$	7.791	7.863
$\dfrac{\|u^* - u\|_{L^2(\Sigma)}}{\|u\|_{L^2(\Sigma)}}$	2.50×10^{-3}	1.21×10^{-3}
$\|f^*\|_{H_0^1(\Omega)}$	6.07	6.041
$\dfrac{\|f^* - f\|_{H_0^1(\Omega)}}{\|f\|_{H_0^1(\Omega)}}$	2.44×10^{-2}	2.85×10^{-2}
$\dfrac{\|f^* - f\|_{L^2(\Omega)}}{\|f\|_{L^2(\Omega)}}$	6.53×10^{-3}	7.02×10^{-3}

On Figures 2 and 3 we have compared $y_T(x_1, 0.5)$ (...) and $y_T^*(x_1, 0.5)$ (—) for $x_1 \in (0,1)$ and $h = \Delta t = 1/32$, $h = \Delta t = 1/64$, respectively; we recall that $y_T^* = y_h^{\Delta t}(T)$ and that our methodology forces $y_h^{\Delta t}(T)$ to belong to H_{0h}^1, explaining the observed behaviour of the above function in the neighbourhood of Γ. On Figures 4 and 5 we have represented the functions $t \to \|u(t)\|_{L^2(\Gamma)}$ (...) and $t \to \|u^*(t)\|_{L^2(\Gamma)}$ (—) for $t \in (0,T)$ and $h = \Delta t = 1/32$, $h = \Delta t = 1/64$, respectively. Finally, on Figures 6 and 7 we have compared $f(x_1, 0.5)$ (...) and $f^*(x_1, 0.5)$ (—) for $x_1 \in (0,1)$ and $h = \Delta t = 1/32$, $h = \Delta t = 1/64$, respectively. Comparing these two figures shows that $h = \Delta t = 1/32$ provides a (slightly) better approximation than $h = \Delta t = 1/64$; this is in agreement with the results in Table 1.

The results obtained here compared favourably with those in Carthel *et al.* (1994) where the same test problem was solved by other methods, including a second-order accurate time discretization method close to that discussed in Section 1.8.6 for distributed control problems (see also Carthel (1994) for further results and comments).

2.6.2. *Second test problem*

If one uses the notation of Section 2.6.1 we have for this test problem $\Omega = (0,1) \times (0,1)$, $T = 1$, $s = 0$, $y_0 = 0$, $y_T(x_1, x_2) = \min(x_1, x_2, 1 - x_1, 1 - x_2)$ and $\nu = 1/2\pi^2$; unlike the test problem of Section 2.6.1, for which $\Gamma_0 = \Gamma$, we have here $\Gamma_0 \neq \Gamma$ since

$$\Gamma_0 = \{\{x_1, x_2\} \mid 0 < x_1 < 1, \ x_2 = 0\}.$$

The function y_T is *Lipschitz continuous*, but not smooth enough to have (see the discussion in Carthel *et al.* (1994, Section 2.3.3)) exact controllability.

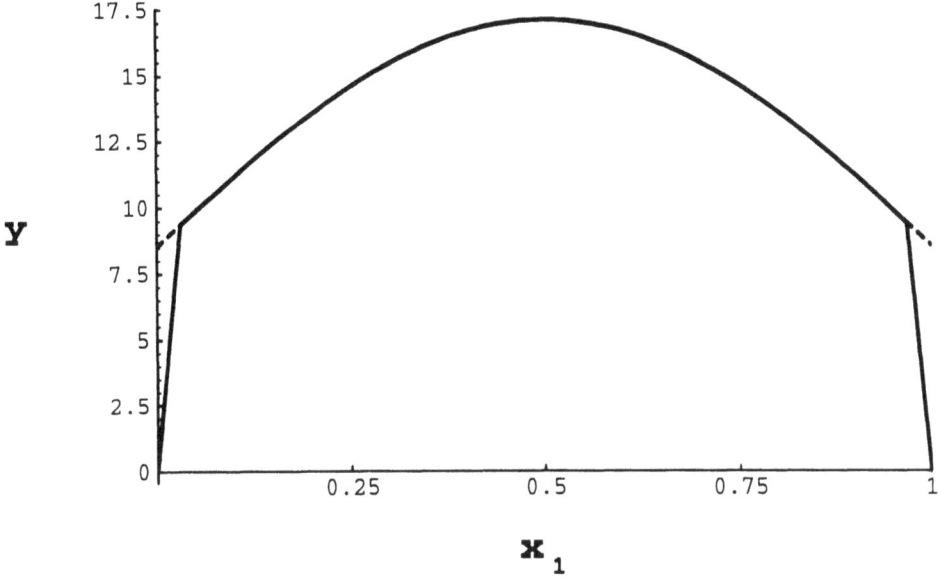

Fig. 2. Comparison between y_T (...) and y_T^* (—) ($h = \Delta t = 1/32$).

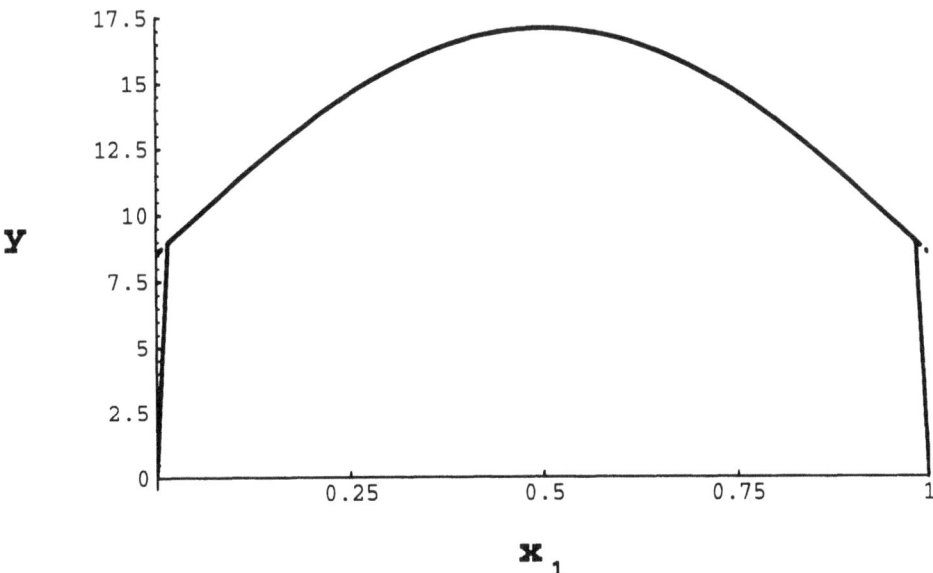

Fig. 3. Comparison between y_T (...) and y_T^* (—) ($h = \Delta t = 1/64$).

This implies that problem (2.21) has no solution if $k = +\infty$; on the other hand, problems (2.11), (2.12), (2.21), (2.25) are *well-posed* for any finite positive value of k or β. Focusing on the solution of problem (2.21) we have used the same space and time discretization methods as for the first test problem, with $h = \Delta t = 1/32$ and $h = \Delta t = 1/64$. We have taken $k = 10^5$

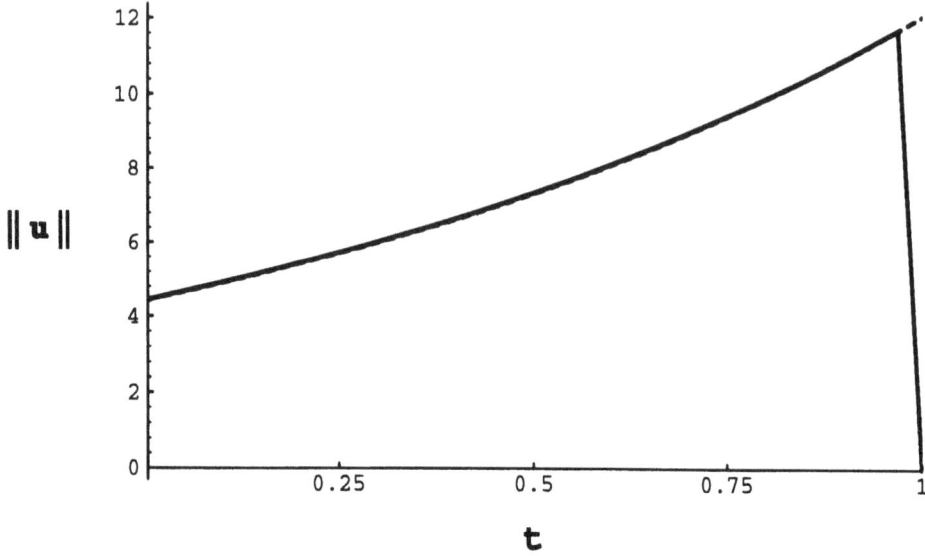

Fig. 4. Comparison between $\|u(t)\|_{L^2(\Gamma)}$ (...) and $\|u^*(t)\|_{L^2(\Gamma)}$ (—) ($h = \Delta t = 1/32$).

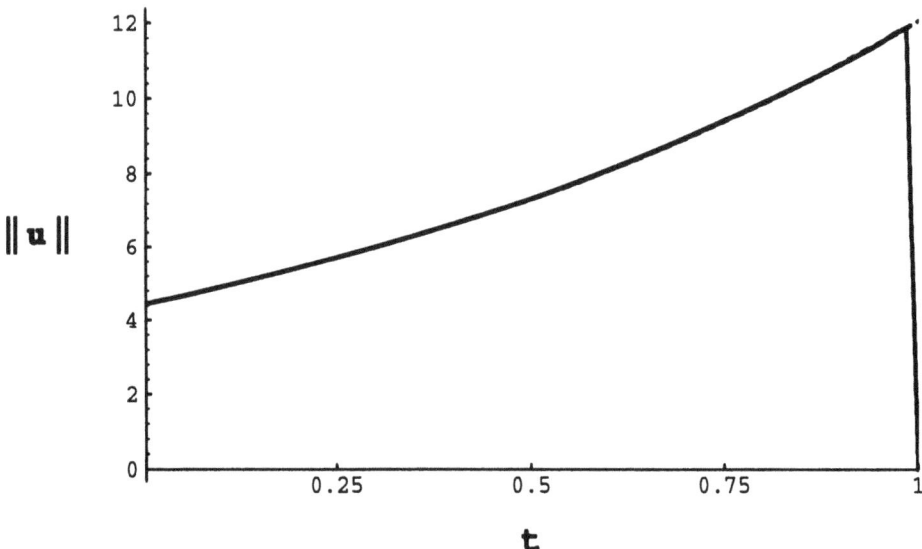

Fig. 5. Comparison between $\|u(t)\|_{L^2(\Gamma)}$ (...) and $\|u^*(t)\|_{L^2(\Gamma)}$ (—) ($h = \Delta t = 1/64$).

and 10^7 for the penalty parameter and used $\epsilon = 10^{-3}$ for the stopping criterium of the conjugate gradient algorithm (2.125)–(2.151) (which has been initialized with $f^0 = 0$).

The numerical results have been summarized in Table 2.

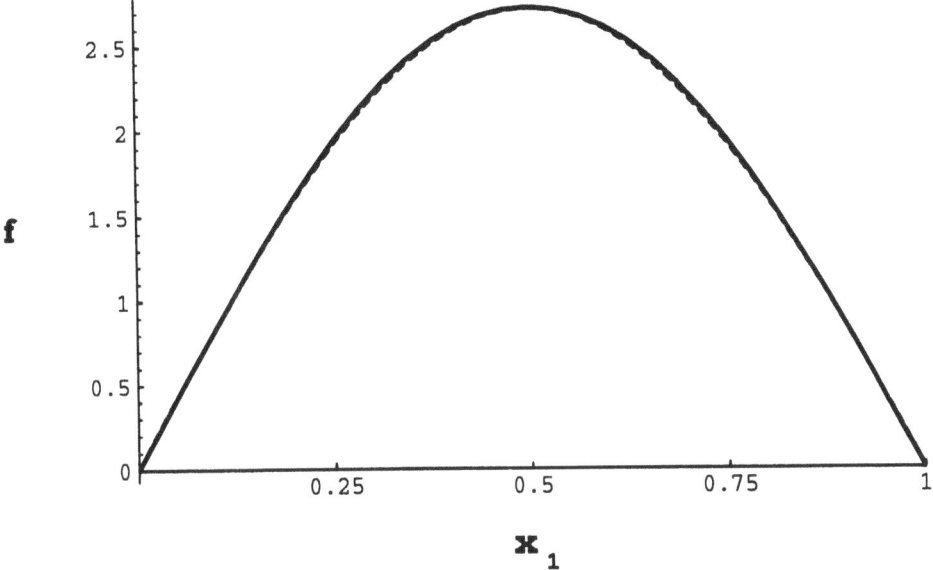

Fig. 6. Comparison between f (...) and f^* (—) ($h = \Delta t = 1/32$).

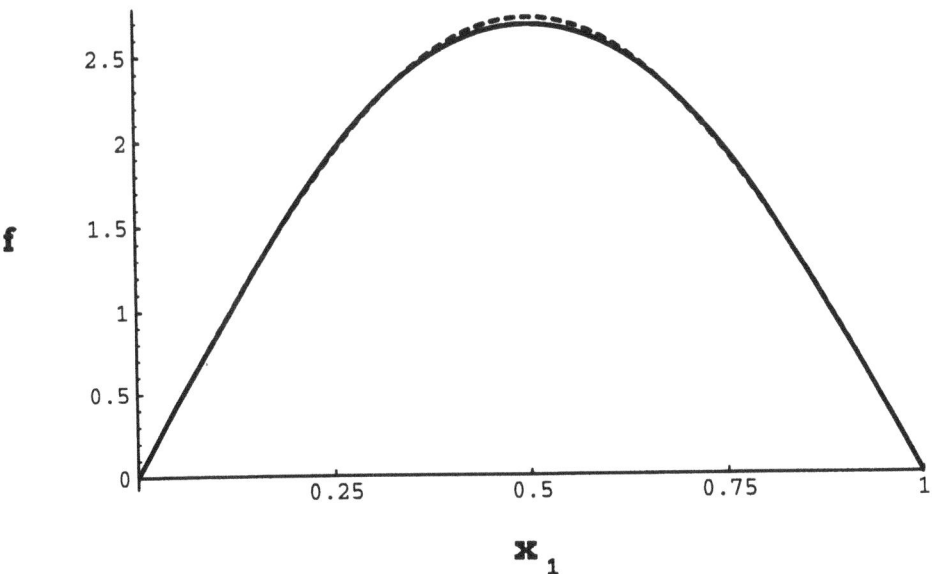

Fig. 7. Comparison between f (...) and f^* (—) ($h = \Delta t = 1/64$).

The above results suggest the following comments: first, we observe that $\|y_T - y_h^{\Delta t}(T)\|_{-1}$ varies like $k^{-1/4}$, approximately. Second, we observe that the number of iterations necessary for convergence, increases as h, Δt and k^{-1} decrease; there is no mystery here, since – from Section 1.8.2, relation

Table 2. *Summary of numerical results.*

$h = \Delta t$	1/32	1/64	1/32	1/64
k	10^5	10^5	10^7	10^7
Number of iterations	56	60	292	505
$\dfrac{\|y_T^* - y_T\|_{-1}}{\|y_T\|_{-1}}$	1.31×10^{-1}	1.28×10^{-1}	4.15×10^{-2}	3.93×10^{-2}
$\|u^*\|_{L^2(\Sigma_0)}$	8.18	8.12	25.59	24.78
$\|f^*\|_{H_0^1(\Omega)}$	600.4	584.2	18,960	17,950
$\|f^*\|_{L^2(\Omega)}$	75.95	73.63	1,632	1,525

(1.130) – the key factor controlling the speed of convergence is the *condition number* of the *bilinear form* in the left-hand side of equation (2.124). This condition number, denoted by $\nu_h^{\Delta t}(k)$, is defined by

$$\nu_h^{\Delta t}(k) = \max_{z \in H_{0h}^1 - \{0\}} R_h(z) \Big/ \min_{z \in H_{0h}^1 - \{0\}} R_h(z), \qquad (2.161)$$

where, in (2.161), $R_h(z)$ is the *Rayleigh quotient* defined by

$$R_h(z) = \frac{k^{-1} \int_\Omega |\nabla z|^2 \, dx + \int_\Omega (\Lambda_h^{\Delta t} z) z \, dx}{\int_\Omega |\nabla z|^2 \, dx}; \qquad (2.162)$$

it can be shown that

$$\lim_{\substack{h \to 0 \\ \Delta t \to 0}} \nu_h^{\Delta t}(k) = k \|\Lambda\|_{\mathcal{L}(H_0^1(\Omega); H^{-1}(\Omega))}, \qquad (2.163)$$

implying that for small values of h, Δt and k^{-1}, problem (2.124) is *badly conditioned*. Indeed, we can expect from (2.163) and from Section 1.8.2, relation (1.130), that for h and Δt sufficiently small the number of iterations necessary to obtain convergence will vary like $k^{1/2}$, approximately; this prediction is confirmed by the results in Table 2 (and will be further confirmed by the results in Section 2.6.3, Table 3, concerning our third test problem). Third, and finally, we observe that $\|u^*\|_{L^2(\Sigma_0)}$ (respectively $\|f^*\|_{H_0^1(\Omega)}$) varies like $k^{1/4}$ (respectively $k^{3/4}$); it can be shown that the behaviour of $\|f^*\|_{H_0^1(\Omega)}$ follows from that of $\|y_T - y_T^*\|_{-1}$ since we have (see, e.g., Carthel *et al.* (1994, Remark 4.3))

$$k^{-1} = \frac{\|y_T - y(T)\|_{-1}}{\|f\|_{H_0^1(\Omega)}}. \qquad (2.164)$$

where y is the state function obtained from the optimal control u via (2.1)–(2.3).

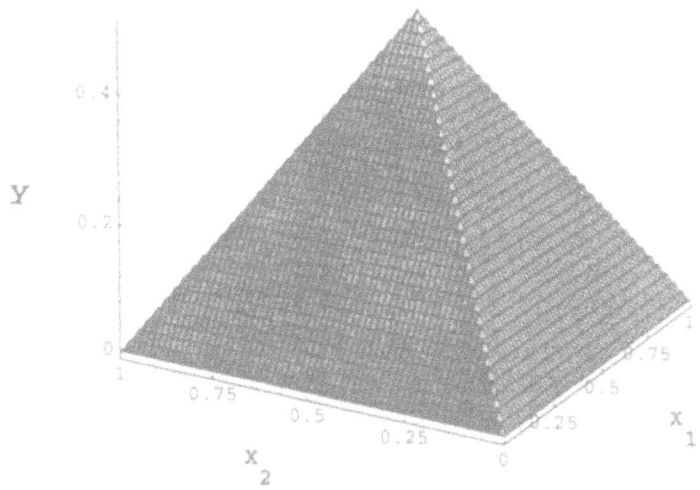

Fig. 8. Graph of the target function y_T $(y_T(x_1, x_2) = \min(x_1, x_2, 1 - x_1, 1 - x_2), 0 \le x_1, x_2 \le 1)$.

On the following figures, we have represented or shown the following information and results.

A view of the target function y_T on Figure 8. On Figures 9(a) to 12(a) (respectively 9(b) to 12(b)) the graph of the function $y_T^*(= y_h^{\Delta t}(T))$ (respectively a comparison between y_T (...) and the actually reached state function y_T^* (—)) for various values of h, Δt and k (we have shown the graphs of the functions $x_2 \to y_T(0.5, x_2)$ and $x_2 \to y_T^*(0.5, x_2)$ for $x_2 \in (0, 1)$). The graphs of the computed solution $f_h^{\Delta t}(= f^*)$ and of the function $x_2 \to f_h^{\Delta t}(0.5, x_2)$ on Figures 13 to 16. On Figures 17 to 20 the graphs of the functions $t \to \|u^*(t)\|_{L^2(\Gamma_0)}$ and $\{x_1, t\} \to u^*(x_1, t)$. Finally, we have visualized on Figures 21 to 24 (using a log-scale) the convergence to zero of the conjugate gradient residual $\|g_m\|_{H_0^1(\Omega)}$; the observed behaviour (highly oscillatory, particularly for $k = 10^7$) is typical of a *badly conditioned* problem.

2.6.3. Third test problem

For this test problem Ω, T, Γ_0, y_0, s, A, ν are as in Section 2.6.2, namely $\Omega = (0.1) \times (0.1)$, $T = 1$, $y_0 = 0$, $s = 0$, $A = -\nu\Delta$ with $\nu = 1/2\pi^2$; the only difference is that this time y_T is the *discontinuous* function defined by

$$y_T(x_1, x_2) = \begin{cases} 1 & \text{if } 1/4 < x_1, x_2 < 3/4, \\ 0 & \text{otherwise.} \end{cases} \tag{2.165}$$

We have applied to this problem the solution methods considered in Section 2.6.2; their behaviour here is essentially the same as that for the test problem of Section 2.6.2 (where y_T was Lipschitz continuous). We have shown in the

Table 3. *Summary of numerical results.*

$h = \Delta t$	1/32	1/64	1/32	1/64
k	10^5	10^5	10^7	10^7
Number of iterations	55	56	361	569
$\dfrac{\|y_T^* - y_T\|_{-1}}{\|y_T\|_{-1}}$	1.64×10^{-1}	1.57×10^{-1}	1.05×10^{-1}	9.88×10^{-2}
$\|u^*\|_{L^2(\Sigma_0)}$	14.68	15.07	56.80	58.53
$\|f^*\|_{H_0^1(\Omega)}$	1,407	1,410	90,010	88,510
$\|f^*\|_{L^2(\Omega)}$	120.7	122.5	5,608	5,566

following Table 3 the results of our numerical experiments (the notation is as in Section 2.6.2).

Comparing to Table 2 we observe that the convergence properties of the conjugate gradient algorithm are essentially the same, despite the fact that y_T is much less smooth here; on the other hand we observe that $\|y_T - y_h^{\Delta t}(T)\|_{-1}$ varies like $k^{-1/3}$, approximately, implying in turn (from (2.164)) that $\|f^*\|_{H_0^1(\Omega)}$ varies like $k^{7/8}$, approximately. The dependence of $\|u^*\|_{L^2(\Sigma_0)}$ is less clear (to us at least); it looks 'faster', however, than $k^{1/4}$.

On Figure 25 we have visualized the graph of the target function y_T, then on Figures 26 and 27 we have compared the function $x_2 \rightarrow y_T(0.5, x_2)$ to $x_2 \rightarrow y_T^*(0.5, x_2)$ (—) for various values of k, h and Δt; on Figures 28 and 29 we have shown the graphs of the corresponding function y_T^*. Finally, for the above values of k, h and Δt, we have shown, on Figures 30 to 35, further information concerning $u_h^{\Delta t}$, $f_h^{\Delta t}$ and the convergence of the conjugate gradient algorithm (2.125)–(2.151).

2.7. Neumann control (I): Formulation of the control problems

We consider again the state equation (2.1) in Q and the initial condition (2.3). We suppose this time that the *boundary control* is of the Neumann's type. To be more precise, the state function y is defined now by

$$\frac{\partial y}{\partial t} + Ay = 0 \text{ in } Q, \quad y(0) = 0, \quad \frac{\partial y}{\partial n_A} = v \text{ on } \Sigma_0, \quad \frac{\partial y}{\partial n_A} = 0 \text{ on } \Sigma \backslash \Sigma_0. \quad (2.166)$$

In (2.166), $\partial/\partial n_A$ denotes the *conormal derivative* operator; if operator A is defined by

$$A\varphi = -\sum_{i=1}^{d} \sum_{j=1}^{d} \frac{\partial}{\partial x_i} a_{ij} \frac{\partial \varphi}{\partial x_j}, \quad (2.167)$$

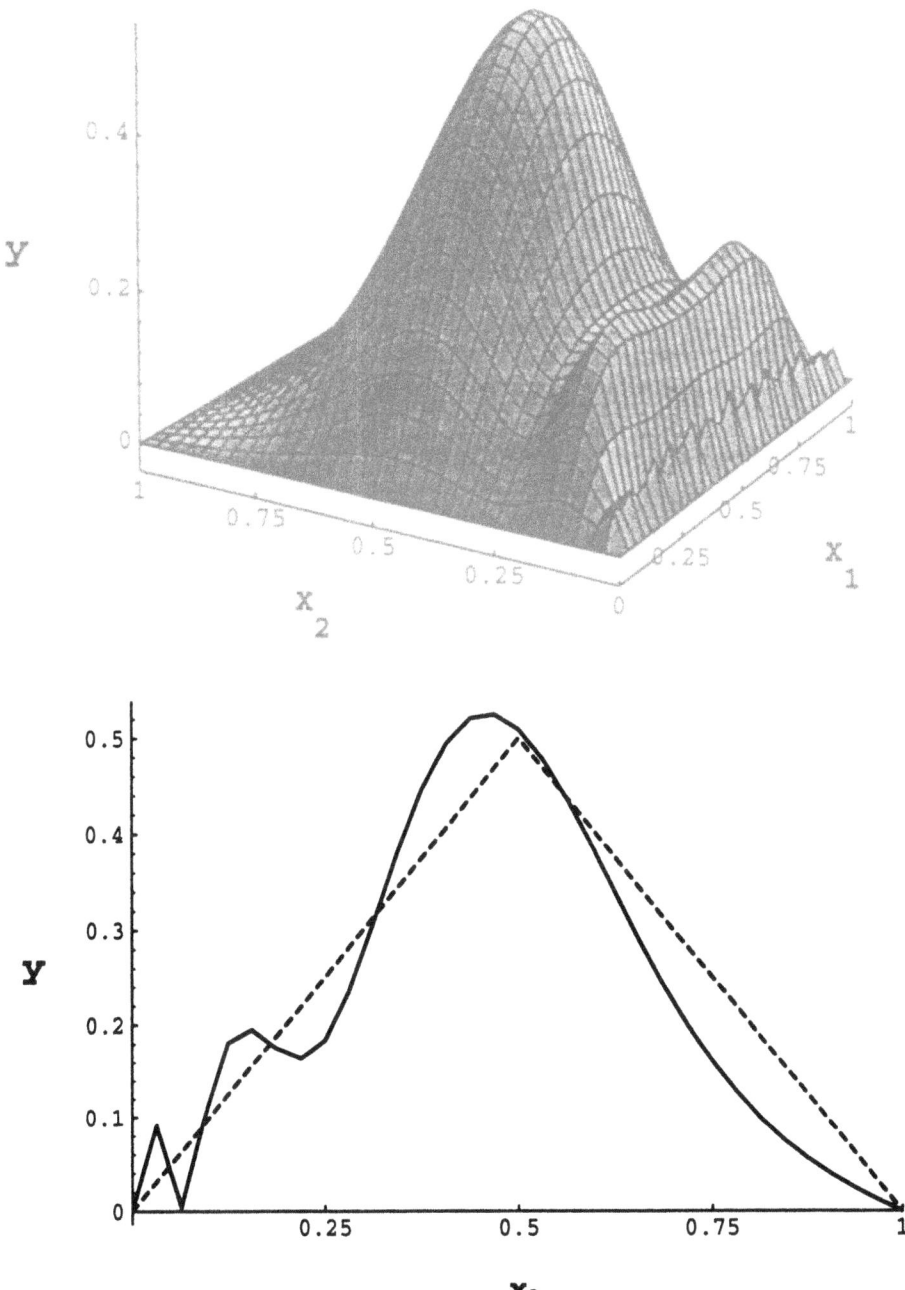

Fig. 9. (a) Graph of the function $y_T^*(k = 10^5,\ h = \Delta t = 1/32)$. (b) Comparison between y_T (...) and y_T^* (—) ($k = 10^5,\ h = \Delta t = 1/32$).

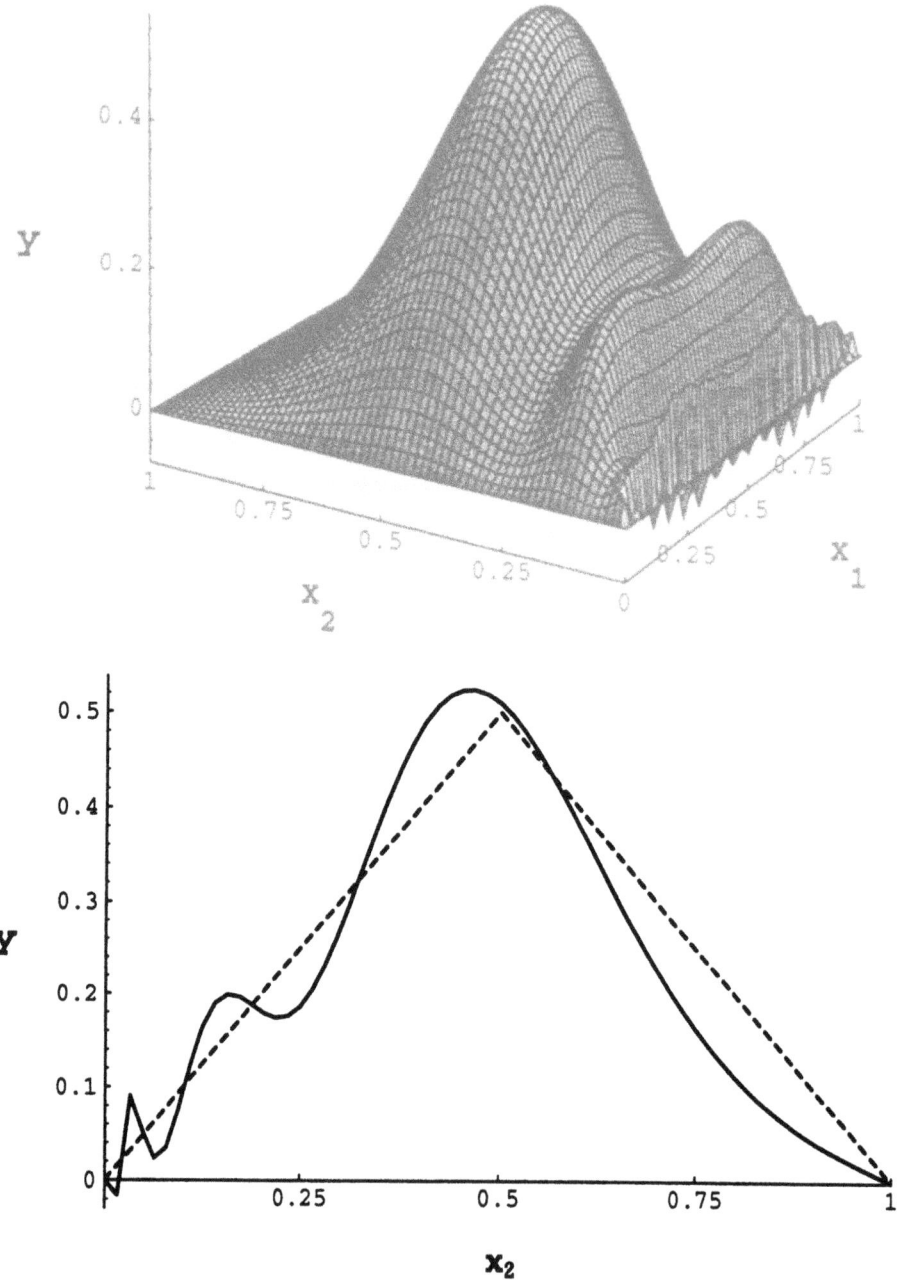

Fig. 10. (a) Graph of the function $y_T^*(k = 10^5, \ h = \Delta t = 1/64)$. (b) Comparison between y_T (...) and y_T^* (—) $(k = 10^5, \ h = \Delta t = 1/64)$.

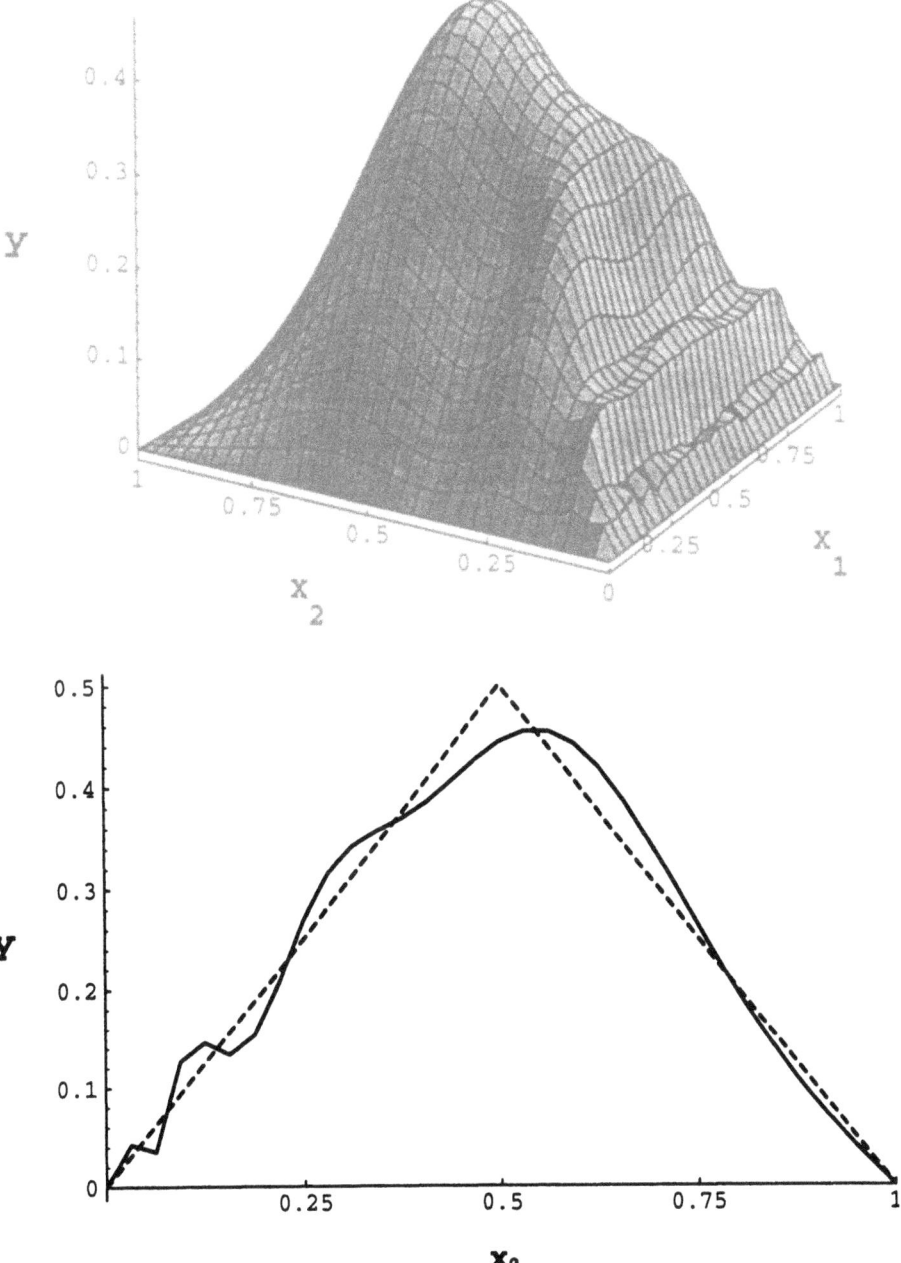

Fig. 11. (a) Graph of the function y_T^* ($k = 10^7$, $h = \Delta t = 1/32$). (b) Comparison between y_T (...) and y_T^* (—) ($k = 10^7$, $h = \Delta t = 1/32$).

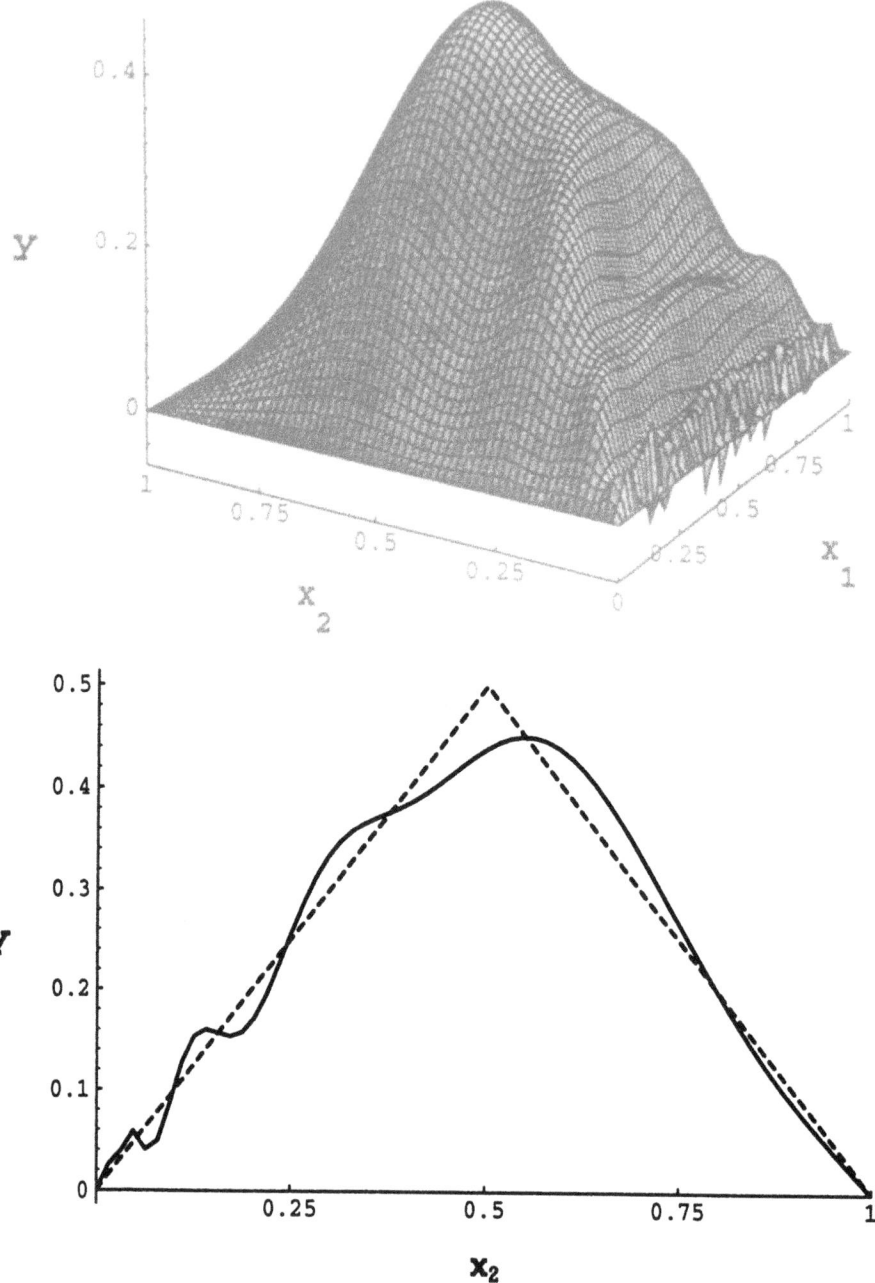

Fig. 12. (a) Graph of the function y_T^* ($k = 10^7$, $h = \Delta t = 1/64$). (b) Comparison between y_T (...) and y_T^* (—) ($k = 10^7$, $h = \Delta t = 1/64$).

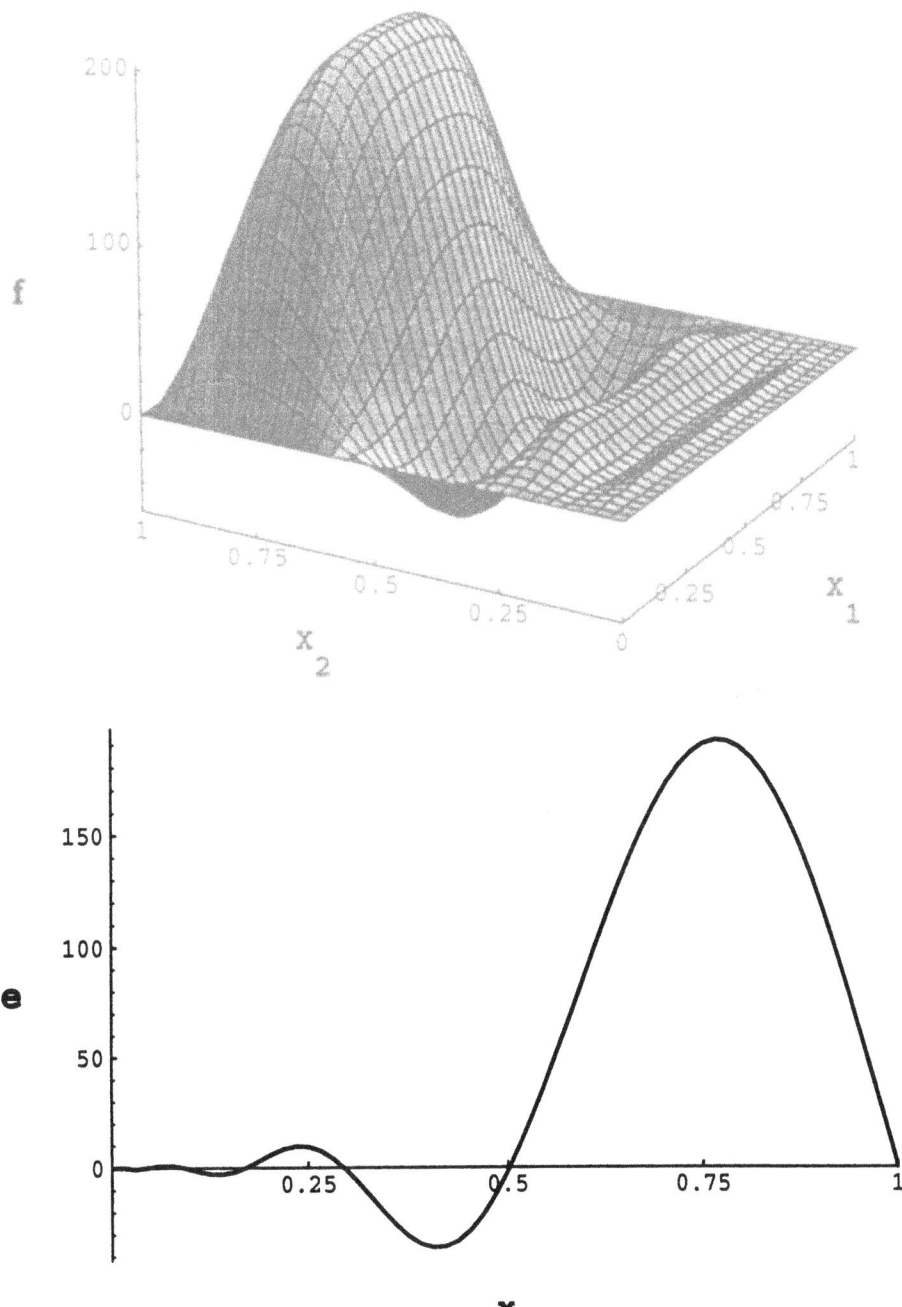

Fig. 13. (a) Graph of the function $f_h^{\Delta t}(k = 10^5, \ h = \Delta t = 1/32)$. (b) Graph of the function $x_2 \to f_h^{\Delta t}(0.5, x_2)(k = 10^5, \ h = \Delta t = 1/32)$.

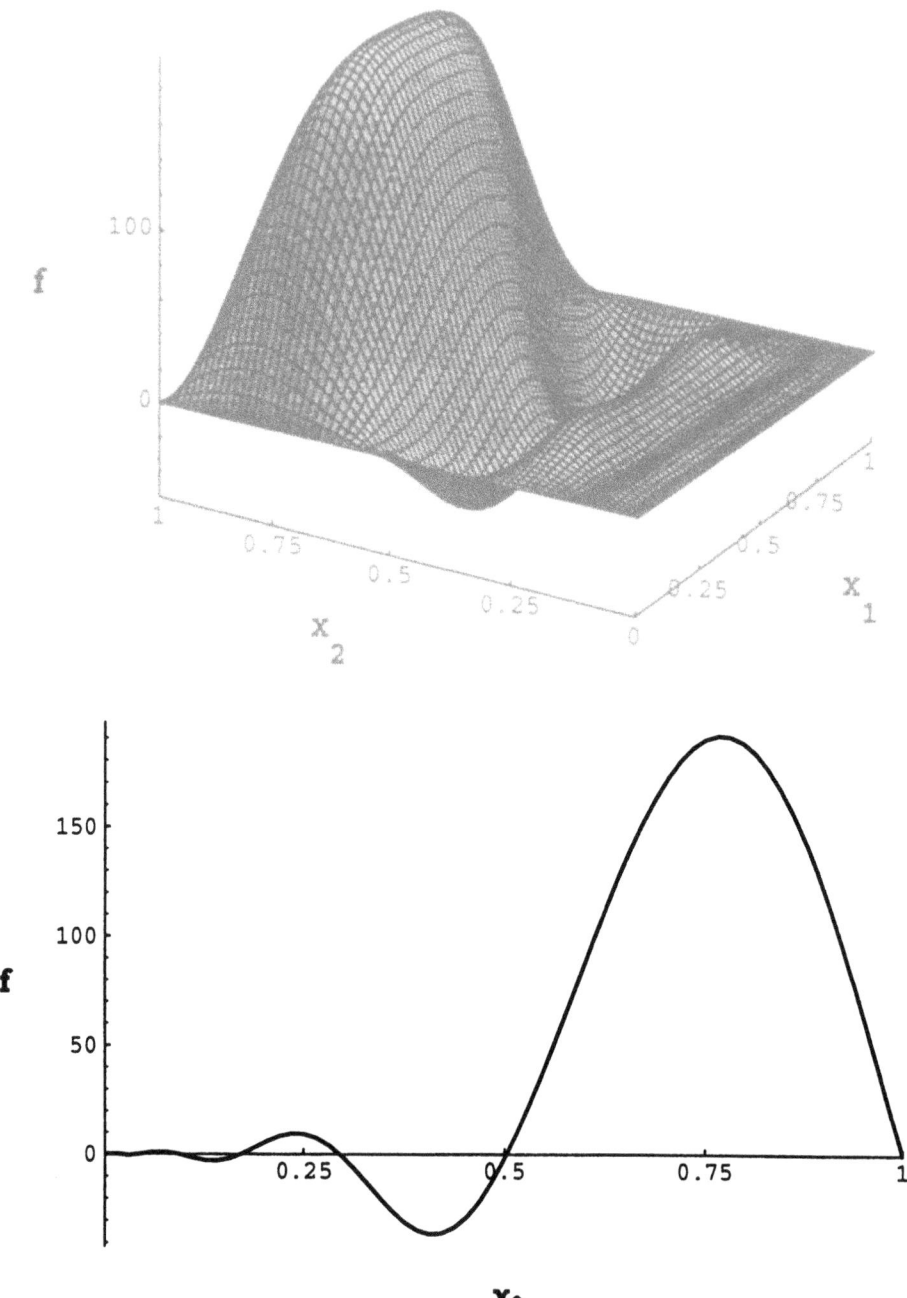

Fig. 14. (a) Graph of the function $f_h^{\Delta t}(k = 10^5, \; h = \Delta t = 1/64)$. (b) Graph of the function $x_2 \to f_h^{\Delta t}(0.5, x_2)(k = 10^5, \; h = \Delta t = 1/64)$.

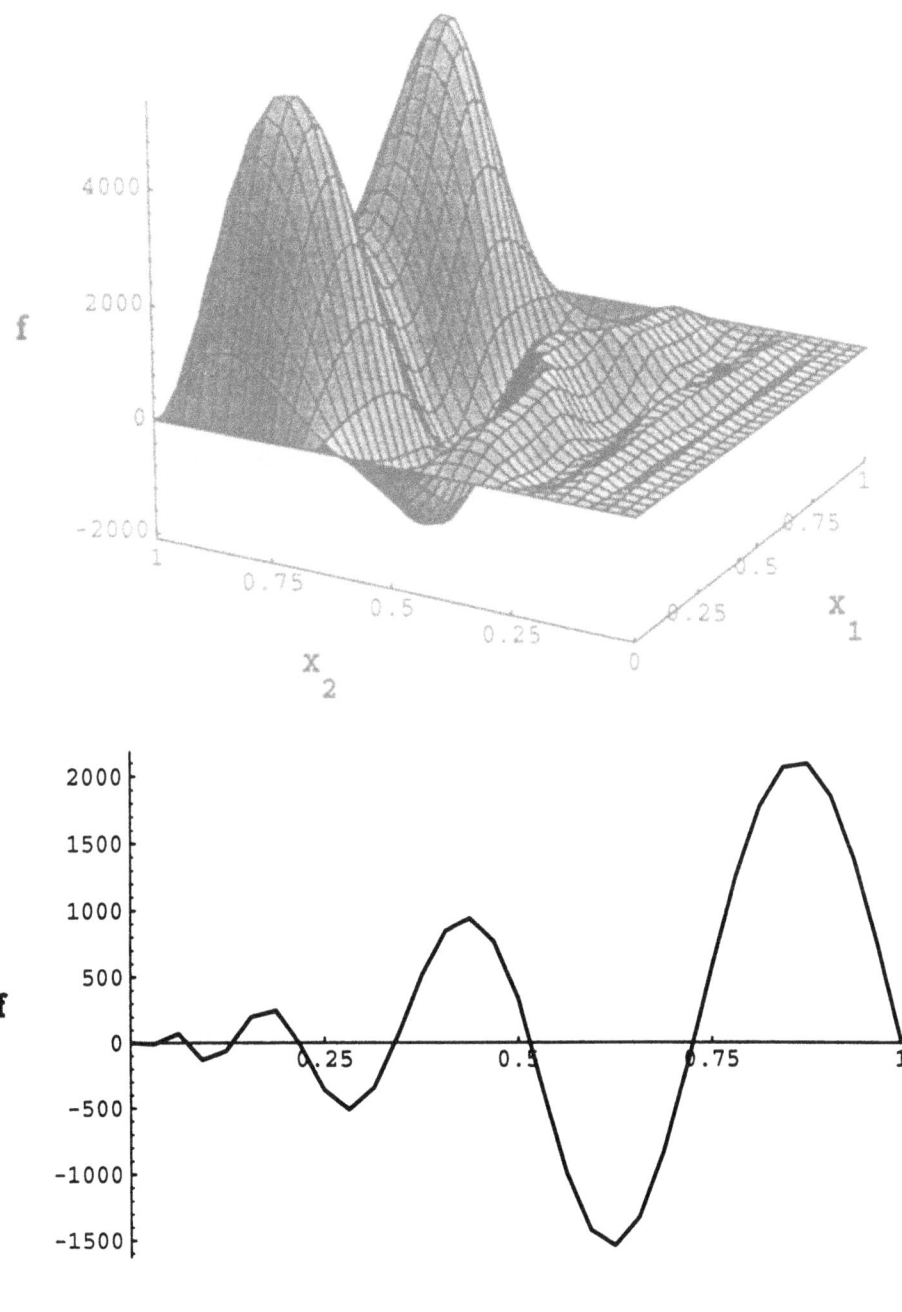

Fig. 15. (a) Graph of the function $f_h^{\Delta t}(k = 10^7,\ h = \Delta t = 1/32)$. (b) Graph of the function $x_2 \to f_h^{\Delta t}(0.5, x_2)(k = 10^7,\ h = \Delta t = 1/32)$.

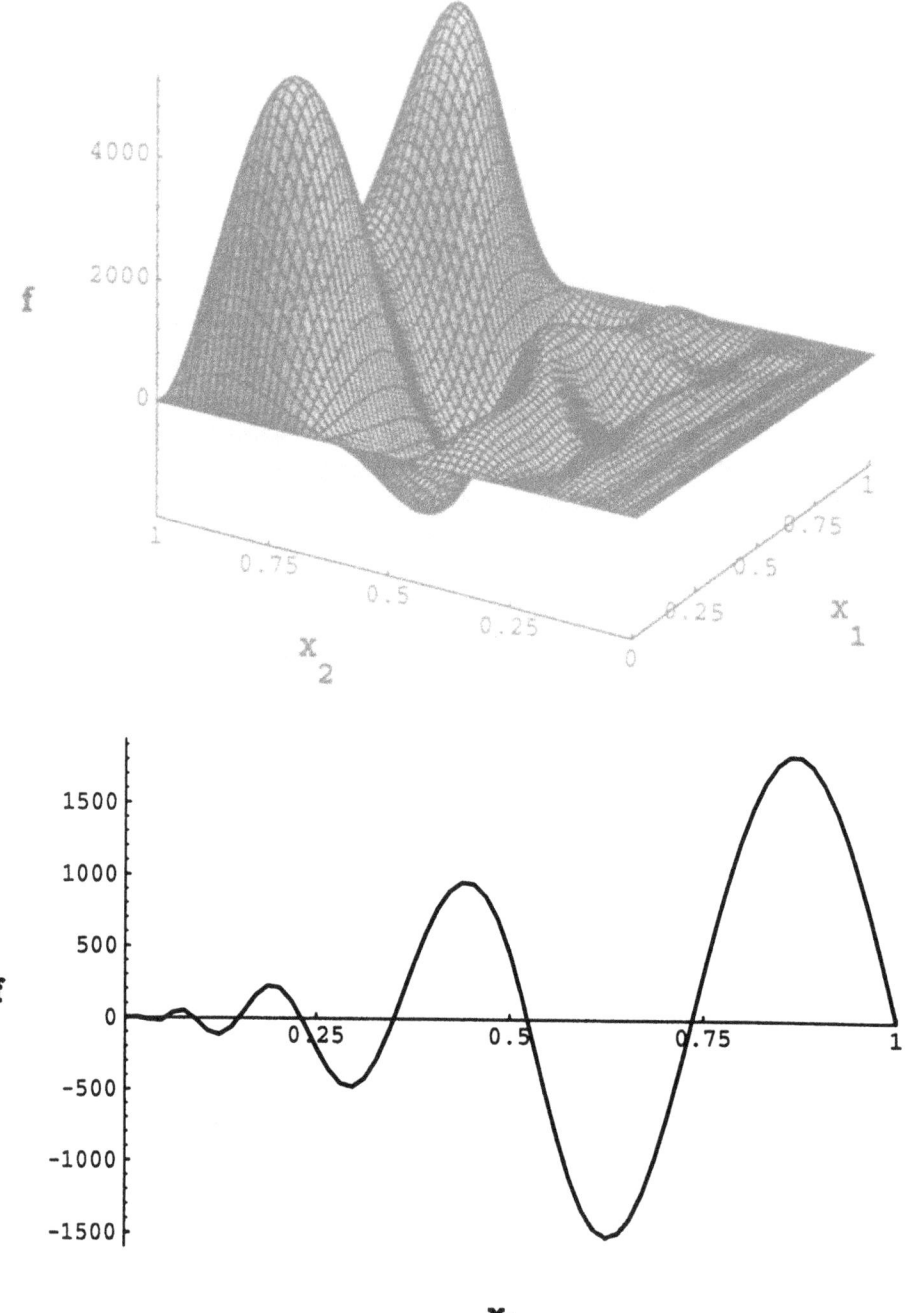

Fig. 16. (a) Graph of the function $f_h^{\Delta t}(k = 10^7,\ h = \Delta t = 1/64)$. (b) Graph of the function $x_2 \to f_h^{\Delta t}(0.5, x_2)(k = 10^7,\ h = \Delta t = 1/64)$.

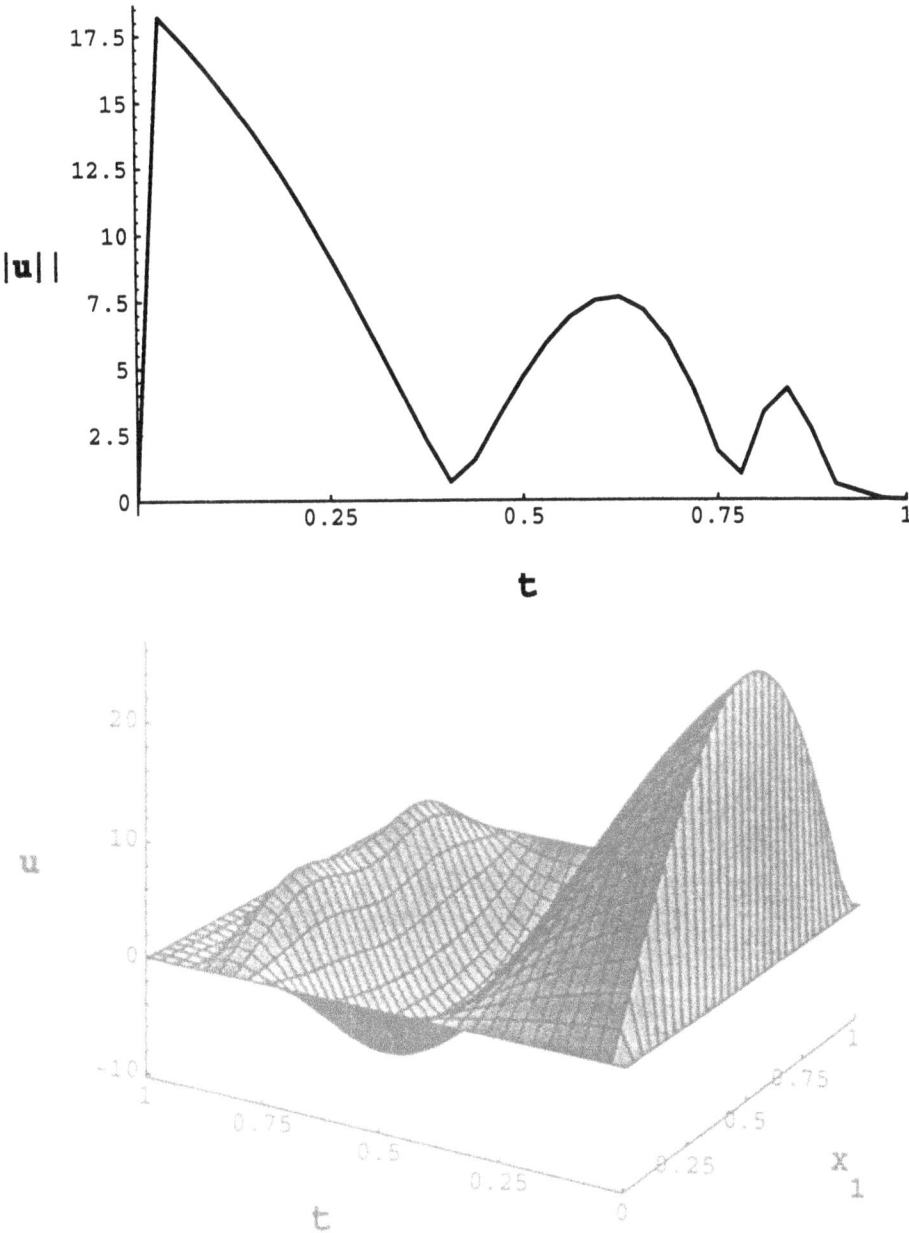

Fig. 17. (a) Graph of $t \to \|u^*(t)\|_{L^2(\Gamma_0)}(k = 10^5, \ h = \Delta t = 1/32)$. (b) Graph of the computed boundary control $(k = 10^5, \ h = \Delta t = 1/32)$.

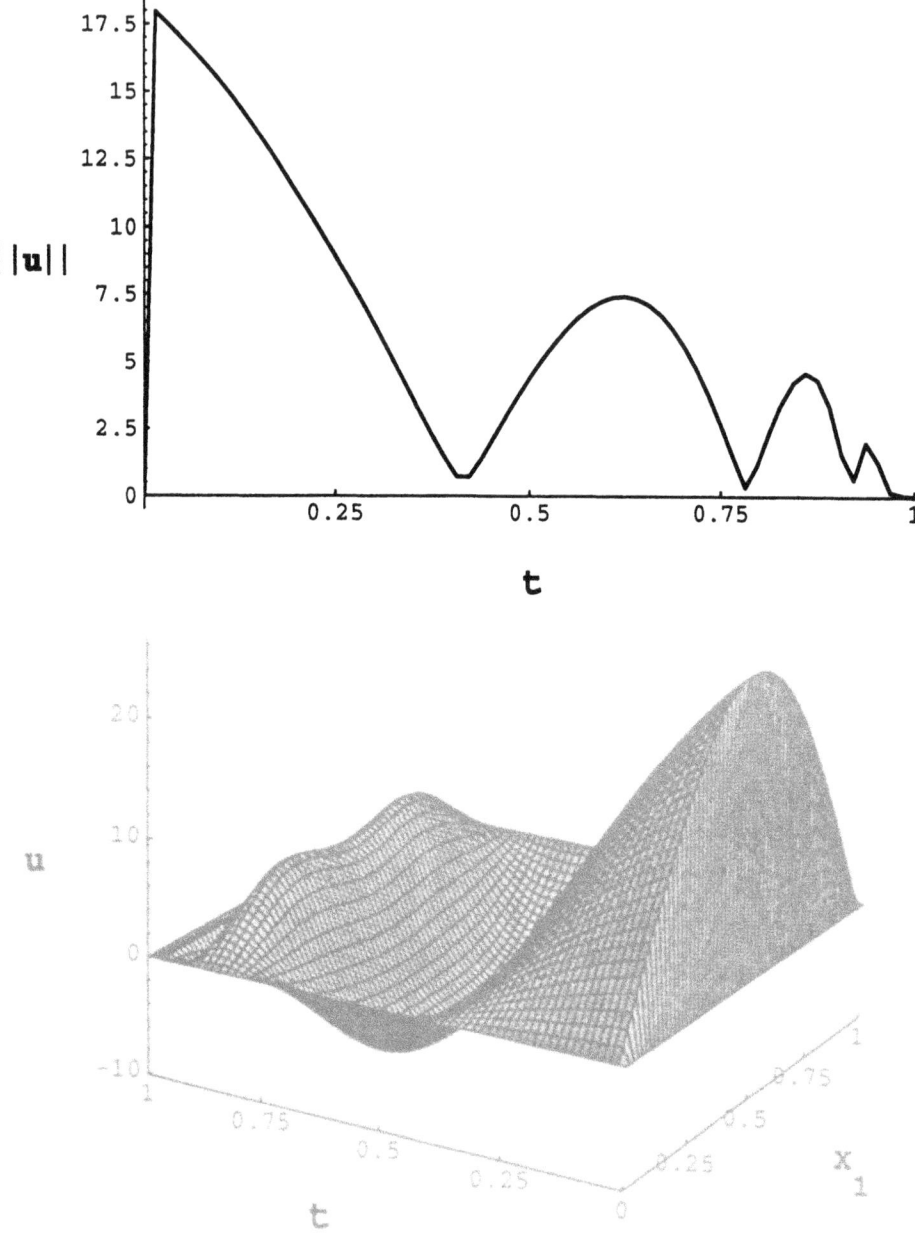

Fig. 18. (a) Graph of $t \rightarrow \|u^*(t)\|_{L^2(\Gamma_0)} (k = 10^5, \ h = \Delta t = 1/64)$. (b) Graph of the computed boundary control $(k = 10^5, \ h = \Delta t = 1/64)$.

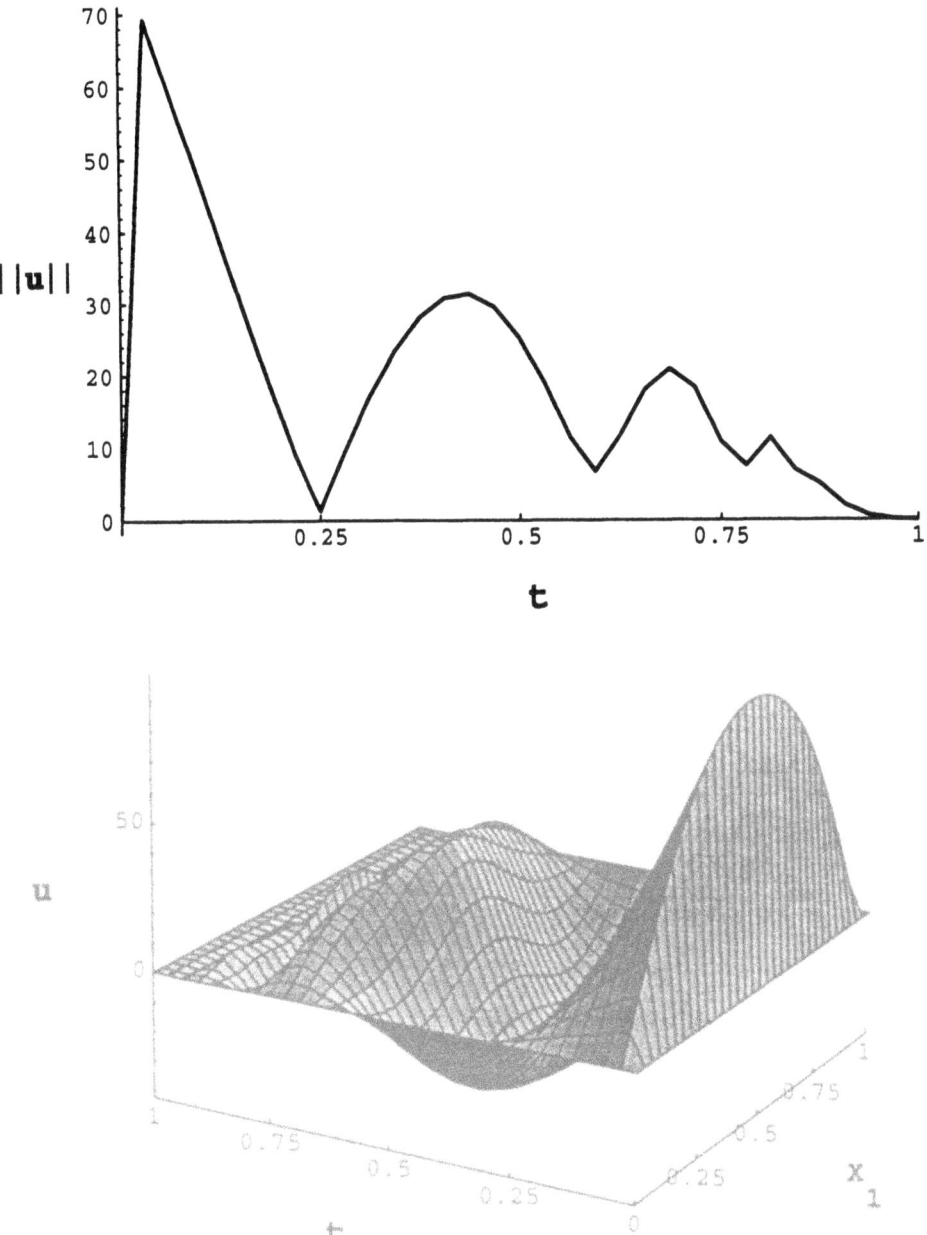

Fig. 19. (a) Graph of $t \to \|u^*(t)\|_{L^2(\Gamma_0)}$ $(k = 10^7,\ h = \Delta t = 1/32)$. (b) Graph of the computed boundary control $(k = 10^7,\ h = \Delta t = 1/32)$.

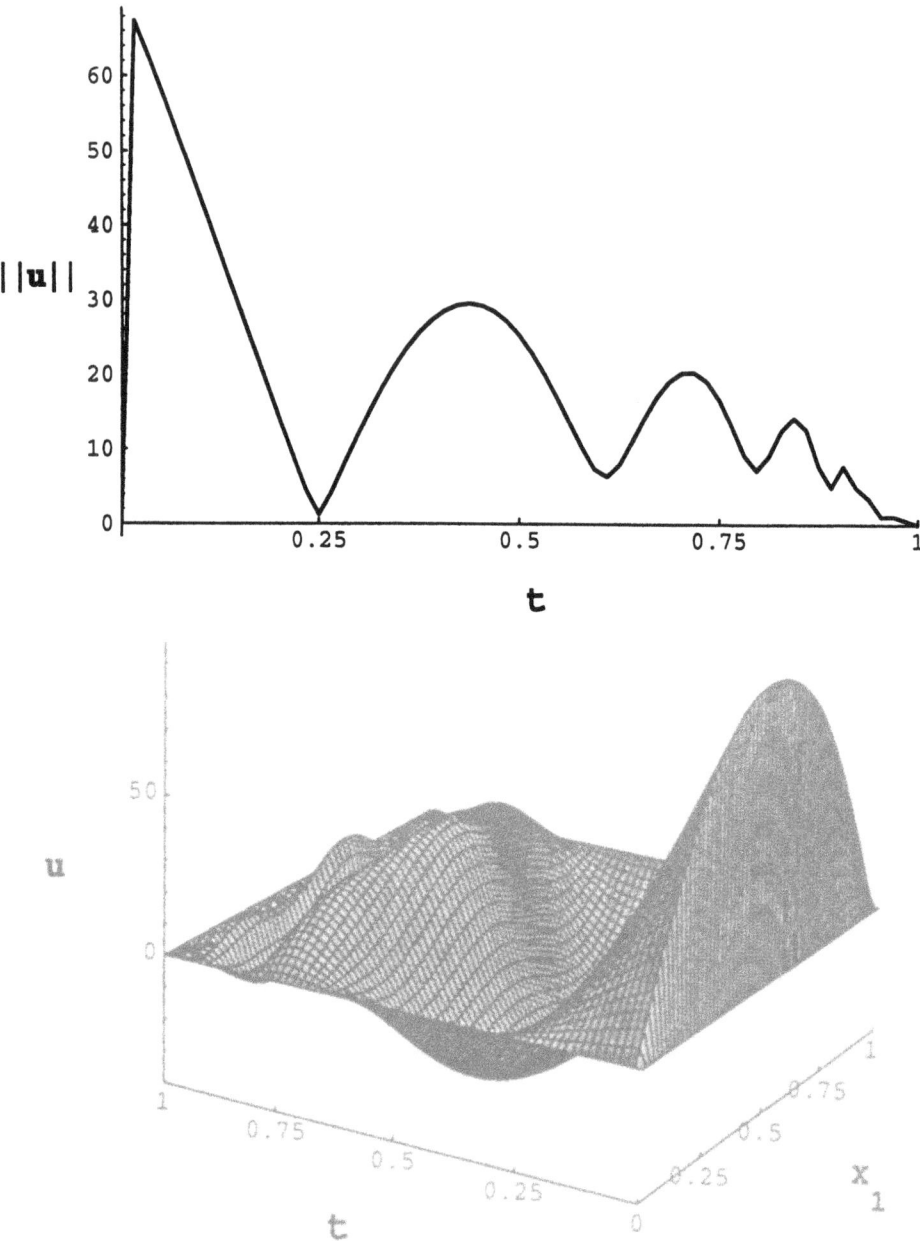

Fig. 20. (a) Graph of $t \to \|u^*(t)\|_{L^2(\Gamma_0)} (k = 10^7, \ h = \Delta t = 1/64)$. (b) Graph of the computed boundary control $(k = 10^7, \ h = \Delta t = 1/64)$.

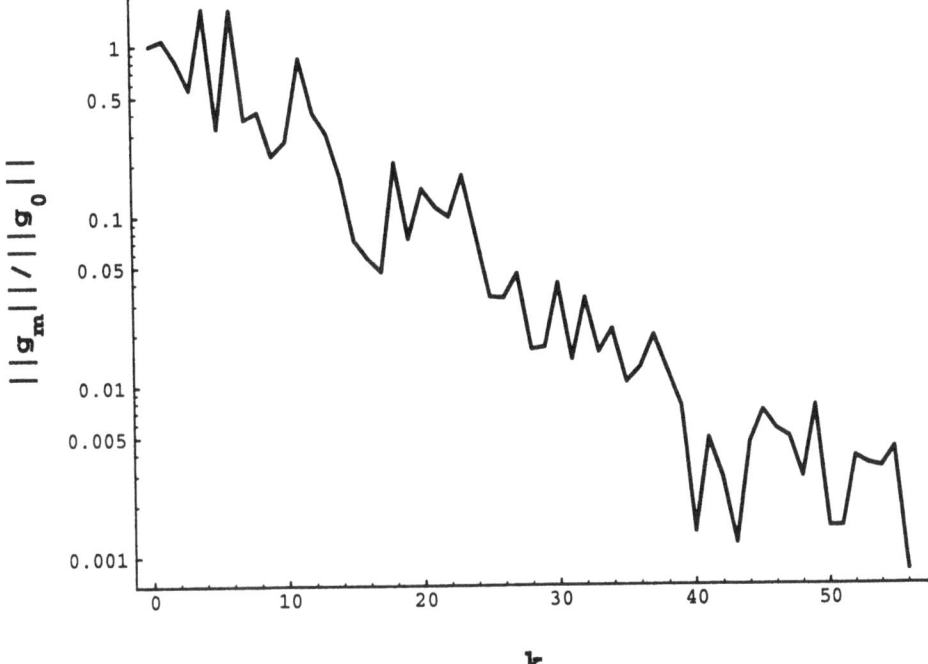

Fig. 21. Variation of $\|g_m\|_{H_0^1(\Omega)}/\|g_0\|_{H_0^1(\Omega)}(k = 10^5, \ h = \Delta t = 1/32)$.

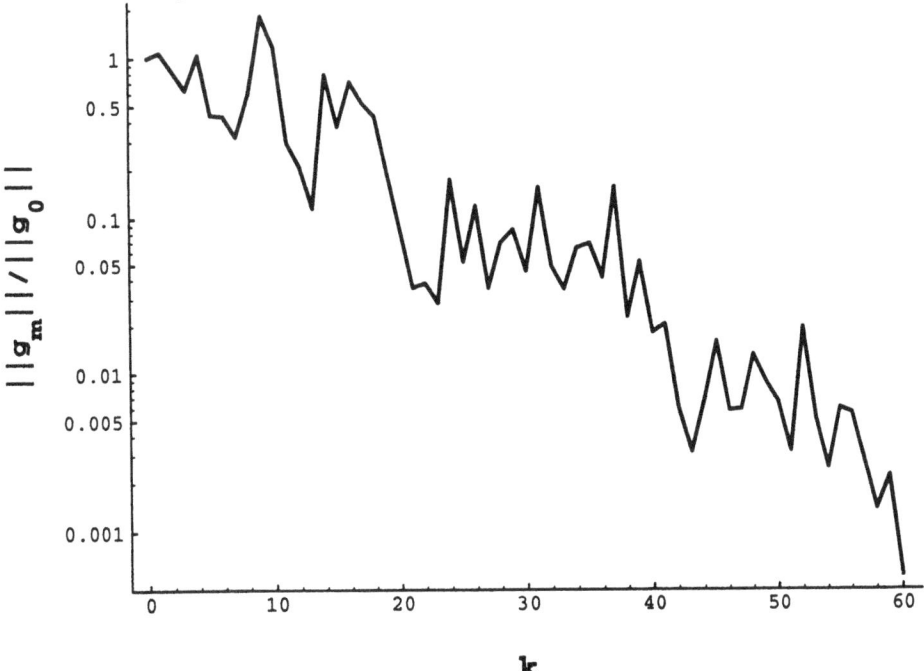

Fig. 22. Variation of $\|g_m\|_{H_0^1(\Omega)}/\|g_0\|_{H_0^1(\Omega)}(k = 10^5, \ h = \Delta t = 1/64)$.

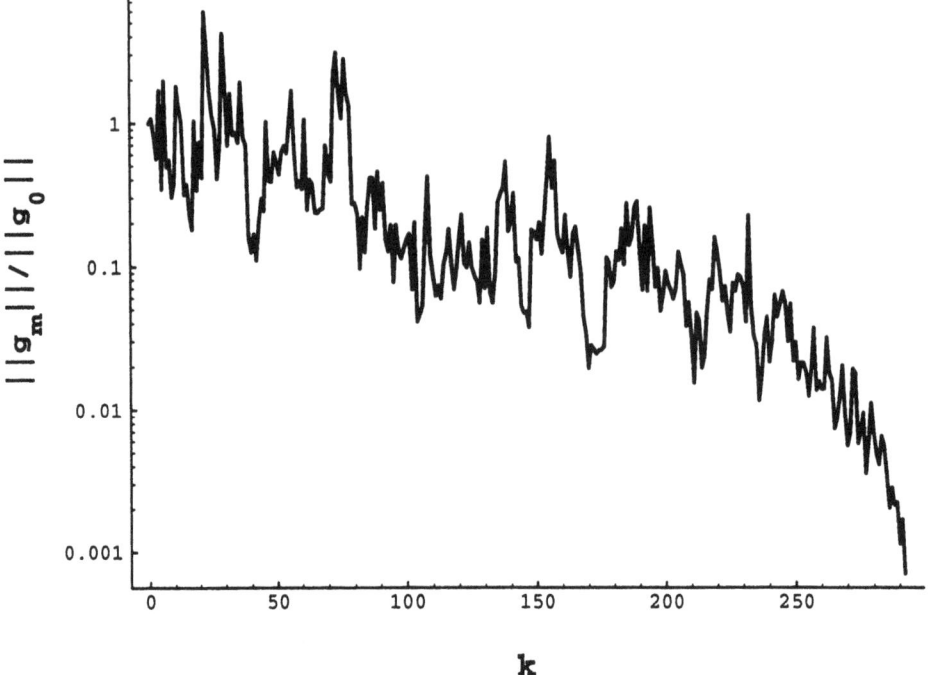

Fig. 23. Variation of $\|g_m\|_{H_0^1(\Omega)}/\|g_0\|_{H_0^1(\Omega)}(k=10^7,\ h=\Delta t=1/32)$.

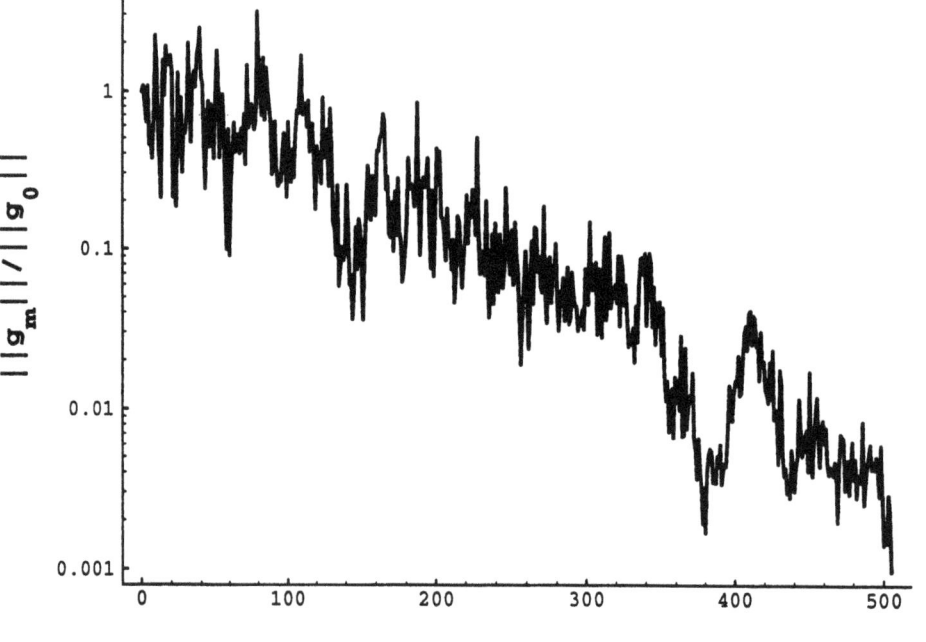

Fig. 24. Variation of $\|g_m\|_{H_0^1(\Omega)}/\|g_0\|_{H_0^1(\Omega)}(k=10^7,\ h=\Delta t=1/64)$.

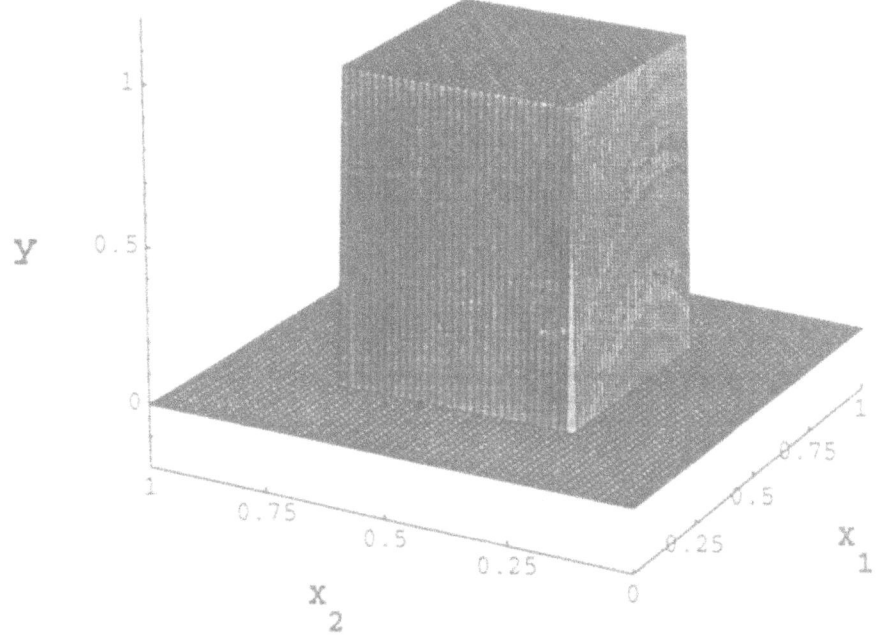

Fig. 25. Graph of the target function y_T (y_T is the characteristic function of the square $(1/4,3/4)^2$).

then $\partial/\partial n_A$ is defined by

$$\frac{\partial \varphi}{\partial n_A} = \sum_{i=1}^{d}\sum_{j=1}^{d} a_{ij}\frac{\partial \varphi}{\partial x_j}n_i, \tag{2.168}$$

where $\mathbf{n} = \{n_i\}_{i=1}^{d}$ is the *unit vector* of the *outward normal* at Γ.
    We assume that

$$v \in L^2(\Sigma_0). \tag{2.169}$$

There are slight (and subtle) technical differences between Neumann and Dirichlet boundary controls. Indeed, suppose that operator A is defined by (2.167) with the following additional properties

$$a_{ij} \in L^\infty(\Omega), \ \forall 1 \le i,j \le d, \tag{2.170}$$

$$\sum_{i=1}^{d}\sum_{j=1}^{d} a_{ij}(x)\xi_i\xi_j \ge \alpha|\boldsymbol{\xi}|^2, \ \forall\boldsymbol{\xi} \in \mathbb{R}^d, \quad \text{a.e. in } \Omega, \text{ with } \alpha > 0 \tag{2.171}$$

(in (2.171), $|\boldsymbol{\xi}|^2 = \sum_{i=1}^{d}|\xi_i|^2$, $\forall\boldsymbol{\xi} = \{\xi_i\}_{i=1}^{d} \in \mathbb{R}^d$); then problem (2.166) can

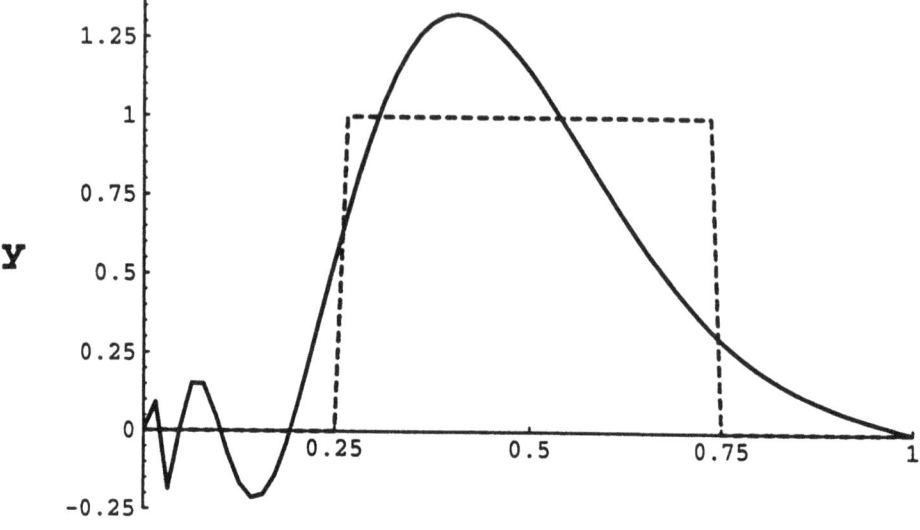

x₂

Fig. 26. Comparison between y_T (...) and y_T^* (—) ($k = 10^5$, $h = \Delta t = 1/64$).

be expressed in *variational* form as follows

$$\begin{cases} \left(\dfrac{\partial y}{\partial t}, \hat{y}\right) + a(y, \hat{y}) = \displaystyle\int_{\Gamma_0} v\hat{y}\, d\Gamma, \ \forall \hat{y} \in H^1(\Omega), \\ y(t) \in H^1(\Omega) \ \text{ a.e. on } (0, T), \ y(0) = 0, \end{cases} \tag{2.172}$$

where

$$a(y, \hat{y}) = \sum_{i=1}^{d} \sum_{j=1}^{d} \int_{\Omega} a_{ij} \frac{\partial y}{\partial x_j} \frac{\partial \hat{y}}{\partial x_i}\, dx \tag{2.173}$$

(actually, all this applies to the case where the coefficients a_{ij} depend on x and t and verify $a_{ij}(x, t) \in L^\infty(Q)$ and

$$\sum_{i=1}^{d} \sum_{j=1}^{d} a_{ij}(x, t)\xi_i\xi_j \geq \alpha|\boldsymbol{\xi}|^2, \ \forall \boldsymbol{\xi} \in \mathbb{R}^d, \ \text{a.e. in } Q$$

with $\alpha > 0$). Therefore, without any further hypothesis on the coefficients a_{ij}, problem (2.166) admits a unique solution $y(v) = y(x, t; v)$ such that

$$y(v) \in L^2(0, T; H^1(\Omega)) \cap C^0([0, T]; L^2(\Omega)). \tag{2.174}$$

To obtain the *approximate controllability* property we shall assume further *regularity* properties for the a_{ij}'s, more specifically we shall assume that

$$a_{ij} \in C^1(\bar{\Omega}), \ \forall 1 \leq i, j \leq d. \tag{2.175}$$

We have then the following

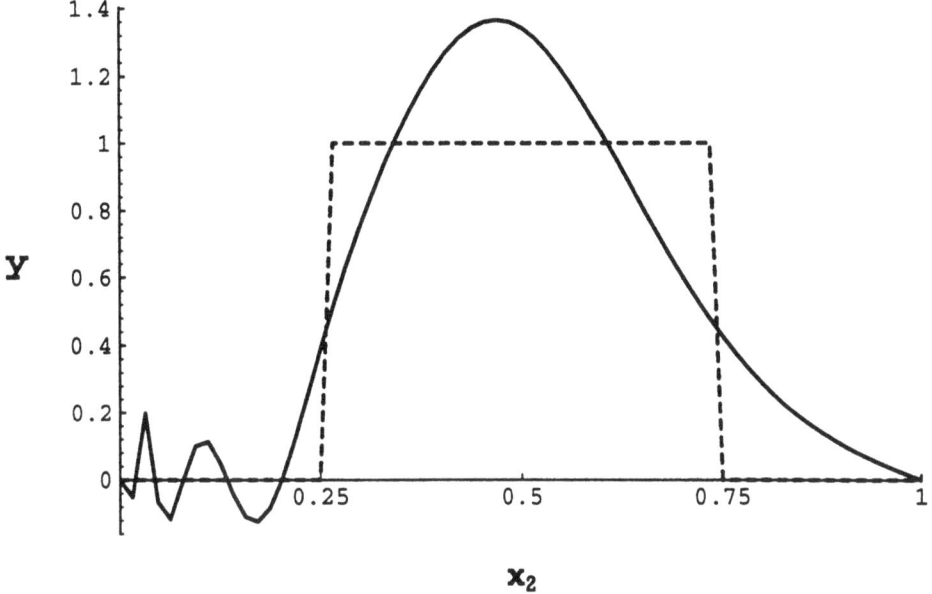

Fig. 27. Comparison between y_T (...) and y_T^* (—) ($k = 10^7$, $h = \Delta t = 1/64$).

Proposition 2.3 *Suppose that coefficients a_{ij} verify* (2.171) *and* (2.175). *Then $y(T; v)$ spans a dense subset of $L^2(\Omega)$ when v spans $L^2(\Sigma_0)$.*

Proof. The proof is similar to the proof of Proposition 2.1 (see Section 2.1). Let us assume therefore that $f \in L^2(\Omega)$ satisfies

$$\int_\Omega y(T; v) f \, dx = 0, \ \forall v \in L^2(\Sigma_0). \tag{2.176}$$

We introduce ψ as the solution of

$$-\frac{\partial \psi}{\partial t} + A^* \psi = 0 \text{ in } Q, \quad \psi(T) = f, \quad \frac{\partial \psi}{\partial n_{A^*}} = 0 \text{ on } \Sigma; \tag{2.177}$$

then (2.176) is equivalent to

$$\psi = 0 \text{ on } \Sigma_0. \tag{2.178}$$

Thanks to the regularity hypothesis (2.175) we can use the *Mizohata's uniqueness theorem* (Mizohata, 1958) (see also Saut and Schoerer (1987)): it follows then from (2.177) and (2.178) that $\psi = 0$, hence $f = 0$ and the proof is completed.

Remark 2.8 The applicability of the Mizohata uniqueness theorem under the only assumption that $a_{ij} \in L^\infty(Q)$ does not seem to have been completely settled, yet.

We can state two basic *controllability problems* both closely related to problems (2.11) and (2.12) in Section 2.1.

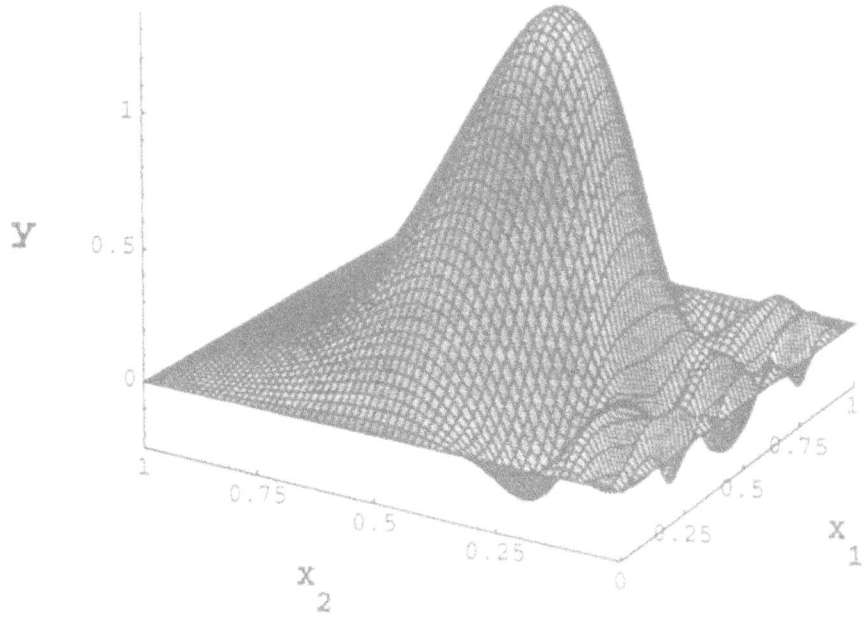

Fig. 28. Graph of the function y_T^* ($k = 10^5$, $h = \Delta t = 1/64$).

The *first* Neumann control problem that we consider is defined by

$$\inf_v \frac{1}{2} \int_{\Sigma_0} v^2 \, \mathrm{d}\Sigma, \tag{2.179}$$

where v is subjected to

$$y(T; v) \in y_T + \beta B; \tag{2.180}$$

in (2.180), $y(t; v)$ is the solution of problem (2.166), the target function y_T belongs to $L^2(\Omega)$, B denotes the closed unit ball of $L^2(\Omega)$ and β is a positive number, arbitrarily small.

The *second* Neumann control problem to be considered is defined by

$$\inf_v \left[\frac{1}{2} \int_{\Sigma_0} v^2 \, \mathrm{d}\Sigma + \frac{1}{2} k \|y(T; v) - y_T\|_{L^2(\Omega)}^2 \right], \tag{2.181}$$

where k is a positive number, arbitrarily large.

Both problems (2.179) and (2.181) admit a *unique* solution. There is however a technical difference between these two problems since problem (2.181) admits a unique solution under the only hypothesis $a_{ij} \in L^\infty(\Omega)$ (and of course the ellipticity property (2.171)), while the existence of a solution for problem (2.179), with β arbitrarily small, requires, so far, some regularity property (such as (2.175)) for the a_{ij}'s. In the following we shall assume that property (2.175) holds, even if this hypothesis is not always necessary.

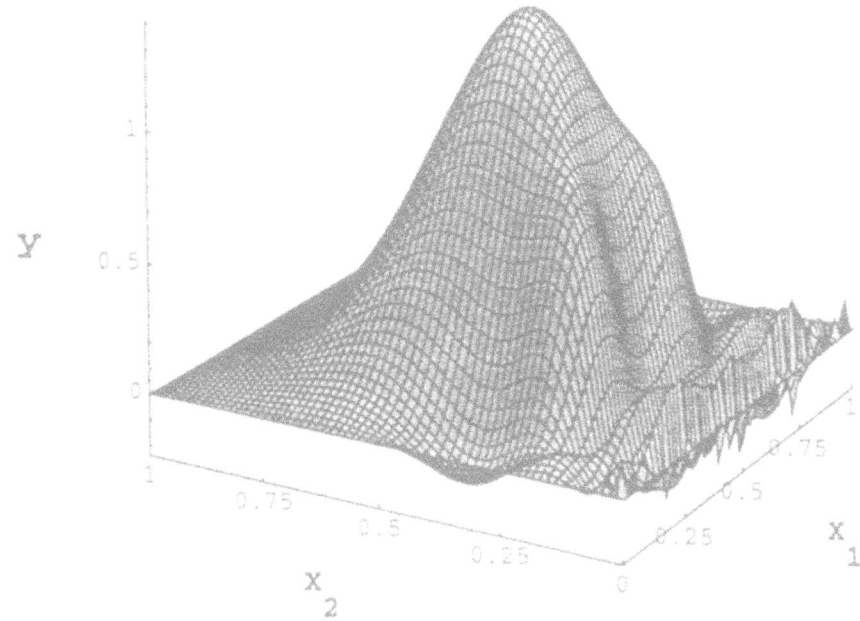

Fig. 29. Graph of the function $y_T^*(k = 10^7, \ h = \Delta t = 1/64)$.

2.8. Neumann control (II): Optimality conditions and dual formulations

The *optimality* system for problem (2.181) is obtained by arguments which are fairly classical (see, e.g., Lions (1968)), as recalled in Section 2.2. Following, precisely, the approach taken in Section 2.2, we introduce the functional $J_k : L^2(\Sigma_0) \to \mathbb{R}$ defined by

$$J_k(v) = \frac{1}{2} \int_{\Sigma_0} v^2 \, d\Sigma + \frac{1}{2} k \|y(T;v) - y_T\|^2_{L^2(\Omega)}. \qquad (2.182)$$

We can show that the derivative J_k' of J_k is defined by

$$(J_k'(v), w)_{L^2(\Sigma_0)} = \int_{\Sigma_0} (v + p)w \, d\Sigma, \ \forall v, w \in L^2(\Sigma_0), \qquad (2.183)$$

where, in (2.183), the *adjoint state function* p is obtained from v via the solution of (2.166) and of the *adjoint state equation*

$$-\frac{\partial p}{\partial t} + A^* p = 0 \text{ in } Q, \quad \frac{\partial p}{\partial n_{A^*}} = 0 \text{ on } \Sigma, \ p(T) = k(y(T) - y_T). \quad (2.184)$$

Suppose now that u is *the* solution of the control problem (2.181); since $J_k'(u) = 0$, we have then the following optimality system satisfied by u and by the corresponding state and adjoint state functions:

$$u = -p|_{\Sigma_0}, \qquad (2.185)$$

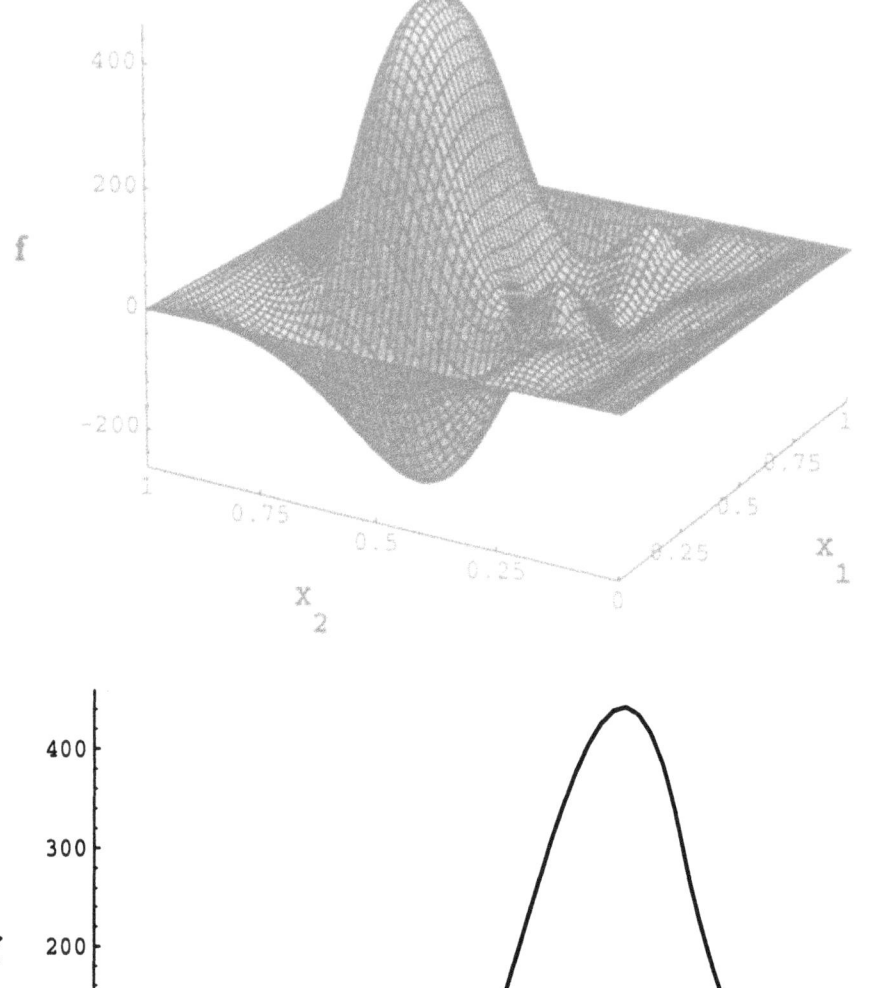

Fig. 30. (a) Graph of $f_h^{\Delta t}(k = 10^5,\ h = \Delta t = 1/64)$. (b) Graph of $x_2 \to f_h^{\Delta t}(0.5, x_2)(k = 10^5,\ h = \Delta t = 1/64)$.

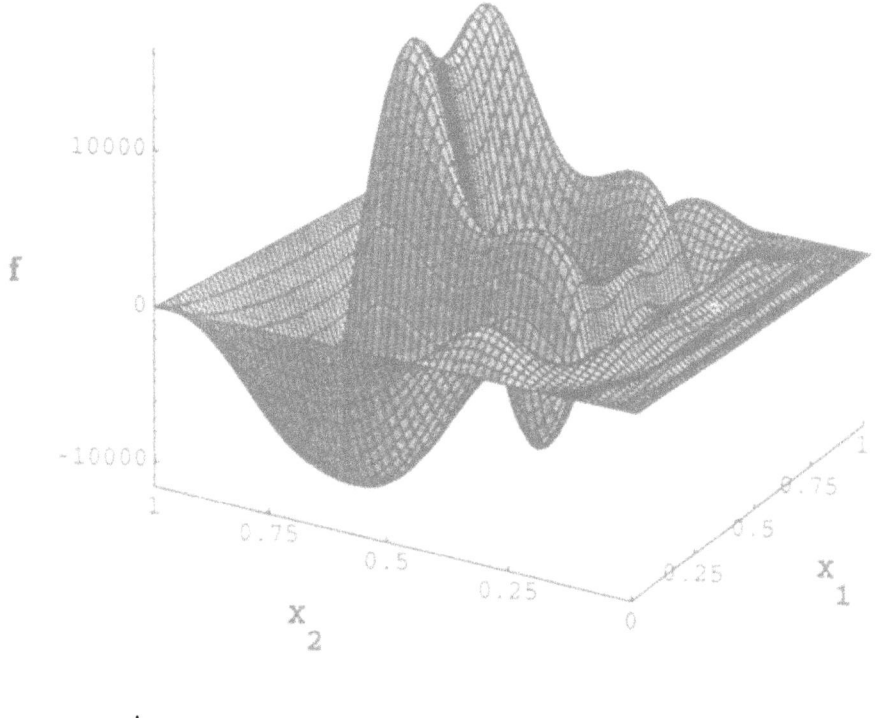

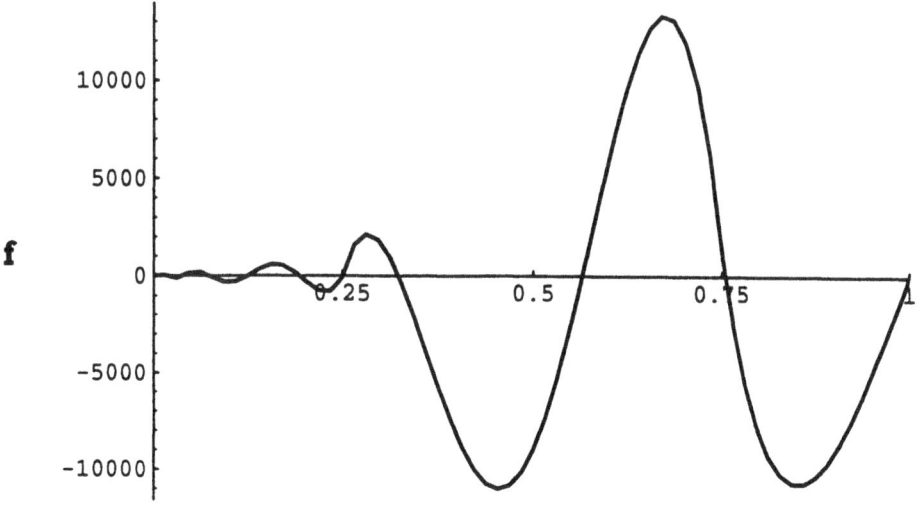

Fig. 31. (a) Graph of $f_h^{\Delta t}(k = 10^7,\ h = \Delta t = 1/64)$. (b) Graph of $x_2 \rightarrow f_h^{\Delta t}(0.5, x_2)(k = 10^7,\ h = \Delta t = 1/64)$.

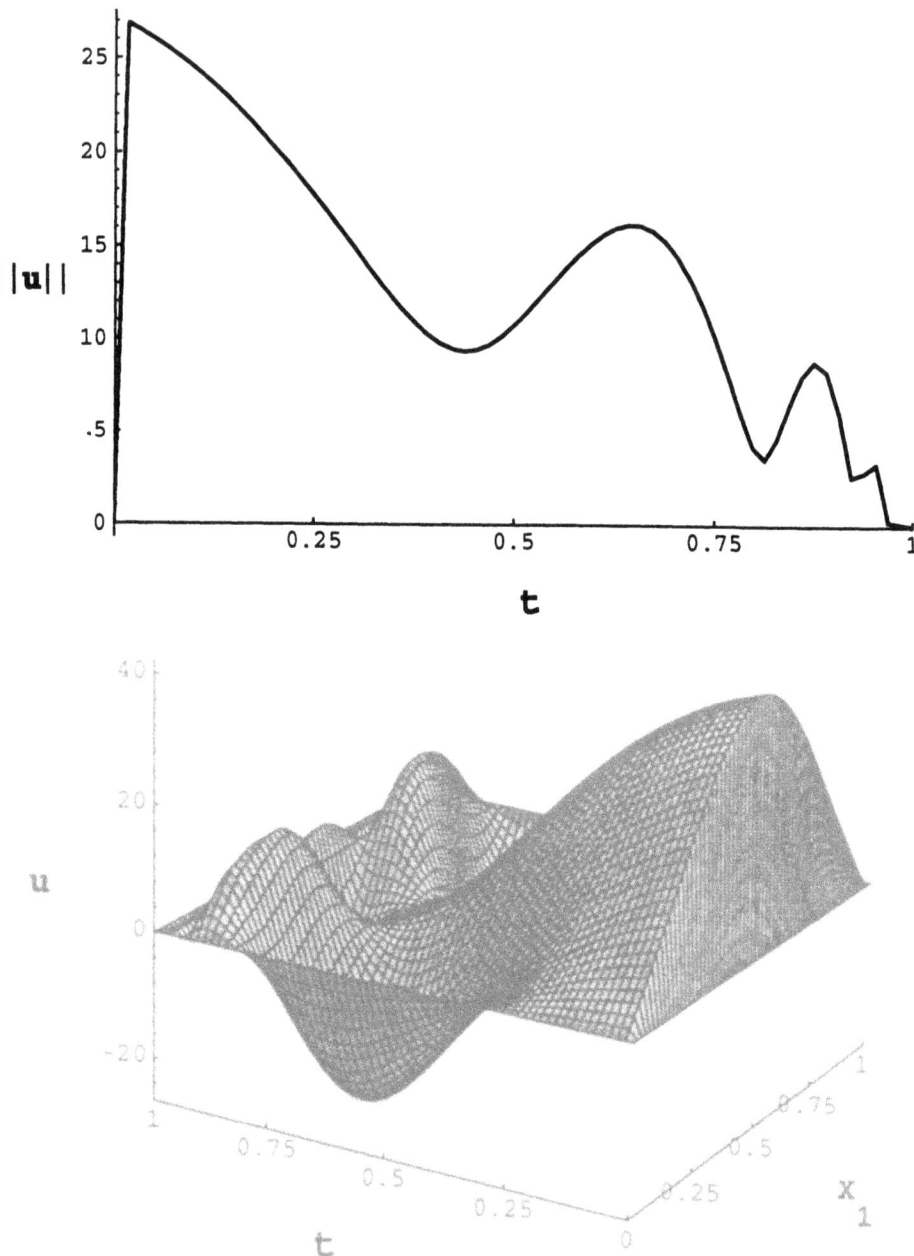

Fig. 32. (a) Graph of $t \to \|u^*(t)\|_{L^2(\Gamma_0)} (k = 10^5, \ h = \Delta t = 1/64)$. (b) Graph of the computed boundary control $(k = 10^5, \ h = \Delta t = 1/64)$.

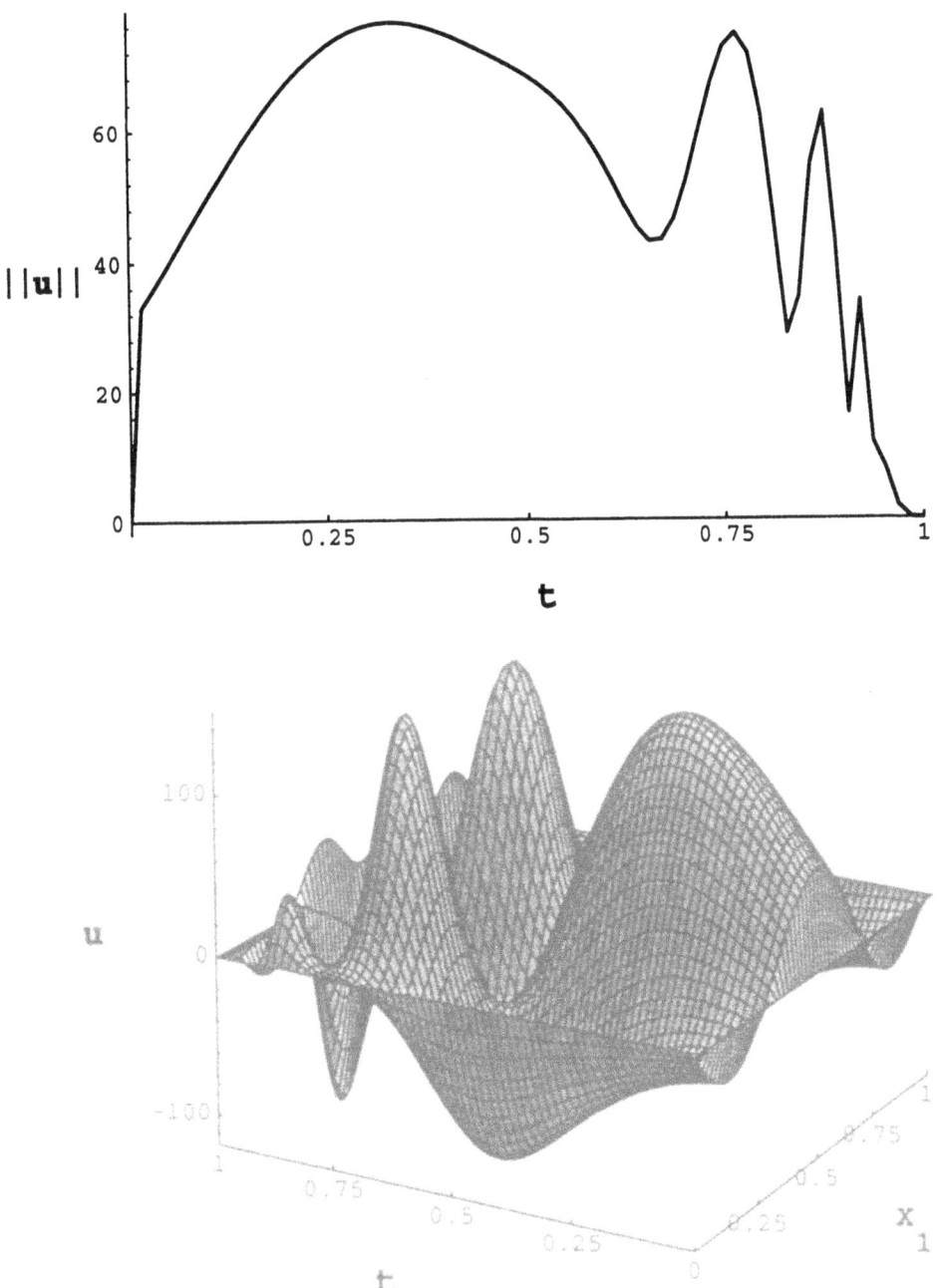

Fig. 33. (a) Graph of $t \to \|u^*(t)\|_{L^2(\Gamma_0)}$ $(k = 10^7, \; h = \Delta t = 1/64)$. (b) Graph of the computed boundary control $(k = 10^7, \; h = \Delta t = 1/64)$.

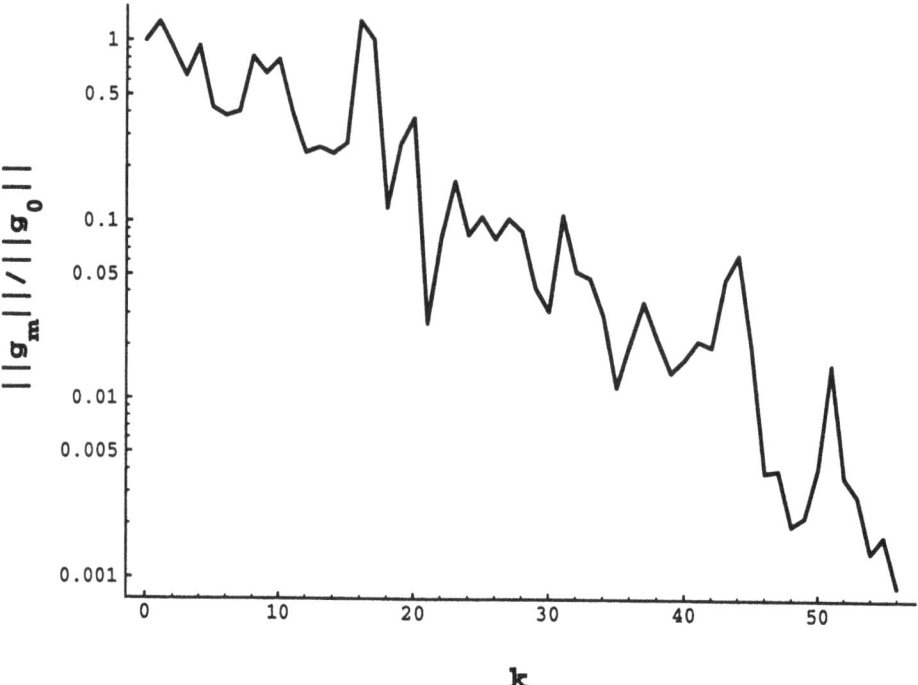

Fig. 34. Variation of $\|g_m\|_{H_0^1(\Omega)}/\|g_0\|_{H_0^1(\Omega)}$ ($k = 10^5$, $h = \Delta t = 1/64$).

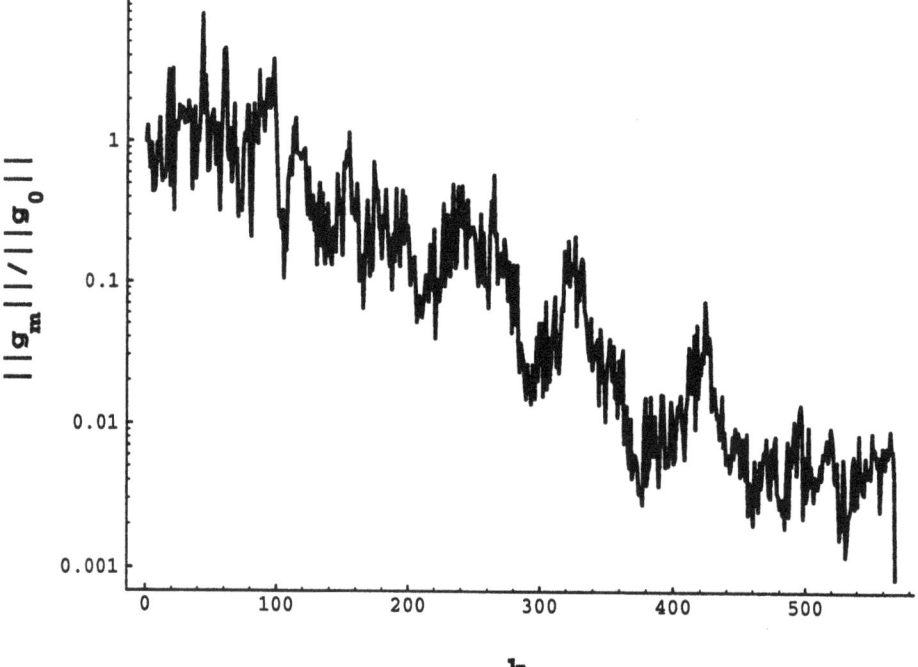

Fig. 35. Variation of $\|g_m\|_{H_0^1(\Omega)}/\|g_0\|_{H_0^1(\Omega)}$ ($k = 10^7$, $h = \Delta t = 1/64$).

$$\frac{\partial y}{\partial t} + Ay = 0 \text{ in } Q, \quad y(0) = 0, \quad \frac{\partial y}{\partial n_A} = 0 \text{ on } \Sigma \backslash \Sigma_0, \quad \frac{\partial y}{\partial n_A} = -p \text{ on } \Sigma_0,$$
$$(2.186)$$

$$-\frac{\partial p}{\partial t} + A^* p = 0 \text{ in } Q, \quad \frac{\partial p}{\partial n_{A^*}} = 0 \text{ on } \Sigma, \quad p(T) = k(y(T) - y_T). \quad (2.187)$$

In order to identify the dual problem of (2.181) we proceed essentially as in Section 2.2. We introduce therefore the operator $\Lambda \in \mathcal{L}(L^2(\Omega); L^2(\Omega))$ defined by

$$\Lambda \hat{f} = -\hat{\varphi}(T), \quad \forall \hat{f} \in L^2(\Omega), \quad (2.188)$$

where, in (2.188), $\hat{\varphi}$ is obtained from $\hat{f}$ as follows.
Solve first

$$-\frac{\partial \hat{\psi}}{\partial t} + A^* \hat{\psi} = 0 \text{ in } Q, \quad \frac{\partial \hat{\psi}}{\partial n_{A^*}} = 0 \text{ on } \Sigma, \quad \hat{\psi}(T) = \hat{f}, \quad (2.189)$$

and then

$$\frac{\partial \hat{\varphi}}{\partial t} + A\hat{\varphi} = 0 \text{ in } Q, \quad \hat{\varphi}(0) = 0, \quad \frac{\partial \hat{\varphi}}{\partial n_A} = 0 \text{ on } \Sigma \backslash \Sigma_0, \quad \frac{\partial \hat{\varphi}}{\partial n_A} = -\hat{\psi} \text{ on } \Sigma_0.$$
$$(2.190)$$

We can easily show that (with obvious notation)

$$\int_\Omega (\Lambda f_1) f_2 \, dx = \int_{\Sigma_0} \psi_1 \psi_2 \, d\Sigma, \quad \forall f_1, f_2 \in L^2(\Omega). \quad (2.191)$$

It follows from (2.191) that operator Λ is *symmetric* and *positive semi-definite*; indeed, it follows from the *Mizohata's uniqueness theorem* that operator Λ is *positive definite* (if (2.175) holds, at least). However, operator Λ is not an isomorphism from $L^2(\Omega)$ onto $L^2(\Omega)$ (implying that, in general, we do not have here exact boundary controllability).

Back to (2.188), we observe that, if we denote by f the function $p(T)$ in (2.187), it follows from the definition of operator Λ that we have

$$k^{-1}f + \Lambda f = -y_T. \quad (2.192)$$

Problem (2.192) is the dual problem of (2.181). From the properties of the operator $k^{-1}I + \Lambda$, problem (2.192) can be solved by a *conjugate gradient algorithm* operating in the space $L^2(\Omega)$; we shall return to this issue in Section 2.9.

The dual problem (2.192) has been obtained by a fairly simple method. Obtaining the dual problem of (2.179) is more complicated. We can use – as already done in previous sections – the *Fenchel–Rockafellar duality theory*; however, in order to introduce (possibly) our readers to other duality techniques we shall derive the dual problem of (2.179) through a *Lagrangian* approach (which is indeed closely related to the Fenchel–Rockafellar method, as shown in, e.g., Rockafellar (1970) and Ekeland and Temam (1974)).

Our starting point is to observe that problem (2.179) is equivalent to

$$\inf_{\{v,z\}} \frac{1}{2} \int_{\Sigma_0} v^2 \, d\Sigma, \tag{2.193}$$

where, in (2.193), the pair $\{v, z\}$ satisfies

$$v \in L^2(\Sigma_0), \tag{2.194}$$

$$z \in y_T + \beta B, \tag{2.195}$$

$$y(T) - z = 0, \tag{2.196}$$

$y(T)$ being obtained from v via the solution of (2.166). The idea here is to 'dualize' the linear constraint (2.196) via an appropriate *Lagrangian functional* and then to compute the corresponding *dual functional*. A Lagrangian functional naturally associated with problem (2.193)–(2.196) is defined by

$$\mathcal{L}(v, z; \mu) = \frac{1}{2} \int_{\Sigma_0} v^2 \, d\Sigma + \int_{\Omega} \mu(y(T) - z) \, dx. \tag{2.197}$$

The dual problem associated with (2.193)–(2.197) is defined by

$$\inf_{\mu \in L^2(\Omega)} J^*(\mu), \tag{2.198}$$

where, in (2.198), the dual functional J^* is defined by

$$J^*(\mu) = - \inf_{\{v,z\}} \mathcal{L}(v, z; \mu), \tag{2.199}$$

where $\{v, z\}$ still satisfies (2.194), (2.195). We clearly have

$$\inf_{\{v,z\}} \mathcal{L}(v, z; \mu) = \inf_{v \in L^2(\Sigma_0)} \left[\frac{1}{2} \int_{\Sigma_0} v^2 \, d\Sigma + \int_{\Omega} y(T)\mu \, dx \right] - \sup_{z \in y_T + \beta B} \int_{\Omega} \mu z \, dx, \tag{2.200}$$

and then

$$\sup_{z \in y_T + \beta B} \int_{\Omega} \mu z \, dx = \sup_{z \in y_T + \beta B} \left[\int_{\Omega} \mu(z - y_T) \, dx + \int_{\Omega} \mu y_T \, dx \right]$$

$$= \beta \|\mu\|_{L^2(\Omega)} + \int_{\Omega} \mu y_T \, dx. \tag{2.201}$$

It remains to evaluate

$$\inf_{v \in L^2(\Sigma_0)} \left[\frac{1}{2} \int_{\Sigma_0} v^2 \, d\Sigma + \int_{\Omega} y(T)\mu \, dx \right]; \tag{2.202}$$

indeed, solving the (linear) control problem (2.202) is quite easy since its unique solution u_μ is characterized (see, e.g., Lions (1968)) by the existence of $\{y_\mu, p_\mu\}$ such that

$$u_\mu = -p_\mu|_{\Sigma_0}, \tag{2.203}$$

$$\frac{\partial y_\mu}{\partial t} + Ay_\mu = 0 \text{ in } Q, \quad y_\mu(0) = 0, \quad \frac{\partial y_\mu}{\partial n_A} = -p_\mu \text{ on } \Sigma_0, \quad \frac{\partial y_\mu}{\partial n_A} = 0 \text{ on } \Sigma \backslash \Sigma_0,$$
(2.204)

$$-\frac{\partial p_\mu}{\partial t} + A^* p_\mu = 0 \text{ in } Q, \quad \frac{\partial p_\mu}{\partial n_{A^*}} = 0 \text{ on } \Sigma, \quad p_\mu(T) = \mu.$$
(2.205)

We have then from (2.202)–(2.205) and from the definition and properties of the operator Λ

$$\inf_{v \in L^2(\Sigma_0)} \left[\frac{1}{2} \int_{\Sigma_0} v^2 \, d\Sigma + \int_\Omega y(T)\mu \, dx \right] = \frac{1}{2} \int_{\Sigma_0} p_\mu^2 \, d\Sigma + \int_\Omega y_\mu(T)\mu \, dx$$
$$= -\frac{1}{2} \int_\Omega (\Lambda\mu)\mu \, dx. \quad (2.206)$$

Combining (2.199), (2.200), (2.201) to (2.206) implies that

$$J^*(\mu) = \frac{1}{2} \int_\Omega (\Lambda\mu)\mu \, dx + \beta \|\mu\|_{L^2(\Omega)} + \int_\Omega y_T \mu \, dx.$$
(2.207)

The dual problem to (2.179) is defined then by

$$\inf_{\hat{f} \in L^2(\Omega)} \left[\frac{1}{2} \int_\Omega (\Lambda\hat{f})\hat{f} \, dx + \beta \|\hat{f}\|_{L^2(\Omega)} + \int_\Omega y_T \hat{f} \, dx \right],$$
(2.208)

or, equivalently, by the following *variational inequality*

$$\begin{cases} f \in L^2(\Omega), \\ \int_\Omega (\Lambda f)(\hat{f} - f) \, dx + \beta \|\hat{f}\|_{L^2(\Omega)} - \beta \|f\|_{L^2(\Omega)} \\ \quad + \int_\Omega y_T(\hat{f} - f) \, dx \geq 0, \ \forall \hat{f} \in L^2(\Omega). \end{cases}$$
(2.209)

Once f is known, obtaining the solution u of problem (2.179) is quite easy, since

$$u = -p|_{\Sigma_0},$$
(2.210)

where, in (2.210), p is the solution of

$$-\frac{\partial p}{\partial t} + A^* p = Q, \quad \frac{\partial p}{\partial n_{A^*}} = 0 \text{ on } \Sigma, \quad p(T) = f.$$
(2.211)

The numerical solution of problem (2.208), (2.209) will be discussed in Section 2.10.

Remark 2.9 Proving *directly* the existence and uniqueness of the solution f of problem (2.208), (2.209) is not obvious. Actually, proving it without some *regularity* hypothesis on the a_{ij}'s (like (2.175)) is still an *open* question.

2.9. Neumann control (III): Conjugate gradient solution of the dual problem (2.192)

We shall address in this section the *iterative solution* of the control problem (2.181), via the solution of its *dual problem* (2.192). From the properties of Λ (*symmetry* and *positive definiteness*) problem (2.192) can be solved by a *conjugate gradient algorithm* operating in the space $L^2(\Omega)$. Such an algorithm is given below; we will use there a *variational description* in order to facilitate *finite element implementations* of the algorithm.

Description of the algorithm

$$f^0 \text{ is given in } L^2(\Omega); \tag{2.212}$$

solve

$$\begin{cases} -\displaystyle\int_\Omega \frac{\partial \psi^0}{\partial t}(t)z\,\mathrm{d}x + a(z, \psi^0(t)) = 0, \ \forall z \in H^1(\Omega), \\ \psi^0(t) \in H^1(\Omega), \quad \text{a.e. on } (0, T), \end{cases} \tag{2.213}_1$$

$$\psi^0(T) = f^0, \tag{2.213}_2$$

and then

$$\begin{cases} \displaystyle\int_\Omega \frac{\partial \varphi^0}{\partial t}(t)z\,\mathrm{d}x + a(\varphi^0(t), z) = -\int_{\Gamma_0} \psi^0(t)z\,\mathrm{d}\Gamma, \ \forall z \in H^1(\Omega), \\ \varphi^0(t) \in H^1(\Omega), \quad \text{a.e. on } (0, T), \end{cases} \tag{2.214}_1$$

$$\varphi^0(0) = 0. \tag{2.214}_2$$

Solve next

$$\begin{cases} g^0 \in L^2(\Omega), \\ \displaystyle\int_\Omega g^0 v\,\mathrm{d}x = k^{-1}\int_\Omega f^0 v\,\mathrm{d}x + \int_\Omega (y_T - \varphi^0(T))v\,\mathrm{d}x, \ \forall v \in L^2(\Omega), \end{cases} \tag{2.215}$$

and set

$$w^0 = g^0. \tag{2.216}$$

Then, for $n \geq 0$, assuming that f^n, g^n, w^n are known, compute f^{n+1}, g^{n+1}, w^{n+1} as follows.
Solve

$$\begin{cases} -\displaystyle\int_\Omega \frac{\partial \bar{\psi}^n}{\partial t}(t)z\,\mathrm{d}x + a(z, \bar{\psi}^n(t)) = 0, \ \forall z \in H^1(\Omega), \\ \bar{\psi}^n(t) \in H^1(\Omega), \quad \text{a.e. on } (0, T) \quad, \end{cases} \tag{2.217}_1$$

$$\bar{\psi}^n(T) = w^n, \tag{2.217}_2$$

and then

$$\begin{cases} \int_{\Omega} \frac{\partial \bar{\varphi}^n}{\partial t}(t)z \, dx + a(\bar{\varphi}^n(t), z) = -\int_{\Gamma_0} \bar{\psi}^n(t)z \, d\Gamma, \ \forall z \in H^1(\Omega), \\ \bar{\varphi}^n(t) \in H^1(\Omega), \quad \text{a.e. on } (0,T) \quad , \end{cases} \tag{2.218}_1$$

$$\bar{\varphi}^n(0) = 0. \tag{2.218}_2$$

Solve next

$$\begin{cases} \bar{g}^n \in L^2(\Omega), \\ \int_{\Omega} \bar{g}^n v \, dx = k^{-1} \int_{\Omega} w^n v \, dx - \int_{\Omega} \bar{\varphi}^n(T)v \, dx, \ \forall v \in L^2(\Omega), \end{cases} \tag{2.219}$$

and compute

$$\rho_n = \int_{\Omega} |g^n|^2 \, dx \Big/ \int_{\Omega} \bar{g}^n w^n \, dx. \tag{2.220}$$

Set then

$$f^{n+1} = f^n - \rho_n w^n, \tag{2.221}$$

$$g^{n+1} = g^n - \rho_n \bar{g}^n. \tag{2.222}$$

If $\|g^{n+1}\|_{L^2(\Omega)}/\|g^0\|_{L^2(\Omega)} \leq \epsilon$, *take* $f = f^{n+1}$; *else, compute*

$$\gamma_n = \int_{\Omega} |g^{n+1}|^2 \, dx \Big/ \int_{\Omega} |g^n|^2 \, dx \tag{2.223}$$

and update w^{n+1} *via*

$$w^{n+1} = g^{n+1} + \gamma_n w^n. \tag{2.224}$$

Do $n = n + 1$ *and go to* (2.217).

In (2.212)–(2.224), the bilinear form $a(\cdot, \cdot)$ is defined by (2.173).

It is fairly easy to derive a fully discrete analogue of algorithm (2.212)–(2.224), obtained by combining finite elements for the space discretization and finite differences for the time discretization. We shall then obtain a variation of algorithm (2.125)–(2.151) (see Section 2.5), which is itself the fully discrete analogue of algorithm (2.42)–(2.54) (see Section 2.3). Actually, algorithm (2.212)–(2.224) is easier to implement than (2.42)–(2.54) since it operates in $L^2(\Omega)$, instead of $H_0^1(\Omega)$; no preconditioning is required, thus.

2.10. Neumann control (IV): Iterative solution of the dual problem (2.208), (2.209)

Problem (2.208), (2.209) can also be formulated as

$$-y_T \in \Lambda f + \beta \partial j(f), \tag{2.225}$$

which is a *multivalued* equation in $L^2(\Omega)$. In (2.225), $\partial j(\cdot)$ is the subgradient of the convex functional $j(\cdot)$ defined by

$$j(\hat{f}) = \left(\int_\Omega |\hat{f}|^2 \, dx \right)^{1/2}, \ \forall \hat{f} \in L^2(\Omega).$$

As done in preceding sections we associate with the *elliptic equation* (2.225) the *initial value problem*

$$\begin{cases} \dfrac{\partial f}{\partial \tau} + \Lambda f + \beta \partial j(f) = -y_T, \\ f(0) = f_0. \end{cases} \tag{2.226}$$

To obtain the steady state solution of (2.226), i.e. the solution of (2.225), we shall use the following algorithm obtained, from (2.226), by application of the *Peaceman–Rachford time discretization scheme* (where $\Delta\tau(> 0)$ is a pseudo-time discretization step)

$$f^0 = f_0; \tag{2.227}$$

then, for $n \geq 0$, compute $f^{n+1/2}$ and f^{n+1}, from f^n, via

$$\dfrac{f^{n+1/2} - f^n}{\Delta\tau/2} + \Lambda f^n + \beta \partial j(f^{n+1/2}) = -y_T, \tag{2.228}$$

$$\dfrac{f^{n+1} - f^{n+1/2}}{\Delta\tau/2} + \Lambda f^{n+1} + \beta \partial j(f^{n+1/2}) = -y_T. \tag{2.229}$$

Problem (2.229) can be reformulated as

$$\dfrac{f^{n+1} - 2f^{n+1/2} + f^n}{\Delta\tau/2} + \Lambda f^{n+1} = \Lambda f^n; \tag{2.230}$$

problem (2.230) being a simple variation of problem (2.192) can be solved by an algorithm similar to (2.212)–(2.224). On the other hand, problem (2.228) can be (easily) solved by the methods used in Section 1.8.8 to solve problems (1.240), (1.243), (1.245) which are simple variants of problems (2.228).

3. CONTROL OF THE STOKES SYSTEM

3.1. Generalities. Synopsis

The control problems and methods which were discussed in Section 2 were mostly concerned with systems governed by *linear diffusion* equations of the *parabolic type*, associated with *second-order elliptic operators*. Indeed, these methods have been applied, in, e.g., Berggren (1992) and Berggren and Glowinski (1994), to the solution of *approximate boundary controllability* problems for systems governed by *strongly* advection dominated *linear advection–diffusion equations.* These methods can also be applied to *systems* of linear advection–diffusion equations and to *higher-order parabolic equations* (or *systems* of such equations). Motivated by the solution of controllability problems for the *Navier–Stokes equations* modelling *incompressible* viscous flow, we will now discuss controllability issues for a system of partial differential equations *which is not of the Cauchy–Kowalewski type,* namely the classical *Stokes* system.

3.2. Formulation of the Stokes system. A fundamental controllability result

In the following, we equip the Euclidian space $\mathbb{R}^d(d \geq 2)$ with its classical scalar product and with the corresponding norm, i.e.

$$\mathbf{a} \cdot \mathbf{b} = \sum_{i=1}^{d} a_i b_i, \ \forall \mathbf{a} = \{a_i\}_{i=1}^d, \ \mathbf{b} = \{b_i\}_{i=1}^d \in \mathbb{R}^d; \ |\mathbf{a}| = (\mathbf{a} \cdot \mathbf{a})^{1/2}, \ \forall \mathbf{a} \in \mathbb{R}^d.$$

We suppose from now on that the control $\mathbf{v}$ is distributed over Ω, with its support in $\bar{\mathcal{O}} \subset \Omega$ (as in Sections 1.1 to 1.8, whose notation is kept). The *state equation* is given by

$$\begin{cases} \dfrac{\partial \mathbf{y}}{\partial t} - \Delta \mathbf{y} = \mathbf{v}\chi_{\mathcal{O}} - \nabla \pi \text{ in } Q, \\ \nabla \cdot \mathbf{y} = 0 \text{ in } Q, \end{cases} \tag{3.1}$$

subjected to the following *initial* and *boundary* conditions

$$\mathbf{y}(0) = \mathbf{0}, \quad \mathbf{y} = \mathbf{0} \text{ on } \Sigma. \tag{3.2}$$

In (3.1) we shall assume that

$$\mathbf{v} \in \mathcal{V} = \text{ closed subpsace of } L^2(\mathcal{O} \times (0,T))^d. \tag{3.3}$$

To fix ideas we shall take $d = 3$, and consider the following cases for $\mathcal{V}$:

$$\mathcal{V} = L^2(\mathcal{O} \times (0,T))^3, \tag{3.4}$$

$$\mathcal{V} = \{v_1, v_2, 0\}, \quad \{v_1, v_2\} \in L^2(\mathcal{O} \times (0,T))^2, \tag{3.5}$$

$$\mathcal{V} = \{v_1, 0, 0\}, \quad v_1 \in L^2(\mathcal{O} \times (0,T)). \tag{3.6}$$

Problem (3.1), (3.2) has a *unique* solution, such that (in particular)

$$\begin{cases} \mathbf{y}(t; \mathbf{v}) \in L^2(0,T; (H_0^1(\Omega))^3), \ \nabla \cdot \mathbf{y} = 0, \\ \dfrac{\partial \mathbf{y}}{\partial t}(t; \mathbf{v}) \in L^2(0,T; V'), \end{cases} \tag{3.7}$$

where V' is the dual space of V with

$$V = \{\varphi \mid \varphi \in (H_0^1(\Omega))^3, \nabla \cdot \varphi = 0\}. \tag{3.8}$$

It follows from (3.7) that

$$t \to \mathbf{y}(t; \mathbf{v}) \text{ belongs to } C^0([0, T]; H), \tag{3.9}$$

where

$$H = \text{ closure of } V \text{ in } (L^2(\Omega))^3$$
$$= \{\varphi \mid \varphi \in (L^2(\Omega))^3, \ \nabla \cdot \varphi = 0, \ \varphi \cdot \mathbf{n} = 0 \text{ on } \Gamma\} \tag{3.10}$$

(where $\mathbf{n}$ denotes the outward unit normal vector at Γ).

We are now going to prove the following

Proposition 3.1 *If V is defined by either (3.4) or (3.5), then the space spanned by $\mathbf{y}(T; \mathbf{v})$ is dense in H.*

Proof. It suffices to prove the above results for the case (3.5). Let us therefore consider $\mathbf{f} \in H$ such that

$$\int_\Omega \mathbf{y}(T; \mathbf{v}) \cdot \mathbf{f} \, dx = 0, \ \forall \mathbf{v} \in \mathcal{V}. \tag{3.11}$$

To $\mathbf{f}$ we associate the solution ψ of the following *backward* Stokes problem

$$\begin{cases} -\dfrac{\partial \psi}{\partial t} - \Delta \psi = -\nabla \sigma \text{ in } Q, \\ \nabla \cdot \psi = 0 \text{ in } Q, \end{cases} \tag{3.12}$$

$$\psi(T) = \mathbf{f}, \ \psi = \mathbf{0} \text{ on } \Sigma. \tag{3.13}$$

Multiplying by $\mathbf{y} = \mathbf{y}(\mathbf{v})$ the first equation in (3.12) and integrating by parts we find that

$$\iint_{\mathcal{O}\times(0,T)} \psi \cdot \mathbf{v} \, dx \, dt = 0, \ \forall \mathbf{v} \in \mathcal{V}. \tag{3.14}$$

Therefore

$$\psi_1 = \psi_2 = 0 \text{ in } \mathcal{O} \times (0, T). \tag{3.15}$$

But ψ is (among other things) *continuous* in t and *real analytic* in x in $\Omega \times (0, T)$, so that (3.15) implies that

$$\psi_1 = \psi_2 = 0 \text{ in } \Omega \times (0, T). \tag{3.16}$$

Since $\nabla \cdot \psi = 0$, it follows from (3.16) that $\partial \psi_3 / \partial x_3 = 0$ in $\Omega \times (0, T)$, and since $\psi_3 = 0$ on Σ, then $\psi_3 = 0$ in $\Omega \times (0, T)$, so that $\mathbf{f} = \mathbf{0}$, which completes the proof. □

Remark 3.1 The above density result does not always hold if $\mathcal{V}$ is defined by (3.6), as proven by I. Diaz and Fursikov (1994).

Remark 3.2 Proposition 3.1 was proved in the lectures of the second author at *Collège de France* in 1990/91. Other results along these lines are due to Fursikov (1992).

The *density* result in Proposition 3.1 implies (at least) *approximate controllability*. Thus, we shall formulate and discuss, in the following sections, two approximate controllability problems.

3.3. Two approximate controllability problems

The *first problem* is defined by

$$\min_{\mathbf{v}\in\mathcal{U}_f} \frac{1}{2} \iint_{\mathcal{O}\times(0,T)} |\mathbf{v}|^2 \, dx \, dt, \tag{3.17}$$

where

$$\mathcal{U}_f = \{\mathbf{v} \mid \mathbf{v} \in \mathcal{V}, \{\mathbf{v}, \mathbf{y}\} \text{ verifies } (3.1), (3.2) \text{ and } \mathbf{y}(T) \in \mathbf{y}_T + \beta B_H\}; \tag{3.18}$$

in (3.18), $\mathbf{y}_T$ is given in H, β is an arbitrary small positive number, B_H is the closed unit ball of H and – to fix ideas – $\mathcal{V}$ is defined by (3.5).

The *second problem* is obtained by *penalization* of the final condition $\mathbf{y}(T) = \mathbf{y}_T$; we have then

$$\min_{\mathbf{v}\in\mathcal{V}} \left[\frac{1}{2} \iint_{\mathcal{O}\times(0,T)} |\mathbf{v}|^2 \, dx \, dt + \frac{1}{2} k \int_\Omega |\mathbf{y}(T) - \mathbf{y}_T|^2 \, dx \right], \tag{3.19}$$

where, in (3.19), k is an arbitrary large positive number, $\mathbf{y}$ is obtained from $\mathbf{v}$ via (3.1), (3.2) and $\mathcal{V}$ is as above.

It follows from Proposition 3.1 that both control problems (3.17) and (3.19) *have a unique solution*.

3.4. Optimality conditions and dual problems

We start with problem (3.19), since it is simpler than problem (3.17). If we denote by J_k the cost functional in (3.19), we have

$$\lim_{\substack{\theta\to 0 \\ \theta\neq 0}} \frac{J_k(\mathbf{v}+\theta\mathbf{w}) - J_k(\mathbf{v})}{\theta} = (J_k'(\mathbf{v}), \mathbf{w}) = \iint_{\mathcal{O}\times(0,T)} (\mathbf{v}-\mathbf{p})\cdot\mathbf{w} \, dx \, dt, \tag{3.20}$$

where, in (3.20), the *adjoint velocity field* $\mathbf{p}$ is solution of the following *backward Stokes problem*

$$\begin{cases} -\dfrac{\partial \mathbf{p}}{\partial t} - \Delta\mathbf{p} + \nabla\sigma = \mathbf{0} \text{ in } Q, \\ \nabla\cdot\mathbf{p} = 0 \text{ in } Q, \end{cases} \tag{3.21}$$

$$\mathbf{p} = \mathbf{0} \text{ on } \Sigma, \quad \mathbf{p}(T) = k(\mathbf{y}_T - \mathbf{y}(T)). \tag{3.22}$$

Suppose now that $\mathbf{u}$ is the unique solution of problem (3.19); it is *characterized* by

$$\begin{cases} \mathbf{u} \in \mathcal{V}, \\ (J_k'(\mathbf{u}), \mathbf{w}) = 0, \ \forall \mathbf{w} \in \mathcal{V}, \end{cases} \tag{3.23}$$

which implies in turn that the *optimal triple* $\{\mathbf{u}, \mathbf{y}, \mathbf{p}\}$ is characterized by

$$u_1 = p_1|_{\mathcal{O}}, \ u_2 = p_2|_{\mathcal{O}}, \ u_3 = 0, \tag{3.24}$$

$$\begin{cases} \dfrac{\partial \mathbf{y}}{\partial t} - \Delta \mathbf{y} + \boldsymbol{\nabla}\pi = \mathbf{u}\chi_{\mathcal{O}} \text{ in } Q, \\ \boldsymbol{\nabla} \cdot \mathbf{y} = 0 \text{ in } Q, \end{cases} \tag{3.25}$$

$$\mathbf{y}(0) = \mathbf{0}, \ \mathbf{y} = \mathbf{0} \text{ on } \Sigma, \tag{3.26}$$

to be completed by (3.21), (3.22).

To obtain the *dual problem* of (3.19) from the above optimality conditions we proceed as in the preceding sections by introducing an operator $\Lambda \in \mathcal{L}(H; H)$ defined as follows:

$$\Lambda\hat{\mathbf{f}} = \hat{\boldsymbol{\varphi}}(T), \ \forall \hat{\mathbf{f}} \in H, \tag{3.27}$$

where to obtain $\hat{\boldsymbol{\varphi}}(T)$ we solve first

$$\begin{cases} -\dfrac{\partial \hat{\boldsymbol{\psi}}}{\partial t} - \Delta \hat{\boldsymbol{\psi}} + \boldsymbol{\nabla}\hat{\sigma} = \mathbf{0} \text{ in } Q, \\ \boldsymbol{\nabla} \cdot \hat{\boldsymbol{\psi}} = 0 \text{ in } Q, \end{cases} \tag{3.28}$$

$$\hat{\boldsymbol{\psi}}(T) = \hat{\mathbf{f}}, \ \hat{\boldsymbol{\psi}} = \mathbf{0} \text{ on } \Sigma, \tag{3.29}$$

and then

$$\begin{cases} \dfrac{\partial \hat{\boldsymbol{\varphi}}}{\partial t} - \Delta \hat{\boldsymbol{\varphi}} + \boldsymbol{\nabla}\hat{\pi} = \{\hat{\psi}_1, \hat{\psi}_2, 0\}\chi_{\mathcal{O}} \text{ in } Q, \\ \boldsymbol{\nabla} \cdot \hat{\boldsymbol{\varphi}} = 0 \text{ in } Q, \end{cases} \tag{3.30}$$

$$\hat{\boldsymbol{\varphi}}(0) = \mathbf{0}, \ \hat{\boldsymbol{\varphi}} = \mathbf{0} \text{ on } \Sigma \tag{3.31}$$

(the two above *Stokes problems* are *well-posed*).

Integrating by parts in time and using Green's formula we can show that (with obvious notation) we have

$$\int_\Omega (\Lambda\hat{\mathbf{f}}) \cdot \hat{\mathbf{f}}' \, dx = \iint_{\mathcal{O}\times(0,T)} (\hat{\psi}_1\hat{\psi}_1' + \hat{\psi}_2\hat{\psi}_2') \, dx \, dt, \ \forall \hat{f}, \ \hat{f}' \in H. \tag{3.32}$$

It follows from relation (3.32) that the operator Λ is *symmetric* and *positive semi-definite* over H; indeed using the approach taken in Section 3.2 to prove Proposition 3.1, we can show that Λ is *positive definite* over H.

Back to the *optimality conditions*, let us denote by $\mathbf{f}$ the function $\mathbf{p}(T)$; it follows then from (3.22) and from the definition of Λ that $\mathbf{f}$ satisfies

$$k^{-1}\mathbf{f} + \Lambda\mathbf{f} = \mathbf{y}_T \tag{3.33}$$

which is precisely the dual problem of (3.19).

From the symmetry of Λ, problem (3.33) can be solved by a *conjugate gradient algorithm* operating in the space H.

Consider control problem (3.17); applying the *Fenchel–Rockafellar duality theory* it can be shown that the unique solution **u** of problem (3.17) can be obtained via

$$u_1 = p_1 \chi_\mathcal{O}, \quad u_2 = p_2 \chi_\mathcal{O}, \quad u_3 = 0, \tag{3.34}$$

where, in (3.34), **p** is the solution of the backward Stokes problem

$$\begin{cases} -\dfrac{\partial \mathbf{p}}{\partial t} - \Delta \mathbf{p} + \nabla \sigma = \mathbf{0} \text{ in } Q, \\ \nabla \cdot \mathbf{p} = 0 \text{ in } Q, \end{cases} \tag{3.35}$$

$$\mathbf{p}(T) = \mathbf{f}, \quad \mathbf{p} = \mathbf{0} \text{ on } \Sigma, \tag{3.36}$$

where, in (3.36), **f** is *the* solution of the following *variational inequality*

$$\begin{cases} \mathbf{f} \in H; \ \forall \hat{\mathbf{f}} \in H, \quad \text{we have} \\ \displaystyle\int_\Omega (\Lambda \mathbf{f}) \cdot (\hat{\mathbf{f}} - \mathbf{f}) \, d\mathbf{x} + \beta \|\hat{\mathbf{f}}\|_H - \beta \|\mathbf{f}\|_H \geq \displaystyle\int_\Omega \mathbf{y}_T \cdot (\hat{\mathbf{f}} - \mathbf{f}) \, d\mathbf{x} \end{cases} \tag{3.37}$$

where $\|\mathbf{f}\|_H = (\int_\Omega |\mathbf{f}|^2 \, d\mathbf{x})^{1/2}$.

Problem (3.37) can be viewed as the *dual* of problem (3.17).

3.5. Iterative solution of the control problem

The various *primal* or *dual* control problems considered in Sections 3.3 and 3.4 can be solved by variants of the algorithms which have been used to solve their scalar diffusion analogues; these algorithms have been described in Section 1.8. Here we shall focus on the *direct solution* of the control problem (3.19), by a *conjugate gradient algorithm*, since we used this approach to solve the test problem discussed in Section 3.7. The unique solution **u** of the control problem (3.19) is *characterized* as also being the unique solution of the linear variational problem (3.23). From the properties of the functional J_k, this problem is a particular case of problem (1.121) in Section 1.8.2; applying thus algorithm (1.122)–(1.129) to problem (3.23) we obtain:

$$\mathbf{u}^0 \text{ chosen in } V; \tag{3.38}$$

solve

$$\begin{cases} \dfrac{\partial \mathbf{y}^0}{\partial t} - \Delta \mathbf{y}^0 + \nabla \pi^0 = \mathbf{u}^0 \chi_\mathcal{O} \text{ in } Q, \\ \nabla \cdot \mathbf{y}_0 = 0 \text{ in } Q, \end{cases} \tag{3.39}$$

$$\mathbf{y}^0(0) = \mathbf{0}, \quad \mathbf{y}^0 = \mathbf{0} \text{ on } \Sigma, \tag{3.40}$$

and then

$$\begin{cases} -\dfrac{\partial \mathbf{p}^0}{\partial t} - \Delta \mathbf{p}^0 + \nabla \sigma^0 = \mathbf{0} \text{ in } Q, \\ \nabla \cdot \mathbf{p}^0 = 0 \text{ in } Q, \end{cases} \tag{3.41}$$

$$\mathbf{p}^0 = \mathbf{0} \text{ on } \Sigma, \ \mathbf{p}^0(T) = k(\mathbf{y}_T - \mathbf{y}^0(T)). \tag{3.42}$$

Solve now

$$\begin{cases} \mathbf{g}^0 \in V, \\ \iint_{\mathcal{O} \times (0,T)} \mathbf{g}^0 \cdot \mathbf{v} \, dx \, dt = \iint_{\mathcal{O} \times (0,T)} (\mathbf{u}^0 - \mathbf{p}^0) \cdot \mathbf{v} \, dx \, dt, \ \forall \mathbf{v} \in V, \end{cases} \tag{3.43}$$

and set

$$\mathbf{w}^0 = \mathbf{g}^0. \tag{3.44}$$

Then for $n \geq 0$, assuming that $\mathbf{u}^n$, $\mathbf{g}^n$, $\mathbf{w}^n$ are known, we obtain $\mathbf{u}^{n+1}$, $\mathbf{g}^{n+1}$, $\mathbf{w}^{n+1}$ as follows.
  Solve

$$\begin{cases} \dfrac{\partial \bar{\mathbf{y}}^n}{\partial t} - \Delta \bar{\mathbf{y}}^n + \nabla \bar{\pi}^n = \mathbf{w}^n \chi_{\mathcal{O}} \text{ in } Q, \\ \nabla \cdot \bar{\mathbf{y}}^n = 0 \text{ in } Q, \end{cases} \tag{3.45}$$

$$\bar{\mathbf{y}}^n(0) = \mathbf{0}, \ \bar{\mathbf{y}}^n = \mathbf{0} \text{ on } \Sigma, \tag{3.46}$$

and then

$$\begin{cases} -\dfrac{\partial \bar{\mathbf{p}}^n}{\partial t} - \Delta \bar{\mathbf{p}}^n + \nabla \bar{\sigma}^n = \mathbf{0} \text{ in } Q, \\ \nabla \cdot \bar{\mathbf{p}}^n = 0 \text{ in } Q, \end{cases} \tag{3.47}$$

$$\bar{\mathbf{p}}^n = \mathbf{0} \text{ on } \Sigma, \ \bar{\mathbf{p}}^n(T) = -k \bar{\mathbf{y}}^n(T). \tag{3.48}$$

Solve now

$$\begin{cases} \bar{\mathbf{g}}^n \in V, \\ \iint_{\mathcal{O} \times (0,T)} \bar{\mathbf{g}}^n \cdot \mathbf{v} \, dx \, dt = \iint_{\mathcal{O} \times (0,T)} (\bar{\mathbf{u}}^n - \bar{\mathbf{p}}^n) \cdot \mathbf{v} \, dx \, dt, \ \forall \mathbf{v} \in V, \end{cases} \tag{3.49}$$

and compute

$$\rho_n = \iint_{\mathcal{O} \times (0,T)} |\mathbf{g}^n|^2 \, dx \, dt \bigg/ \iint_{\mathcal{O} \times (0,T)} \bar{\mathbf{g}}^n \cdot \mathbf{w}^n \, dx \, dt, \tag{3.50}$$

$$\mathbf{u}^{n+1} = \mathbf{u}^n - \rho_n \mathbf{w}^n, \tag{3.51}$$

$$\mathbf{g}^{n+1} = \mathbf{g}^n - \rho_n \bar{\mathbf{g}}^n. \tag{3.52}$$

If $\|\mathbf{g}^{n+1}\|_{L^2(\mathcal{O} \times (0,T))^d}/\|\mathbf{g}^0\|_{L^2(\mathcal{O} \times (0,T))^d} \leq \epsilon$ take $\mathbf{u} = \mathbf{u}^{n+1}$; else, compute

$$\gamma_n = \iint_{\mathcal{O} \times (0,T)} |\mathbf{g}^{n+1}|^2 \, dx \, dt \bigg/ \iint_{\mathcal{O} \times (0,T)} |\mathbf{g}^n|^2 \, dx \, dt \tag{3.53}$$

and update $\mathbf{w}^n$ *by*

$$\mathbf{w}^{n+1} = \mathbf{g}^{n+1} + \gamma_n \mathbf{w}^n. \tag{3.54}$$

Do $n = n + 1$ *and go to* (3.45).

Remark 3.3 For a given value of ϵ the number of iterations necessary to obtain the convergence of algorithm (3.38)–(3.54) varies like $k^{1/2}$ (as before for closely related algorithms).

Remark 3.4 The implementation of algorithm (3.38)–(3.54) requires *efficient Stokes solvers*, for solving problems (3.39) (3.40), (3.41) (3.42), (3.45) (3.46), (3.47) and (3.48). Such solvers can be found in, e.g., Glowinski and Le Tallec (1989), Glowinski (1991; 1992a); actually, this issue is fully addressed in the related article by Berggren and Glowinski (1994), for *more general* boundary conditions than Dirichlet.

3.6. Time discretization of the control problem (3.19)

The practical implementation of algorithm (3.38)–(3.54) requires space and time approximations of the control problem (3.19). Focusing on the *time discretization* only (the space discretization will be addressed in Berggren and Glowinski (1994)) we introduce a *time discretization step* $\Delta t = T/N$ (with N a *positive* integer), denote by $\mathbf{v}$ the vector $\{\mathbf{v}^n\}_{n=1}^N$ and approximate problem (3.19) by

$$\min_{\mathbf{v} \in \mathcal{V}^{\Delta t}} \left[\frac{\Delta t}{2} \sum_{n=1}^N \int_{\mathcal{O}} |\mathbf{v}^n|^2 \, \mathrm{d}x + \frac{k}{2} \int_{\Omega} |\mathbf{y}^N - \mathbf{y}_T|^2 \, \mathrm{d}x \right], \qquad (3.55)$$

where, by analogy with (3.4)–(3.6), $\mathcal{V}^{\Delta t}$ is defined by either

$$\mathcal{V}^{\Delta t} = \left\{ \{\mathbf{v}^n\}_{n=1}^N \mid \mathbf{v}^n = \{v_1^n, v_2^n, v_3^n\} \in (L^2(\mathcal{O}))^3, \ \forall n = 1, \dots, N \right\}$$

or

$$\mathcal{V}^{\Delta t} = \left\{ \{\mathbf{v}^n\}_{n=1}^N \mid \mathbf{v}^n = \{v_1^n, v_2^n, 0\}, \{v_1^n, v_2^n, \} \in (L^2(\mathcal{O}))^2, \ \forall n = 1, \dots, N \right\},$$

$$\mathcal{V}^{\Delta t} = \left\{ \{\mathbf{v}^n\}_{n=1}^N \mid \mathbf{v}^n = \{v_1^n, 0, 0\}, v_1^n \in L^2(\mathcal{O}), \ \forall n = 1, \dots, N \right\},$$

and where $\mathbf{y}^n$ is obtained from $\mathbf{v}$ via

$$\mathbf{y}^0 = \mathbf{0}; \qquad (3.56)$$

for $n = 1, \dots, N$, we obtain $\{\mathbf{y}^n, \pi^n\}$ from $\mathbf{y}^{n-1}$ by solving the following steady Stokes type problem

$$\begin{cases} \dfrac{\mathbf{y}^n - \mathbf{y}^{n-1}}{\Delta t} - \Delta \mathbf{y}^n + \nabla \pi^n = \mathbf{v}^n \chi_{\mathcal{O}} \text{ in } \Omega, \\ \nabla \cdot \mathbf{y}^n = 0 \text{ in } \Omega, \end{cases} \qquad (3.57)$$

$$\mathbf{y}^n = \mathbf{0} \text{ on } \Gamma. \qquad (3.58)$$

The above scheme is nothing but a *backward Euler time discretization* of problem (3.1), (3.2). Efficient algorithms for solving problem (3.57), (3.58) (and finite element approximations of it) can be found in, e.g., Glowinski and

Le Tallec (1989), Glowinski (1991; 1992a) (see also Berggren and Glowinski (1994)).

The discrete control problem (3.55) has a *unique* solution; for the optimality conditions and a discrete analogue of the conjugate gradient algorithm (3.38)–(3.54) see Berggren and Glowinski (1994) (see also the above reference for a discussion of the *full discretization* of problem (3.19) and solution methods for the fully discrete problem).

3.7. Numerical experiments

Following Berggren and Glowinski (1994), we (briefly) consider the practical solution of the following variant of problem (3.19):

$$\min_{\mathbf{v} \in \mathcal{V}} \left[\frac{1}{2} \iint_{\mathcal{O} \times (0,T)} |\mathbf{v}|^2 \, dx \, dt + \frac{k}{2} \int_{\Omega} |\mathbf{y}(T) - \mathbf{y}_T|^2 \, dx \right], \tag{3.59}$$

where, in (3.59), $\mathcal{O} \subset \Omega \subset \mathbb{R}^2$, $\mathbf{v} = \{v_1, 0\}$, $\mathcal{V} = \{\mathbf{v} \mid \mathbf{v} = \{v_1, 0\}, v_1 \in L^2(\mathcal{O} \times (0,T))\}$, where $\mathbf{y}(T)$ is obtained from $\mathbf{v}$ via the solution of the following *Stokes problem*

$$\begin{cases} \dfrac{\partial \mathbf{y}}{\partial t} - \nu \Delta \mathbf{y} + \nabla \pi = \mathbf{v} \chi_{\mathcal{O}} \text{ in } Q, \\ \nabla \cdot \mathbf{y} = 0 \text{ in } Q, \end{cases} \tag{3.60}$$

$$\mathbf{y}(0) = \mathbf{y}_0, \text{ with } \mathbf{y}_0 \in (L^2(\Omega))^2, \ \nabla \cdot \mathbf{y}_0 = 0, \ \mathbf{y}_0 \cdot \mathbf{n} = 0 \text{ on } \Sigma_0 (= \Gamma_0 \times (0,T)), \tag{3.61}$$

$$\mathbf{y} = \mathbf{g}_0 \text{ on } \Sigma_0, \tag{3.62}$$

$$\nu \frac{\partial \mathbf{y}}{\partial n} - \mathbf{n}\pi = \mathbf{g}_1 \text{ on } \Sigma_1 (= \Gamma_1 \times (0,T)), \tag{3.63}$$

and where the target function $\mathbf{y}_T$ is given in $(L^2(\Omega))^2$. In (3.60)–(3.63) $\nu (> 0)$ is a viscosity parameter and $\Gamma_0 \cap \Gamma_1 = \emptyset$, closure of $\Gamma_0 \cup \Gamma_1 = \Gamma$. Actually the boundary condition (3.63) is not particularly physical, but it can be used to implement downstream boundary conditions for flow in unbounded regions.

The test problem that we consider is the particular problem (3.59) where:

1 $\Omega = (0,2) \times (0,1)$, $\mathcal{O} = (1/2, 3/2) \times (1/4, 3/4)$, $T = 1$;
2 $\Gamma_0 = \{\{x_i\}_{i=1}^2 \mid x_2 = 0 \text{ or } 1, \ 0 < x_1 < 2\}$, $\Gamma_1 = \{\{x_i\}_{i=1}^2 \mid x_1 = 0 \text{ or } 2, \ 0 < x_2 < 1\}$;
3 $\mathbf{g}_0 = \mathbf{0}$, $\mathbf{g}_1 = \mathbf{0}$;
4 $\mathbf{y}_T = \mathbf{0}$, $k = 20$;
5 $\nu = 5 \times 10^{-2}$;
6 $\mathbf{y}_0$ corresponds to a plane *Poiseuille flow* of maximum velocity equal to 1, i.e.

$$\mathbf{y}_0(x) = \{4x_2(1 - x_2), 0\}, \ \forall x \in \Omega.$$

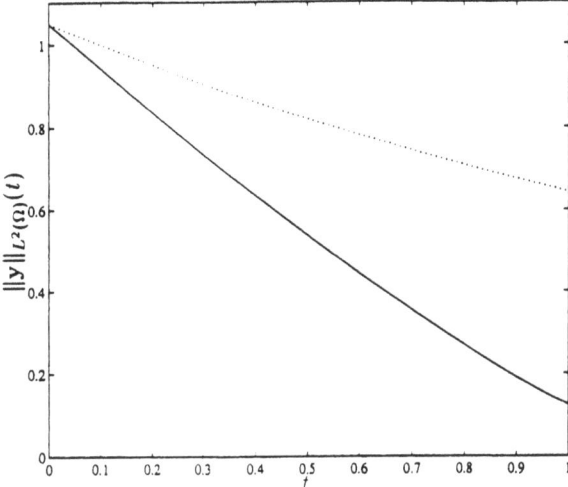

Fig. 36. Variation of $\|\mathbf{y}_h^{\Delta t}(t)\|_{(L^2(\Omega))^2}$ with (—) and without (...) control.

Integrating equations (3.60)–(3.63) with $\mathbf{v} = \mathbf{0}$ will lead to a solution that decays in time with a rate determined by the size of the viscosity parameter ν. The problem here is to find – via (3.59) – a control that will speed up this decay as much as possible at time T.

The *time discretization* has been obtained through a variant of scheme (3.56)–(3.58), using $\Delta t = 1/50$; the space discretization was achieved using a *finite element* approximation associated with a 32×16 (respectively (16×8)) regular grid for the *velocity* (respectively the *pressure*) (see Berggren and Glowinski (1994), for details). A *fully discrete* variant of the conjugate gradient algorithm (3.38)–(3.54) was used to compute the *approximate optimal control* $\mathbf{u}_h^{\Delta t}$ and the associated *velocity field* $\mathbf{y}_h^{\Delta t}$.

On Figure 36 we compare the decays between $t = 0$ and $t = T = 1$ of the noncontrolled flow velocity (...) and of the controlled flow velocity (—) (we have shown the values of $(\int_\Omega |\mathbf{y}(t)|^2 \, d\mathbf{x})^{1/2}$; remember that $\frac{1}{2} \int_\Omega |\mathbf{y}(t)|^2 \, d\mathbf{x}$ is the *flow kinetic* energy). On Figure 37, we have compared, at time T, the *kinetic energy* distributions of the controlled flow (lower graph) and of the noncontrolled one (upper graph). Control has been effective to reduce the flow kinetic energy, particularly on the support $\mathcal{O}$ of the optimal control (according to Figure 37, at least). The results displayed on the following figures were obtained after 70 iterations.

Finally, we have shown on Figure 38 the graph of the first component of the computed optimal control $\mathbf{u}_h^{\Delta t}$ at various values of t.

For further details and comments about these computations see Berggren and Glowinski (1994), where further numerical experiments are also discussed.

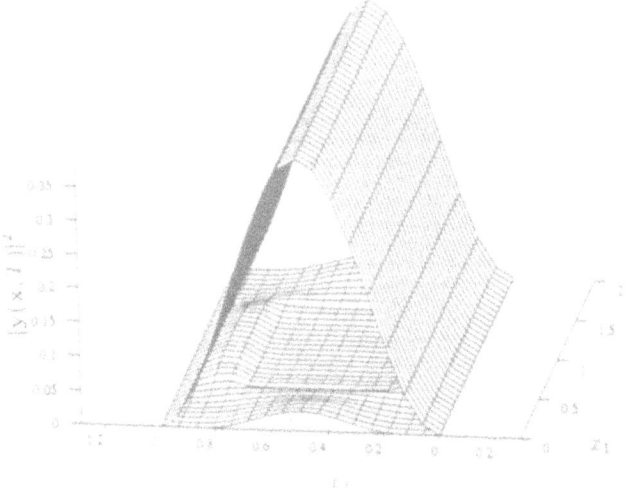

Fig. 37. Kinetic energy distribution of the controlled flow (lower graph) and noncontrolled flow (upper graph). Kinetic energy distribution of the controlled flow (lower graph) and noncontrolled flow (upper graph).

4. CONTROL OF NONLINEAR DIFFUSION SYSTEMS

4.1. Generalities. Synopsis

The various *controllability problems* which have been discussed so far have all been associated with systems governed by linear diffusion equations.

In this section we briefly address the *nonlinear* situation and would like to show that *nonlinearity* may bring *noncontrollability* (as seen in Section 4.2) and also to discuss (in Section 4.3) the solution of *pointwise control problems* for the *viscous Burgers equation*.

Further information is given in V. Komornik (1994), J.L. Lions (1991a), I. Lasiecka (1992), I. Lasiecka and R. Tataru (1994), E. Zuazua (1988) and the references therein.

4.2. An example of a noncontrollable nonlinear system

In this section, we want to emphasize that *approximate controllability is very unstable under 'small' nonlinear perturbations.*

Let us consider again the state equation

$$\frac{\partial y}{\partial t} - \Delta y = v\chi_\mathcal{O} \text{ in } Q, \quad y(0) = 0, \ y = 0 \text{ on } \Sigma, \tag{4.1}$$

which is the same equation as in Section 1.1, but where we take $A = -\Delta$ to make things as simple as possible.

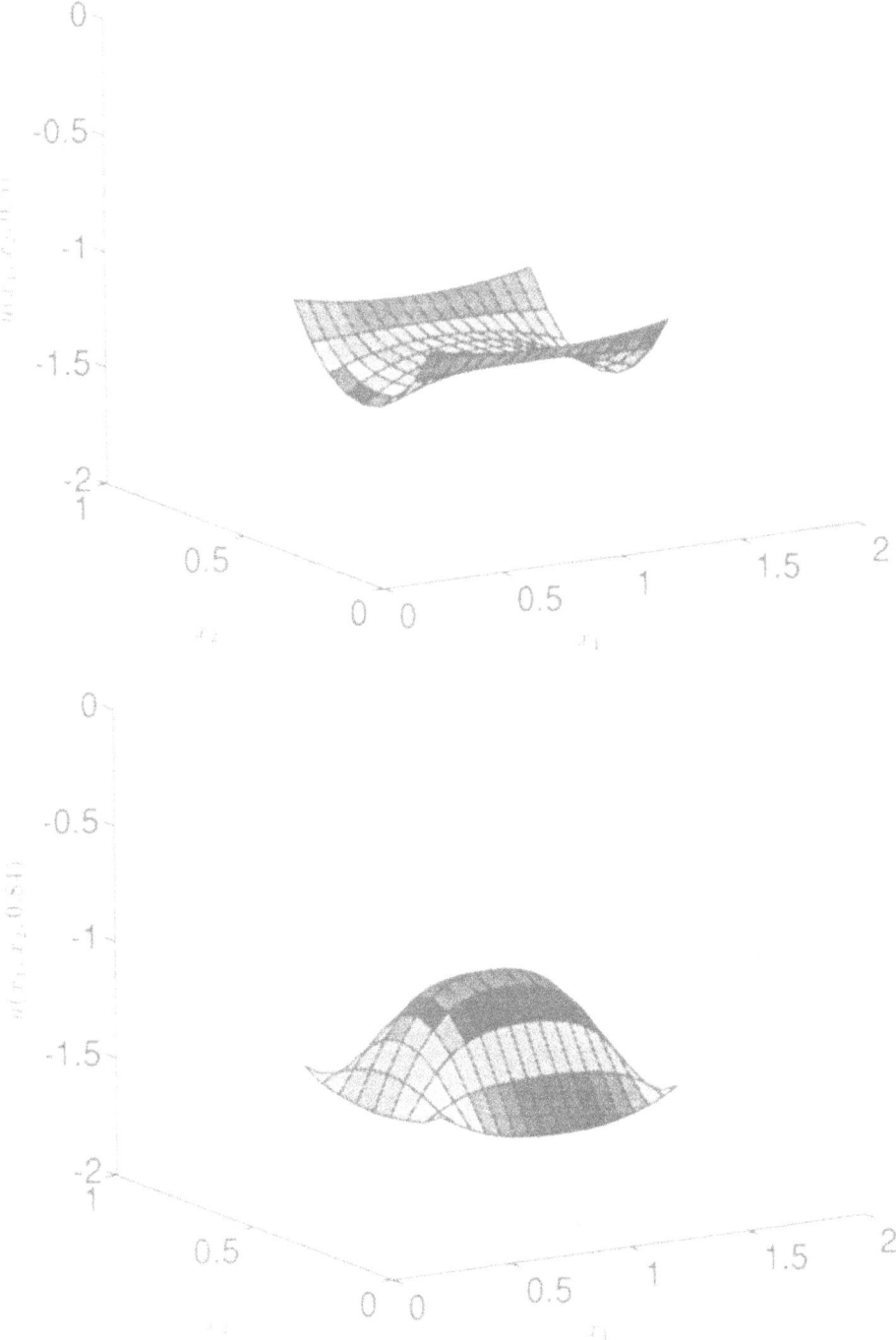

Fig. 38. (a) Graph of the computed optimal control $(t = 0.5)$. (b) Graph of the computed optimal control $(t = 0.84)$.

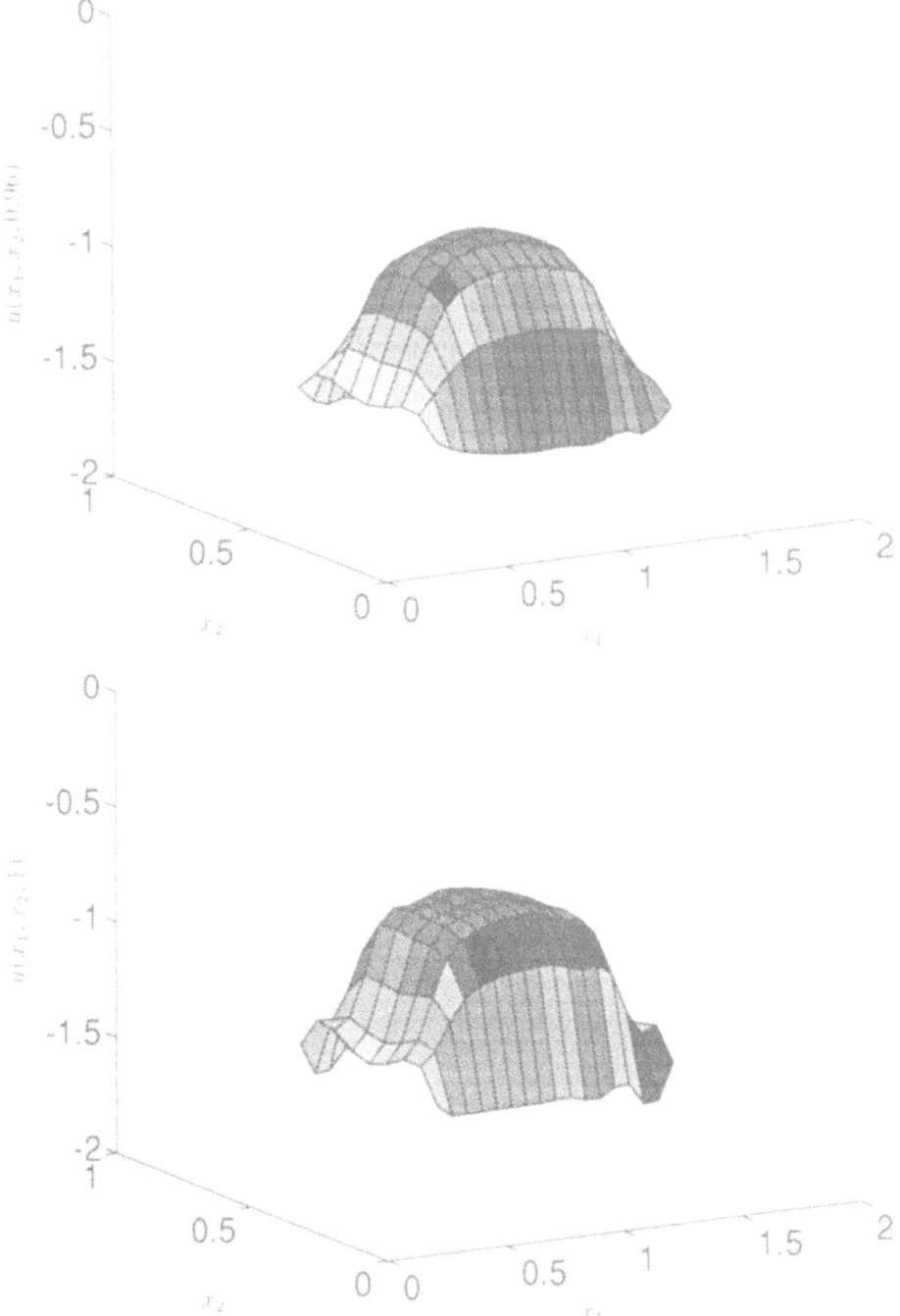

Fig. 38 (*cont.*) (c) Graph of the computed optimal control ($t = 0.96$). (d) Graph of the computed optimal control ($t = 1.0$).

We consider now the *nonlinear* partial differential equation

$$\frac{\partial y}{\partial t} - \Delta y + \alpha y^3 = v\chi_{\mathcal{O}} \text{ in } Q, \quad y(0) = 0, \ y = 0 \text{ on } \Sigma, \qquad (4.2)$$

where α is *positive*, otherwise arbitrarily small. Problem (4.2) has a *unique* solution (see, e.g., Lions (1969)). Contrary to what happens for (4.1), *the set described by $y(T; v)$ ($y(v)$ is the solution of (4.2)) when v spans $L^2(\mathcal{O} \times (0, T))$ is far from being dense in $L^2(\Omega)$.*

There are several proofs of this result, some of them based on *maximum principles*. The following one is due to A. Bamberger (1977) and is reported in the PhD thesis of Henry (1978). It is based on a simple *energy estimate*. One multiplies (4.2) by my, where $m(x) \geq 0$, $m \equiv 0$ near $\mathcal{O}$, $m \in C^1(\bar{\Omega})$. Then

$$\frac{1}{2}\frac{d}{dt}\int_\Omega my^2\,dx + \int_\Omega m|\nabla y|^2\,dx + \int_\Omega y\nabla y\cdot\nabla m\,dx + \alpha\int_\Omega my^4\,dx = 0. \quad (4.3)$$

Let us write

$$\int_\Omega y\nabla y\cdot\nabla m\,dx = \int_\Omega m^{1/4}y(m^{1/2}\nabla y)\cdot(m^{-3/4}\nabla m)\,dx$$

so that there exists a constant C such that

$$\left|\int_\Omega y\nabla y\cdot\nabla m\,dx\right| \leq \alpha\int_\Omega my^4\,dx + \int_\Omega m|\nabla y|^2\,dx + C\int_\Omega m^{-3}|\nabla m|^4\,dx.$$
$$(4.4)$$

Combining (4.3) and (4.4) gives

$$\frac{d}{dt}\frac{1}{2}\int_\Omega my^2\,dx \leq C\int_\Omega m^{-3}|\nabla m|^4\,dx$$

so that

$$\frac{1}{2}\int_\Omega m(x)|y(x, T; v)|^2\,dx \leq CT\int_\Omega m^{-3}|\nabla m|^4\,dx \qquad (4.5)$$

no matter how v is chosen, since the right-hand side of (4.5) *does not depend on v.* Of course, this calculation assumes that we can choose m as above and such that $\int_\Omega m^{-3}|\nabla m|^4\,dx < +\infty$; such functions m are easy to construct.

Remark 4.1 Examples and counter examples of controllability for nonlinear diffusion type equations are given in Diaz (1991).

4.3. Pointwise control of the viscous Burgers equation

4.3.1. Motivation

The *inviscid* or *viscous Burgers equations* have, for many years, attracted the attention of many investigators, from both the theoretical and numerical points of view. There are several reasons for this 'popularity', one of them being certainly that the Burgers equations provide not too unrealistic simplifications of the *Euler* and *Navier–Stokes* equations of *Fluid Dynamics*; among the features, common with those more complicated equations, *nonlinearity* is certainly the most important single one. It is not surprising, therefore, that the Burgers equations have also attracted the attention of the *Control Community* (see, e.g., Burns and Kang (1991), Burns and Marrekchi (1993)). The present section is another contribution in that direction: we shall address here the solution of controllability problems for the *viscous Burgers equation* via *pointwise controls*; from that point of view this section can be seen as a generalization of Section 1.10 where we addressed the pointwise control of linear diffusion systems (the viscous Burgers equation considered here belongs to the class of the *nonlinear advection–diffusion* systems whose most celebrated representative is the *Navier–Stokes equation* system).

4.3.2. Formulation of the control problems

As in Berggren and Glowinski (1994) (see also Berggren (1992) and Dean and Gubernatis (1991)) we can consider the following *pointwise control problem* for the *viscous Burgers equation*

$$\min_{\mathbf{v} \in \mathcal{U}} \left[\frac{1}{2} \|\mathbf{v}\|_{\mathcal{U}}^2 + \frac{1}{2} k \|y(T) - y_T\|_{L^2(0,1)}^2 \right], \tag{4.6}$$

where, in (4.6), we have:

1    $\mathbf{v} = \{v_m\}_{m=1}^M$, $\mathcal{U} = L^2(0,T;\mathbb{R}^M)$, $\|\mathbf{v}\|_{\mathcal{U}} = (\sum_{m=1}^M \int_0^T |v_m|^2 \, dt)^{1/2}$;
2    $k > 0$, arbitrarily large;
3    $y_T \in L^2(0,1)$ and $y(T)$ is obtained from $\mathbf{v}$ via the solution of the *viscous Burgers equation*, below

$$\frac{\partial y}{\partial t} - \nu \frac{\partial^2 y}{\partial x^2} + y \frac{\partial y}{\partial x} = f + \sum_{m=1}^M v_m \delta(x - a_m) \text{ in } Q(= (0,1) \times (0,T)), \tag{4.7}$$

$$\frac{\partial y}{\partial x}(0,t) = 0, y(1,t) = 0 \quad \text{a.e. on } (0,T), \tag{4.8}$$

$$y(0) = y_0 (\in L^2(0,1)); \tag{4.9}$$

in (4.7), $\nu(> 0)$ is a *viscosity parameter*, f a *forcing term*, $a_m \in (0,1)$, $\forall m = 1, \ldots, M$ and $x \to \delta(x - a_m)$ denotes the *Dirac measure* at a_m.

Let us denote by V the (Sobolev) space defined by

$$V = \{z \mid z \in H^1(0,1), z(1) = 0\}, \tag{4.10}$$

and suppose that $f \in L^2(0,T;V')$ (V': dual space of V); it follows then from Lions (1969) that for $\mathbf{v}$ given in $\mathcal{U}$ the Burgers system (4.7)–(4.9) has a *unique solution* in $L^2(0,T;V) \cap C^0([0,T];L^2(0,1))$. From this result, we can show that the control problem (4.6) *has a solution* (not *necessarily unique*, due to the *nonconvexity* of the functional $J : \mathcal{U} \to \mathbb{R}$, where J is the *cost function* in (4.6)).

Remark 4.2 In Glowinski and Berggren (1994), we have discussed the solution of the variant of problem (4.6) where the location on $(0,1)$ of the a_m's is *unknown* (a_m: 'support' of the mth pointwise control). The solution methods described in the following can easily be modified to accommodate this more complicated situation (see the above reference for details and numerical results).

4.3.3. Optimality conditions for problem (4.6)

To compute a control **u** solution of problem (4.6) we shall derive first *necessary optimality conditions* and use them (in Section 4.3.4) through a *conjugate gradient algorithm* to obtain the above solution.

The derivative $J'(\mathbf{v})$ of J at $\mathbf{v}$ can be obtained from

$$(J'(\mathbf{v}), \mathbf{w})_{\mathcal{U}} = \lim_{\substack{\theta \to 0 \\ \theta \neq 0}} \frac{J(\mathbf{v} + \theta\mathbf{w}) - J(\mathbf{v})}{\theta}. \tag{4.11}$$

Actually, instead of (4.11), we shall use a (formal) perturbation analysis to obtain $J'(\mathbf{v})$:

First, we have

$$\delta J(\mathbf{v}) = (J'(\mathbf{v}), \delta\mathbf{v})_{\mathcal{U}} = \sum_{m=1}^{M} \int_0^T v_m \delta v_m \, dt + k \int_0^1 (y(T) - y_T)\delta y(T) \, dx \tag{4.12}$$

where (from (4.7)–(4.9)) $\delta y(T)$ is obtained from $\delta\mathbf{v}$ via the solution of the following variational problem

$$\delta y(t) \in V \quad \text{a.e. on } (0,T); \quad \forall z \in V \quad \text{we have a.e. on } (0,T)$$

$$\left\langle \frac{\partial}{\partial t}\delta y, z \right\rangle + \nu \int_0^1 \frac{\partial}{\partial x}\delta y \frac{\partial z}{\partial x} \, dx + \int_0^1 \delta y \frac{\partial y}{\partial x} z \, dx + \int_0^1 y \frac{\partial}{\partial x}\delta y z \, dx$$

$$= \sum_{m=1}^{m} \delta v_m z(a_m), \tag{4.13}$$

$$\delta y(0) = 0; \tag{4.14}$$

in (4.13), $\langle \cdot, \cdot \rangle$ denotes the *duality pairing* between V' and V.

Consider now $p \in L^2(0,T;V) \cap C^0([0,T]; L^2(0,1))$ such that $\partial p/\partial t \in L^2(0,T;V')$; taking $z \in p(t)$ in (4.13) we obtain

$$\int_0^T \left\langle \frac{\partial}{\partial t}\delta y, p \right\rangle dt + \nu \int_0^T \int_0^1 \frac{\partial}{\partial x}\delta y \frac{\partial p}{\partial x} \, dx \, dt$$

$$+ \int_0^T \int_0^1 \left(\delta y \frac{\partial y}{\partial x} + y \frac{\partial}{\partial x}\delta y \right) p \, dx \, dt$$

$$= \sum_{m=1}^{M} \int_0^T p(a_m, t)\delta v_m \, dt. \tag{4.15}$$

Integrating by parts over $(0, T)$ it follows from (4.14), (4.15) that

$$
\int_0^1 p(T)\delta y(T)\, dx - \int_0^T \left\langle \frac{\partial p}{\partial t}, \delta y \right\rangle dt + \nu \int_0^T \int_0^1 \frac{\partial}{\partial x}\delta y \frac{\partial p}{\partial x}\, dx\, dt
$$
$$
+ \int_0^T \int_0^1 \left(\delta y \frac{\partial y}{\partial x} + y \frac{\partial}{\partial x}\delta y \right) p\, dx\, dt
$$
$$
= \sum_{m=1}^M \int_0^T p(a_m, t)\delta v_m\, dt. \tag{4.16}
$$

Suppose now that p satisfies also

$$
-\left\langle \frac{\partial p}{\partial t}, z \right\rangle + \nu \int_0^1 \frac{\partial p}{\partial x}\frac{\partial z}{\partial x}\, dx + \int_0^1 p\left(\frac{\partial y}{\partial x}z + y\frac{\partial z}{\partial x} \right) dx = 0,
$$
$$
\forall z \in V, \text{ a.e. on } (0, T), \tag{4.17}
$$

and

$$
p(T) = k(y_T - y(T)); \tag{4.18}
$$

it follows then from (4.16) that

$$
k \int_0^1 (y(T) - y_T)\delta y(T)\, dx = -\sum_{m=1}^M \int_0^T p(a_m, t)\delta v_m\, dt,
$$

which combined with (4.12) implies in turn that

$$
(J'(\mathbf{v}), \delta \mathbf{v})_{\mathcal{U}} = \sum_{m=1}^M \int_0^T (v_m(t) - p(a_m, t))\delta v_m(t)\, dt.
$$

We have thus 'proved' that, $\forall \mathbf{v}, \mathbf{w} \in \mathcal{U}$

$$
(J'(\mathbf{v}), \mathbf{w})_{\mathcal{U}} = \sum_{m=1}^M \int_0^T (v_m(t) - p(a_m, t))w_m(t)\, dt. \tag{4.19}
$$

Remark 4.3 Starting from (4.11) we can give a rigorous proof of (4.19).

Suppose now that $\mathbf{u}$ is a solution of problem (4.6); we have then $J'(\mathbf{u}) = \mathbf{0}$ which provides the following *optimality system*

$$
u_m(t) = p(a_m, t), \; \forall m = 1, \ldots, M, \text{ on } (0, T), \tag{4.20}
$$

$$
\frac{\partial y}{\partial t} - \nu \frac{\partial^2 y}{\partial x^2} + y\frac{\partial y}{\partial x} = f + \sum_{m=1}^M u_m\delta(x - a_m) \text{ in } Q, \tag{4.21}
$$

$$
\frac{\partial y}{\partial x}(0, t) = 0, \; y(1, t) = 0 \text{ on } (0, T), \tag{4.22}
$$

$$
y(0) = y_0 \tag{4.23}
$$

$$-\frac{\partial p}{\partial t} - \nu\frac{\partial^2 p}{\partial x^2} - y\frac{\partial p}{\partial x} = 0 \text{ in } Q, \tag{4.24}$$

$$\nu\frac{\partial p}{\partial x}(0, t) + y(0, t)p(0, t) = 0, \ p(1, t) = 0 \text{ on } (0, T), \tag{4.25}$$

$$p(T) = k(y_T - y(T)). \tag{4.26}$$

4.3.4. Iterative solution of the control problem (4.6)

Conjugate gradient algorithms are particularly attractive for large scale nonlinear problems since their applications requires only – in principle – *first derivative information* (see, e.g., Daniel (1970), Polack (1971) and Nocedal (1992) for further comments and convergence proofs). Problem (4.6) is a particular case of the minimization problem

$$\begin{cases} u \in H, \\ j(u) \le j(v), \ \forall v \in H, \end{cases} \tag{4.27}$$

where H is a real Hilbert space for the scalar product $(\cdot, \cdot)$ and the corresponding norm $\|\cdot\|$ and where the functional $j : H \to \mathbb{R}$ is *differentiable*; we denote by $j'(v)$ ($\in H'; H'$: dual space of H) the differential of j at v.

A *conjugate gradient* algorithm for solving (4.27) is defined as follows:

$$u^0 \text{ is given in } H; \tag{4.28}$$

solve

$$\begin{cases} g^0 \in H, \\ (g^0, v) = \langle j'(u^0), v\rangle, \ \forall v \in H, \end{cases} \tag{4.29}$$

and set

$$w^0 = g^0. \tag{4.30}$$

For $n \ge 0$, assuming that u^n, g^n, w^n are known, compute u^{n+1}, g^{n+1}, w^{n+1} by

$$\begin{cases} \text{Find } \rho_n \in \mathbb{R} \text{ such that} \\ j(u^n - \rho_n w^n) \le j(u^n - \rho w^n), \ \forall \rho \in \mathbb{R}, \end{cases} \tag{4.31}$$

set

$$u^{n+1} = u^n - \rho_n w^n, \tag{4.32}$$

and solve

$$\begin{cases} g^{n+1} \in H, \\ (g^{n+1}, v) = \langle j'(u^{n+1}), v\rangle, \ \forall v \in H. \end{cases} \tag{4.33}$$

If $\|g^{n+1}\|/\|g^0\| \le \epsilon$ take $u = u^{n+1}$; else compute either

$$\gamma_n = \frac{\|g^{n+1}\|^2}{\|g^n\|^2} \quad \text{(Fletcher–Reeves update)} \tag{4.34}_1$$

or

$$\gamma_n = \frac{(g^{n+1}, g^{n+1} - g^n)}{\|g^n\|^2} \quad \text{(Polack–Ribière update)} \tag{4.34}_2$$

and then

$$w^{n+1} = g^{n+1} + \gamma_n w^n. \tag{4.35}$$

Do $n = n + 1$ *and go to* (4.31).

We observe that each iteration requires the solution of a *linear problem* ((4.29) for $n = 0$, (4.33) for $n \geq 1$) and the *line search* (4.31). In most applications the Polack–Ribière variant of algorithm (4.28)–(4.35) is faster than the Fletcher–Reeves one (see, e.g,, Powell (1976) for an explanation of this fact).

Application to problem (4.6). Problem (4.6) is a particular case of (4.27) where $H = \mathcal{U} = L^2(0, T; \mathbb{R}^M)$; combining (4.19) and (4.28)–(4.35) we obtain the following solution method for problem (4.6):

$$u^0 \text{ is given in } \mathcal{U}; \tag{4.36}$$

solve

$$\begin{cases} \dfrac{\partial y^0}{\partial t} - \nu \dfrac{\partial^2 y^0}{\partial x^2} + y^0 \dfrac{\partial y^0}{\partial x} = f + \displaystyle\sum_{m=1}^{M} u_m^0 \delta(x - a_m) \text{ in } Q, \\[4mm] \dfrac{\partial y^0}{\partial x}(0, t) = 0, y^0(1, t) = 0 \text{ on } (0, T), \ y^0(0) = y_0, \end{cases} \tag{4.37}$$

and

$$\begin{cases} -\dfrac{\partial p^0}{\partial t} - \nu \dfrac{\partial^2 p^0}{\partial x^2} - y^0 \dfrac{\partial p^0}{\partial x} = 0 \text{ in } Q, \\[4mm] \nu \dfrac{\partial p^0}{\partial x}(0, t) + y^0(0, t)p^0(0, t) = 0, \ p^0(1, t) = 0 \text{ on } (0, T), \end{cases} \tag{4.38}_1$$

$$p^0(T) = k(y_T - y^0(T)). \tag{4.38}_2$$

Solve then

$$\begin{cases} g^0 \in \mathcal{U}; \ \forall v \in \mathcal{U}, \text{ we have} \\[2mm] \displaystyle\int_0^T g^0 \cdot v \, dt = \sum_{m=1}^{M} \int_0^T (u_m^0(t) - p^0(a_m, t))v_m(t) \, dt, \end{cases} \tag{4.39}$$

and set

$$w^0 = g^0. \tag{4.40}$$

Then for $n \geq 0$, assuming that u^n, g^n, w^n are known compute u^{n+1}, g^{n+1}, w^{n+1} as follows.

Solve the following one-dimensional minimization problem

$$\begin{cases} \rho_n \in \mathbb{R}, \\ J(\mathbf{u}^n - \rho_n \mathbf{w}^n) \leq J(\mathbf{u}^n - \rho \mathbf{w}^n), \ \forall \rho \in \mathbb{R}, \end{cases} \tag{4.41}$$

and update $\mathbf{u}^n$ *by*

$$\mathbf{u}^{n+1} = \mathbf{u}^n - \rho_n \mathbf{w}^n. \tag{4.42}$$

Next, solve

$$\begin{cases} \dfrac{\partial y^{n+1}}{\partial t} - \nu \dfrac{\partial^2 y^{n+1}}{\partial x^2} = f + \displaystyle\sum_{m=1}^{M} u_m^{n+1} \delta(x - a_m) \text{ in } Q, \\[4mm] \dfrac{\partial y^{n+1}}{\partial x}(0, t) = 0, \, y^{n+1}(1, t) = 0 \text{ on } (0, T), \ y^{n+1}(0) = y_0, \end{cases} \tag{4.43}$$

and

$$\begin{cases} -\dfrac{\partial p^{n+1}}{\partial t} - \nu \dfrac{\partial^2 p^{n+1}}{\partial x^2} - y^{n+1} \dfrac{\partial p^{n+1}}{\partial x} = 0 \text{ in } Q, \\[4mm] \nu \dfrac{\partial p^{n+1}}{\partial x}(0, t) + y^{n+1}(0, t) p^{n+1}(0, t) = 0, \, p^{n+1}(1, t) = 0 \text{ on } (0, T), \end{cases}$$

$$\tag{4.44}_1$$

$$p^{n+1}(T) = k(y_T - y^{n+1}(T)). \tag{4.44}_2$$

Solve then

$$\begin{cases} \mathbf{g}^{n+1} \in \mathcal{U}; \ \forall \mathbf{v} \in \mathcal{U}, \ \text{ we have} \\[2mm] \displaystyle\int_0^T \mathbf{g}^{n+1} \cdot \mathbf{v} \, dt = \displaystyle\sum_{m=1}^{M} \int_0^T (u_m^{n+1}(t) - p^{n+1}(a_m, t)) v_m(t) \, dt. \end{cases} \tag{4.45}$$

If $\|\mathbf{g}^{n+1}\|_{\mathcal{U}} / \|\mathbf{g}^0\|_{\mathcal{U}} \leq \epsilon$ *take* $\mathbf{u} = \mathbf{u}^{n+1}$; *else compute either*

$$\gamma_n = \int_0^T |\mathbf{g}^{n+1}|^2 \, dt \Big/ \int_0^T |\mathbf{g}^n|^2 \, dt \ \text{(Fletcher–Reeves)} \tag{4.46}_1$$

or

$$\gamma_n = \int_0^T \mathbf{g}^{n+1} \cdot (\mathbf{g}^{n+1} - \mathbf{g}^n) \, dt \Big/ \int_0^T |\mathbf{g}^n|^2 \, dt \ \text{(Polack–Ribière)} \tag{4.46}_2$$

and update $\mathbf{w}^n$ *by*

$$\mathbf{w}^{n+1} = \mathbf{g}^{n+1} + \gamma_n \mathbf{w}^n. \tag{4.47}$$

Do $n = n + 1$ *and go to* (4.41).

The practical implementation of algorithm (4.36)–(4.47) will rely on the *numerical integration* of the parabolic problems (4.37), (4.38), (4.43), (4.44) (to be discussed in Section 4.3.5) and on the efficiency and accuracy of the *line search* (4.41); actually, to solve the nonlinear problem (4.41) we have employed the *cubic backtracking* strategy advocated in Dennis and Schnabel (1983, Ch. 6).

4.3.5. Space–time discretization of the control problem (4.6). Optimality conditions

We shall use a combination of *finite element* and *finite difference* methods for the space-time discretization of problem (4.6); for simplicity, we shall consider *uniform meshes* for both discretizations. We consider therefore two positive integers I and Δt (to be 'large' in practice) and define the discretization steps h and Δt by $h = 1/I$, $\Delta t = T/N$. Next, we define $x_i = ih$, $i = 0, 1, \ldots, I$ and approximate $L^2(0,1)$ and $H^1(0,1)$ by

$$H_h^1 = \{z_h \mid z_h \in C^0[0,1], z_h|_{[x_{i-1},x_i]} \in P_1, \forall i = 1, \ldots, I\},$$

where P_1 denotes the space of the polynomials in one variable of degree less than or equal to one. The space V in (4.10) is approximated by

$$V_h = \{z_h \mid z_h \in H_h^1, z_h(1) = 0\} \ (= V \cap H_h^1),$$

while the *control space* $\mathcal{U}(= L^2(0,T;\mathbb{R}^M))$ in (4.6) is approximated by

$$\mathcal{U}^{\Delta t} = (\mathbb{R}^M)^N = \left\{\mathbf{v} \mid \mathbf{v} = \{\{v_m^n\}_{m=1}^M\}_{n=1}^N\right\}, \tag{4.48}$$

to be equipped with the following scalar product

$$(\mathbf{v}, \mathbf{w})_{\Delta t} = \Delta t \sum_{n=1}^N \sum_{m=1}^M v_m^n w_m^n, \ \forall \mathbf{v}, \mathbf{w} \in \mathcal{U}^{\Delta t}.$$

We approximate then the control problem (4.6) by

$$\min_{\mathbf{v} \in \mathcal{U}^{\Delta t}} J_h^{\Delta t}(\mathbf{v}), \tag{4.49}$$

where the functional $J_h^{\Delta t} : \mathcal{U}^{\Delta t} \to \mathbb{R}$ is defined by

$$J_h^{\Delta t}(\mathbf{v}) = \tfrac{1}{2}(\mathbf{v}, \mathbf{v})_{\Delta t} + \tfrac{1}{2}k\|y^N - y_T\|_{L^2(0,1)}^2, \tag{4.50}$$

with y^N defined from $\mathbf{v}$ via the solution of the following *discrete Burgers equation*:

$$y^0 = y_{0h} \in H_h^1 \text{ such that } \lim_{h \to 0} \|y_{0h} - y_0\|_{L^2(0,1)} = 0; \tag{4.51}$$

for $n = 1, \ldots, N$ we obtain y^n from y^{n-1} via the solution of the following discrete linear (elliptic) variational problem

$$\left\{ \begin{array}{l} y^n \in V_h; \ \forall z \in V_h \text{ we have} \\[2mm] \displaystyle\int_0^1 \frac{y^n - y^{n-1}}{\Delta t} z \, dx + \nu \int_0^1 \frac{dy^n}{dx}\frac{dz}{dx} \, dx + \int_0^1 y^{n-1}\frac{dy^{n-1}}{dx} z \, dx \\[4mm] \displaystyle = \int_0^1 fz \, dx + \sum_{m=1}^M v_m^n z(a_m). \end{array} \right. \tag{4.52}$$

Scheme (4.51), (4.52) is *semi-implicit* since the nonlinear term $y(dy/dx)$ is treated *explicitly*; we can expect therefore that Δt has to satisfy a *stability* condition. It is easily verified that obtaining y^n from y^{n-1} is *equivalent* to solving a *linear system* for a matrix (the discrete analogue of operator $(I/\Delta t) - \nu d^2/dx^2$) which is *tridiagonal, symmetric* and *positive definite*. If Δt is constant over the time interval $(0, T)$ this matrix being independent of n can be *Cholesky* factored once for all.

The *approximate control problem* (4.49) *has at least one solution* $\mathbf{u}_h^{\Delta t} = \{\{u_m^n\}_{m=1}^M\}_{n=1}^N$. Any solution of problem (4.49) satisfies the (necessary) *optimality condition*

$$\nabla J_h^{\Delta t}(\mathbf{u}_h^{\Delta t}) = 0, \qquad (4.53)$$

where $\nabla J_h^{\Delta t}$ is the *gradient* of the functional $J_h^{\Delta t}$.

Following the approach taken in Section 4.3.3 for the continuous problem (4.6) we can show that

$$(\nabla J_h^{\Delta t}(\mathbf{v}), \mathbf{w}) = \Delta t \sum_{n=1}^N \sum_{m=1}^M (v_m^n - p^n(a_m))w_m^n, \ \forall \mathbf{v}, \mathbf{w} \in \mathcal{U}^{\Delta t}, \qquad (4.54)$$

where $\{p^n\}_{n=1}^N$ is obtained from $\mathbf{v}$ via the solution of the discrete Burgers equation (4.51), (4.52), followed by the solution of the *discrete adjoint* equation, below.

Compute

$$p^{N+1} \in V_h \text{ such that } \int_0^1 p^{N+1}z \, dx = k \int_0^1 (y_T - y^N)z \, dx, \ \forall z \in V_h, \quad (4.55)$$

and then for $n = N, N-1, \ldots, 1$, p^n is obtained from p^{n+1} via the solution of the discrete elliptic problem

$$\left\{ \begin{array}{l} p^n \in V_h; \text{ we have} \\ \int_0^1 \dfrac{p^n - p^{n+1}}{\Delta t} z \, dx + \nu \int_0^1 \dfrac{dp^n}{dx} \dfrac{dz}{dx} \, dx + \int_0^1 p^{n+1} \left(y^n \dfrac{dz}{dx} + \dfrac{dy^n}{dx} z \right) dx = 0. \end{array} \right.$$

$$(4.56)$$

The comments concerning the calculation of y^n from y^{n-1} still apply here; actually, the linear systems to be solved at each time step to obtain p^n from p^{n+1} have the same matrix as those encountered in the calculation of y^n from y^{n-1}.

From (4.54), we can derive a fully discrete variant of algorithm (4.36)–(4.47) to solve the approximate control problem (4.49) via the optimality conditions (4.53); such an algorithm is discussed in Berggren and Glowinski (1994).

4.3.6. Numerical experiments

Following Berggren and Glowinski (1994) (see also Dean and Gubernatis (1991), Glowinski (1991)) we consider particular cases of problem (4.6) which

Table 4. *Summary of numerical results.*

a	1/5	2/3
Number of iterations	89	47
$\dfrac{\|y_h^{\Delta t}(T) - y_T\|_{L^2(0,1)}}{\|y_T\|_{L^2(0,1)}}$	2×10^{-1}	9×10^{-2}
$\|u_h^{\Delta t}\|_{L^2(0,T)}$	0.11	0.11

have in common:

$$T = 1, \ \nu = 10^{-2}, \ k = 8, \ y_0 = 0,$$

$$f(x,t) = \begin{cases} 1 \text{ if } \{x,t\} \in (0, 1/2) \times (0,T), \\ 2(1-x) \text{ if } \{x,t\} \in (1/2, 1) \times (0,T), \end{cases}$$

$$y_T(x) = 1 - x^2, \text{ if } x \in (0,1).$$

To discretize the corresponding control problems, we have used the methods described in Section 4.3.5 with $h = 1/128$ and $\Delta t = 1/256$. The discrete control problems (4.49) have been solved by the fully discrete variant of algorithm (4.36)–(4.47) mentioned in Section 4.3.5, using $\mathbf{u}^0 = \mathbf{0}$ as an *initial guess* and $\epsilon = 10^{-5}$ as the *stopping criterium*.

First, several experiments were performed with a *single* control point ($M = 1$) for different values of $a(= a_1)$. In Table 4 we have summarized some of the numerical results concerning the *computed optimal control* $u_h^{\Delta t}$ and the corresponding *discrete state function* $y_h^{\Delta t}$:

For $a = 1/5$ (respectively $a = 2/3$) we have visualized on Figure 39(a) (respectively Figure 40(a)) the *computed optimal control* $u_h^{\Delta t}$ while on Figure 39(a) (respectively Figure 40(b)) we have compared the *target function* y_T (...) with the computed approximation $y_h^{\Delta t}(T)$ (—) of $y(T)$.

For $a = 2/3$ a good fit *downstream* from the control point can be noticed, while the solution seems to be *close to uncontrollable upstream*. The positive sign of the solution implies that the *convection* is directed towards the increasing values of x, which is why it seems reasonable that the system is at least locally controllable in that direction. The only way of controlling the system upstream is through the diffusion term, which is small here ($\nu = 10^{-2}$) compared with the convection term. For the case $a = 1/5$ there are clearly problems with controllability far downstream of the controller (recall that there is a distributed, uncontrolled forcing term, f, which affects the solution).

Figure 41 shows the target and the final state when *two control points*

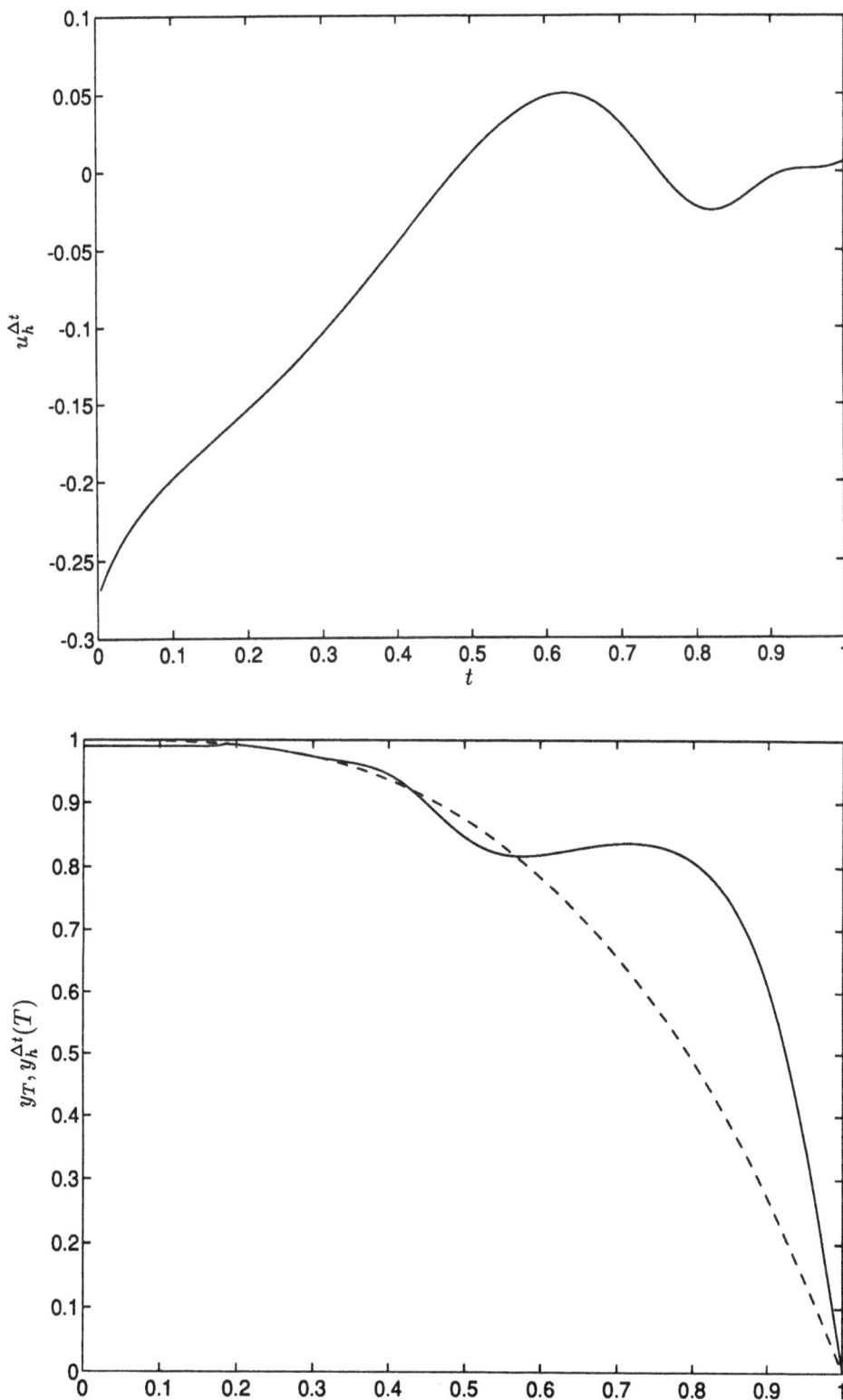

Fig. 39. (a) Graph of the computed control $u_h^{\Delta t}(a = 1/5)$. (b) Comparison between y_T (...) and $y_h^{\Delta t}(T)$ (—) $(a = 1/5)$.

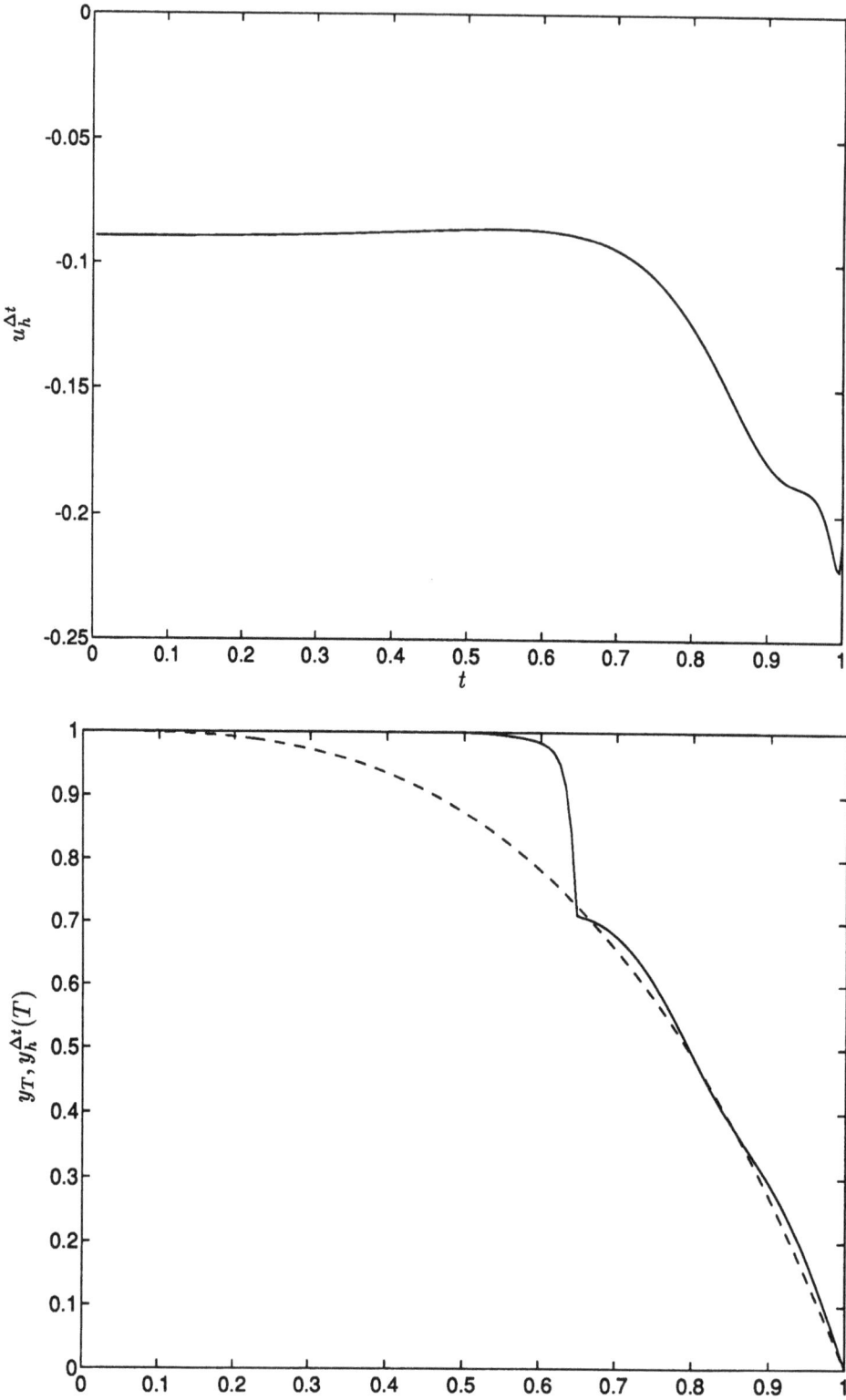

Fig. 40. (a) Graph of the computed control $u_h^{\Delta t}(a = 2/3)$. (b) Comparison between y_T (...) and $y_h^{\Delta t}(T)$ (—) $(a = 2/3)$.

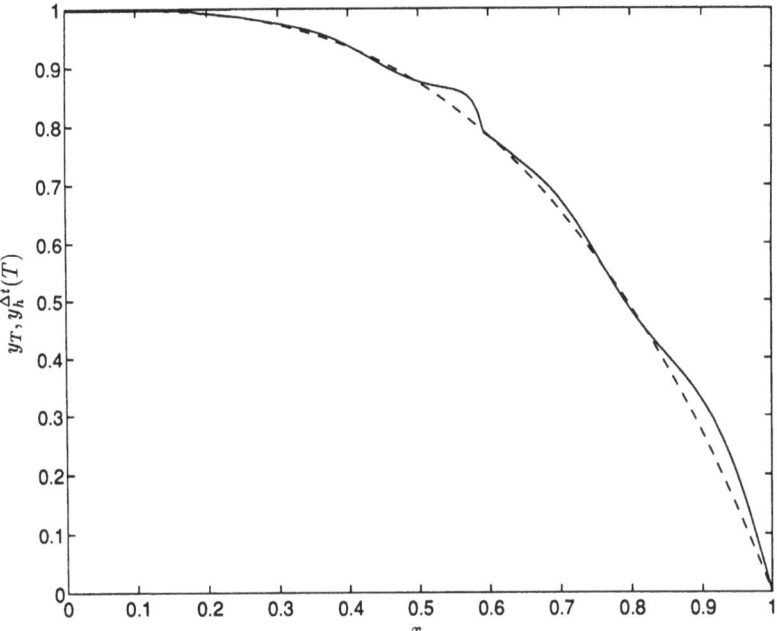

Fig. 41. Comparison between y_T (...) and $y_h^{\Delta t}(T)$ (—) ($\mathbf{a} = \{1/5, 3/5\}$).

Table 5. *Summary of numerical results.*

a	$\{1/5, 3/5\}$	$\{0.1, 0.3, 0.5, 0.7, 0.9\}$
Number of iterations	86	82
$\dfrac{\|y_h^{\Delta t}(T) - y_T\|_{L^2(0,1)}}{\|y_T\|_{L^2(0,1)}}$	2.5×10^{-2}	8.5×10^{-3}

are used, namely $a_1 = 1/5$ and $a_2 = 3/5$; the results are significantly better. Actually the results become 'very good' (as shown on Figure 42) when one uses the *five* control points $a_1 = 0.1$, $a_2 = 0.3$, $a_3 = 0.5$, $a_4 = 0.7$ and $a_5 = 0.9$; in that case we are 'close' to a control distributed over the whole interval (0,1). Some further results are summarized in Table 5.

Remark 4.4 Concerning the convergence of the *conjugate gradient algorithm* used to solve the approximate control problems (4.49) let us mention that

(i) The *Fletcher–Reeves* variant seems to have here a faster convergence than the *Polak–Ribière* one.

(ii) The computational time does not depend too much on the number M of control points. For example the CPU time (user time on a SUN

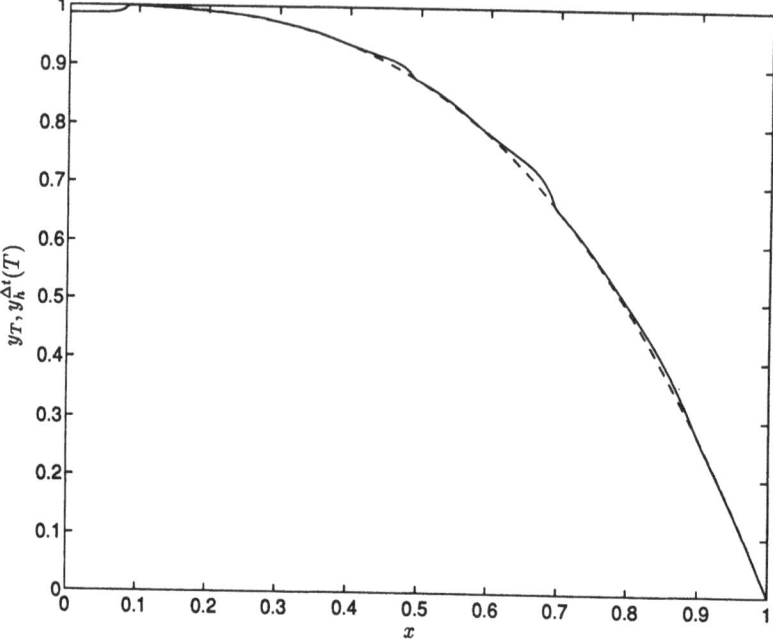

Fig. 42. Comparison between y_T (...) and $y_h^{\Delta t}(T)$ (—) ($\mathbf{a} = \{0.1, 0.3, 0.5,$ $0.7, 0.9\}$).

Workstation SPARC10) was about 22 s for the case with *one* control point at $a = 1/5$, to be compared with 27 s for the *five* control points test problem. Thus, the time-consuming part is the solution of the discrete state and adjoint state equations and not the manipulation of the control vectors (see Berggren and Glowinski (1994) for further details).

Remark 4.5 In Berggren and Glowinski (1994) we have also addressed and solved the more complicated problem where the control $\mathbf{u}$ *and* the location $\mathbf{a}$ of the controllers are unknown; this new problem can also be solved by a *conjugate gradient algorithm* operating in $L^2(0, T; \mathbb{R}^M) \times \mathbb{R}^M$; compared with the case where $\mathbf{a}$ is fixed the convergence of the new algorithm is much (about 4 times) slower (see, Berggren and Glowinski (1994) for the computational aspects and for numerical results).

4.3.7. Controllability and the Navier–Stokes equations

  Flow control is an important part of Engineering and from that point of view has been around for many years. However the corresponding *mathematical problems* are quite difficult and most of them are still open; it is therefore our opinion that a survey on the *numerical aspects* of these problems is still premature.

  It is nevertheless worth mentioning that a most important issue in that direction is the *control of turbulence* motivated, for example, by *drag reduc-*

tion (see, e.g., Buschnell and Hefner (1990) and Sellin and Moses (1989)). Another important issue concerns the control of *turbulent combustion* as discussed in, e.g., McManus, Poinsot and Candel (1993) and Samaniego, Yip, Poinsot and Candel (1993).

Despite the lack of theoretical results there is an enormous amount of literature on flow control topics (see, e.g., the four above publications and the references therein). Focusing on recent work in the spirit of the present article, let us mention Abergel and Temam (1990), Lions (1991a), Glowinski (1991), this list being far from complete. In the following we shall give further references; they concern the application of *Dynamic Programming* to the control of system governed by the Navier–Stokes equations.

5. DYNAMIC PROGRAMMING FOR LINEAR DIFFUSION EQUATIONS

5.1. Introduction. Synopsis

We address in this section the *'real time'* aspect of the controllability problems. We proceed in a largely formal fashion. The content of this section is based on Lions (1991b)

We consider again the *state equation*

$$\frac{\partial y}{\partial t} + Ay = v\chi_{\mathcal{O}}, \tag{5.1}$$

now in the time interval $(s, T], 0 \leq s \leq T$; the *'initial' condition* is

$$y(s) = h, \tag{5.2}$$

where h is an *arbitrary* function in $L^2(\Omega)$; the *boundary* condition is

$$y = 0 \text{ on } \Sigma_s = \Gamma \times (s, T). \tag{5.3}$$

Consider now the following *control problem*

$$\inf \frac{1}{2} \iint_{\mathcal{O} \times (s,T)} v^2 \, dx \, dt, \ v \in L^2(\mathcal{O} \times (s,T)) \text{ so that } y(T;v) \in y_T + \beta B, \tag{5.4}$$

where in (5.4), $\beta > 0$, B is the closed unit ball of $L^2(\Omega)$ centred at 0, $y_T \in L^2(\Omega)$ and $t \to y(t;v)$ is the solution of (5.1)–(5.3).

The *minimum* in (5.4) is now a function of h and s, we define $\phi(h, s)$ by

$$\phi(h, s) = \text{ minimal value of the cost function in (5.4)}. \tag{5.5}$$

We now derive the *Hamilton–Jacobi–Bellman* (HJB) *equation* satisfied by ϕ on $\times(0, T)$.

5.2. Derivation of the Hamilton–Jacobi–Bellman equation

As we said above, we shall proceed in a largely formal fashion. We take

$$v(x, t) = w(x) \text{ in } (s, s + \varepsilon), \ \varepsilon > 0 \text{ 'very small'}. \tag{5.6}$$

With this choice of v, the state function $y(t)$ moves during the time interval $(s, s + \varepsilon)$ from h to an element 'very close' to

$$h_\varepsilon = h - \varepsilon A h + \varepsilon w \chi_{\mathcal{O}}, \tag{5.7}$$

assuming that $h \in H^2(\Omega) \cap H_0^1(\Omega)$ (h_ε is obtained from h by the *explicit Euler scheme*).

On the time interval $(s + \varepsilon, T)$ we consider the whole process starting from h_ε at time $s + \varepsilon$. The *optimality principle* leads to

$$\phi(h, s) = \inf_w \left[\frac{\varepsilon}{2} \int_{\mathcal{O}} w^2 \, dx + \phi(h_\varepsilon, s + \varepsilon) \right] + \text{'negligible terms'}. \tag{5.8}$$

Taking now the ε-expansion of the function $\phi(h_\varepsilon, s + \varepsilon)$ we obtain

$$
\begin{aligned}
\phi(h_\varepsilon, s + \varepsilon) &= \phi(h - \varepsilon A h + \varepsilon w \chi_{\mathcal{O}}, s + \varepsilon), \\
&= \phi(h, x) - \varepsilon \left(\frac{\partial \phi}{\partial h}(h, s), A h \right) + \varepsilon \left(\frac{\partial \phi}{\partial h}(h, s), w \chi_{\mathcal{O}} \right) \\
&\quad + \varepsilon \frac{\partial \phi}{\partial s}(h, s) + \text{ higher-order terms}, \tag{5.9}
\end{aligned}
$$

where

$$\left(\frac{\partial \phi}{\partial h}(h, s), \hat{h} \right) = \lim_{\lambda \to 0} \frac{d}{d\lambda} \phi(h + \lambda \hat{h}, s),$$

with h and $\hat{h}$ in $L^2(\Omega)$ and, actually, smooth enough so that h and $\hat{h}$ belong to $H^2(\Omega) \cap H_0^1(\Omega)$.

Combining (5.8) to (5.9), dividing by ε, and letting $\varepsilon \to 0$, we obtain

$$\inf_w \left[\frac{1}{2} \int_{\mathcal{O}} w^2 \, dx - \left(\frac{\partial \phi}{\partial h}(h, s), A h \right) + \left(\frac{\partial \phi}{\partial h}(h, s), w \chi_{\mathcal{O}} \right) + \frac{\partial \phi}{\partial s}(h, s) \right] = 0; \tag{5.10}$$

hence it follows that

$$-\frac{\partial \phi}{\partial s}(h, s) + \left(\frac{\partial \phi}{\partial h}(h, s), A h \right) + \frac{1}{2} \int_{\mathcal{O}} \left(\frac{\partial \phi}{\partial h}(h, s) \right)^2 dx = 0. \tag{5.11}$$

The functional equation (5.11) is the *Hamilton–Jacobi–Bellmann* equation. It is a *partial differential equation in infinite dimensions* since $h \in L^2(\Omega)$ (in fact, $h \in H^2(\Omega) \cap H_0^1(\Omega)$), and where $s \in (0, T)$.

We have to add an '*initial*' condition, here for $t = T$, since we integrate (5.11) *backward in time*.

When $s \to T$, we have less and less time to 'correct' the trajectory. Therefore (this is again formal but it can be made precise without difficulty)

$$\phi(h, T) = \begin{cases} 0 \text{ if } h \in y_T + \beta B, \\ +\infty \text{ otherwise.} \end{cases} \tag{5.12}$$

5.3. Some remarks

Remark 5.1 The 'solution' of equations (5.11) and (5.12) should be defined in the framework of the *viscosity solutions* of Crandall and P.L. Lions (1985; 1986a,b; 1990; 1991), which was generalized by those authors to the infinite-dimensional case, which is the present situation.

Remark 5.2 Let h be given in $L^2(\Omega)$ and let y_h be the solution of

$$\frac{\partial y_h}{\partial t} + Ay_h = 0 \text{ in } \Omega \times (s, T), \ y_h(s) = h, \ y_h = 0 \text{ on } \Sigma_s \tag{5.13}$$

(i.e. we choose $v = 0$ in (5.1)–(5.3)). Let us denote by E_s the set of those functions h in (5.13) such that

$$y_h(T) \in y_T + \beta B. \tag{5.14}$$

We clearly have (from (5.11), (5.12))

$$\phi(h, s) = 0 \text{ if } h \in E_s. \tag{5.15}$$

We can – formally – draw the picture of Figure 43.

Remark 5.3 As usual in the *dynamic programming* approach, the *best decision* at time s corresponds to the element w in $L^2(\mathcal{O})$ which achieves the minimum in (5.10), namely

$$u(s) = -\frac{\partial \phi}{\partial h}(h, s)\chi_{\mathcal{O}}; \tag{5.16}$$

This is the 'real time' optimal policy – provided we know how to compute $(\partial\phi/\partial h)(h, s)$ – a formidable task indeed!

Remark 5.4 The Duality formulas of Section 1.4 can of course be applied. We obtain

$$\phi(h, s) = -\inf_{\hat{f}\in L^2(\Omega)} \left[\frac{1}{2} \iint_{\mathcal{O}\times(s,T)} \hat{\psi}^2 \, dx \, dt - \int_\Omega \hat{f}(y_T - y_h(T)) \, dx + \beta\|\hat{f}\|_{L^2(\Omega)} \right], \tag{5.17}$$

where y_h is defined by (5.13) and where $\hat{\psi}$ is defined by

$$-\frac{\partial\hat{\psi}}{\partial t} + A^*\hat{\psi} = 0 \text{ in } \Omega \times (s, T), \ \hat{\psi}(T) = \hat{f}, \ \hat{\psi} = 0 \text{ on } \Sigma_s. \tag{5.18}$$

Remark 5.5 *Dynamic programming* has been applied to the *closed loop* control of the *Navier–Stokes equations for incompressible viscous flow* in Sritharan (1991a,b).

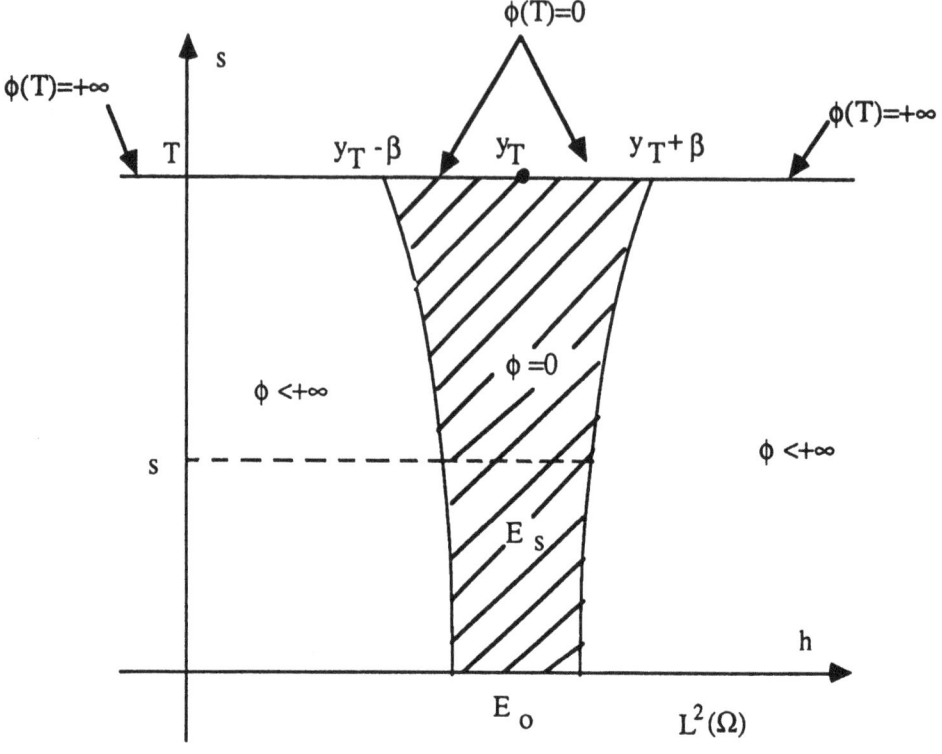

Fig. 43. Distribution of ϕ in the set $L^2(\Omega) \times (0,T)$.

6. WAVE EQUATIONS

6.1. Wave equations: Dirichlet boundary control

Let Ω be a *bounded* open set in $\mathbb{R}^d$, with a smooth boundary Γ. In $Q = \Omega \times (0,T)$, we consider the *wave equation*

$$\frac{\partial^2 y}{\partial t^2} + Ay = 0, \tag{6.1}$$

where A is a *second-order elliptic operator*, with *smooth coefficients*, and such that

$$A = A^*. \tag{6.2}$$

A classical case is

$$A = -\Delta \left(= -\nabla^2 = -\sum_{i=1}^{d} \frac{\partial^2}{\partial x_i^2} \right). \tag{6.3}$$

We assume that

$$y(0) = 0, \quad \frac{\partial y}{\partial t}(0) = 0, \tag{6.4}$$

and we suppose that the *control is applied on a part of the boundary*. More precisely, let Γ_0 be a 'smooth' subset of Γ. Then

$$y = \begin{cases} v \text{ on } \Sigma_0 = \Gamma_0 \times (0,T), \\ 0 \text{ on } \Sigma\backslash\Sigma_0, \Sigma = \Gamma \times (0,T). \end{cases} \tag{6.5}$$

We denote by $y(v) : t \to y(t; v)$ the solution of the wave problem (6.1), (6.4), (6.5), assuming that the control v satisfies 'some' further properties. Indeed, we shall assume that

$$v \in L^2(\Sigma_0), \tag{6.6}$$

since this is – as far as the control itself is concerned! – certainly the simplest possible choice. However, a few preliminary remarks are necessary here.

Remark 6.1. Even assuming that Γ, Γ_0 and the coefficients of operator A are very smooth, once the choice (6.6) has been made, one *has* to deal with *weak solutions* of (6.1), (6.4), (6.5). In fact (cf. Lions (1988a) and (1988b, Vol. 1)) the (unique) solution $y(v)$ of (6.1), (6.4), (6.5) satisfies the following properties

$$y(v) \text{ is continuous from } [0,T] \text{ to } L^2(\Omega), \tag{6.7}$$

$$y_t(v) \text{ is continuous from } [0,T] \text{ to } H^{-1}(\Omega), \tag{6.8}$$

where, in (6.8) and in the following, we have set

$$\varphi_t = \frac{\partial\varphi}{\partial t}, \quad \varphi_{tt} = \frac{\partial^2\varphi}{\partial t^2}.$$

The solution $y = y(v)$ is defined by *transposition* as in Lions and Magenes (1968). If we consider the adjoint equation

$$\begin{cases} \varphi_{tt} + A\varphi = f \text{ in } Q, \\ \varphi(T) = \varphi_t(T) = 0, \quad \varphi = 0 \text{ on } \Sigma, \end{cases} \tag{6.9}$$

where $f \in L^1(0,T; L^2(\Omega))$, then y is defined by

$$\int_Q yf \, dx \, dt = -\int_{\Sigma_0} \frac{\partial\varphi}{\partial n_A} v \, d\Sigma, \tag{6.10}$$

where $\partial/\partial n_A$ denotes the *normal derivative* associated with A (it is the usual normal derivative if $A = -\Delta$). The linear form

$$f \to -\int_{\Sigma_0} \frac{\partial\varphi}{\partial n_A} v \, d\Sigma$$

is *continuous* over $L^1(0,T; L^2(\Omega))$; this is the *key point* since we can show that

$$\left\| \frac{\partial\varphi}{\partial n_A} \right\|_{L^2(\Sigma)} \le C\|f\|_{L^1(0,T;L^2(\Omega))}. \tag{6.11}$$

One uses then the restriction of $\partial\varphi/\partial n_A$ to Σ_0 and therefore

$$y \in L^\infty(0, T; L^2(\Omega)).$$

One proceeds then to obtain (6.7), (6.8).

Remark 6.2. The original proof (Lions (1983)) assumes that Γ is *smooth*. Strangely enough, it took ten years – and a nontrivial technical proof – to generalize (6.11) to *Lipschitz boundaries* (in the sense of Nečas (1967)); this was done by Chiara (1993).

We now want to study the *controllability* for systems modelled by (6.1), (6.4), (6.5), i.e., *given*

$$T(0 < T < +\infty), \quad given \ \{z^0, z^1\} \in L^2(\Omega) \times H^{-1}(\Omega),$$

can we find v such that

$$\begin{cases} y(T; v) = z^0 \text{ or } y(T; v) \text{ 'very close' to } z^0, \\ y_t(T; v) = z^1 \text{ or } y_t(T; v) \text{'very close' to } z^1. \end{cases} \qquad (6.12)$$

There is a *fundamental difference* between the present situation and those discussed in Sections 1 and 2 for *diffusion* equations, due to the *finite propagation velocity* of the waves (or singularities) the solution is made of, whereas this velocity is *infinite* for *diffusion* equations (and for *Petrowsky's* type equations as well). It follows from this property that

Conditions (6.12) may be possible only if T is sufficiently large.   (6.13)

This will be made precise in the following sections.

6.2. Approximate controllability

For technical reasons, we shall always consider the mapping

$$L : v \rightarrow \{-y_t(T; v), y(T; v)\} \qquad (6.14)$$

(which is a *continuous linear* mapping from $L^2(\Sigma_0)$ into $H^{-1}(\Omega) \times L^2(\Omega)$), instead of the mapping

$$v \rightarrow \{y(T; v), y_t(T; v)\}$$

(but this does not go beyond simplifying – we hope – some formulae).

Let us first discuss the range $R(L)$ of operator L; we consider thus $\mathbf{f} = \{f^0, f^1\}$ such that

$$\mathbf{f} \in H_0^1(\Omega) \times L^2(\Omega) \qquad (6.15)$$

and

$$\langle Lv, \mathbf{f} \rangle = 0, \quad \forall v \in L^2(\Sigma_0), \qquad (6.16)$$

i.e.

$$-\langle y_t(T;v), f^0\rangle + \int_\Omega y(T;v)f^1\,dx = 0, \ \forall v \in L^2(\Sigma_0); \tag{6.16}'$$

in (6.16) (respectively (6.16)'), $\langle\cdot,\cdot\rangle$ denotes the duality pairing between $H^{-1}(\Omega) \times L^2(\Omega)$ and $H_0^1(\Omega) \times L^2(\Omega)$ (respectively $H^{-1}(\Omega)$ and $H_0^1(\Omega)$).

We introduce ψ solution of

$$\psi_{tt} + A\psi = 0 \text{ in } Q = \Omega \times (0,T), \quad \psi(T) = f^0, \ \psi_t(T) = f^1, \ \psi = 0 \text{ on } \Sigma. \tag{6.17}$$

It is a smooth solution, which satisfies in particular

$$\frac{\partial\psi}{\partial n_A} \in L^2(\Sigma), \left\|\frac{\partial\psi}{\partial n_A}\right\|_{L^2(\Sigma)} \le C(\|f^0\|_{H_0^1(\Omega)} + \|f^1\|_{L^2(\Omega)}). \tag{6.18}$$

Multiplying the first equation in (6.17) by y and integrating by parts, we obtain

$$\langle Lv, \mathbf{f}\rangle = \int_{\Sigma_0} \frac{\partial\psi}{\partial n_A} v\,d\Sigma. \tag{6.19}$$

Thus (6.16) is equivalent to

$$\frac{\partial\psi}{\partial n_A} = 0 \text{ on } \Sigma_0. \tag{6.20}$$

Therefore the *Cauchy data* are zero for ψ on Σ_0. According to the *Holmgren's Uniqueness Theorem* (cf. Hörmander (1976)) it follows that

$$\text{If } T > 2(\text{diameter of } \Omega), \text{ then } \{y(T;v), y_t(T;v)\}$$
$$\text{describes a dense subspace of } L^2(\Omega) \times H^{-1}(\Omega) \tag{6.21}$$

(in (6.21), the 'diameter' of Ω is related to the geodetic distance associated with A. It is the usual geodetic distance if $A = -\Delta$).

Indeed, according to *Holmgren's Theorem*, we have $\psi \equiv 0$ so that $f = 0$ (see, for example, Lions (1988b, Vol. 1)).

Remark 6.3. Holmgren's theorem applies with the conditions

$$\psi = 0 \quad \text{and} \quad \frac{\partial\psi}{\partial n_A} = 0 \text{ on } \Sigma_0,$$

without having necessarily $\psi = 0$ on $\Sigma\backslash\Sigma_0$. The fact that in the present situation we have $\psi = 0$ on the whole Σ provides some more flexibility to obtain uniqueness results. We shall return to this later on.

6.3. Formulation of the approximate controllability problem

We shall make the following hypothesis:

$$\Sigma_0 \text{ allows the application of the Holmgren's Uniqueness Theorem.} \tag{6.22}$$

Then, $\{z^0, z^1\}$ being given in $L^2(\Omega) \times H^{-1}(\Omega)$, there always exist controls v (actually an infinite number of them) such that

$$y(T; v) \in z^0 + \beta_0 B, \quad y_t(T; v) \in z^1 + \beta_1 B_{-1}, \tag{6.23}$$

where B (respectively B_{-1}) denotes the unit ball of $L^2(\Omega)$ (respectively $H^{-1}(\Omega)$), and where β_0, β_1 are given positive members, arbitrarily small.

The *optimal control* problem that we consider is

$$\inf_v \frac{1}{2} \int_{\Sigma_0} v^2 \, d\Sigma, \quad v \text{ satisfying } (6.23). \tag{6.24}$$

Remark 6.4. *Exact controllability* corresponds to $\beta_0 = \beta_1 = 0$.

6.4. Dual problems

We proceed essentially as in Section 1.4. We introduce therefore

$$F_1(v) = \frac{1}{2} \int_{\Sigma_0} v^2 \, d\Sigma, \quad \forall v \in L^2(\Sigma_0), \tag{6.25}$$

and then $F_2 : H^{-1}(\Omega) \times L^2(\Omega) \to \mathbb{R} \cup \{+\infty\}$ by

$$F_2(\hat{f}) = F_2(\hat{f}^0, \hat{f}^1) = \begin{cases} 0 \text{ if } \hat{f}^0 \in -z^1 + \beta_1 B_{-1} \text{ and } \hat{f}^1 \in z^0 + \beta_0 B, \\ +\infty \text{ otherwise.} \end{cases} \tag{6.26}$$

With this notation, the control problem (6.20) can be formulated as

$$\inf_{v \in L^2(\Sigma_0)} [F_1(v) + F_2(Lv)]. \tag{6.27}$$

Using, as in Section 1.4, *Duality Theory* we obtain

$$\int_{v \in L^2(\Sigma_0)} [F_1(v) + F_2(Lv)] + \inf_{\hat{f} \in H_0^1(\Omega) \times L^2(\Omega)} [F_1^*(L^*\hat{f}) + F_2^*(-\hat{f})] = 0, \tag{6.28}$$

where

$$F_1^*(v) = \frac{1}{2} \int_{\Sigma_0} v^2 \, d\Sigma, \quad \forall v \in L^2(\Sigma_0), \tag{6.29}_1$$

$$\begin{cases} F_2^*(\hat{f}) = -\langle z^1, \hat{f}^0 \rangle + \int_\Omega z^0 \hat{f}^1 \, dx + \beta_1 \|\hat{f}^0\|_{H_0^1(\Omega)} + \beta_0 \|\hat{f}^1\|_{L^2(\Omega)}, \\ \forall \hat{f} = \{\hat{f}^0, \hat{f}^1\} \in H_0^1(\Omega) \times L^2(\Omega). \end{cases} \tag{6.29}_2$$

Using (6.19) we have

$$L^*\hat{f} = \frac{\partial \hat{\psi}}{\partial n_A} \text{ on } \Sigma_0, \tag{6.30}$$

where $\hat{\psi}$ is given by (6.17) (with $f = \hat{f}$). We have therefore the following

Theorem 6.1 *We suppose that (6.22) holds true. For β_0 and β_1 given arbitrarily small, problem (6.24) has a unique solution such that*

$$\inf_v \frac{1}{2} \int_{\Sigma_0} v^2 \, d\Sigma = -\inf_{\hat{\mathbf{f}}} \left[\frac{1}{2} \int_{\Sigma_0} \left(\frac{\partial \hat{\psi}}{\partial n_A} \right)^2 d\Sigma + \langle z^1, \hat{f}^0 \rangle \right.$$

$$\left. - \int_\Omega z^0 \hat{f}^1 \, dx + \beta_1 \|\hat{f}^0\|_{H_0^1(\Omega)} + \beta_0 \|\hat{f}^1\|_{L^2(\Omega)} \right],$$

(6.31)

where, in (6.31), $v \in L^2(\Sigma_0)$ and verifies (6.23), $\hat{\mathbf{f}} \in H_0^1(\Omega) \times L^2(\Omega)$, and where $\hat{\psi}$ is given by (6.17), with $\mathbf{f} = \hat{\mathbf{f}}$.

The *dual problem* is the minimization problem in the right-hand side of (6.31). If $\mathbf{f}$ is the solution of the dual problem and if ψ is the corresponding solution of (6.17) then the optimal control, i.e. the solution u of problem (6.24) is given by

$$u = \frac{\partial \psi}{\partial n_A} \text{ on } \Sigma_0. \tag{6.32}$$

6.5. Direct solution of the dual problem

One can formulate the dual problem in an equivalent fashion which will be useful when β_0 and β_1 converge to zero, and also for numerical calculations.
  To this effect, we introduce the following operator Λ:
  Given $\hat{\mathbf{f}} = \{\hat{f}^0, \hat{f}^1\} \in H_0^1(\Omega) \times L^2(\Omega)$, we define $\hat{\psi}$ and $\hat{y}$ by

$$\hat{\psi}_{tt} + A\hat{\psi} = 0 \text{ in } Q, \quad \hat{\psi}(T) = \hat{f}^0, \ \hat{\psi}_t(T) = \hat{f}^1, \ \hat{\psi} = 0 \text{ on } \Sigma, \tag{6.33}_1$$

$$\hat{y}_{tt} + A\hat{y} = 0 \text{ in } Q, \quad \hat{y}(0) = \hat{y}_t(0) = 0, \ \hat{y} = \frac{\partial \hat{\psi}}{\partial n_A} \text{ on } \Sigma_0, \ \hat{y} = 0 \text{ on } \Sigma \backslash \Sigma_0,$$

$$(6.33)_2$$

and we set

$$\Lambda \hat{\mathbf{f}} = \{-\hat{y}_t(T), \hat{y}(T)\}. \tag{6.34}$$

We define in this way an operator Λ such that

$$\Lambda \in \mathcal{L}(H_0^1(\Omega) \times L^2(\Omega); H^{-1}(\Omega) \times L^2(\Omega)). \tag{6.35}$$

If we multiply both sides of the first equation in $(6.33)_2$ by $\hat{\psi}'$ (which corresponds to $\hat{f}' \in H_0^1(\Omega) \times L^2(\Omega)$) and if we integrate by parts we obtain (with obvious notation):

$$\langle \Lambda \hat{\mathbf{f}}, \hat{\mathbf{f}}' \rangle = \int_{\Sigma_0} \frac{\partial \hat{\psi}}{\partial n_A} \frac{\partial \hat{\psi}'}{\partial n_A} \, d\Sigma. \tag{6.36}$$

It follows from (6.36) that the operator Λ is *self-adjoint* and *positive semi-definite*.

The *dual problem* is then *equivalent* to

$$\inf_{\hat{\mathbf{f}}} \left[\frac{1}{2}\langle \Lambda \hat{\mathbf{f}}, \hat{\mathbf{f}} \rangle + \langle z^1, \hat{f}^0 \rangle - \int_\Omega z^0 \hat{f}^1 \, dx + \beta_1 \|\hat{f}^0\|_{H_0^1(\Omega)} + \beta_0 \|\hat{f}^1\|_{L^2(\Omega)} \right], \quad (6.37)$$

where, in (6.37), $\hat{\mathbf{f}} \in H_0^1(\Omega) \times L^2(\Omega)$.

Assuming that the condition (6.22) holds true, problem (6.37) has a *unique* solution for β_0 and $\beta_1 > 0$, arbitrarily small. If we denote by $\mathbf{f}$ the solution of problem (6.37) it is also the solution of the following *variational inequality*

$$\mathbf{f} \in H_0^1(\Omega) \times L^2(\Omega); \ \forall \hat{\mathbf{f}} \in H_0^1(\Omega) \times L^2(\Omega) \text{ we have}$$

$$\langle \Lambda \mathbf{f}, \hat{\mathbf{f}} - \mathbf{f} \rangle + \langle z^1, \hat{f}^0 - f^0 \rangle - \int_\Omega z^0 (\hat{f}^1 - f^1) \, dx$$

$$+ \beta_1 (\|\hat{f}^0\|_{H_0^1(\Omega)} - \|f^0\|_{H_0^1(\Omega)})$$

$$+ \beta_0 (\|\hat{f}^1\|_{L^2(\Omega)} - \|f^1\|_{L^2(\Omega)}) \geq 0. \quad (6.38)$$

Remark 6.5. Problems (6.37), (6.38) are equivalent to the minimization problem in the right-hand side of (6.31), but they are better suited for the solution of the dual problem.

Remark 6.6. The operator Λ is the same as the one introduced in the *Hilbert Uniqueness Method* (HUM). This is made more precise in the following section (see also Lions (1986; 1988a,b)).

Remark 6.7. Relation (6.36) makes sense, because there exists a constant C such that

$$\int_{\Sigma_0} \left| \frac{\partial \hat{\psi}}{\partial n_A} \right|^2 d\Sigma \leq C(\|\hat{f}^0\|_{H_0^1(\Omega)}^2 + \|\hat{f}^1\|_{L^2(\Omega)}), \quad (6.39)$$

where, in (6.39), $\hat{\psi}$ and $\hat{\mathbf{f}} = \{\hat{f}^0, \hat{f}^1\}$ are related by (6.33).

6.6. Exact controllability and new functional spaces

Let us now consider problem (6.37), (6.38) with the idea of letting β_0 and β_1 converge to zero. We introduce on $H_0^1(\Omega) \times L^2(\Omega)$ the following new functional

$$[\hat{f}] = (\langle \Lambda \hat{f}, \hat{f} \rangle)^{1/2}. \quad (6.40)$$

Since we assume that the condition (6.22) holds true, the functional $[\cdot]$ is in fact a *norm*, of a *pre-Hilbertian* nature. We introduce then

$$E = \text{ Completion of } H_0^1(\Omega) \times L^2(\Omega) \text{ for the norm } [\cdot]; \quad (6.41)$$

with this notation we can state that

$$\Lambda \text{ is an isomorphism from } E \text{ onto } E'. \quad (6.42)$$

If $\beta_0 = \beta_1 = 0$, problem (6.37), (6.38) is *equivalent* to

$$\inf_{\hat{\mathbf{f}}} \left[\frac{1}{2} [\hat{f}]^2 + \langle z^1, \hat{f}^0 \rangle - \int_\Omega z^0 \hat{f}^1 \, dx \right], \quad \hat{\mathbf{f}} \in H_0^1(\Omega) \times L^2(\Omega). \qquad (6.43)$$

Problem (6.44) has a unique solution if and only if

$$\{-z^1, z^0\} \in E'. \qquad (6.44)$$

If we denote by $\mathbf{f}_\beta = \{f_\beta^0, f_\beta^1\}$ the solution of problem (6.37), (6.38), where $\beta = \{\beta_0, \beta_1\}$, then

$$\lim_{\beta \to 0} \mathbf{f}_\beta = \mathbf{f}_0 = \text{the solution of (6.43)} \qquad (6.45)$$

if and only if condition (6.44) holds true.

Remark 6.8. The method of solution that we have just presented is what is called HUM (*Hilbert Uniqueness Method*) (cf. Lions (1986; 1988a,b)) since the key element is the introduction of the new Hilbert space E based on a *uniqueness* property.

Remark 6.9. Problems (6.38) or (6.43) give a *constructive* approach to approximate or exact controllability; we shall make this more precise in the next sections.

Remark 6.10. Condition (6.44) means that *exact controllability is possible if and only if* z^0 *and* z^1 *are taken in a convenient Hilbert space.*

Remark 6.11. The approach taken in the present section is closely related to the one followed in Section 1.5. With the notation of Section 1.5, Remark 1.14, we would have

$$E = \widehat{H_0^1(\Omega) \times L^2(\Omega)}.$$

There is, however, a very important technical difference between the two situations, since for the *diffusion* problems discussed in Section 1 the space $\widehat{L^2(\Omega)}$ is *never* a 'simple' *distribution space* (except for the case $\mathcal{O} = \Omega$, i.e. the control is distributed over the whole domain Ω). For the *wave equation* the situation is quite different, as we shall see in the next section.

6.7. On the structure of space E

We follow here Bardos, Lebeau and Rauch (1988).

We shall say that Σ_0 enjoys the *geometrical control condition* if any ray, starting from any point of Ω at $t = 0$, reaches eventually (after geometrical reflexions on Γ) the set Γ_0 before time $t = T$.

The main result is then

If Σ_0 satisfies the geometrical control condition, then $E = H_0^1(\Omega) \times L^2(\Omega)$.

$$(6.46)$$

Actually the geometrical control condition is also *necessary* in order (6.46) to be true. The inequality corresponding to (6.46) is the *reverse* of inequality (6.39): there exists a constant $C_1 > 0$ such that

$$\int_{\Sigma_0} \left| \frac{\partial \hat{\psi}}{\partial n_A} \right|^2 d\Sigma \geq C_1 (\|\hat{f}^0\|^2_{H^1_0(\Omega)} + \|\hat{f}^1\|^2_{L^2(\Omega)}) \tag{6.47}$$

if and only if Σ_0 satisfies the geometrical control condition.

Remark 6.12. We refer to Lions (1988b) for the various contributions, by many authors, which led to the fundamental inequality (6.47).

Remark 6.13. If Σ_0 *does not* satisfy the geometrical control condition, but *does* satisfy the conditions for the Holmgren's Uniqueness Theorem, then

$$[\hat{f}] = \left(\int_{\Sigma_0} \left| \frac{\partial \hat{\psi}}{\partial n_A} \right|^2 d\Sigma \right)^{1/2}$$

is a norm, *strictly weaker* than the $H^1_0(\Omega) \times L^2(\Omega)$ norm. In that case E is a *new* Hilbert space, such that

$$H^1_0(\Omega) \times L^2(\Omega) \subset E, \quad \text{strictly,} \tag{6.48}$$

and the exact structure of E is far from being simple, since the space E can *contain elements which are not distributions on* Ω.

6.8. Numerical methods for the Dirichlet boundary controllability of the wave equation

6.8.1. Generalities. Synopsis

In this section which is largely inspired by Dean, Glowinski and Li (1989), Glowinski *et al.* (1990), Glowinski (1992a) we shall discuss the *numerical solution* of the *exact* and *approximate Dirichlet boundary controllability problems* considered in the preceding sections.

To make it simpler we shall assume that Σ_0 satisfies the *geometrical control condition* (see the above section), so that

$$E = H^1_0(\Omega) \times L^2(\Omega), \tag{6.49}$$

and the operator Λ defined in Section 6.5 is an *isomorphism* from E onto $E'(= H^{-1}(\Omega) \times L^2(\Omega))$. The properties of Λ (*symmetry* and *strong ellipticity*) will make the solution of the exact controllability problem possible by a *conjugate gradient algorithm* operating in the space E. We shall describe next the *time* and *space discretizations* of the exact controllability problem by a combination of *finite difference* (FD) and *finite element* (FE) methods and then discuss the iterative solution of the corresponding approximate problem. Finally we shall describe solution methods for the *approximate boundary controllability problem* (6.24).

Both exact and approximate controllability problems will be solved using their *dual formulation* since the corresponding control problems are *easier* to solve than their primal counterparts.

The results of numerical experiments obtained by the methods described in the present section will be reported in Section 6.9, hereafter.

Remark 6.14. A *spectral* method – still based on HUM – for solving *directly* (i.e. non*iteratively*) the *exact* Dirichlet boundary controllability problem is discussed in Bourquin (1993), where numerical results are also presented.

6.8.2. Dual formulation of the exact controllability problem. Further properties of Λ

To obtain the *dual problem* corresponding to *exact controllability* it suffices to take $\beta_0 = \beta_1 = 0$ in formulations (6.37), (6.38); we obtain then

$$\Lambda \mathbf{f} = \{-z^1, z^0\}. \tag{6.50}$$

Since we supposed (see Section 6.8.1) that the *geometrical control condition* holds, we know (from Section 6.6) that

$$\Lambda \text{ is an isomorphism from } E \text{ onto } E', \tag{6.51}$$

with $E = H_0^1(\Omega) \times L^2(\Omega)$, $E' = H^{-1}(\Omega) \times L^2(\Omega)$. Problem (6.50) has therefore a *unique* solution, $\forall \{z^0, z^1\} \in L^2(\Omega) \times H^{-1}(\Omega)$. The solution $\mathbf{f}$ of (6.50) is also *the* solution of the following *linear variational* problem

$$\begin{cases} \mathbf{f} \in E; \; \forall \hat{\mathbf{f}} = \{\hat{f}^0, \hat{f}^1\} \in E \text{ we have} \\ \langle \Lambda \mathbf{f}, \hat{\mathbf{f}} \rangle = -\langle z^1, \hat{f}^0 \rangle + \int_\Omega z^0 \hat{f}^1 \, dx. \end{cases} \tag{6.52}$$

Since (from Sections 6.5 to 6.7) Λ is *continuous*, *self-adjoint* and *strongly elliptic* (in the sense that there exists $C > 0$ such that

$$\langle \Lambda \hat{\mathbf{f}}, \hat{\mathbf{f}} \rangle \geq C \|\hat{\mathbf{f}}\|_E^2, \quad \forall \hat{\mathbf{f}} \in E)$$

the *bilinear functional*

$$\{\hat{\mathbf{f}}, \hat{\mathbf{f}}'\} \to \langle \Lambda \hat{\mathbf{f}}, \hat{\mathbf{f}}' \rangle : E \times E \to \mathbb{R}$$

is *continuous, symmetric* and *E-elliptic* over $E \times E$. On the other hand, the *linear* functional in the right-hand side of (6.52) is clearly *continuous* over E, implying (cf. Section 1.8.2) that problem (6.50), (6.52) can be solved by a *conjugate gradient algorithm* operating in the space E. such an algorithm will be described in the following section.

Remark 6.15. We suppose here that $\Gamma_0 = \Gamma$ and that $A = -\Delta$; we suppose also that there exists $x_0 \in \Omega$ and $C > 0$ such that

$$\overrightarrow{x_0 M} \cdot \mathbf{n} = C, \quad \forall M \in \Gamma, \tag{6.53}$$

with $\mathbf{n}$ the unit vector of the outward normal at Γ, at M. Domains satisfying

(6.53) are easy to characterize geometrically, simple cases being disks and squares. Now let us denote by Λ_T the operator Λ associated with T. It has been shown by J.L. Lions (unpublished result) and Bensoussan (1990) that

$$\lim_{T \to +\infty} \frac{\Lambda_T}{T} = \frac{1}{C} \begin{bmatrix} -\Delta & 0 \\ 0 & I \end{bmatrix}. \tag{6.54}$$

Result (6.54) is quite important for the validation of the numerical methods described hereafter, since it easily provides

$$\lim_{T \to +\infty} T\mathbf{f}_T = \{\chi^0, \chi^1\}, \tag{6.55}$$

where, from (6.54),

$$\Delta \chi^0 = C z^1 \text{ in } \Omega, \quad \chi^0 = 0 \text{ on } \Gamma, \tag{6.56}$$

$$\chi^1 = C z^0. \tag{6.57}$$

6.8.3. Conjugate gradient solution of problem (6.50), (6.52).

Assuming that the *geometrical control condition* holds, it follows from Section 6.8.2. that we can apply the general *conjugate gradient* algorithm (1.122)–(1.129) to the solution of problem (6.50), (6.52); indeed, it suffices to take

$$V = E, a(\cdot, \cdot) = \langle \Lambda \cdot, \cdot \rangle, \quad L : \hat{\mathbf{f}} \to -\langle z^1, \hat{f}^0 \rangle + \int_\Omega z^0 \hat{f}^1 \, dx.$$

On E, we shall use as scalar product

$$\{\mathbf{v}, \mathbf{w}\} \to \int_\Omega (\nabla v^0 \cdot \nabla w^0 + v^1 w^1) \, dx, \quad \forall \mathbf{v}, \mathbf{w} \in E.$$

We obtain then the following algorithm

Algorithm. Step 0: Initialization

$$f_0^0 \in H_0^1(\Omega) \quad \text{and} \quad f_0^1 \in L^2(\Omega) \text{ are given;} \tag{6.58}$$

solve then

$$\begin{cases} \dfrac{\partial^2 \psi_0}{\partial t^2} + A\psi_0 = 0 \text{ in } Q, \quad \psi_0 = 0 \text{ on } \Sigma, \\[2mm] \psi_0(T) = f_0^0, \quad \dfrac{\partial \psi_0}{\partial t}(T) = f_0^1, \end{cases} \tag{6.59}$$

and

$$\begin{cases} \dfrac{\partial^2 \varphi_0}{\partial t^2} + A\varphi_0 = \text{ in } Q, \quad \varphi_0 = \dfrac{\partial \psi_0}{\partial n_A} \text{ on } \Sigma_0, \quad \varphi_0 = 0 \text{ on } \Sigma \backslash \Sigma_0, \\[2mm] \varphi_0(0) = 0, \quad \dfrac{\partial \varphi_0}{\partial t}(0) = 0. \end{cases} \tag{6.60}$$

Compute $\mathbf{g}_0 = \{g_0^0, g_0^1\} \in E$ *by*

$$-\Delta g_0^0 = z^1 - \frac{\partial \varphi_0}{\partial t}(T) \text{ in } \Omega, g_0^0 = 0 \text{ on } \Gamma, \tag{6.61}$$

$$g_0^1 = \varphi_0(T) - z^0, \tag{6.62}$$

respectively. Set then

$$\mathbf{w}_0 = \mathbf{g}_0. \tag{6.63}$$

Now, for $n \geq 0$, assuming that $\mathbf{f}_n, \mathbf{g}_n, \mathbf{w}_n$ are known, compute $\mathbf{f}_{n+1}, \mathbf{g}_{n+1}, \mathbf{w}_{n+1}$ as follows.
  Step 1: Descent

  Solve

$$\begin{cases} \dfrac{\partial^2 \bar{\psi}_n}{\partial t^2} + A\bar{\psi}_n = 0 \text{ in } Q, \quad \bar{\psi}_n = 0 \text{ on } \Sigma, \\ \bar{\psi}_n(T) = w_n^0, \quad \dfrac{\partial \bar{\psi}_n}{\partial t}(T) = w_n^1, \end{cases} \tag{6.64}$$

$$\begin{cases} \dfrac{\partial^2 \bar{\varphi}_n}{\partial t^2} + A\bar{\varphi}_n = 0 \text{ in } Q, \quad \bar{\varphi}_n = \dfrac{\partial \bar{\psi}^n}{\partial n_A} \text{ on } \Sigma_0, \quad \bar{\varphi}_n = 0 \text{ on } \Sigma \backslash \Sigma_0, \\ \bar{\varphi}_n(0) = 0, \quad \dfrac{\partial \bar{\varphi}_n}{\partial t}(0) = 0, \end{cases}$$

$$\tag{6.65}$$

$$\Delta \bar{g}_n^0 = \dfrac{\partial \bar{\varphi}_n}{\partial t}(T) \text{ in } \Omega, \bar{g}_n^0 = 0 \text{ on } \Gamma, \tag{6.66}$$

and set

$$\bar{g}_n^1 = \bar{\varphi}_n(T). \tag{6.67}$$

Compute now

$$\rho_n = \int_\Omega (|\nabla g_n^0|^2 + |g_n^1|^2) \, dx \Big/ \int_\Omega (\nabla \bar{g}_n^0 \cdot \nabla w_n^0 + \bar{g}_n^1 w_n^1) \, dx, \tag{6.68}$$

$$\mathbf{f}_{n+1} = \mathbf{f}_n - \rho_n \mathbf{w}_n, \tag{6.69}$$

$$\mathbf{g}_{n+1} = \mathbf{g}_n - \rho_n \bar{\mathbf{g}}_n. \tag{6.70}$$

Step 2: Test of the convergence and construction of the new descent direction. If $\mathbf{g}_{n+1} = \mathbf{0}$, or is sufficiently small (i.e.

$$\int_\Omega (|\nabla g_{n+1}^0|^2 + |g_{n+1}^1|^2) \, dx \Big/ \int_\Omega (|\nabla g_0^0|^2 + |g_0^1|^2) \, dx \leq \epsilon^2) \tag{6.71}$$

take $\mathbf{f} = \mathbf{f}_{n+1}$; if not, compute

$$\gamma_n = \int_\Omega (|\nabla g_{n+1}^0|^2 + |g_{n+1}^1|^2) \, dx \Big/ \int_\Omega (|\nabla g_n^0|^2 + |g_n^1|^2) \, dx, \tag{6.72}$$

and set

$$\mathbf{w}_{n+1} = \mathbf{g}_{n+1} + \gamma_n \mathbf{w}_n. \tag{6.73}$$

Do $n = n + 1$ and go to (6.64).

Remark 6.16. It appears at first glance that algorithm (6.58)–(6.73) is quite memory demanding since it seems to require the storage of $\partial \bar{\psi}_n / \partial n_A |_{\Sigma_0}$ (in practice the storage of $\partial \bar{\psi}_n / \partial n_A$ over a discrete – but still large – subset of Σ_0). In fact, we can avoid this storage problem by observing that since the wave equation in (6.64) is *reversible* we can integrate *simultaneously, from 0 to T*, the wave equations (6.65) and

$$\begin{cases} \dfrac{\partial^2 \bar{\psi}_n}{\partial t^2} + A \bar{\psi}_n = 0 \text{ in } Q, \quad \bar{\psi}_n = 0 \text{ on } \Sigma, \\[2mm] \bar{\psi}_n(0) \text{ and } \dfrac{\partial \bar{\psi}_n}{\partial t}(0) \text{ known from the integration of (6.64) from } T \text{ to } 0. \end{cases} \tag{6.74}$$

In the particular case where an *explicit scheme* is used for solving the wave equations (6.64), (6.65) and (6.74), the extra cost associated with the solution of (6.74) is *negligible* compared with the saving due to not storing $\partial \bar{\psi}_n / \partial n_A$ on Σ_0.

Remark 6.17. Once the solution $\mathbf{f}$ of the dual problem (6.50), (6.52) is known it suffices to integrate the wave equation (6.33)$_1$ with $\hat{\mathbf{f}} = \mathbf{f}$ to obtain ψ. The optimal control u, solution of the exact controllability problem is given then by

$$u = \left. \frac{\partial \psi}{\partial n_A} \right|_{\Sigma_0} \tag{6.75}$$

6.8.4. Finite difference approximation of the dual problem (6.50), (6.52)

6.8.4.1. Generalities. An FE/FD approximation of problem (6.50), (6.52) will be discussed in Section 6.8.7 (see also Glowinski et al. (1990), Glowinski (1992a), and the references therein). At the present moment, we shall concentrate on the case where

$$\Omega = (0,1)^2, \quad A = -\Delta, \quad \Gamma_0 = \Gamma,$$

and where FD methods are used both for the space and time discretizations. Indeed, these approximations can also be obtained via space discretizations associated with FE grids like the one shown on Figure 1 of Section 2.6 (we should use, as shown in Glowinski et al. (1990), piecewise linear approximations and numerical integration by the trapezoidal rule).

Let I and N be positive integers; we define h (*space discretization step*) and Δt (*time discretization step*) by

$$h = \frac{1}{(I+1)}, \quad \Delta t = \frac{T}{N}, \tag{6.76}$$

respectively, and then denote by M_{ij} the point $\{ih, jh\}$.

6.8.4.2. Approximation of the wave equation $(6.33)_1$. Let us first discuss the discretization of the following wave problem

$$\begin{cases} \psi_{tt} - \Delta\psi = 0 \text{ in } Q, \quad \psi = 0 \text{ on } \Sigma, \\ \psi(T) = f^0, \quad \psi_t(T) = f^1. \end{cases} \tag{6.77}$$

With ψ_{ij}^n an approximation of $\psi(M_{ij}, n\Delta t)$, we approximate (6.77) by the following *explicit* FD scheme

$$\begin{cases} \dfrac{\psi_{ij}^{n-1} + \psi_{ij}^{n+1} - 2\psi_{ij}^n}{|\Delta t|^2} - \dfrac{\psi_{i+1j}^n + \psi_{i-1j}^n + \psi_{ij+1}^n + \psi_{ij-1}^n - 4\psi_{ij}^n}{h^2} = 0, \\ 1 \le i, j \le I, \quad 0 \le n \le N, \end{cases} \tag{6.78}_1$$

$$\psi_{kl}^n = 0 \quad \text{if } M_{kl} \in \Gamma, \tag{6.78}_2$$

$$\psi_{ij}^N = f^0(M_{ij}), \quad \psi_{ij}^{N+1} - \psi_{ij}^{N-1} = 2\Delta t f^1(M_{ij}), \quad 1 \le i, j \le I. \tag{6.78}_3$$

To be *stable*, the above scheme has to satisfy the following (*stability*) condition

$$\Delta t \le h/\sqrt{2}. \tag{6.79}$$

6.8.4.3. Approximation of $(\partial\psi/\partial n)|_\Sigma$. Suppose that we want to approximate $\partial\psi/\partial n$ at $M \in \Gamma$, as shown in Figure 44. Suppose that ψ is known at E; we shall then approximate $\partial\psi/\partial n$ at M by

$$\frac{\partial\psi}{\partial n}(M) \approx \frac{\psi(E) - \psi(W)}{2h}. \tag{6.80}$$

In fact, $\psi(E)$ is not known since $E \notin \bar{\Omega}$. However – formally at least – $\psi = 0$ on Σ implies $\psi_{tt} = 0$ on Σ, which combined with $\psi_{tt} - \Delta\psi = 0$ implies $\Delta\psi = 0$ on Σ; discretizing this last relation at M yields

$$\frac{\psi(W) + \psi(E) + \psi(N) + \psi(S) - 4\psi(M)}{h^2} = 0. \tag{6.81}$$

Since N, M, S belong to Γ, (6.81) reduces to

$$\psi(W) = -\psi(E), \tag{6.82}$$

which combined with (6.80) implies that

$$\frac{\partial\psi}{\partial n}(M) \approx -\frac{\psi(W)}{h} = \frac{0 - \psi(W)}{h} = \frac{\psi(M) - \psi(W)}{h}. \tag{6.83}$$

In that particular case, the *centred* approximation (6.80) (which is *second-order accurate*) coincides with the *one-sided* one in (6.83) (which is only *first-order accurate*, in general). In the sequel, we shall use, therefore, (6.83) to approximate $\partial\psi/\partial n$ at M and we shall denote by $\delta_{kl}\psi$ the corresponding approximation of $\partial\psi/\partial n$ at $M_{kl} \in \Gamma$.

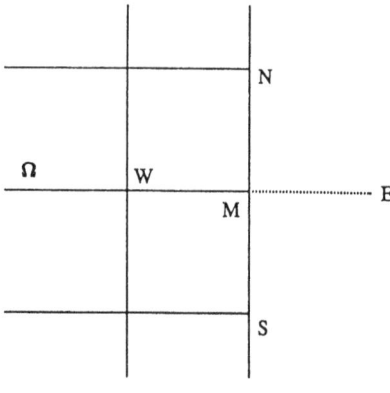

Fig. 44.

6.8.4.4. Approximation of the wave problem (6.33)₂. Similarly to (6.33)₁, the wave problem (6.33)₂, namely here,

$$\begin{cases} \varphi_{tt} - \Delta\varphi = 0 \text{ in } Q, \quad \varphi = \dfrac{\partial\psi}{\partial n} \text{ on } \Sigma, \\ \varphi(0) = 0, \quad \varphi_t(0) = 0 \end{cases} \tag{6.84}$$

will be approximated by

$$\begin{cases} \dfrac{\varphi_{ij}^{n+1} + \varphi_{ij}^{n-1} - 2\varphi_{ij}^{n}}{|\Delta t|^2} - \dfrac{\varphi_{i+1j}^{n} + \varphi_{i-1j}^{n} + \varphi_{ij+1}^{n} + \varphi_{ij-1}^{n} - 4\varphi_{ij}^{n}}{h^2} = 0, \\ 1 \leq i,j \leq I, \quad 0 \leq n \leq N, \end{cases} \tag{6.85}_1$$

$$\varphi_{kl}^{n} = \delta_{kl}\psi^{n} \quad \text{if } M_{kl} \in \Gamma, \tag{6.85}_2$$

$$\varphi_{ij}^{0} = 0, \quad \dfrac{\varphi_{ij}^{1} - \varphi_{ij}^{-1}}{2\Delta t} = 0, \quad 1 \leq i,j \leq I. \tag{6.85}_3$$

6.8.4.5. Approximation of Λ. Starting from

$$\mathbf{f}_h = \left\{ \{ f_{ij}^0, f_{ij}^1 \} \right\}_{1 \leq i,j \leq I}$$

and via the solution of the discrete wave equations (6.78), (6.85) we approximate Λf by

$$\Lambda_h^{\Delta t}\mathbf{f}_h = \left\{ \left\{ -\dfrac{\varphi_{ij}^{N+1} - \varphi_{ij}^{N-1}}{2\Delta t}, \varphi_{ij}^{N} \right\} \right\}_{1 \leq i,j \leq I} \tag{6.86}$$

It is proved in Glowinski *et al.* (1990, pp. 17–19) that we have (with obvious notation)

$$\langle \Lambda_h^{\Delta t} \mathbf{f}_h, \hat{\mathbf{f}}_h \rangle_{h,\Delta t} = h^2 \sum_{1 \le i,j \le I} \left[\varphi_{ij}^N \hat{f}_{ij}^1 - \left(\frac{\varphi_{ij}^{N+1} - \varphi_{ij}^{N-1}}{2\Delta t} \right) \hat{f}_{ij}^0 \right]$$

$$= h \Delta t \sum_{n=0}^{N} \alpha_n \sum_{M_{kl} \in \Gamma^*} \delta_{kl} \psi^n \delta_{kl} \hat{\psi}^n, \tag{6.87}$$

where, in (6.87), $\alpha_0 = \alpha_N = \frac{1}{2}$, $\alpha_n = 1, \forall n = 1, \ldots, N-1$, and where $\Gamma^* = \Gamma$ minus the four corners $\{0,0\}$, $\{0,1\}$, $\{1,0\}$, $\{1,1\}$. It follows from (6.87) that $\Lambda_h^{\Delta t}$ is *symmetric* and *positive semi-definite*. Actually, it is proved in Glowinski *et al.* (1990, Section 6.2) that $\Lambda_h^{\Delta t}$ is *positive definite* if $T > T_{min} \approx \Delta t/h$. This property implies that if $T(> 0)$ is given, it suffices to take $\Delta t/h$ sufficiently small to have exact boundary controllability for the discrete wave equation. This property is in contradiction with the continuous case where the exact boundary controllability property is lost if T is too small ($T < 1$ here). The reason for this discrepancy will be discussed in the following.

6.8.4.6. Approximation of the dual problem (6.50), (6.52). With $\mathbf{z}_h$ a convenient approximation of $\mathbf{z} = \{z^0, z^1\}$ we approximate problem (6.50), (6.52) by

$$\Lambda_h^{\Delta t} \mathbf{f}_h^{\Delta t} = \sigma \mathbf{z}_h, \tag{6.88}$$

where, in (6.88), σ denotes the matrix $\begin{pmatrix} 0 & -1 \\ 1 & 0 \end{pmatrix}$. In Glowinski *et al.* (1990, Section 6.3), one may find a discrete variation of the conjugate gradient algorithm (6.58)–(6.73) which can be used to solve the approximate problem (6.88).

6.8.5. Numerical solution of a test problem; ill-posedness of the discrete problem (6.88)

Following Glowinski *et al.* (1990, Section 7), Dean *et al.* (1989, Section 2.7), Glowinski (1992a, Section 2.7) we still consider the case $\Omega = (0,1)^2$, $\Gamma_0 = \Gamma$, $A = -\Delta$; we take $T = 3.75/\sqrt{2}$ (> 1, so that the exact controllability property holds) and $\mathbf{f} = \{f^0, f^1\}$ defined by

$$f^0(x_1, x_2) = \sin \pi x_1 \sin \pi x_2, \quad f^1 = -\pi \sqrt{2} f^0. \tag{6.89}$$

It is shown in Glowinski *et al.* (1990, Section 7) that using *separation of variables methods* we can compute a *Fourier Series* expansion of $\Lambda \mathbf{f}$. The corresponding functions z^0 and z^1 (both computed by *Fast Fourier Transform*) have been visualized on Figures 45 and 46, respectively (the graph on Figure 46 is the plot of $-z^1$).

From the above figures, z^0 is a *Lipschitz continuous function* which is not C^1; similarly, z^1 is *bounded* but *discontinuous*. On Figure 47, we have shown

Table 6. *Summary of numerical results (n.c.: no convergence)*

h	$\frac{1}{8}$	$\frac{1}{16}$	$\frac{1}{32}$	$\frac{1}{64}$	$\frac{1}{128}$
Number of conjugate gradient iterations	20	38	84	363	n.c.
$\|f^0 - f^0_*\|_{L^2(\Omega)}$	0.42×10^{-1}	0.18×10^{-1}	0.41×10^{-1}	3.89	n.c.
$\|f^0 - f^0_*\|_{H^1_0(\Omega)}$	0.65	0.54	2.54	498.1	n.c.
$\|f^1 - f^1_*\|_{L^2(\Omega)}$	0.20	0.64×10^{-1}	1.18	170.6	n.c.
$\|u - u_*\|_{L^2(\Sigma)}$	0.51	0.24	0.24	1.31	n.c.
$\|u_*\|_{L^2(\Sigma)}$	7.320	7.395	7.456	7.520	n.c.

the plot of the function $t \to \|\partial\psi/\partial n(t)\|_{L^2(\Gamma)}$ where ψ, given by

$$\psi(x, t) = \sqrt{2}\cos \pi\sqrt{2}\left(t - \frac{7}{2\sqrt{2}}\right)\sin \pi x_1 \sin \pi x_2,$$

is the solution of the wave equation (6.77) when f^0 and f^1 are given by (6.89); we recall that $\partial\psi/\partial n|_\Sigma (= u)$ is precisely the optimal Dirichlet control for which we have exact boundary controllability.

The numerical methods described in Sections 6.8.3 and 6.8.4 have been applied to the solution of the above test problem taking $\Delta t = h/\sqrt{2}$. Interestingly enough, the numerical results deteriorate as h and Δt converge to zero; moreover, taking Δt twice smaller, i.e. $\Delta t = h/2\sqrt{2}$, does not improve the situation. Also, the number of conjugate gradient iterations necessary to achieve convergence increases as h and Δt decrease. Results of the numerical experiments are reported on Table 6. In Table 6, f^0_*, f^1_* and u_*, are the computed values of f^0, f^1 and u respectively.

The most striking fact coming from Table 6 is the deterioration in the numerical results as h and Δt tend to zero; indeed, for $h = 1/128$, convergence was not achieved after 1000 iterations. To illustrate this deterioration further as h and $\Delta t \to 0$ we have compared, in Figures 48 to 51, f^0 and f^1 with their computed approximations f^0_* and f^1_*, for $h = 1/32$ and $1/64$; we observe that for $h = 1/64$ the variations in f^0_* and f^1_* are so large that we have been obliged to use a very large scale to be able to picture them (indeed we have plotted $-f^1, -f^1_*$)

If, for the same values of h, one takes Δt smaller than $h/\sqrt{2}$, the results remain practically the same. In Section 6.8.6, we shall try to analyse the reasons for this deterioration in the numerical results as $h \to 0$ and also to cure it. To conclude this section we observe that the error $\|u - u_*\|_{L^2(\Sigma)}$ deteriorates much more slowly as $h \to 0$ than the errors $f^0 - f^0_*, f^1 - f^1_*$; in fact, the approximate values $\|u_*\|_{L^2(\Sigma)}$ of $\|u\|_{L^2(\Sigma)}$ are quite good, even for

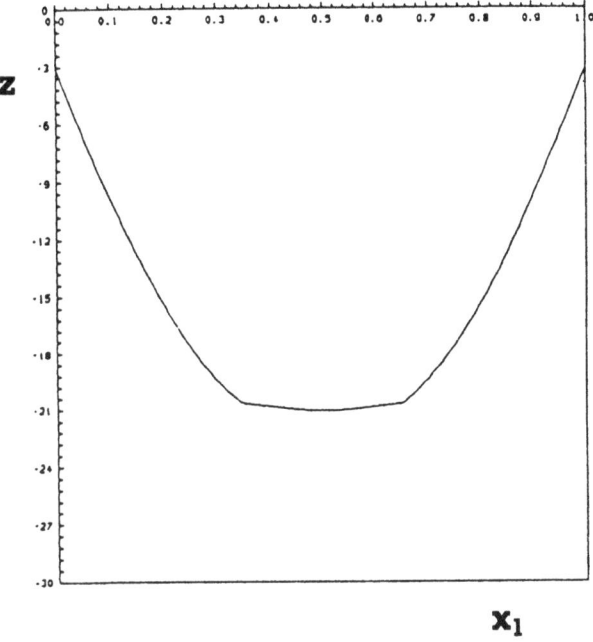

Fig. 45. $z^0(x_1, .5)$.

$h = 1/64$ if one realizes that the exact value of $\|u\|_{L^2(\Sigma)}$ is 7.386 68... For further illustrations and more details see Glowinski (1992a, Section 2.7) and the references therein.

6.8.6. Analysis and cures of the ill-posedness of the approximate problem (6.88)

It follows from the numerical results discussed in Section 6.8.5, that when h decreases to zero, the *ill-posedness* of the discrete problem gets worse. From the oscillatory results shown in Figures 48 to 51 it is quite clear that the trouble lies with the *high-frequency components* of the discrete solution or, to be more precise, with the way in which the discrete operator $\Lambda_h^{\Delta t}$ acts on the *short-wavelength* components of $\mathbf{f}_h$. Before analysing the mechanism producing these unwanted oscillations let us introduce a vector basis of $\mathbb{R}^{I \times I}$, well suited to the following discussion. This basis $\mathcal{B}_h$ is defined by

$$\mathcal{B}_h = \{w_{pq}\}_{1 \le p,q \le I}, \tag{6.90}$$

$$w_{pq} = \{\sin p\pi ih \times \sin q\pi jh\}_{1 \le i,j \le I}; \tag{6.91}$$

we recall that $h = 1/(I+1)$.

From the oscillatory results described in Section 6.8.5 it is reasonable to assume that the discrete operator $\Lambda_h^{\Delta t}$ damps too strongly those components of $\mathbf{f}_h^{\Delta t}$ with *large wavenumbers* p and q; in other words, we can expect that

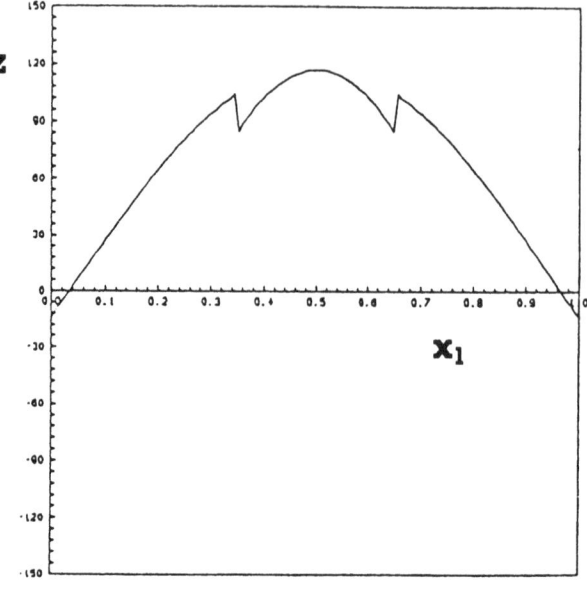

Fig. 46. $z^1(x_1, .5)$.

if p and/or q are large then $\Lambda_h^{\Delta t}\{w_{pq}, 0\}$ or $\Lambda_h^{\Delta t}\{0, w_{pq}\}$ will be quite small implying in turn (this is typical of ill-posed problems) that small perturbations of the right-hand side of the discrete problem (6.88) can produce very large variations in the corresponding solution.

Operator $\Lambda_h^{\Delta t}$ is fairly complicated (see Section 6.8.4 for its precise definition) and we can wonder which stage in it in particular acts as a *low pass filter* (i.e. selectively damping the large wavenumber components of the discrete solutions). Starting from the observation that the ill-posedness persists if, for a fixed h, we decrease Δt, it is then natural (and much simpler) to consider the *semi-discrete* case, where only the space derivatives have been discretized.

In such a case, problem (6.77) is discretized as follows (with $\dot{\psi} = \partial\psi/\partial t$, $\ddot{\psi} = \partial^2\psi/\partial t^2$) if $\Omega = (0, 1)^2$ as in Sections 6.8.4, 6.8.5:

$$\ddot{\psi}_{ij} - \frac{\psi_{i+1j} + \psi_{i-1j} + \psi_{ij+1} + \psi_{ij-1} - 4\psi_{ij}}{h^2} = 0, \quad 1 \le i, j \le I, \quad (6.92)_1$$

$$\psi_{kl} = 0 \text{ if } \{kh, lh\} \in \Gamma, \quad (6.92)_2$$

$$\psi_{ij}(T) = f_h^0(ij, jh), \dot{\psi}_{ij}(T) = f_h^1(ih, jh), \quad 1 \le i, j \le I. \quad (6.92)_3$$

Consider now the particular case where

$$f_h^0 = w_{pq}, \quad f_h^1 = 0. \quad (6.93)$$

Since the vectors w_{pq} are for $1 \le p, q \le I$ the *eigenvectors* of the discrete

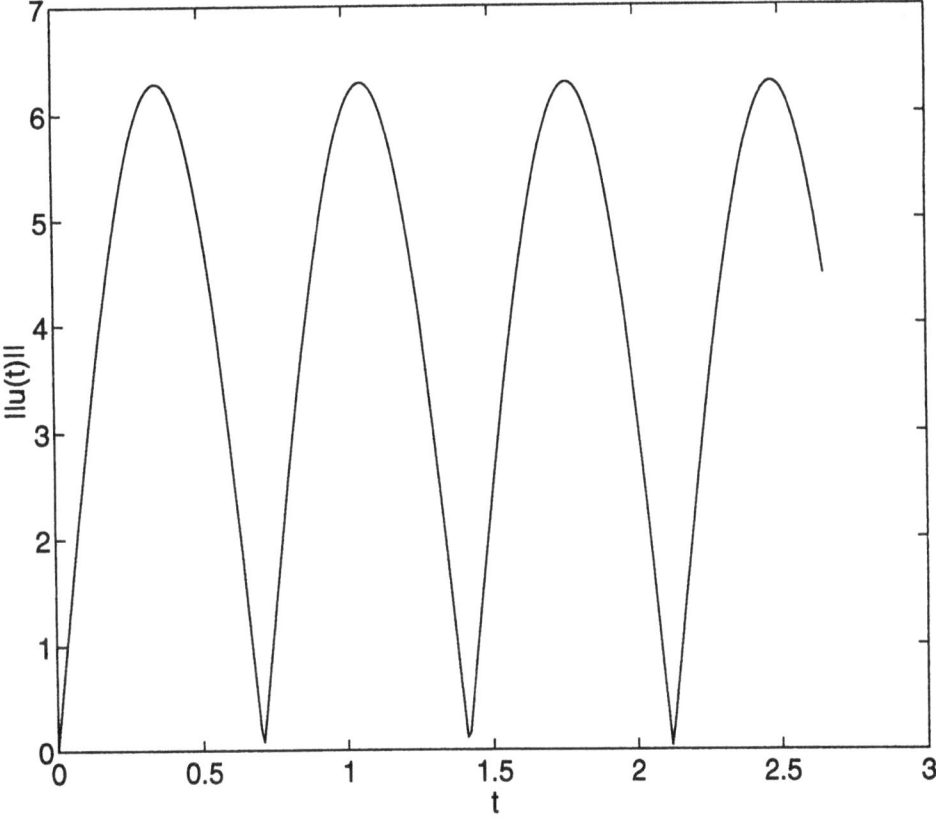

Fig. 47. $\|(\partial\psi/\partial n)(t)\|_{L^2(\Gamma)}$.

Laplace operator occurring in $(6.92)_1$ and that the corresponding eigenvalues $\lambda_{pq}(h)$ are given by

$$\lambda_{pq}(h) = \frac{4}{h^2}\left(\sin^2 p\pi\frac{h}{2} + \sin^2 q\pi\frac{h}{2}\right), \tag{6.94}$$

we can easily prove that the solution of (6.92), (6.93) is given by

$$\psi_{ij}(t) = \sin p\pi ih \sin q\pi jh \cos\left(\sqrt{\lambda_{pq}(h)}(T-t)\right), \quad 0 \leq i,j \leq I+1. \tag{6.95}$$

Next, we use (6.83) (see Section 6.8.4.3) to compute, from (6.95), the approximation of $\partial\psi/\partial n$ at the boundary point $M_{0j} = \{0, jh\}$, with $1 \leq j \leq I$; thus at time t, $\partial\varphi/\partial n$ is approximated at M_{0j} by

$$\delta\psi_h(M_{0j}, t) = -\frac{1}{h}\sin p\pi h \sin q\pi jh \cos\left(\sqrt{\lambda_{pq}(h)}(T-t)\right). \tag{6.96}$$

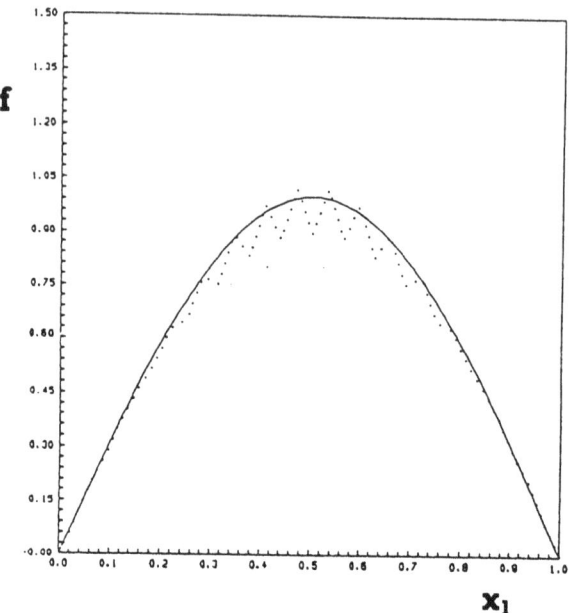

Fig. 48. Variations of $f^0(x_1, .5)$ (——) and $f^0_*(x_1, .5)$ (·····) ($h = 1/32$).

If $1 \leq p \ll I$, the coefficient $K_h(p)$ defined by

$$K_h(p) = \frac{\sin p\pi h}{h} \tag{6.97}$$

is an approximation of $p\pi$ which is second-order accurate (with respect to h); now if $p \sim I/2$ we have $K_h(p) \sim I$ and if $p = I$ we have (since $h = 1/(I+1))K_h(I) \sim \pi$.

Back to the *continuous problem*, it is quite clear that (6.92), (6.93) is in fact a semi-discrete approximation of the wave problem

$$\psi_{tt} - \Delta\psi = 0 \text{ in } Q, \quad \psi = 0 \text{ on } \Sigma, \tag{6.98}_1$$

$$\psi(x, T) = \sin p\pi x_1 \sin q\pi x_2, \quad \psi_t(x, T) = 0. \tag{6.98}_2$$

The solution of (6.98) is given by

$$\psi(x, t) = \sin p\pi x_1 \sin q\pi x_2 \cos\left(\pi\sqrt{p^2 + q^2}(T - t)\right). \tag{6.99}$$

Computing $(\partial\psi/\partial n)|_\Sigma$ we obtain

$$\frac{\partial\psi}{\partial n}(M_{0j}, t) = -p\pi \sin q\pi jh \cos\left(\pi\sqrt{p^2 + q^2}(T - t)\right). \tag{6.100}$$

We observe that if $p \ll I$ and $q \ll I$, then $(\partial\psi/\partial n)(M_{0j}, t)$ and $\delta\psi_h(M_{0j}, t)$ are close quantities. Now, if the wavenumber p is large, then the coefficient $K(p) = \pi p$ in (6.100) is much larger than the corresponding coefficient $K_h(p)$

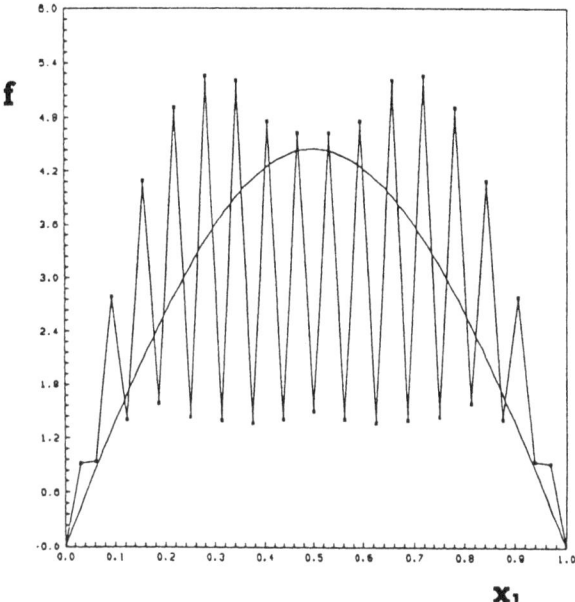

Fig. 49. Variations of $f^1(x_1, .5)$ (——) and $f_*^1(x_1, .5)$ ($\cdots\cdots$) ($h = 1/32$).

in (6.97); we have, in fact,

$$\frac{K(I/2)}{K_h(I/2)} \approx \frac{\pi}{2}, \quad \frac{K(I)}{K_h(I)} \approx I.$$

Figure 52, (where we have visualized, with an appropriate scaling, the function $p\pi \to p\pi$ and its discrete analogue, namely the function $p\pi \to \sin p\pi h/h$) shows that for $p, q > (I+1)/2$, the approximate normal derivative operator introduces a very strong damping. We would have obtained similar results by considering, instead of (6.93), initial conditions such as

$$f_h^0 = 0, \quad f_h^1 = w_{pq}. \tag{6.101}$$

From the above analysis it appears that the approximation of $(\partial\psi/\partial n)|_\Sigma$, which is used to construct operator $\Lambda_h^{\Delta t}$, introduces very strong damping of the *large wavenumber components* of $\mathbf{f}_h$. Possible cures for the ill-posedness of the discrete problem have been discussed in Glowinski *et al.* (1990), Dean *et al.* (1989), Glowinski (1992a). The first reference, in particular, contains a detailed discussion of a *biharmonic Tychonoff regularization procedure*, where problem (6.50) is approximated by a discrete version of

$$\varepsilon\mathbf{M}\mathbf{f}_\varepsilon + \Lambda\mathbf{f}_\varepsilon = \begin{pmatrix} -z^1 \\ z^0 \end{pmatrix} \text{ in } \Omega, \tag{6.102}_1$$

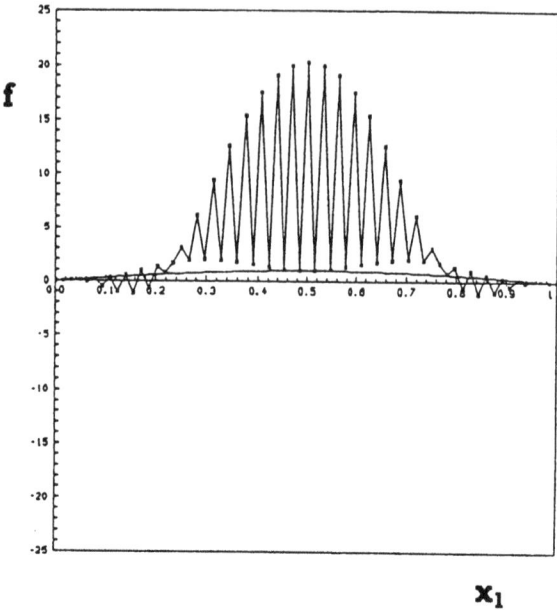

f

x₁

Fig. 50. Variations of $f^0(x_1, .5)$ (——) and $f_*^0(x_1, .5)$ ($\cdots\cdots$) ($h = 1/64$).

$$\Delta f_\varepsilon^0 = f_\varepsilon^0 = f_\varepsilon^1 = 0 \text{ on } \Gamma,$$

where, in (6.102), $\varepsilon > 0$, $\mathbf{f}_\varepsilon = \{f_\varepsilon^0, f_\varepsilon^1\}$, and where operator $\mathbf{M}$ is defined by

$$\mathbf{M} = \begin{pmatrix} \Delta^2 & 0 \\ 0 & -\Delta \end{pmatrix}. \tag{6.102}_2$$

Various theoretical and numerical issues associated with (6.102) are discussed in Glowinski *et al.* (1990), including the choice of ε as a function of h; indeed elementary *boundary layer* considerations show that ε has to be of the order of h^2. The numerical results presented in Glowinski *et al.* (1990) and Dean *et al.* (1989) validate convincingly the above regularization approach. Also in Glowinski *et al.* (1990, p. 42) we suggest that *mixed* FE approximations (see, e.g. Roberts and Thomas (1991), Brezzi and Fortin (1991) for introductions to mixed FE methods) may improve the quality of the numerical results; one of the reasons for this potential improvement is that mixed FE methods are known to provide accurate approximations of derivatives and also that derivative values at selected nodes (including boundary ones) are natural degrees of freedom for these approximations. As shown in Glowinski, Kinton and Wheeler (1989) and Dupont, Glowinski, Kinton and Wheeler (1992) this approach substantially reduces the unwanted oscillations, since *without* any regularization good numerical results have been obtained using mixed FE implementation of HUM. The main drawback of

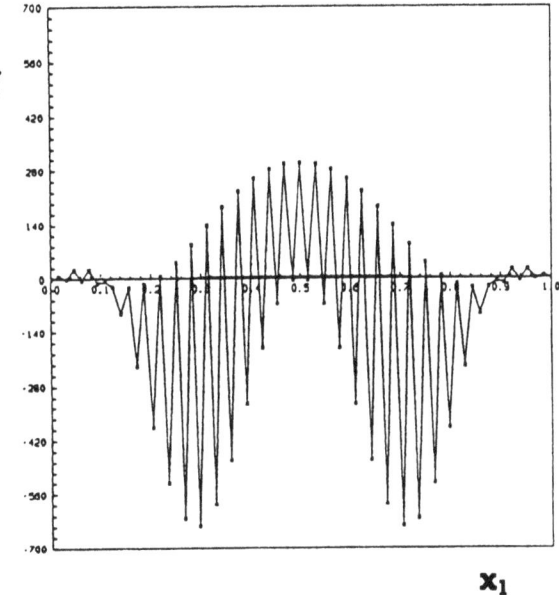

Fig. 51. Variations of $f^1(x_1, .5)$ (———) and $f^1_*(x_1, .5)$ ($\cdots\cdots$) ($h = 1/64$).

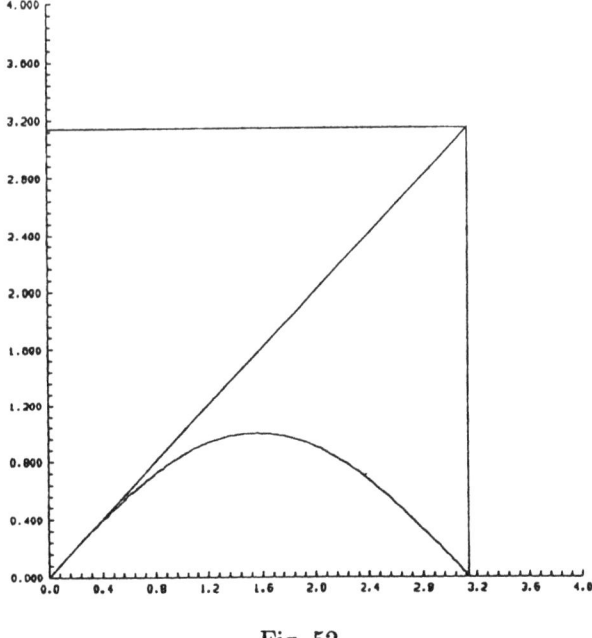

Fig. 52.

the mixed FE approach is that (without regularization) the number of conjugate gradient iterations necessary to achieve convergence increases (slowly) with h (in fact, roughly, as $h^{-1/2}$); it seems, also, on the basis of numerical

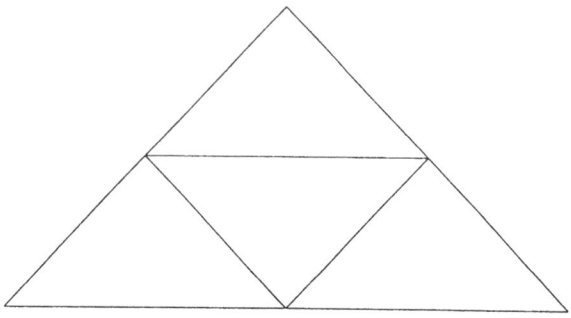

Fig. 53. Triangles of $\mathcal{T}_h$ and $\mathcal{T}_{h/2}$.

experiments, that the level of unwanted oscillations increases (slowly, again) with T.

Another cure for spurious oscillations has been introduced in Glowinski and Li (1990) (see also Glowkinski (1992a, Section 3)); this (simple) cure, suggested by Figure 52, consists of eliminating the *short-wavelength* components of $\mathbf{f}_h$ with wavenumbers p and q larger than $(I+1)/2$; to achieve this radical filtering it suffices to define $\mathbf{f}_h$ on an FD grid of step size $\geq 2h$. An FE implementation of the above filtering technique is discussed in Section 6.8.7; also, for the calculations described in Section 6.9 we have defined $\mathbf{f}_h$ over a grid of step size $2h$.

6.8.7. An FE implementation of the filtering technique of Section 6.8.6

6.8.7.1. Generalities. We go back to the case where (possibly) $\Omega \neq (0,1)^2$, $\Gamma_0 \neq \Gamma$ and $A \neq -\Delta$; the most natural fashion of combining HUM and the filtering technique discussed in Section 6.8.6 is to use *finite elements* for the space approximation; in fact, as shown in Glowinski et al. (1990, Section 6.2), special triangulations (like the one shown in Figure 1 of Section 2.6.1) will give back FD approximations closely related to the one discussed in Section 6.8.6. For simplicity, we suppose that Ω is a polygonal domain of $\mathbb{R}^2$; we then introduce a triangulation $\mathcal{T}_h$ of Ω such that $\bar{\Omega} = \bigcup_{T \in \mathcal{T}_h} T$, with h the length of the largest edge (s) of $\mathcal{T}_h$. From $\mathcal{T}_h$, we define $\mathcal{T}_{h/2}$ by joining (see Figure 53), the midpoints of the edges of the triangles of $\mathcal{T}_h$.

With P_1 the space of the polynomials in two variables of degree ≤ 1, we define the spaces H_h^1 and H_{0h}^1 by

$$H_h^1 = \{v \mid \in C^0(\bar{\Omega}), v|_T \in P_1, \forall T \in \mathcal{T}_h\}, \quad H_{0h}^1 = \{v \mid v \in H_h^1, v|_\Gamma = 0\};$$
$$(6.103)$$

similarly, we define $H_{h/2}^1$ and $H_{0h/2}^1$ by replacing h by $h/2$ in (6.103). We observe that $H_h^1 \subset H_{h/2}^1$, $H_{0h}^1 \subset H_{0h/2}^1$. We then approximate the $L^2(\Omega)$-

scalar product over H_h^1 by

$$(v, w)_h = \frac{1}{3} \sum_Q \omega_Q v(Q) w(Q), \quad \forall v, w \in H_h^1, \qquad (6.104)$$

where, in (6.104), Q describes the set of the vertices of $\mathcal{T}_h$ and where ω_Q is the area of the polygonal domain, union of those triangles of $\mathcal{T}_h$, with Q as a common vertex. Similarly, we define $(\cdot, \cdot)_{h/2}$ by substituting $h/2$ to h in (6.104).

Finally, assuming that the points at the interface of Γ_0 and $\Gamma \backslash \Gamma_0$ are vertices of $\mathcal{T}_{h/2}$, we define $V_{0h/2}$ by

$$V_{0h/2} = \{ v \mid v \in H_{h/2}^1, \ v = 0 \text{ on } \Gamma_0 \backslash \Gamma \}. \qquad (6.105)$$

6.8.7.2. *Approximation of problem (6.50).* We approximate the fundamental equation $\Lambda f = \{-z^1, z^0\}$ by the following linear variational problem in $H_{0h}^1 \times H_{0h}^1$:

$$\begin{cases} \mathbf{f}_h^{\Delta t} \in H_{0h}^1 \times H_{0h}^1, \\ \lambda_h^{\Delta t}(\mathbf{f}_h, \mathbf{v}) = -\langle z^1, v^0 \rangle + \int_\Omega z^0 v^1 \, dx, \ \forall \mathbf{v} = \{v^0, v^1\} \in H_{0h}^1 \times H_{0h}^1. \end{cases} \qquad (6.106)$$

In (6.106), $\langle \cdot, \cdot \rangle$ denotes the duality pairing between $H^{-1}(\Omega)$ and $H_0^1(\Omega)$, and the bilinear form $\lambda_h^{\Delta t}(\cdot, \cdot)$ is defined as follows.

(i) Take $\hat{\mathbf{f}}_h = \{\hat{f}_h^0, \hat{f}_h^1\} \in H_{0h}^1 \times H_{0h}^1$ and solve, for $n = N, \ldots, 0$, the *discrete variational problem*

$$\begin{cases} \hat{\psi}_h^{n-1} \in H_{0h/2}^1, \\ (\hat{\psi}_h^{n-1} + \hat{\psi}_h^{n+1} - 2\hat{\psi}_h^n, v)_{h/2} + |\Delta t|^2 a(\hat{\psi}_h^n, v) = 0, \ \forall v \in H_{0h/2}^1, \end{cases} \qquad (6.107)$$

with the final conditions

$$\hat{\psi}_h^N = \hat{f}_h^0, \quad \hat{\psi}_h^{N+1} - \hat{\psi}_h^{N-1} = 2\Delta t \hat{f}_h^1; \qquad (6.108)$$

we recall that $a(\cdot, \cdot)$ denotes the bilinear form defined by

$$a(v, w) = \langle Av, w \rangle, \ \forall v \in H^1(\Omega), \quad w \in H_0^1(\Omega).$$

(ii) To approximate $\partial \psi / \partial n_A$ over Σ_0, first introduce the complementary subspace $M_{h/2}$ of $H_{0h/2}^1$ defined by

$$\begin{cases} M_{h/2} \oplus H_{0h/2}^1 = V_{0h/2}, \\ v \in M_{h/2} \Longrightarrow v|_T = 0, \ \forall T \in \mathcal{T}_{h/2} \text{ such that } T \cap \Gamma = \emptyset; \end{cases} \qquad (6.109)$$

we observe that $M_{h/2}$ is isomorphic to the space $\gamma V_{0h/2}$ of the traces over Γ_0 of the functions of $V_{0h/2}$. The approximation of $(\partial \hat{\psi}/\partial n_A)|_{\Gamma_0}$ at $t = n\Delta t$ is then defined (cf. Glowinski *et al.* (1990) and Section 2.4.3) by solving the linear variational problem

$$\begin{cases} \delta\hat{\psi}_h^n \in \gamma V_{0h/2}, \\ \int_{\Gamma_0} \delta\hat{\psi}_h^n v \, d\Gamma = a(\hat{\psi}_h^n, v), \ \forall v \in M_{h/2}. \end{cases} \quad (6.110)$$

Variants of (6.110), leading to linear systems with diagonal matrices are given in Glowinski et al. (1990).

(iii) Now, for $n = 0, \dots, N$; solve the discrete variational problem

$$\begin{cases} \hat{\varphi}_h^{n+1} \in V_{0h/2}; \hat{\varphi}_h^{n+1} = \delta\hat{\psi}_h^{n+1} \text{ on } \Gamma_0, \\ (\hat{\varphi}_h^{n+1} + \hat{\varphi}_h^{n-1} - 2\hat{\varphi}_h^n, v)_{h/2} + |\Delta t|^2 a(\hat{\varphi}_h^n, v) = 0, \ \forall v \in H_{0h/2}^1 \end{cases} \quad (6.111)$$

initialized via

$$\hat{\varphi}_h^0 = 0, \quad \hat{\varphi}_h^1 - \hat{\varphi}_h^{-1} = 0. \quad (6.112)$$

(iv) Finally, define $\lambda_h^{\Delta t}(\cdot, \cdot)$ by

$$\lambda_h^{\Delta t}(\hat{\mathbf{f}}_h, \mathbf{v}) = (\hat{\lambda}_h^0, v^0)_{h/2} + (\hat{\lambda}_h^1, v^1)_{h/2}, \ \forall \mathbf{v} = \{v^0, v^1\} \in H_{0h}^1 \times H_{0h}^1, \quad (6.113)$$

where, in (6.113), $\hat{\lambda}_h^0$ and $\hat{\lambda}_h^1$ both belong to $H_{0h/2}^1$ and satisfy

$$\hat{\lambda}_h^0(P) = -\frac{\hat{\varphi}_h^{N+1}(P) - \hat{\varphi}_h^{N-1}(P)}{2\Delta t},$$
$$\hat{\lambda}_h^1(P) = \hat{\varphi}_h^N(P), \ \forall P \text{ interior vertex of } \mathcal{T}_{h/2}. \quad (6.114)$$

Following Glowinski et al. (1990, Section 6) we can prove that

$$\lambda_h^{\Delta t}(\mathbf{f}_{1h}, \mathbf{f}_{2h}) = \Delta t \sum_{n=0}^{N} \alpha_n \int_{\Gamma_0} \delta\psi_{1h}^n \delta\psi_{2h}^n \, d\Gamma, \ \forall f_{1h}, f_{2h} \in H_{0h}^1 \times H_{0h}^1, \quad (6.115)$$

where, in (6.115), $\alpha_0 = \alpha_N = 1/2$, and $\alpha_n = 1$ if $0 < n < N$.

It follows from (6.115) that the bilinear form $\lambda_h^{\Delta t}(\cdot, \cdot)$ is symmetric and positive semi-definite. As in Glowinski et al. (1990, Section 6.2), we should prove that $\lambda_h^{\Delta t}(\cdot, \cdot)$ is positive definite if T is sufficiently large and if Ω is a square (or a rectangle) and $\mathcal{T}_h$, $\mathcal{T}_{h/2}$ regular triangulations of Ω. From the properties of $\lambda_h^{\Delta t}(\cdot, \cdot)$ the linear variational problem (6.106) (which approximates problem (6.50)) can be solved by a conjugate gradient algorithm operating in $H_{0h}^1 \times H_{0h}^1$. This algorithm is described in Section 6.8.7.3.

6.8.7.3. Conjugate gradient solution of the approximate problem (6.106). The conjugate gradient algorithm for solving problem (6.106) is an FE implementation of algorithm (6.58)–(6.73) (see Section 6.8.3).

Description of the Conjugate Gradient Algorithm

Step 0: Initialization

$$f_0^0 \in H_{0h}^1, \quad f_0^1 \in H_{0h}^1 \text{ are given;} \quad (6.116)$$

solve then, for $n = N, N - 1, \ldots, 0$, the discrete linear variational problem

$$\begin{cases} \psi_0^{n-1} \in H_{0h/2}^1, \\ \left(\dfrac{\psi_0^{n-1} + \psi_0^{n+1} - 2\psi_0^n}{|\Delta t|^2}, v \right)_{h/2} + a(\psi_0^n, v) = 0, \ \forall v \in H_{0h/2}^1, \end{cases} \tag{6.117}$$

initialized by

$$\psi_0^N = f_0^0, \quad \psi_0^{N+1} - \psi_0^{N-1} = 2\Delta t f_0^1, \tag{6.118}$$

and store ψ_0^0, ψ_0^{-1}.

Then for $n = 0, 1, \ldots, N$, compute $\psi_0^n, \delta\psi_0^n, \varphi_0^{n+1}$ by forward (discrete) time integration, as follows.

1   *If $n = 0$, compute $\delta\psi_0^0$ from ψ_0^0 using (6.110). If $n > 0$, compute first ψ_0^n by solving*

$$\begin{cases} \psi_0^n \in H_{0h/2}^1, \\ \left(\dfrac{\psi_0^n + \psi_0^{n-2} - 2\psi_0^{n-1}}{|\Delta t|^2}, v \right)_{h/2} + a(\psi_0^{n-1}, v) = 0, \ \forall v \in H_{0h/2}^1 \end{cases}$$

$$\tag{6.119}$$

and then $\delta\psi_0^n$ by using (6.110).

2   *Take $\varphi_0^n = \delta\psi_0^n$ on Γ_0 and use*

$$\left(\dfrac{\varphi_0^{n+1} + \varphi_0^{n-1} - 2\varphi_0^n}{|\Delta t|^2}, v \right)_{h/2} + a(\varphi_0^n, v) = 0, \ \forall v \in H_{0h/2}^1, \tag{6.120}$$

to compute the values taken by $\varphi_0^{n+1} (\in V_{0h/2})$ at the interior vertices of $\mathcal{T}_{h/2}$. These calculations are initialized by

$$\varphi_0^0(P) = 0, \varphi_0^1(P) - \varphi_0^{-1}(P) = 0, \ \forall P \ \text{interior vertex of } \mathcal{T}_{h/2}. \tag{6.121}$$

Compute then $\mathbf{g}_0 = \{g_0^0, g_0^1\} \in H_{0h}^1 \times H_{0h}^1$ by solving the following discrete Dirichlet problem

$$\begin{cases} g_0^0 \in H_{0h}^1, \\ \displaystyle\int_\Omega \nabla g_0^0 \cdot \nabla v \, dx = \langle z^1, v \rangle - \left(\dfrac{\varphi_0^{N+1} - \varphi_0^{N-1}}{2\Delta t}, v \right)_{h/2}, \ \forall v \in H_{0h}^1, \end{cases}$$

$$\tag{6.122}$$

and then

$$\begin{cases} g_0^1 \in H_{0h}^1, \\ (g_0^1, v)_h = (\varphi_0^N, v)_{h/2} - \displaystyle\int_\Omega z^0 v \, dx, \ \forall v \in H_{0h}^1. \end{cases} \tag{6.123}$$

If $\mathbf{g}_0 = \mathbf{0}$, or is 'small', take $\mathbf{f}_h^{\Delta t} = \mathbf{f}_0$; if not, set

$$\mathbf{w}_0 = \mathbf{g}_0. \tag{6.124}$$

Then for $k \geq 0$, assuming that $\mathbf{f}_k$, $\mathbf{g}_k$, $\mathbf{w}_k$ are known, compute $\mathbf{f}_{k+1}$, $\mathbf{g}_{k+1}$, $\mathbf{w}_{k+1}$ as follows.

Step 1: Descent
For $n = N, N-1, \ldots, 0$, solve the discrete backward wave equation

$$\begin{cases} \bar{\psi}_k^{n-1} \in H_{0h/2}^1, \\ \left(\dfrac{\bar{\psi}_k^{n-1} + \bar{\psi}_k^{n+1} - 2\bar{\psi}_k^n}{|\Delta t|^2}, v \right)_{h/2} + a(\bar{\psi}_k^n, v) = 0, \ \forall v \in H_{0h/2}^1, \end{cases} \tag{6.125}$$

initialized by

$$\bar{\psi}_k^N = w_k^0, \quad \bar{\psi}_k^{N+1} - \bar{\psi}_k^{N-1} = 2\Delta t w_k^1, \tag{6.126}$$

and store $\bar{\psi}_k^0$, $\bar{\psi}_k^{-1}$.

Then for $n = 0, 1, \ldots, N$, compute $\bar{\psi}_k^n$, $\delta\bar{\psi}_k^n$, $\bar{\varphi}_k^{n+1}$ by forward time integration as follows.

1   If $n = 0$, compute $\delta\bar{\psi}_k^0$ from $\bar{\psi}_k^0$ using (6.110).
If $n > 0$, compute first $\bar{\psi}_k^n$ by solving

$$\begin{cases} \bar{\psi}_k^n \in H_{0h/2}^1, \\ \left(\dfrac{\bar{\psi}_k^n + \bar{\psi}_k^{n-2} - 2\bar{\psi}_k^{n-1}}{|\Delta t|^2}, v \right)_{h/2} + a(\bar{\psi}_k^{n-1}, v) = 0, \ \forall v \in H_{0h/2}^1, \end{cases}$$
$$\tag{6.127}$$

and then $\delta\bar{\psi}_k^n$ by using (6.110).

2   Take $\bar{\varphi}_k^n = \delta\bar{\psi}_k^n$ on Γ_0 and use

$$\left(\dfrac{\bar{\varphi}_k^{n+1} + \bar{\varphi}_k^{n-1} - 2\bar{\varphi}_k^n}{|\Delta t|^2}, v \right)_{h/2} + a(\bar{\varphi}_k^n, v) = 0, \ \forall v \in H_{0h/2}^1, \tag{6.128}$$

to compute the values taken by $\bar{\varphi}_k^{n+1} (\in V_{0h/2})$ at the interior vertices of $\mathcal{T}_{h/2}$. These calculations are initialized by

$$\bar{\varphi}_k^1(P) - \bar{\varphi}_k^{-1}(P) = \bar{\varphi}_k^0(P) = 0, \ \forall P \text{ interior vertex of } \mathcal{T}_{h/2}. \tag{6.129}$$

Compute now $\mathbf{g}_k(= \{g_k^0, g_k^1\}) \in H_{0h}^1 \times H_{0h}^1$ by

$$\begin{cases} \bar{g}_k^0 \in H_{0h}^1, \\ \displaystyle\int_\Omega \nabla\bar{g}_k^0 \cdot \nabla v \, dx = -\left(\dfrac{\bar{\varphi}_k^{N+1} - \bar{\varphi}_k^{N-1}}{2\Delta t}, v \right)_{h/2}, \ \forall v \in H_{0h}^1, \end{cases} \tag{6.130}$$

$$\begin{cases} \bar{g}_k^1 \in H_{0h}^1, \\ (\bar{g}_k^1, v)_h = (\bar{\varphi}_k^N, v)_{h/2}, \ \forall v \in V_{0h}, \end{cases} \tag{6.131}$$

and then ρ_k by

$$\rho_k = \int_\Omega |\nabla g_k^0|^2 \, dx + (g_k^1, g_k^1)_h \Big/ \int_\Omega \nabla \bar{g}_k^0 \cdot \nabla w_k^0 \, dx + (\bar{g}_k^1, w_k^1)_h. \quad (6.132)$$

Once ρ_k is known, compute

$$\mathbf{f}_{k+1} = \mathbf{f}_k - \rho_k \mathbf{w}_k, \quad (6.133)$$

$$\mathbf{g}_{k+1} = \mathbf{g}_k - \rho_k \bar{\mathbf{g}}_k. \quad (6.134)$$

Step 2. Test of the convergence and construction of the new descent direction

If $\mathbf{g}_{k+1} = \mathbf{0}$, or is 'small', take $\mathbf{f}_h^{\Delta t} = \mathbf{f}_{k+1}$; if not, compute

$$\gamma_k = \int_\Omega |\nabla g_{k+1}^0|^2 \, dx + (g_{k+1}^1, g_{k+1}^1)_h \Big/ \int_\Omega |g_k^0|^2 \, dx + (g_k^1, g_k^1)_h, \quad (6.135)$$

and set

$$\mathbf{w}_{k+1} = \mathbf{g}_{k+1} + \gamma_k \mathbf{w}_k. \quad (6.136)$$

Do $k = k + 1$ and go to (6.125).

Remark 6.18 The above algorithm may seem a little bit complicated at first glance (21 statements); in fact, it is fairly easy to implement, since the only nontrivial part of it is the solution (on the coarse grid) of the discrete Dirichlet problems (6.122) and (6.130). An interesting feature of algorithm (6.116)–(6.136) is that the *forward integration* of the discrete wave equations (6.117) and (6.125) provides a very substantial computer memory saving. To illustrate this claim, let us consider the case where $\Omega = (0,1) \times (0,1)$, $\Gamma_0 = \Gamma$, $T = 2\sqrt{2}$, $h = 1/64$, $\Delta t = h/2\sqrt{2} = \sqrt{2}/256$; we have then – approximately – $(512)^2$ discretization points on Σ, therefore in that specific case, using algorithm (6.116)–(6.136) avoids the storage of 2.62×10^5 real numbers. The saving would be even more substantial for larger T and would be an absolute necessity for three-dimensional problems. In fact, the above storage-saving strategy which is based on the *time reversibility* of the *wave equation* (6.1) cannot be applied to the control problems discussed in Sections 1 and 2 since they concern systems modelled by *diffusion* equations which are, unfortunately, *time irreversible*.

Remark 6.19 The above remark shows the interest of solving the *dual problem* from a computational point of view. In the original control problem, the unknown is the control u which is defined over Σ_0; for the dual problem the unknown is then the solution $\mathbf{f}$ of problem (6.50). If one considers again the particular case of Remark 6.18, i.e. $\Omega = (0,1) \times (0,1)$, $\Gamma_0 = \Gamma$, $T = 2\sqrt{2}$, $h = 1/64$, $\Delta t = h/2\sqrt{2}$ the unknown u will be approximated by a finite dimensional vector $u_h^{\Delta t}$ with 2.62×10^5 components, while $\mathbf{f}$ is approximated by $\mathbf{f}_h^{\Delta t}$ of dimension $2 \times (63)^2 = 7.938 \times 10^3$, a substantial saving indeed. Also, the dimension of $\mathbf{f}_h^{\Delta t}$ remains the same as T increases, while the dimension of $u_h^{\Delta t}$ is proportional to T.

Numerical results obtained using algorithm (6.116)–(6.136) will be discussed in Section 6.9.

6.8.8. Solution of the approximate boundary controllability problem (6.24)

Following the approach advocated for the exact boundary controllability problem, we shall address the *numerical solution* of the *approximate boundary controllability* problem (6.24) via the solution of its *dual problem*, namely problem (6.37), (6.38). This can also be formulated as

$$\Lambda \mathbf{f} + \partial j(\mathbf{f}) = \begin{pmatrix} -z^1 \\ z^0 \end{pmatrix}, \tag{6.137}$$

where, in (6.137), the *convex functional* $j : H_0^1(\Omega) \times L^2(\Omega) \to \mathbb{R}$ is defined by

$$j(\hat{\mathbf{f}}) = \beta_1 \|\hat{f}^0\|_{H_0^1(\Omega)} + \beta_0 \|\hat{f}^1\|_{L^2(\Omega)}, \ \forall \hat{\mathbf{f}} = \{\hat{f}^0, \hat{f}^1\} \in H_0^1(\Omega) \times L^2(\Omega). \tag{6.138}$$

Following a strategy already used in preceding sections (see, e.g. Section 1.8.8) we associate with the 'elliptic' problem (6.137) the following *initial value problem*

$$\begin{cases} \begin{pmatrix} -\Delta & 0 \\ 0 & I \end{pmatrix} \dfrac{\partial \mathbf{f}}{\partial \tau} + \Lambda \mathbf{f} + \partial j(\mathbf{f}) = \begin{pmatrix} -z^1 \\ z^0 \end{pmatrix}, \\ \mathbf{f}(0) = \mathbf{f}_0, \end{cases} \tag{6.139}$$

where τ is a pseudo-time. The particular form of problem (6.139) clearly suggests time integration by *operator splitting* (see, again, Section 1.8.8). Concentrating on the *Peaceman–Rachford scheme*, we obtain – with $\Delta\tau(>0)$ a pseudo-time step – the following algorithm to compute the steady-state solution of problem (6.139), i.e. the solution of problem (6.37), (6.38), (6.137):

$$\mathbf{f}^0 = \mathbf{f}_0; \tag{6.140}$$

then, for $k \geq 0$, assuming that $\mathbf{f}^k$ is known, we compute $\mathbf{f}^{k+1/2}$ and $\mathbf{f}^{k+1}$ via the solution of

$$\begin{pmatrix} -\Delta & 0 \\ 0 & I \end{pmatrix} \frac{\mathbf{f}^{k+1/2} - \mathbf{f}^k}{\Delta\tau/2} + \Lambda\mathbf{f}^k + \partial j(\mathbf{f}^{k+1/2}) = \begin{pmatrix} -z^1 \\ z^0 \end{pmatrix}, \tag{6.141}$$

$$\begin{pmatrix} -\Delta & 0 \\ 0 & I \end{pmatrix} \frac{\mathbf{f}^{k+1} - \mathbf{f}^{k+1/2}}{\Delta\tau/2} + \Lambda\mathbf{f}^{k+1} + \partial j(\mathbf{f}^{k+1/2}) = \begin{pmatrix} -z^1 \\ z^0 \end{pmatrix}. \tag{6.142}$$

Let us discuss the solution of the subproblems (6.141), (6.142):
(i) Assuming that (6.141) has been solved, equation (6.142) can be for-

mulated as

$$\begin{pmatrix} -\Delta & 0 \\ 0 & I \end{pmatrix} \frac{\mathbf{f}^{k+1} - 2\mathbf{f}^{k+1/2} + \mathbf{f}^k}{\Delta\tau/2} + \Lambda\mathbf{f}^{k+1} = \Lambda\mathbf{f}^k,$$

i.e.

$$\begin{pmatrix} -\Delta & 0 \\ 0 & I \end{pmatrix} \mathbf{f}^{k+1} + \frac{\Delta\tau}{2}\Lambda\mathbf{f}^{k+1} = \begin{pmatrix} -\Delta & 0 \\ 0 & I \end{pmatrix}(2\mathbf{f}^{k+1/2} - \mathbf{f}^k) + \frac{\Delta\tau}{2}\Lambda\mathbf{f}^k. \quad (6.143)$$

Problem (6.143) is a variant of problem (6.50) (a regularized one, in fact) and can be solved by a conjugate gradient algorithm closely related to algorithm (6.58)–(6.73) (we have to replace the bilinear form

$$\{\mathbf{f}_1, \mathbf{f}_2\} \to \langle \Lambda\mathbf{f}_1, \mathbf{f}_2 \rangle : (H_0^1(\Omega) \times L^2(\Omega))^2 \to \mathbb{R}$$

by

$$\{\mathbf{f}_1, \mathbf{f}_2\} \to \int_\Omega \nabla f_1^0 \cdot \nabla f_2^0 \, dx + \int_\Omega f_1^1 f_2^1 \, dx + \frac{1}{2}\Delta\tau \langle \Lambda\mathbf{f}_1, \mathbf{f}_2 \rangle)$$

(ii) Concerning the solution of problem (6.141), we shall take advantage of the fact that operator $\partial j(\cdot)$ is *diagonal* from $H_0^1(\Omega) \times L^2(\Omega)$ into $H^{-1}(\Omega) \times L^2(\Omega)$; solving problem (6.141) is then equivalent to solving the two following *uncoupled* minimization problems (where the notation is fairly obvious):

$$\min_{\hat{f}^0 \in H_0^1(\Omega)} \left[\frac{1}{2}\int_\Omega |\nabla\hat{f}^0|^2 \, dx + \beta_1\frac{\Delta\tau}{2}\left(\int_\Omega |\nabla\hat{f}^0|^2 \, dx\right)^{1/2} \right.$$
$$\left. + \frac{\Delta\tau}{2}\langle z^1 + (\Lambda\mathbf{f}^k)^0, \hat{f}^0 \rangle - \int_\Omega \nabla f^{0,k} \cdot \nabla\hat{f}^0 \, dx \right], \quad (6.144)$$

$$\min_{\hat{f}^1 \in L^2(\Omega)} \left[\frac{1}{2}\int_\Omega |\hat{f}^1|^2 \, dx + \beta_0\frac{\Delta\tau}{2}\|\hat{f}^1\|_{L^2(\Omega)} - \frac{\Delta\tau}{2}\int_\Omega (z^0 - (\Lambda\mathbf{f}^k)^1)\hat{f}^1 \, dx \right.$$
$$\left. - \int_\Omega f^{1,k}\hat{f}^1 \, dx \right]. \quad (6.145)$$

Both problems (6.144), (6.145) have *closed form* solutions which can be obtained as in Section 1.8.8 for the solution of problem (1.115). The solution of problem (6.144) (respectively (6.145)) clearly provides the first (respectively the second) component of $\mathbf{f}^{k+1/2}$, i.e. the one in $H_0^1(\Omega)$ (respectively in $L^2(\Omega)$).

6.9. Experimental validation of the filtering procedure of Section 6.8.7 via the solution of the test problem of Section 6.8.5

We consider in this section the solution of the test problem of Section 6.8.5. The filtering technique discussed in Section 6.8.7 is applied with $\mathcal{T}_h$ a regular

triangulation like the one shown on Figure 1 of Section 2.6; we recall that
$\mathcal{T}_h$ is used to approximate $\mathbf{f}_h^{\Delta t}$, while ψ and φ are approximated on $\mathcal{T}_{h/2}$ as
shown in Section 6.8.7. Instead of taking h to be equal to the length of the
largest edges of $\mathcal{T}_h$, it is convenient here to take h as the length of the edges
adjacent to the right angles of $\mathcal{T}_h$. The approximate problems (6.106) have
been solved by the conjugate gradient algorithm (6.116)–(6.136) of Section
6.8.7.3. This algorithm has been *initialized* with $f_0^0 = f_0^1 = 0$ and we have
used

$$\int_\Omega |\nabla g_k^0|^2 \, dx + (g_k^1, g_k^1)_h \Big/ \int_\Omega |\nabla g_0^0|^2 \, dx + (g_0^1, g_0^1)_h \leq 10^{-14} \qquad (6.146)$$

as the *stopping criterion* (for calculations on a CRAY X-MP).

Let us also mention that the functions z^0, z^1, u of the test problem in
Section 6.8.5, satisfy

$$\|z^0\|_{L^2(\Omega)} = 12.92\ldots, \quad \|z^1\|_{H^{-1}(\Omega)} = 11.77\ldots, \quad \|u\|_{L^2(\Sigma)} = 7.386\,68\ldots.$$

In the following we shall denote by $\|\cdot\|_{0,\Omega}, |\cdot|_{1,\Omega}, \|\cdot\|_{-1,\Omega}, \|\cdot\|_{0,\Sigma}$ the $L^2(\Omega)$,
$H_0^1(\Omega)$, $H^{-1}(\Omega)$, $L^2(\Sigma)$ norms, respectively (here $|v|_{1,\Omega} = (\int_\Omega |\nabla v|^2 \, dx)^{1/2}$
and $\|v\|_{-1,\Omega} = |w|_{1,\Omega}$ where $w \in H_0^1(\Omega)$ is the solution of the Dirichlet
problem $-\Delta w = v$ in Ω, $w = 0$ on Γ).

To approximate problem (6.50) by the discrete problem (6.106) we have
been using $h = 1/4, 1/8, 1/16, 1/32, 1/64$ and $\Delta t = h/2\sqrt{2}$ (since the *wave
equations* are solved on a space/time grid of step size $h/2$ for the space
discretization and $h/2\sqrt{2}$ for the time discretization); we recall that $T =
15/4\sqrt{2}$.

Results of our numerical experiments have been summarized in Table 7.
In this table f_*^0, f_*^1, u_* are defined as in Section 6.8.5, and the new quantities
z_*^0, z_*^1 are the discrete analogues of $y(T)$ and $y_t(T)$, where y is the solution
of $(6.33)_2$, associated via $(6.33)_1$, to the solution $\mathbf{f}$ of problem (6.50).

Remark 6.20 In Table 7 we have taken $h/2$ as discretization parameter
to make easier comparisons with the results of Table 6 and Glowinski *et al.*
(1990, Section 10).

Comparing the above results to those in Table 6, the following facts appear
quite clearly.

1   The *filtering method* described in Section 6.8.7 has been a *very effective*
     cure to the *ill-posedness* of the approximate problem (6.88).
2   The number of *conjugate gradient* iterations necessary to achieve the
     convergence is (for h sufficiently small) essentially *independent* of h; in
     fact, if one realizes that for $h = 1/64$ the number of unknowns is $2 \times$
     $(63)^2 = 7938$, converging in 12 iterations is a fairly good performance.
3   The target functions z^0 and z^1 have been reached within a fairly high
     accuracy.

Table 7. *Table 2.1. Summary of numerical results.* [a] *indicates the number of conjugate gradient iterations.*

	$\frac{1}{8}$	$\frac{1}{16}$	$\frac{1}{32}$	$\frac{1}{64}$	$\frac{1}{128}$
			$h/2$		
a	7	10	12	12	12
CPU time(s) CRAY X-MP	0.1	0.6	2.8	14.8	83.9
$\frac{\|f^0 - f_*^0\|_{0,\Omega}}{\|f^0\|_{0,\Omega}}$	9.6×10^{-2}	2.6×10^{-2}	2.2×10^{-2}	6.4×10^{-3}	1.5×10^{-3}
$\frac{|f^0 - f_*^0|_{1,\Omega}}{|f^0|_{1,\Omega}}$	3.5×10^{-1}	1.8×10^{-1}	9×10^{-2}	4.4×10^{-2}	2.2×10^{-2}
$\frac{\|f^1 - f_*^1\|_{0,\Omega}}{\|f^1\|_{0,\Omega}}$	1×10^{-1}	2.6×10^{-2}	1.5×10^{-2}	7×10^{-3}	3.2×10^{-3}
$\frac{\|z^0 - z_*^0\|_{0,\Omega}}{\|z^0\|_{0,\Omega}}$	2.4×10^{-8}	3×10^{-8}	6×10^{-8}	8.3×10^{-8}	6.6×10^{-8}
$\frac{\|z^1 - z_*^1\|_{-1,\Omega}}{\|z^1\|_{-1,\Omega}}$	6.9×10^{-7}	4.6×10^{-7}	9.4×10^{-6}	2×10^{-5}	8.5×10^{-5}
$\frac{\|u - u_*\|_{0,\Sigma}}{\|u\|_{0,\Sigma}}$	1.2×10^{-1}	4.3×10^{-2}	2×10^{-2}	7.6×10^{-3}	3.4×10^{-3}
$\|u_*\|_{0,\Sigma}$	7.271	7.386	7.453	7.405	7.381

The results of Table 6 compare favourably with those displayed in Tables 10.3 and 10.4 of Glowinski *et al.* (1990, pp. 58, 59) which were obtained using the Tychonoff regularization procedure briefly recalled in Section 6.8.6; in fact, fewer iterations are needed here, implying a smaller CPU time (actually the CPU time seems to be a *sublinear* function of h^{-3} which is – modulo a multiplicative constant – the number of points of the space/time discretization grid). Table 7 also shows that the approximation errors (roughly) satisfy

$$\|f^0 - f_*^0\|_{L^2(\Omega)} = 0(h^2), \quad \|f^0 - f_*^0\|_{H_0^1(\Omega)} = 0(h), \quad \|f^1 - f_*^1\|_{L^2(\Omega)} = 0(h), \tag{6.147}$$

$$\|u - u_*\|_{L^2(\Sigma)} = 0(h). \tag{6.148}$$

Estimates (6.147) are of *optimal order* with respect to h in the sense that they have the order that we can expect when one approximates the solution of a boundary value problem, for a second-order elliptic operator, by piecewise linear FE approximations; this result is not surprising since (from Section 6.8.2, relation (6.54)) the operator Λ associated with $\Omega = (0,1) \times (0,1)$ behaves for T *sufficiently large* like

$$2T \begin{pmatrix} -\Delta & 0 \\ 0 & I \end{pmatrix} \tag{6.149}$$

(we have here $x_0 = \{1/2, 1/2\}$ and $C = \frac{1}{2}$).

In order to visualize the influence of h we have plotted for $h = 1/4$, 1/8, 1/16, 1/32, 1/64 and $\Delta t = h/2\sqrt{2}$ the exact solutions f^0, f^1 and

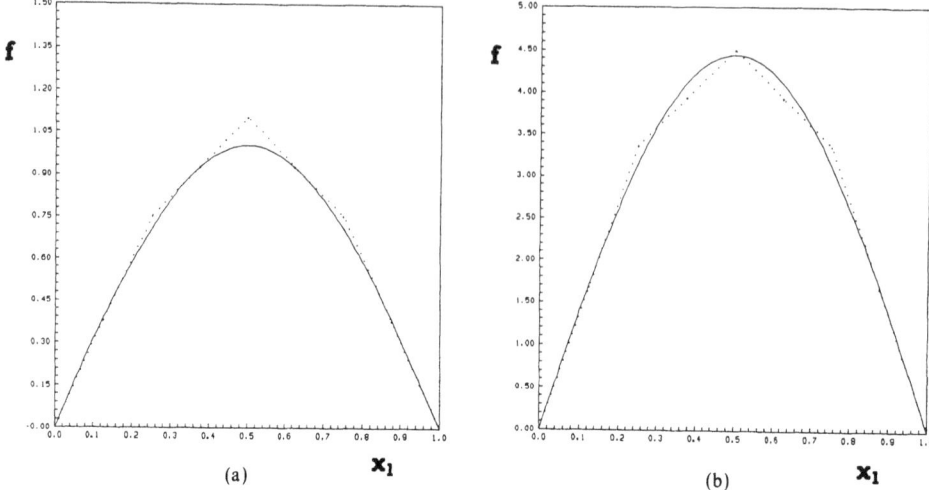

(a) x_1 (b) x_1

Fig. 54. ($h = 1/4$, $\Delta t = h/2\sqrt{2}$). (a) Variation of $f^0(x_1, 1/2)$ (——) and $f^0_*(x_1, 1/2)$ ($\cdots\cdots$); (b) Variation of $-f^1(x_1, 1/2)$ (——) and $-f^1_*(x_1, 1/2)$ ($\cdots\cdots$).

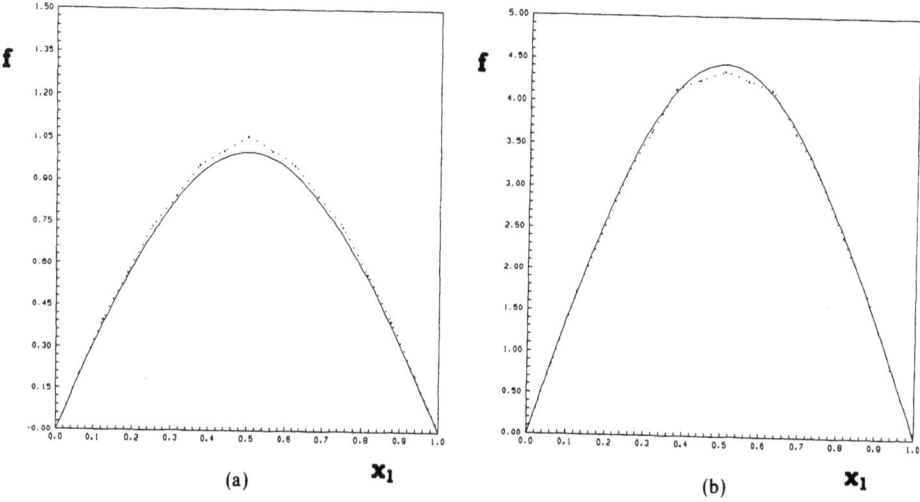

(a) x_1 (b) x_1

Fig. 55. ($h = 1/8$, $\Delta t = h/2\sqrt{2}$). (a) Variation of $f^0(x_1, 1/2)$ (——) and $f^0_*(x_1, 1/2)$ ($\cdots\cdots$). (b) Variation of $-f^1(x_1, 1/2)$ (——) and $-f^1_*(x_1, 1/2)$ ($\cdots\cdots$).

the corresponding computed solutions f^0_*, f^1_*. To be more precise, we have shown the plots of the functions $x_1 \to f^0(x_1, 1/2)$, $x_1 \to -f^1(x_1, 1/2)$ (full curves) and of the corresponding computed functions (dotted curves). These results have been reported in Figures 54 to 58 and the captions there are self-explanatory.

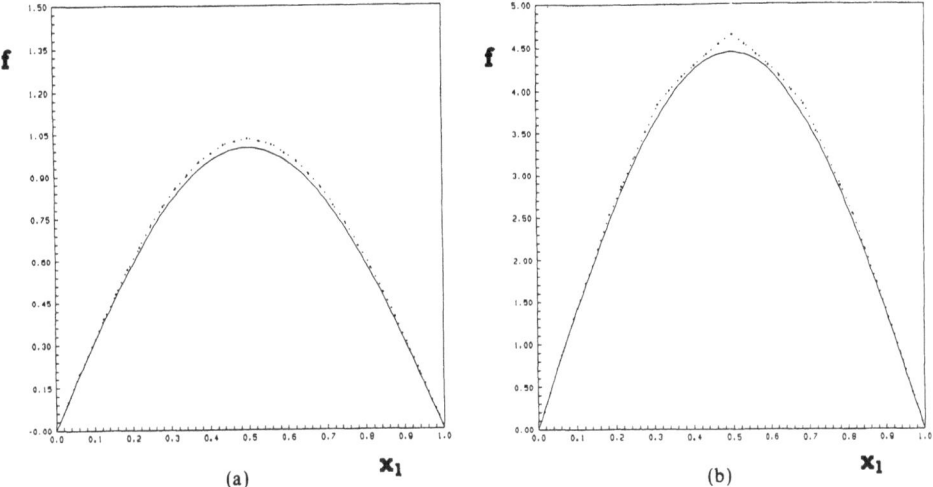

(a) $\mathbf{x_1}$

(b) $\mathbf{x_1}$

Fig. 56. ($h = 1/16$, $\Delta t = h/2\sqrt{2}$). (a) Variation of $f^0(x_1, 1/2)$ (———) and $f^0_*(x_1, 1/2)$ ($\cdots\cdots$). (b) Variation of $-f^1(x_1, 1/2)$ (———) and $-f^1_*(x_1, 1/2)$ ($\cdots\cdots$).

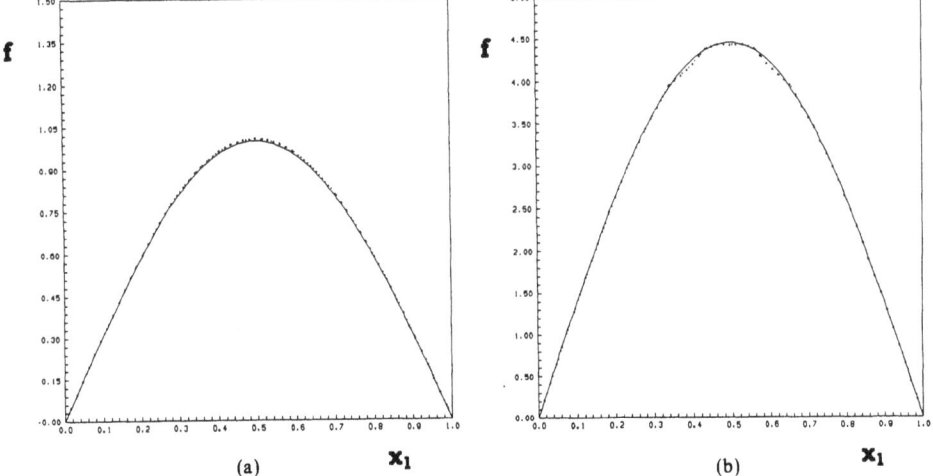

(a) $\mathbf{x_1}$

(b) $\mathbf{x_1}$

Fig. 57. ($h = 1/32$, $\Delta t = h/2\sqrt{2}$). (a) Variation of $f^0(x_1, 1/2)$ (———) and $f^0_*(x_1, 1/2)$ ($\cdots\cdots$). (b) Variation of $-f^1(x_1, 1/2)$ (———) and $-f^1_*(x_1, 1/2)$ ($\cdots\cdots$).

The above numerical experiments have been done with $T = 15/4\sqrt{2}$; in order to study the influence of T we have kept z^0 and z^1 as in the above experiments and taken $T = 28.2843$. For $h = 1/64$ and $\Delta t = h/2\sqrt{2}$ we need just 10 iterations of algorithm (6.116)–(6.136) to achieve convergence, the corresponding CRAY X-MP CPU time being then 800s (!) (the number

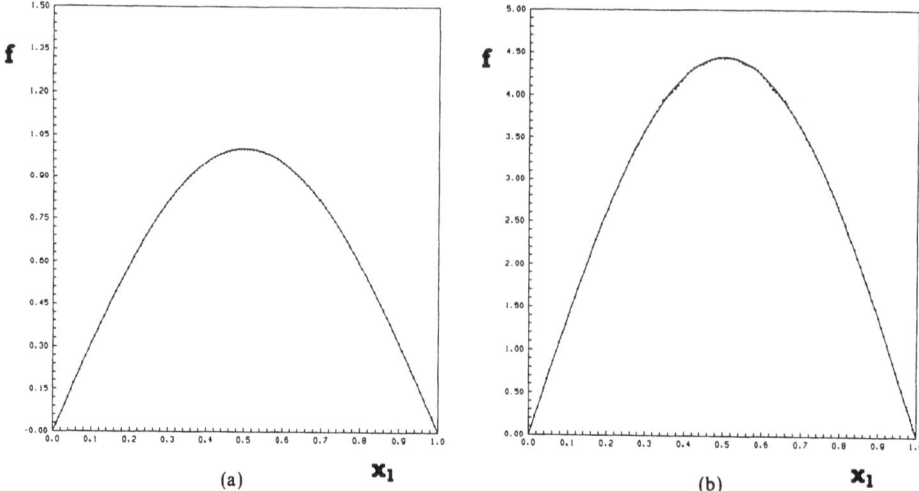

Fig. 58. ($h = 1/64$, $\Delta t = h/2\sqrt{2}$). (a) Variation of $f^0(x_1, 1/2)$ (———) and $f_*^0(x_1, 1/2)$ ($\cdots\cdots$). (b) Variation of $-f^1(x_1, 1/2)$ (———) and $-f_*^1(x_1, 1/2)$ ($\cdots\cdots$).

of grid points for the space/time discretization is now $\approx 86 \times 10^6$). We have $\|u_*\|_{L^2(\Sigma)} = 2.32$, $\|z^0 - z_*^0\|_{L^2(\Omega)} = 5.8 \times 10^{-6}$, $\|z^1 - z_*^1\|_{-1,\Omega} = 1.6 \times 10^{-5}$. The most interesting results are the ones reported on Figures 59(a) and (b). There, we have compared Tf_*^0 and Tf_*^1 (for $T = 28.2843$) with the corresponding theoretical limits χ^0 and χ^1 which, according to Section 6.8.2, relations (6.55)–(6.57), are given by

$$\Delta\chi^0 = z^1/2 \text{ in } \Omega, \quad \chi^0 = 0 \text{ on } \Gamma, \tag{6.150}$$

$$\chi^1 = z^0/2. \tag{6.151}$$

The *full* curves represent the variations of $x_1 \to \chi^0(x_1, 1/2)$ and of $x_1 \to -\chi^1(x_1, 1/2)$, while the *dotted* curves represent the variations of $x_1 \to Tf_*^0(x_1, 1/2)$ and $x_1 \to -Tf_*^1(x_1, 1/2)$.

In our opinion the above figures provide an excellent *numerical verification* of the convergence result (6.55) of Section 6.8.2 (we observe at $x_1 = 0$ and $x_1 = 1$ a (numerical) *Gibbs phenomenon* associated with the L^2 convergence of Tf_*^1 to χ^1). Conversely, these results provide a *validation* of the numerical methodology discussed here; they show that this methodology is particularly robust, accurate, nondissipative and perfectly able to handle very long time intervals $[0, T]$. In fact, numerical experiments have shown that the above-mentioned qualities of the numerical methods discussed here persist for target functions z^0 and z^1 much rougher than those considered in this section.

Additional results can be found in Glowinski *et al.* (1990, Section 4).

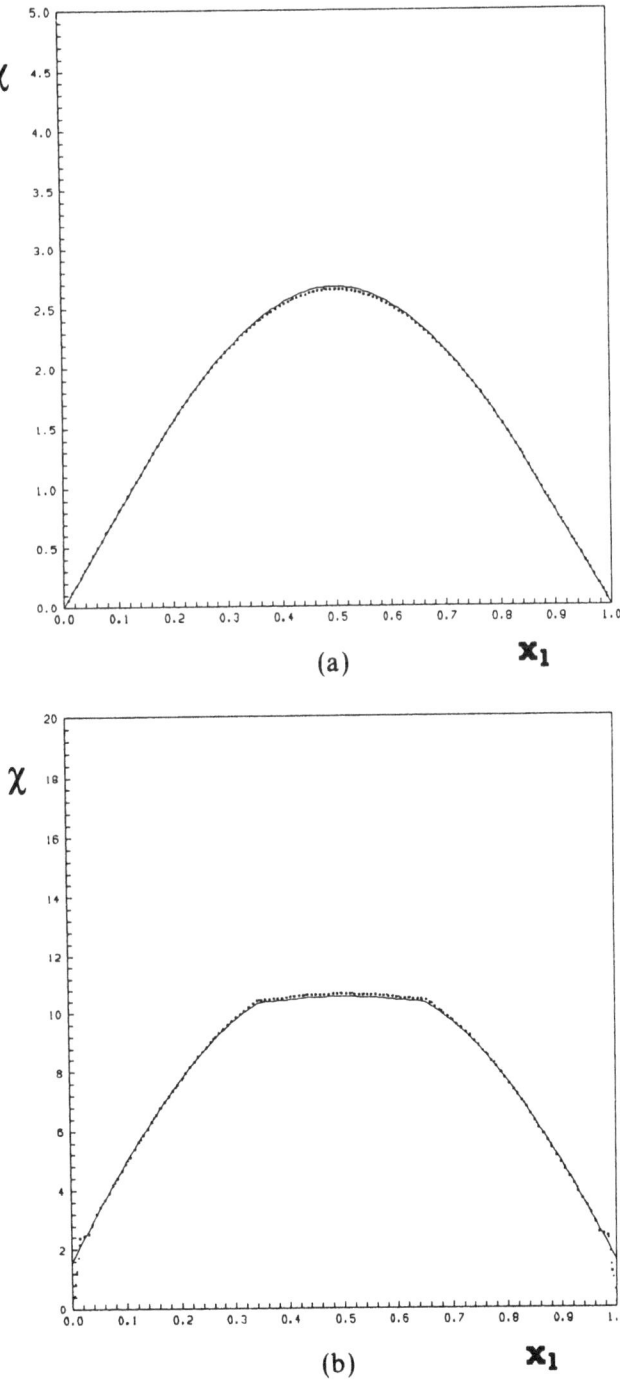

Fig. 59. ($h = 1/64$, $\Delta t = h/2\sqrt{2}$, $T = 28.2843$). (a) Variation of $\chi^0(x_1, 1/2)$ (——) and $Tf_*^0(x_1, 1/2)$ ($\cdots\cdots$). (b) Variation of $-\chi^1(x_1, 1/2)$ (——) and $-Tf_*^1(x_1, 1/2)$ ($\cdots\cdots$).

6.10. Other boundary controls

6.10.1. Approximate Neumann boundary controllability

We consider now problems entirely similar to the previous ones but where we 'exchange' the *Dirichlet conditions* for *Neumann conditions*.

We therefore define the state function $y = y(v)$ by

$$y_{tt} + Ay = 0 \text{ in } Q = \Omega \times (0, T), \tag{6.152}_1$$

$$y(0) = 0, \quad y_t(0) = 0, \tag{6.152}_2$$

$$\frac{\partial y}{\partial n_A} = v \text{ on } \Sigma_0 = \Gamma_0 \times (0, T), \quad \frac{\partial y}{\partial n_A} = 0 \text{ on } \Sigma \backslash \Sigma_0. \tag{6.152}_3$$

To fix ideas, we still assume that

$$v \in L^2(\Sigma_0). \tag{6.153}$$

Using again *transposition*, we can show that problem (6.152) has a *unique* (weak) solution if (6.153) holds. In fact, the solution here is (slightly) smoother than the one in Section 6.1. We can, in any case, define an operator L from $L^2(\Sigma_0)$ into $H^{-1}(\Omega) \times L^2(\Omega)$ by

$$Lv = \{-y_t(T; v), y(T; v)\}. \tag{6.154}$$

If v is smooth, the solution $y(v)$ will also be smooth, assuming of course that the coefficients of A are also smooth.

Let us study *approximate controllability* first.

We suppose that v is *smooth*; indeed, to fix ideas we assume that

$$v, \frac{\partial v}{\partial t} \in L^2(\Sigma_0), \quad v|_t = 0. \tag{6.155}$$

Then, $y(v)$ can be defined by a *variational formulation*, showing that

$$\begin{cases} y \text{ is continuous from } [0, T] \text{ into } H^1(\Omega), \\ y_t \text{ is continuous from } [0, T] \text{ into } L^2(\Omega). \end{cases} \tag{6.156}$$

Then in particular

$$Lv \in L^2(\Omega) \times L^2(\Omega). \tag{6.157}$$

Let us consider now $\mathbf{f}$ belonging to the orthogonal of the range of L, i.e.

$$\begin{cases} \mathbf{f} = \{f^0, f^1\} \in L^2(\Omega) \times L^2(\Omega), \\ -\int_\Omega f^0 y_t(T) \, dx + \int_\Omega f^1 y(T) \, dx = 0, \ \forall v \text{ satisfying (6.155)}. \end{cases} \tag{6.158}$$

We introduce ψ defined by

$$\psi_{tt} + A\psi = 0 \text{ in } Q, \quad \psi(T) = f^0, \ \psi_t(T) = f^1, \ \frac{\partial \psi}{\partial n_A} = 0 \text{ on } \Sigma. \tag{6.159}$$

Multiplying the wave equation in (6.159) by $y = y(v)$ and integrating by parts, we obtain

$$\langle Lv, \mathbf{f} \rangle = - \int_{\Sigma_0} \psi v \, d\Sigma. \tag{6.160}$$

Therefore (6.158) is equivalent to

$$\psi = 0 \text{ on } \Sigma_0. \tag{6.161}$$

If we assume (as in Section 6.3, relation (6.22)) that

$$\Sigma_0 \text{ allows the application of the Holmgren's uniqueness theorem} \tag{6.162}$$

then (6.159), (6.161) imply that $\psi = 0$ in Q, so that $f = 0$; we have proved thus that

$$\text{assuming (6.162) the range of } L \text{ is dense in } L^2(\Omega) \times L^2(\Omega), \tag{6.163}$$

which implies, in turn, *approximate controllability*.

Remark 6.20 Suppose that Γ is a C^∞ manifold. Then we can take

$$v \in \mathcal{D}(\Sigma_0) \text{ (the space of the } C^\infty \text{ functions with compact support in } \Sigma_0), \tag{6.164}$$

and the range of L, for v describing $\mathcal{D}(\Sigma_0)$, is still dense in $L^2(\Omega) \times L^2(\Omega)$.

We can now state the following control problem

$$\inf_v \frac{1}{2} \int_{\Sigma_0} v^2 \, d\Sigma; y(T) \in z^0 + \beta_0 B, \quad y_t(T) \in z^1 + \beta_1 B, \tag{6.165}$$

where, in (6.165), $\{y, v\}$ satisfies (6.152), (6.153), $\{z^0, z^1\} \in L^2(\Omega) \times L^2(\Omega)$, and where B denotes the closed unit ball of $L^2(\Omega)$.

Remark 6.21 We do *not* introduce $H^{-1}(\Omega)$ here for two reasons:

1. in the present context the $H^{-1}(\Omega)$ space (which is not the dual of $H^1(\Omega)$) is not natural;
2. the choice of the same norm, in (6.165), for both $y(T)$ and $y_t(T)$ shows the flexibility of the methodology.

Remark 6.22 Problem (6.165) has a unique solution. *Uniqueness* follows from the *strict convexity*. As far as *existence* is concerned let $\{u_n\}_{n\geq 0}$ be a *minimizing sequence*. Then $\{u_n\}_{n\geq 0}$ is *bounded* in $L^2(\Sigma_0)$. Let us set $y_n = y(u_n)$. By definition, $\{\{y_n(T), \partial y_n/\partial t(T)\}\}_{n\geq 0}$ remains in a bounded set of $L^2(\Omega) \times L^2(\Omega)$. We can therefore extract from $\{u_n\}_{n\geq 0}$ a subsequence, still denoted by $\{u_n\}_{n\geq 0}$, such that

$$u_n \to u \text{ weakly in } L^2(\Sigma_0), \tag{6.166}$$

$$\left\{ y_n(T), \frac{\partial y_n}{\partial t}(T) \right\} \to \{\xi_0, \xi_1\} \text{ weakly in } L^2(\Omega) \times L^2(\Omega). \tag{6.167}$$

However,

$$\{y(T; u_n), y_t(T; u_n)\} \to \{y(T; u), y_t(T; u)\}$$

weakly in $L^2(\Omega) \times H^{-1}(\Omega)$, so that $\xi_0 = y(T; u), \xi_1 = y_t(T; u)$, which proves the existence of a solution namely u to problem (6.165).

The uniqueness of the solution implies that the *whole* minimizing sequence converges to u.

Remark 6.23 The use of β_0 and $\beta_1 > 0$ allows the introduction of new (and complicated) function spaces to be avoided. Unfortunately, these spaces cannot be avoided if we let β_0 and $\beta_1 \to 0$, as we shall see below.

6.10.2. Duality results: exact Neumann boundary controllability

Now, we use *duality*, as in previous sections. We then introduce functionals F_1 and F_2 by

$$F_1(v) = \frac{1}{2} \int_{\Sigma_0} v^2 \, d\Sigma, \tag{6.168}$$

$$F_2(\hat{\mathbf{f}}) = \begin{cases} 0 \text{ if } \hat{f}^0 \in -z^1 + \beta_1 B \quad \hat{f}^1 \in z^0 + \beta_0 B, \\ +\infty \text{ otherwise on } L^2(\Omega) \times L^2(\Omega). \end{cases} \tag{6.169}$$

It follows then from (6.154) that problem (6.165) is equivalent to

$$\inf_{v \in L^2(\Sigma_0)} [F_1(v) + F_2(Lv)]. \tag{6.170}$$

Using *convex duality* arguments, we obtain

$$\inf_{v \in L^2(\Sigma_0)} [F_1(v) + F_2(Lv)] = - \inf_{\hat{\mathbf{f}} \in L^2(\Omega) \times L^2(\Omega)} [F_1^*(L^*\hat{\mathbf{f}}) + F_2^*(-\hat{\mathbf{f}})], \tag{6.171}$$

where we use L^* with L thought of as an unbounded operator.

By virtue of (6.171), we have

$$L^*\hat{\mathbf{f}} = -\hat{\psi}|_{\Sigma_0}, \tag{6.172}$$

where $\hat{\psi}$ is the solution of (6.159) when $\hat{\mathbf{f}}$ replaces $\mathbf{f}$.

We obtain then as *dual problem* of the control problem (6.165)

$$\inf_{\hat{\mathbf{f}} \in L^2(\Omega) \times L^2(\Omega)} \left[\frac{1}{2} \int_{\Sigma_0} \hat{\psi}^2 \, d\Sigma + \int_\Omega (z^1 \hat{f}_0 - z^0 \hat{f}^1) \, dx \right.$$

$$\left. + \beta_1 \|\hat{f}^0\|_{L^2(\Omega)} + \beta_0 \|\hat{f}^1\|_{L^2(\Omega)} \right]. \tag{6.173}$$

Remark 6.23 We shall give an alternative formulation of the dual problem (6.173). This new formulation is particularly useful when $\{\beta_0, \beta_1\} \to \mathbf{0}$. Using the HUM approach, we introduce the operator Λ defined as follows.

The functions $\hat{f}^0$ and $\hat{f}^1$ being given in, say, $L^2(\Omega)$, we define $\hat{\psi}$ and $\hat{y}$ by

$$\hat{\psi}_{tt} + A\hat{\psi} = 0 \text{ in } Q, \quad \hat{\psi}(T) = \hat{f}^0, \ \hat{\psi}_t(T) = \hat{f}^1, \quad \frac{\partial \hat{\psi}}{\partial n_A} = 0 \text{ on } \Sigma, \quad (6.174)$$

$$\begin{cases} \hat{y}_{tt} + A\hat{y} = 0 \text{ in } Q, \quad \hat{y}(0) = \hat{y}_t(0) = 0, \quad \dfrac{\partial \hat{y}}{\partial n_A} = -\hat{\psi} \text{ on } \Sigma_0, \\[2mm] \dfrac{\partial \hat{y}}{\partial n_A} = 0 \text{ on } \Sigma \backslash \Sigma_0. \end{cases} \quad (6.175)$$

We set then (with $\hat{f} = \{\hat{f}^0, \hat{f}^1\}$):

$$\Lambda \hat{f} = \{-\hat{y}_t(T), \hat{y}(T)\}. \quad (6.176)$$

Taking $\hat{f} = f_1$ and f_2, and denoting by ψ_1, ψ_2 the corresponding solutions of (6.174) we obtain from (6.174), (6.175) that

$$\int_\Omega (\Lambda f_1) \cdot f_2 \, dx = \int_{\Sigma_0} \psi_1 \psi_2 \, d\Sigma. \quad (6.177)$$

It follows from (6.177) that

Λ is symmetric and positive semi-definite over $L^2(\Omega) \times L^2(\Omega)$. $\quad (6.178)$

It follows from (6.177) that problem (6.173) is *equivalent* to

$$\inf_{\hat{f} \in L^2(\Omega) \times L^2(\Omega)} \left[\frac{1}{2} \int_\Omega (\Lambda \hat{f}) \cdot \hat{f} \, dx + \int_\Omega (z^1 \hat{f}^0 - z^0 \hat{f}^1) \, dx + \beta_1 \|\hat{f}^0\|_{L^2(\Omega)} \right.$$

$$\left. + \beta_0 \|\hat{f}^1\|_{L^2(\Omega)} \right]. \quad (6.179)$$

In order to discuss the case $\beta_0 = \beta_1 = 0$ in (6.179), we introduce over $L^2(\Omega) \times L^2(\Omega)$ the norm $[\ldots]$ defined by

$$[\hat{f}] = \left(\int_\Omega (\Lambda \hat{f}) \cdot \hat{f} \, dx \right)^{1/2}, \quad \forall \hat{f} \in L^2(\Omega) \times L^2(\Omega). \quad (6.180)$$

We define next the space E by

$$E = \text{ completion of } L^2(\Omega) \times L^2(\Omega) \text{ for the norm } [\ldots]. \quad (6.181)$$

Taking now the limit in (6.179) as $\{\beta_0, \beta_1\} \to \mathbf{0}$ we obtain – *formally* – the dual problem associated with *exact controllability*, namely

$$\inf_{\hat{f} \in E} \left[\frac{1}{2} [\hat{f}]^2 - \langle \sigma z, \hat{f} \rangle \right] \quad (6.182)$$

where, in (6.182), $\langle \ldots, \ldots \rangle$ denotes the duality pairing between E' and E, $z = \{z^0, z^1\}$ and $\sigma = \begin{pmatrix} 0 & -1 \\ 1 & 0 \end{pmatrix}$.

Problem (6.182) has a solution (necessarily unique) if and only if

$$\{-z^1, z^0\} \in E';$$ (6.183)

equivalently, exact controllability is true if and only if condition (6.183) is satisfied.

Remark 6.24 Contrary to the situation in Section 6.6, the space E, as defined by (6.181), has no simple interpretation. For further information concerning space E, we refer to Lions (1988b) and the references therein.

Remark 6.25 It is by now clear that the method followed here is general. It can, in particular, be applied to *other boundary conditions*.

6.10.3. A second approximate Neumann boundary controllability problem

Inspired by Sections 1 and 2, we consider, for $k > 0$, the following control problem

$$\min_{v \in L^2(\Sigma_0)} \left[\frac{1}{2} \int_{\Sigma_0} v^2 \, d\Sigma + \frac{k}{2} (\|y(T) - z^0\|_{L^2(\Omega)}^2 + \|y_t(T) - z^1\|_{L^2(\Omega)}^2) \right],$$
                                                              (6.184)

where, in (6.184), y is – still – defined from v via the wave equation (6.152); problem (6.184) is obtained by *penalization* of the conditions $y(T) = z^0$, $y_t(T) = z^1$.

Using the results of Section 6.10.1 it is quite easy to show that *problem (6.184) has a (necessarily unique) solution* (even if (6.162) does not hold). If we denote by u the solution of problem (6.184), it is characterized by the existence of an *adjoint state* function p such that

$$\begin{cases} y_{tt} + Ay = 0 \text{ in } Q, \quad y(0) = y_t(0) = 0, \\ \dfrac{\partial y}{\partial n_A} = u \text{ on } \Sigma_0, \quad \dfrac{\partial y}{\partial n_A} = 0 \text{ on } \Sigma \backslash \Sigma_0, \end{cases}$$ (6.185)

$$p_{tt} + Ap = 0 \text{ in } Q, \quad \frac{\partial p}{\partial n_A} = 0 \text{ on } \Sigma,$$ (6.186)$_1$

$$p(T) = k(y_t(T) - z^1), \quad p_t(T) = -k(y(T) - z^0),$$ (6.186)$_2$

$$u = -p \text{ on } \Sigma_0.$$ (6.187)

Let us define $\mathbf{f} = \{f^0, f^1\} \in L^2(\Omega) \times L^2(\Omega)$ by

$$f^0 = p(T), \quad f^1 = p_t(T);$$ (6.188)

it follows then from (6.186)$_2$, and from the definition of Λ (see Section 6.10.2) that

$$k^{-1}\mathbf{f} + \Lambda\mathbf{f} = \{-z^1, z^0\}.$$ (6.189)

Problem (6.189) is the *dual* problem of (6.184).

From the properties of Λ (*symmetry* and *positivity*) and from the $(L^2(\Omega))^2$ ellipticity of the bilinear form associated with operator $k^{-1}I + \Lambda$, problem (6.189) can be solved by a *conjugate gradient algorithm* operating in $L^2(\Omega) \times L^2(\Omega)$; such an algorithm will be described in Section 6.10.4.

6.10.4. Conjugate gradient solution of the dual problem (6.189)

We can solve problem (6.189) by the following variant of algorithm (6.58)–(6.73) (see Section 6.8.3):

$$\mathbf{f}_0 = \{f_0^0, f_0^1\} \text{ given in } L^2(\Omega) \times L^2(\Omega); \tag{6.190}$$

solve then

$$\begin{cases} \dfrac{\partial^2 \psi_0}{\partial t^2} + A\psi_0 = 0 \text{ in } Q, \quad \psi_0(T) = f_0^0, \\[2mm] \dfrac{\partial \psi_0}{\partial t}(T) = f_0^1, \quad \dfrac{\partial \psi_0}{\partial n_A} = 0 \text{ on } \Sigma, \end{cases} \tag{6.191}$$

$$\begin{cases} \dfrac{\partial^2 \varphi_0}{\partial t^2} + A\varphi_0 = 0 \text{ in } Q, \quad \varphi_0(0) = \dfrac{\partial \varphi_0}{\partial t}(0) = 0, \\[2mm] \dfrac{\partial \varphi_0}{\partial n_A} = -\psi_0 \text{ on } \Sigma_0, \quad \dfrac{\partial \varphi_0}{\partial n} = 0 \text{ on } \Sigma \backslash \Sigma_0. \end{cases} \tag{6.192}$$

Define $\mathbf{g}_0 = \{g_0^0, g_0^1\} \in L^2(\Omega) \times L^2(\Omega)$ *by*

$$\int_\Omega g_0^0 v \, dx = k^{-1} \int_\Omega f_0^0 v \, dx + \int_\Omega \left(z^1 - \dfrac{\partial \varphi_0}{\partial t}(T) \right) v \, dx, \quad \forall v \in L^2(\Omega), \tag{6.193$_1$}$$

$$\int_\Omega g_0^1 v \, dx = k^{-1} \int_\Omega f_0^1 v \, dx + \int_\Omega (\varphi_0(T) - z^0) v \, dx, \quad \forall v \in L^2(\Omega), \tag{6.193$_2$}$$

and define $\mathbf{w}_0 = \{w_0^0, w_1^0\}$ *by*

$$\mathbf{w}_0 = \mathbf{g}_0. \tag{6.194}$$

Assuming that $\mathbf{f}_n, \mathbf{g}_n, \mathbf{w}_n$ *are known, we obtain* $\mathbf{f}_{n+1}, \mathbf{g}_{n+1}, \mathbf{w}_{n+1}$ *as follows.*
 Solve

$$\begin{cases} \dfrac{\partial^2 \bar\psi_n}{\partial t^2} + A\bar\psi_n = 0 \text{ in } Q, \quad \bar\psi_n(T) = w_n^0, \\[2mm] \dfrac{\partial \bar\psi_n}{\partial t}(T) = w_n^1, \quad \dfrac{\partial \bar\psi_n}{\partial n_A} = 0 \text{ on } \Sigma, \end{cases} \tag{6.195}$$

$$\begin{cases} \dfrac{\partial^2 \bar\varphi_n}{\partial t^2} + A\bar\varphi_n = 0 \text{ in } Q, \quad \bar\varphi_n(0) = \dfrac{\partial \bar\varphi_n}{\partial t}(0) = 0, \\[2mm] \dfrac{\partial \bar\varphi_n}{\partial n_A} = -\bar\psi_n \text{ on } \Sigma_0, \quad \dfrac{\partial \bar\varphi_n}{\partial n_A} = 0 \text{ on } \Sigma \backslash \Sigma_0. \end{cases} \tag{6.196}$$

Define $\bar{\mathbf{g}}_n = \{\bar{g}_n^0, \bar{g}_n^1\} \in L^2(\Omega) \times L^2(\Omega)$ *by*

$$\int_\Omega \bar{g}_n^0 v \, \mathrm{d}x = k^{-1} \int_\Omega w_n^0 v \, \mathrm{d}x - \int_\Omega \frac{\partial \bar{\varphi}_n}{\partial t}(T) v \, \mathrm{d}x, \quad \forall v \in L^2(\Omega), \qquad (6.197)_1$$

$$\int_\Omega \bar{g}_n^1 v \, \mathrm{d}x = k^{-1} \int_\Omega w_n^1 v \, \mathrm{d}x + \int_\Omega \bar{\varphi}_n(T) v \, \mathrm{d}x, \quad \forall v \in L^2(\Omega). \qquad (6.197)_2$$

Compute

$$\rho_n = \int_\Omega (|g_n^0|^2 + |g_n^1|^2) \, \mathrm{d}x \Big/ \int_\Omega (\bar{g}_n^0 w_n^0 + \bar{g}_n^1 w_n^1) \, \mathrm{d}x, \qquad (6.198)$$

and then

$$\mathbf{f}_{n+1} = \mathbf{f}_n - \rho_n \mathbf{w}_n, \qquad (6.199)$$

$$\mathbf{g}_{n+1} = \mathbf{g}_n - \rho_n \bar{\mathbf{g}}_n. \qquad (6.200)$$

If $\|\mathbf{g}_{n+1}\|_{L^2(\Omega) \times L^2(\Omega)} / \|\mathbf{g}_0\|_{L^2(\Omega) \times L^2(\Omega)} \leq \epsilon$ *take* $\mathbf{f} = \mathbf{f}_{n+1}$; *if not compute*

$$\gamma_n = \frac{\|\mathbf{g}_{n+1}\|^2_{L^2(\Omega) \times L^2(\Omega)}}{\|\mathbf{g}_n\|^2_{L^2(\Omega) \times L^2(\Omega)}} \qquad (6.201)$$

and update $\mathbf{w}_n$ *by*

$$\mathbf{w}_{n+1} = \mathbf{g}_{n+1} + \gamma_n \mathbf{w}_n. \qquad (6.202)$$

Do $n = n + 1$ *and go to* (6.195).

Remark 6.26 The FE implementation of the above algorithm is just a variation of the one of algorithm (6.58)–(6.73) (it is in fact simpler). In fact, here too we can take advantage of the *reversibility* of the wave equations to reduce the storage requirements of the discrete analogues of algorithm (6.190)–(6.202).

Remark 6.27 In Glowinski and Li (1990), we can find a discussion of numerical methods for solving exact Neumann boundary controllability problems; the solution method is based on a combination of *finite element* approximations and of a conjugate gradient algorithm closely related to algorithm (6.190)–(6.202). We also discuss, in the above reference, the asymptotic behaviour of the solution $\mathbf{f}$ of the dual problem when $T \to +\infty$; there too the analytical results confirmed the numerical ones, validating therefore the computational methodology.

6.10.5. Application to the solution of the dual problem (6.179)

Assuming that β_0 and β_1 are *positive* the dual problem (6.179) can also be written as

$$\Lambda \mathbf{f} + \partial j(\mathbf{f}) = \begin{pmatrix} -z^1 \\ z^0 \end{pmatrix}, \qquad (6.203)$$

where the functional $j : L^2(\Omega) \times L^2(\Omega) \to \mathbb{R}$ is defined by

$$j(\hat{f}) = \beta_1 \|\hat{f}^0\|_{L^2(\Omega)} + \beta_0 \|\hat{f}^1\|_{L^2(\Omega)}, \quad \forall \hat{f} = \{f^0, f^1\} \in L^2(\Omega) \times L^2(\Omega). \quad (6.204)$$

As in Section 6.8.8, we associate with (6.203) the following *initial value problem*

$$\begin{cases} \dfrac{\partial \mathbf{f}}{\partial \tau} + \Lambda \mathbf{f} + \partial j(\mathbf{f}) = \begin{pmatrix} -z^1 \\ z^0 \end{pmatrix}, \\ \mathbf{f}(0) = \mathbf{f}_0 \end{cases} \quad (6.205)$$

to be discretized, for example, by the following *Peaceman–Rachford scheme*:

$$\mathbf{f}^0 = \mathbf{f}_0; \quad (6.206)$$

then for $k \geq 0$, *assuming that* $\mathbf{f}^k$ *is known, we compute* $\mathbf{f}^{k+1/2}$ *and* $\mathbf{f}^{k+1}$ *via*

$$\frac{\mathbf{f}^{k+1/2} - \mathbf{f}^k}{\Delta\tau/2} + \Lambda f^k + \partial j(\mathbf{f}^{k+1/2}) = \begin{pmatrix} -z^1 \\ z^0 \end{pmatrix}, \quad (6.207)$$

$$\frac{\mathbf{f}^{k+1} - \mathbf{f}^{k+1/2}}{\Delta\tau/2} + \Lambda f^{k+1} + \partial j(\mathbf{f}^{k+1/2}) = \begin{pmatrix} -z^1 \\ z^0 \end{pmatrix}. \quad (6.208)$$

Problem (6.207) is fairly easy to solve (see Section 6.8.8) since the operator $\partial j(\ldots)$ is *diagonal*. On the other hand, once $\mathbf{f}^{k+1/2}$ is known, problem (6.208) is just a particular case of problem (6.189) (with $k = \Delta\tau/2$); it can be solved therefore by the conjugate gradient algorithm (6.190)–(6.202).

6.11. Distributed controls for wave equations

Let us consider $\mathcal{O} \subset \Omega$ and let the *state equation* be

$$y_{tt} + Ay = v\chi_{\mathcal{O}} \text{ in } Q, \quad y(0) = y_t(0) = 0, \quad y = 0 \text{ on } \Sigma. \quad (6.209)$$

We choose

$$v \in L^2(\mathcal{O} \times (0,T)). \quad (6.210)$$

The solution of problem (11.1) is *unique*. It satisfies

$$\{y, y_t\} \text{ is continuous from } [0,T] \text{ into } H_0^1(\Omega) \times L^2(\Omega). \quad (6.211)$$

Let us see when

$$\{y(T), y_t(T)\} \text{ spans a dense subset of } H_0^1(\Omega) \times L^2(\Omega). \quad (6.212)$$

We consider $\mathbf{f} = \{f^0, f^1\} \in L^2(\Omega) \times H^{-1}(\Omega)$ such that

$$-\int_\Omega y_t(T)f^0 \, dx + \langle f^1, y(T)\rangle = 0, \quad \forall v \in L^2(\mathcal{O} \times (0,T)), \quad (6.213)$$

where, in (6.213), $\langle \cdot, \cdot \rangle$ denotes the duality pairing between $H^{-1}(\Omega)$ and $H_0^1(\Omega)$.

We introduce ψ solution of

$$\psi_{tt} + A\psi = 0 \text{ in } Q, \quad \psi(T) = f^0, \ \psi_t(T) = f^1, \ \psi = 0 \text{ on } \Sigma. \tag{6.214}$$

Then

$$-\int_\Omega y_t(T) f^0 \, dx + \langle f^1, y(T) \rangle = \int_{\mathcal{O} \times (0,T)} \psi v \, dx \, dt. \tag{6.215}$$

Therefore (6.213) is equivalent to

$$\psi = 0 \text{ on } \mathcal{O} \times (0, T). \tag{6.216}$$

We shall assume that we can apply Holmgren's uniqueness theorem to $\mathcal{O} \times (0, T)$; then $\psi \equiv 0$ and $f = 0$, so that (6.212) holds true.

We can then consider

$$\inf_v \frac{1}{2} \int \int_{\mathcal{O} \times (0,T)} v^2 \, dx \, dt; \, y(T; v) \in z^0 + \beta_0 B_1, y_t(T; v) \in z^1 + \beta_1 B \tag{6.217}$$

where, in (6.217), $y(v)$ is obtained from v via (6.209), $\{z^0, z^1\}$ is given in $H_0^1(\Omega) \times L^2(\Omega)$, B_1 (respectively B) is the closed unit ball of $H_0^1(\Omega)$ (respectively $L^2(\Omega)$).

Similar considerations to everything which has been said in the previous sections can be adapted to the present situation, from either the *purely mathematical point of view* (see Lions (1988b)) or the *numerical point of view*.

Remark 6.28 One can also consider *pointwise control*, as in

$$\begin{cases} y_{tt} + Ay = v(t)\delta(x - b) \text{ in } Q, \\ y(0) = y_t(0) = 0, \quad y = 0 \text{ on } \Sigma \text{ (to fix ideas)}. \end{cases} \tag{6.218}$$

Control problems for systems modelled by (6.218) have been discussed in Lions (1988b, Volume 1, Chapter 7). Interesting phenomena appear concerning the role of $b \in \Omega$. Methods from harmonic analysis have been used in this respect by Meyer (1989) and further developed by Haraux and Jaffard (1991), I. Joó (1991).

6.12. Dynamic Programming

We are going to apply *Dynamic Programming* to the situations described in Section 6.11. The approach is *formal*, somewhat similar to the one in Section 5.

Remark 6.29. We could have applied dynamic programming to the situations described in Sections 6.1 or 6.10, but the situation is simpler for the control problems described in Section 6.11.

We consider for s given in $(0, T)$

$$\begin{cases} y_{tt} + Ay = v\chi_{\mathcal{O}} \text{ in } \Omega \times (s, T), \\ y(s) = h^0, \quad y_t(s) = h^1, \quad y = 0 \text{ on } \Sigma_s = \Gamma \times (s, T), \end{cases} \tag{6.219}$$

with $\{h^0, h^1\} \in H_0^1(\Omega) \times L^2(\Omega)$.

We introduce

$$\phi(h^0, h^1, s) = \inf_v \int\int_{\mathcal{O}\times(s,T)} v^2 \, dx \, dt \qquad (6.220)$$

where, in (6.220), v is such that $\{y(v), v\}$ satisfies (6.219) and

$$y(T; v) \in z^0 + \beta_0 B_1, \quad y_t(T; v) \in z^1 + \beta_1 B. \qquad (6.221)$$

The quantity $\phi(h^0, h^1, s)$ is finite for every

$$z^0 \in H_0^1(\Omega), \quad z^1 \in L^2(\Omega), \quad \beta_0 > 0, \quad \beta_1 > 0$$

if and only if the Holmgren's uniqueness theorem applies for $\mathcal{O} \times (s, T)$ in $\Omega \times (s, T)$. This is true for $s < s_0$, s_0 a suitable number in $(0, T)$. In that case, the infimum in (6.221) is finite for $s < s_0$, implying that the function ϕ is defined over $H_0^1(\Omega) \times L^2(\Omega) \times (0, s_0)$.

Let us write now the *Hamilton–Jacobi–Bellmann (HJB) equation*; we take

$$v(x, t) = w(x) \text{ in } \mathcal{O} \times (s, s + \varepsilon). \qquad (6.222)$$

With this choice of v, $\{y(t), y_t(t)\}$ 'moves' during the time interval $(s, s + \varepsilon)$ from $\{h^0, h^1\}$ to

$$\{h^0 + \varepsilon h^1, h^1 + \varepsilon w\chi_{\mathcal{O}} - \varepsilon A h^0\} + 0(\varepsilon^2)$$

(assuming that $h^0 \in H_0^1(\Omega) \cap H^2(\Omega)$). Then, according to the optimality principle, we have

$$\phi(h^0, h^1, s) = \inf_w \left[\frac{\varepsilon}{2} \int_{\mathcal{O}} w^2 \, dx + \phi(h^0 + \varepsilon h^1, h^1 + \varepsilon w\chi_{\mathcal{O}} \right.$$
$$\left. - \varepsilon A h^0, s + \varepsilon) \right] + 0(\varepsilon^2). \qquad (6.223)$$

Expanding ϕ we obtain

$$\frac{\partial\phi}{\partial s} + \left(\frac{\partial\phi}{\partial h^0}, h^1\right) - \left(\frac{\partial\phi}{\partial h^1}, A h^0\right) + \inf_{w\in L^2(\mathcal{O})} \left[\frac{1}{2}\int_{\mathcal{O}} w^2 \, dx + \left(\frac{\partial\phi}{\partial h^1}, w\chi_{\mathcal{O}}\right)\right] = 0. \qquad (6.224)$$

This is the *HJB equation*. We have the 'final' condition

$$\phi(h^0, h^1, s_0) = \begin{cases} 0 \text{ if } \{h^0, h^1\} \in E, \\ +\infty \text{ otherwise} \end{cases} \qquad (6.225)$$

where E is the set described by $y(s_0; v), y_t(s_0; v)$ when y satisfies $y_{tt} + Ay = v\chi_{\mathcal{O}}$ in $\Omega \times (s_0, T)$, $v \in L^2(\mathcal{O} \times (s_0, T))$ $y = 0$ on $\Gamma \times (s_0, T)$, and (6.221) holds true. This definition is not constructive. See Remark 6.31 below.

Remark 6.30 We emphasize once more that the above approach is fairly *formal.*

Remark 6.31 The *real time optimal policy* is given at time $t \in (0, s_0)$ by

$$u(t) = -\frac{\partial \phi}{\partial h^1}(h^0, h^1, t)\chi_O. \tag{6.226}$$

How to proceed for $t \in (s_0, T)$ seems to be an open question, even from a conceptual point of view.

6.13. On the application of controllability methods to the solution of the Helmholtz equation at large wave numbers

6.13.1. Introduction

Stealth technologies have enjoyed a considerable growth of interest during this last decade both for aircraft and space applications. Due to the *very high frequencies* used by modern radars the computation of the *Radar Cross Section (RCS)* of a full aircraft using the *Maxwell equations* is still a *great challenge* (see Talflove (1992)). From the fact that *boundary integral methods* are not well suited to general coated materials, *field approaches* seem to provide an alternative which is worth exploring.

In this section, we consider a particular application of *controllability methods* to the solution of the *Helmholtz equations* obtained when looking for the *monochromatic solutions of linear wave problems*. The idea here is to go back to the original wave equation and to apply techniques, inspired by controllability, which find its *time periodic* solutions. Indeed, this method (introduced in Bristeau, Glowinski and Périaux (1993a,b)) is in competition with – and is related to – the one in which the wave equation is integrated from 0 to $+\infty$ in order to obtain asymptotically a time periodic solution; it is well known from Lax and Phillips (1989) that if the scattering body is convex then the solution will converge *exponentially* to the periodic solution. On the other hand, for *non-convex* reflectors (which is quite a common situation) the convergence can be very slow; the method described in this section substantially improves the speed of convergence of the asymptotic one, particularly for stiff problems where internal rays can be trapped by successive reflections.

6.13.2. The Helmholtz equation and its equivalent wave problem

Let us consider a scattering body B, of boundary $\partial B = \gamma$, 'illuminated' by an *incident monochromatic wave* of frequency $f = k/2\pi$ (see Figure 60).

In the case of the wave equation $u_{tt} - \Delta u = 0$ with a periodic solution $u = \text{Re}\,(Ue^{-ikt})$, the associated *Helmholtz equation*, satisfied by the coefficient $U(x)$ of e^{-ikt} is given by

$$\Delta U + k^2 U = 0 \text{ in } \mathbb{R}^d \backslash \bar{B}(d = 2, 3), \tag{6.227}$$

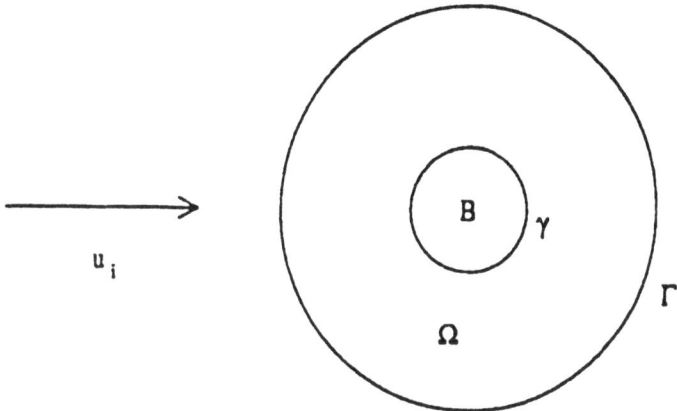

Fig. 60. u_i is the incident field.

$$U = G \text{ on } \gamma. \tag{6.228}$$

In practice, we bound $\mathbb{R}^d \backslash \bar{B}$ by an artificial boundary Γ on which we pre-scribe, for example, an *approximate first-order Sommerfeld condition such as*

$$\frac{\partial U}{\partial n} - ikU = 0 \text{ on } \Gamma; \tag{6.229}$$

now, equation (6.227) is prescribed on Ω only, where Ω is this portion of $\mathbb{R}^d \backslash \bar{B}$ between γ and Γ. In the above equations, U is the *scattered field*, $-G$ is the incident field, U and G are *complex valued functions*.

Remark 6.32 More complicated (and efficient) absorbing conditions than (6.229) have been coupled to the controllability method described hereafter; they allow the use of smaller computational domains. The resulting method-ology will be described in a forthcoming publication.

Systems (6.227)–(6.229) is related to the T-periodic solutions ($T = 2\pi/k$) of the following wave equation and associated boundary conditions

$$u_{tt} - \Delta u = 0 \text{ in } Q(= \Omega \times (0, T)), \tag{6.230}$$

$$u = g \text{ on } \sigma(= \gamma \times (0, T)), \tag{6.231}$$

$$\frac{\partial u}{\partial n} + \frac{\partial u}{\partial t} = 0 \text{ on } \Sigma(= \Gamma \times (0, T)), \tag{6.232}$$

where, in (6.231), $g(x, t)$ is a time periodic function related to G by $g(x, t) = \text{Re}\,(e^{-ikt}G(x))$. If we denote

$$G(x) = G_r(x) + iG_{im}(x),$$

g satisfies

$$g(x, t) = G_r(x) \cos kt + G_{im}(x) \sin kt.$$

The goal, here, is to find *periodic solutions* to system (6.230)–(6.232) without solving the Helmholtz problem (6.227)–(6.229).

In the following, we look for T-periodic solutions to systems such as (6.230)–(6.232); this means solutions satisfying

$$u(0) = u(T), \quad u_t(0) = u_t(T). \tag{6.233}$$

These solutions can be written

$$u(x, t) = \mathrm{Re}\left(e^{-ikt}U(x)\right)$$

(or $u(x, t) = U_\mathrm{r} \cos kt + U_\mathrm{im} \sin kt$) where $U = U_\mathrm{r} + iU_\mathrm{im}$ is the solution of (6.227)–(6.228); so we have

$$u(0) = U_\mathrm{r}, \quad u_t(0) = kU_\mathrm{im}.$$

6.13.3. *Exact controllability methods for the calculation of time periodic solutions to the wave equation*

In order to solve problem (6.230)–(6.233) we advocate the following approach whose main merit is to reduce the above problem to an *exact controllability* one, close to those problems whose solution is discussed in Sections 6.1 to 6.12. Indeed, problem (6.230)–(6.233) is clearly equivalent to the following one:

Find $\mathbf{e} = \{e^0, e^1\}$ *such that*

$$u_{tt} - \Delta u = 0 \text{ in } Q, \tag{6.234}$$

$$u = g \text{ on } \sigma, \tag{6.235}$$

$$\frac{\partial u}{\partial n} + \frac{\partial u}{\partial t} = 0 \text{ on } \Sigma, \tag{6.236}$$

$$u(0) = e^0, \quad u_t(0) = e^1, \quad u(T) = e^0, \quad u_t(T) = e^1. \tag{6.237}$$

Problem (6.234)–(6.237) is an *exact controllability problem* which can be solved by methods directly inspired by those in Sections 6.1 to 6.10. We shall not address here the *existence* and *uniqueness* of solutions to problem (6.234)–(6.237) (these issues are addressed in Bardos and Rauch (1994)); instead we shall focus on the practical calculation of such solutions, assuming they do exist.

6.13.4. *Least-squares formulation of the problem (6.234)–(6.237)*

In order to be able to apply controllability methods to the solution of problem (6.234)–(6.237) the appropriate choice for the space E containing $\mathbf{e} = \{e^0, e^1\}$ is fundamental. We advocate

$$E = V_g \times L^2(\Omega), \tag{6.238}$$

where

$$V_g = \{\varphi \mid \varphi \in H^1(\Omega), \varphi = g(0) \text{ on } \gamma\}. \tag{6.239}$$

In order to solve (6.234)–(6.237), we use the following *least-squares* formulation (where y plays the role of u in (6.234)–(6.237)):

$$\min_{\mathbf{v} \in E} J(\mathbf{v}) \tag{6.240}$$

with

$$J(\mathbf{v}) = \frac{1}{2} \int_\Omega (|\nabla(y(T) - v^0)|^2 + |y_t(T) - v^1|^2) \, dx, \quad \forall \mathbf{v} = \{v^0, v^1\}, \tag{6.241}$$

where, in (3.241), the function y is the solution of

$$y_{tt} - \Delta y = 0 \text{ in } Q, \tag{6.242}$$

$$y = g \text{ on } \sigma, \tag{6.243}$$

$$\frac{\partial y}{\partial n} + \frac{\partial y}{\partial t} = 0 \text{ on } \Sigma, \tag{6.244}$$

$$y(0) = v^0, \quad y_t(0) = v^1. \tag{6.245}$$

The choice of J is directly related to the fact that the *natural energy* $\mathcal{E}(\cdot)$ associated with the system is defined by

$$\mathcal{E}(t) = \frac{1}{2} \int_\Omega (|\nabla y|^2 + |y_t|^2) \, dx. \tag{6.246}$$

Assuming that $\mathbf{e}$ is the solution of problem (6.240), it will satisfy the following equation

$$\langle J'(\mathbf{e}), \mathbf{v} \rangle = 0, \quad \forall \mathbf{v} \in E_0, \tag{6.247}$$

where, in (6.247), $E_0 = V_0 \times L^2(\Omega)$ (with $V_0 = \{\varphi \mid \varphi \in H^1(\Omega), \varphi = 0 \text{ on } \gamma\}$) and where $\langle \cdot, \cdot \rangle$ denotes the duality pairing between E_0' and E_0 (E_0': dual space of E_0). In (6.247), J' denotes the *differential* of J.

Problem (6.247) can be solved by a *conjugate gradient algorithm* (described in Section 6.13.6) operating in E; in order to implement this algorithm, we need to be able to compute $J'(\mathbf{v}), \forall \mathbf{v} \in E$; this is the object of the following section.

6.13.5. Calculation of J'

To compute J' we use a *perturbation analysis*: starting from (6.241), we obtain

$$\begin{aligned} \delta J(\mathbf{v}) &= \langle J'(\mathbf{v}), \delta\mathbf{v} \rangle \\ &= \int_\Omega \nabla(v^0 - y(T)) \cdot \nabla \delta v^0 \, dx + \int_\Omega (v^1 - y_t(T)) \delta v^1 \, dx \end{aligned}$$

$$+ \int_\Omega \nabla(y(T) - v^0) \cdot \nabla \delta y(T) \, dx$$

$$+ \int_\Omega (y_t(T) - v^1) \delta y_t(T) \, dx. \tag{6.248}$$

We also have from (6.242)–(6.245):

$$\delta y_{tt} - \Delta \delta y = 0 \text{ in } Q, \tag{6.249}$$

$$\delta y = 0 \text{ on } \sigma, \tag{6.250}$$

$$\left(\frac{\partial}{\partial n} + \frac{\partial}{\partial t} \right) \delta y = 0 \text{ on } \Sigma, \tag{6.251}$$

$$\delta y(0) = \delta v^0, \quad \delta y_t(0) = \delta v^1. \tag{6.252}$$

Consider now a function p of x and t such that the function $p(t) : x \to p(x, t)$ vanishes on γ; next, multiply both sides of (6.249) by p, integrate on Q and then by parts. It follows then from (6.250), (6.251) that

$$\int_\Omega \delta y_t p \, dx|_0^T - \int_\Omega \delta y p_t \, dx|_0^T + \int_Q p_{tt} \delta y \, dx \, dt + \int_Q \nabla p \cdot \nabla \delta y \, dx \, dt$$

$$+ \int_\Gamma \delta y p \, d\Gamma|_0^T - \int_\Sigma p_t \delta y \, d\Gamma \, dt = 0. \tag{6.253}$$

Suppose that p satisfies

$$\int_\Omega (p_{tt} z + \nabla p \cdot \nabla z) \, dx - \int_\Gamma p_t z \, d\Gamma = 0, \ \forall z \in V_0, p = 0 \text{ on } \sigma, \text{ a.e. on } (0, T);$$
$$\tag{6.254}$$

(6.254) is *equivalent* to

$$p_{tt} - \Delta p = 0 \text{ in } Q, \tag{6.255}$$

$$p = 0 \text{ on } \sigma, \tag{6.256}$$

$$\frac{\partial p}{\partial n} - \frac{\partial p}{\partial t} = 0 \text{ on } \Sigma. \tag{6.257}$$

Using (6.252), equation (6.253) reduces then to

$$\int_\Omega \delta y_t(T) p(T) \, dx - \int_\Omega \delta y(T) p_t(T) \, dx + \int_\Gamma \delta y(T) p(T) \, d\Gamma'$$

$$= \int_\Omega \delta y_t(0) p(0) \, dx - \int_\Omega \delta y(0) p_t(0) \, dx + \int_\Gamma \delta y(0) p(0) \, d\Gamma$$

$$= \int_\Omega p(0) \delta v^1 \, dx - \int_\Omega p_t(0) \delta v^0 \, dx + \int_\Gamma p(0) \delta v^0 \, d\Gamma. \tag{6.258}$$

Let us define $p(T)$ and $p_t(T)$ by

$$p(T) = y_t(T) - v^1 \tag{6.259}$$

and

$$\int_\Omega p_t(T) z \, dx = \int_\Gamma (y_t(T) - v^1) z \, d\Gamma - \int_\Omega \nabla(y(T) - v^0) \cdot \nabla z \, dx, \; \forall z \in V_0,$$
(6.260)

respectively. Finally, using (6.248) and (6.258)–(6.260), with $z = \delta y(T)$, shows that

$$\langle J'(\mathbf{v}), \mathbf{w} \rangle = \int_\Omega \nabla(v^0 - y(T)) \cdot \nabla w^0 \, dx - \int_\Omega p_t(0) w^0 \, dx$$
$$+ \int_\Gamma p(0) w^0 \, d\Gamma + \int_\Omega p(0) w^1 \, dx + \int_\Omega (v^1 - y_t(T)) w^1 \, dx,$$

$$\forall \mathbf{w} = \{w^0, w^1\} \in E_0,$$
(6.261)

where, in (6.261), p is the solution of the *adjoint equation* (6.255)–(6.257), completed by the 'final conditions' (6.259), (6.260).

Remark 6.33 Relations (6.260) and (6.261) are largely formal; however, it is worth mentioning that the discrete variants of these two relations make sense and lead to algorithms with fast convergence properties.

Remark 6.34 The *well-posedness* of problem (6.240) is discussed in Bardos and Rauch (1994), where sufficient conditions for existence and uniqueness are given.

6.13.6. Conjugate gradient solution of the least-squares problem (6.240)
   A *conjugate gradient algorithm* for the solution of the *linear problem* (6.247) (equivalent to problem (6.240)) is given by
   Step 0: Initialization

$$\mathbf{e}_0 = \{e_0^0, e_0^1\} \text{ is given in } E.$$
(6.262)

Solve the following forward wave problem

$$\frac{\partial^2 y_0}{\partial t^2} - \Delta y_0 = 0 \text{ in } Q,$$
(6.263)$_1$

$$y_0 = g \text{ on } \sigma,$$
(6.263)$_2$

$$\frac{\partial y_0}{\partial n} + \frac{\partial y_0}{\partial t} = 0 \text{ on } \Sigma,$$
(6.263)$_3$

$$y_0(0) = e_0^0, \frac{\partial y_0}{\partial t}(0) = e_0^1.$$
(6.263)$_4$

Solve the following backward wave problem

$$\frac{\partial^2 p_0}{\partial t^2} - \Delta p_0 = 0 \text{ in } Q,$$
(6.264)$_1$

$$p_0 = 0 \text{ on } \sigma,$$
(6.264)$_2$

$$\frac{\partial p_0}{\partial n} - \frac{\partial p_0}{\partial t} = 0 \text{ on } \Sigma, \tag{6.264}_3$$

with $p_0(T)$ and $(\partial p_0/\partial t)(T)$ given by

$$p_0(T) = \frac{\partial y_0}{\partial t}(T) - e_0^1, \tag{6.264}_4$$

$$\int_\Omega \frac{\partial p_0}{\partial t}(T) z \, dx = \int_\Gamma p_0(T) z \, d\Gamma - \int_\Omega \nabla(y_0(T) - e_0^0) \cdot \nabla z \, dx, \ \forall z \in V_0, \tag{6.264}_5$$

respectively.

Define next $\mathbf{g}_0 = \{g_0^0, g_0^1\} \in E_0(= V_0 \times L^2(\Omega))$ by

$$\int_\Omega \nabla g_0^0 \cdot \nabla z \, dx = \int_\Omega \nabla(e_0^0 - y_0(T)) \cdot \nabla z \, dx - \int_\Omega \frac{\partial p_0}{\partial t}(0) z \, dx$$
$$+ \int_\Gamma p_0(0) z \, d\Gamma, \ \forall z \in V_0, \tag{6.265}_1$$

$$g_0^1 = p_0(0) + e_0^1 - \frac{\partial y_0}{\partial t}(T), \tag{6.265}_2$$

and then

$$\mathbf{w}^0 = \mathbf{g}^0. \tag{6.266}$$

For $k \geq 0$, suppose that $\mathbf{e}_k$, $\mathbf{g}_k$, $\mathbf{w}_k$ are known, we compute then $\mathbf{e}_{k+1}$, $\mathbf{g}_{k+1}$, $\mathbf{w}_{k+1}$ as follows

Step 1: Descent

$$\frac{\partial^2 \bar{y}_k}{\partial t^2} - \Delta \bar{y}_k = 0 \text{ in } Q, \tag{6.267}_1$$

$$\bar{y}_k = 0 \text{ on } \sigma, \tag{6.267}_2$$

$$\frac{\partial \bar{y}_k}{\partial t} + \frac{\partial \bar{y}_k}{\partial n} = 0 \text{ on } \Sigma, \tag{6.267}_3$$

$$\bar{y}_k(0) = w_k^0, \quad \frac{\partial \bar{y}_k}{\partial t}(0) = w_k^1. \tag{6.267}_4$$

Solve then

$$\frac{\partial^2 \bar{p}_k}{\partial t} - \Delta \bar{p}_k = 0 \text{ in } Q, \tag{6.268}_1$$

$$\bar{p}_k = 0 \text{ on } \sigma, \tag{6.268}_2$$

$$\frac{\partial \bar{p}_k}{\partial t} - \frac{\partial \bar{p}_k}{\partial n} = 0 \text{ on } \Sigma, \tag{6.268}_3$$

with $\bar{p}_k(T)$ and $(\partial \bar{p}_k/\partial t)(T)$ given by

$$\bar{p}_k(T) = \frac{\partial \bar{y}_k}{\partial t}(T) - w_k^1, \tag{6.268}_4$$

$$\int_\Omega \frac{\partial \bar{p}_k}{\partial t}(T)z \, dx = \int_\Gamma \bar{p}_k(T)z \, d\Gamma - \int_\Omega \nabla(\bar{y}_k(T) - w_k^0) \cdot \nabla z \, dx, \quad \forall z \in V_0,$$

$$(6.268)_5$$

respectively. Define next $\bar{\mathbf{g}}_k = \{\bar{g}_k^0, \bar{g}_k^1\} \in V_0 \times L^2(\Omega)$ *by*

$$\int_\Omega \nabla \bar{g}_k^0 \cdot \nabla z \, dx = \int_\Omega \nabla(w_k^0 - \bar{y}_k(T)) \cdot \nabla z \, dx - \int_\Omega \frac{\partial \bar{p}_k}{\partial t}(0)z \, dx$$
$$+ \int_\Gamma \bar{p}_k(0)z \, d\Gamma, \quad \forall z \in V_0,$$

$$(6.269)_1$$

$$\bar{g}_k^1 = \bar{p}_k(0) + w_k^1 - \frac{\partial \bar{y}_k}{\partial t}(T),$$

$$(6.269)_2$$

and then ρ_k *by*

$$\rho_k = \int_\Omega (|\nabla g_k^0|^2 + |g_k^1|^2) \, dx \Big/ \int_\Omega (\nabla \bar{g}_k^0 \cdot \nabla w_k^0 + \bar{g}_k^1 w_k^1) \, dx.$$

$$(6.270)$$

We update then $\mathbf{e}_k$ *and* $\mathbf{g}_k$ *by*

$$\mathbf{e}_{k+1} = \mathbf{e}_k - \rho_k \mathbf{w}_k,$$

$$(6.271)$$

$$\mathbf{g}_{k+1} = \mathbf{g}_k - \rho_k \bar{\mathbf{g}}_k.$$

$$(6.272)$$

Step 2: Test of the convergence and construction of the new descent direction. If $(\int_\Omega (|\nabla g_{k+1}^0|^2 + |g_{k+1}^1|^2) \, dx)^{1/2} / (\int_\Omega (|\nabla g_0^0|^2 + |g_0^1|^2) \, dx)^{1/2} \le \epsilon$ *take* $\mathbf{e} = \mathbf{e}_{k+1}$; *if not, compute*

$$\gamma_k = \int_\Omega (|\nabla g_{k+1}^0|^2 + |g_{k+1}^1|^2) \, dx \Big/ \int_\Omega (|\nabla g_k^0|^2 + |g_k^1|^2) \, dx$$

$$(6.273)$$

and update $\mathbf{w}_k$ *by*

$$\mathbf{w}_{k+1} = \mathbf{g}_{k+1} + \gamma_k \mathbf{w}_k.$$

$$(6.274)$$

Do $k = k + 1$ *and go to* (6.267).

Remark 6.35 Algorithm (6.262)–(6.274) looks complicated at first glance. In fact, it is not that complicated to implement since each iteration requires basically the solution of *two wave equations* such as (6.267) and (6.268) and of an *elliptic problem* such as $(6.269)_1$. The above problems are classical ones for which efficient solution methods already exist.

Remark 6.36 Algorithm (6.262)–(6.274) can be seen as a variation of the *asymptotic method* mentioned in Section 6.13.1; there, we integrate the periodically excited wave equation until we reach a periodic solution (i.e. a *limit cycle*). What algorithm (6.262)–(6.274) does indeed is to *periodically* measure the *lack* (or *defect*) *of periodicity* and use the result of this measure as a *residual* to speed up the convergence to a periodic solution. In fact, a similar idea was used in Auchmuty, Dean, Glowinski and Zhang (1987) to compute the periodic solutions of systems of *stiff* nonlinear differential equations (including cases where the period itself is an unknown parameter of the problem).

6.13.7. An FD/FE implementation

The practical implementation of the previously presented control-based method is straightforward. It is based on a *time discretization* by the *centred second-order accurate explicit FD scheme*, already employed in Sections 6.8 and 6.9. This scheme is combined to *piecewise linear FE approximations* (as in Sections 6.8 and 6.9) for the space variables; we use *mass lumping* – through numerical integration by the *trapezoïdal rule* – to obtain a *diagonal* mass matrix for the acceleration term. The fully discrete scheme has to satisfy a *stability condition* such as $\Delta t \leq Ch$, where C is a constant. To obtain accurate solutions, we need to have h at least *ten times smaller* than the wavelength; in fact, h has to be even smaller ($h \approx \lambda/20$) in those regions where internal rays are trapped by successive reflections. For the *mesh generation*, the advancing front method proposed by George (1971) has been used; this method (implemented at INRIA by George and Seveno) gives *homogeneous* triangulations (see the following figures).

6.13.8. Numerical experiments.

In order to validate the methods discussed in the above sections, we have considered the solution of three test problems of increasing difficulty, namely the scattering of planar incident waves by a *disk*, then by a *nonconvex* reflector which can be seen as a semi-open cavity (a kind of – very – idealized air intake) and finally the scattering of a planar wave by a *two-dimensional aircraft related body*. For these different cases the artificial boundary is located at a 3λ distance from B and we assume that the boundary of the reflector is *perfectly conducting*.

The following results have been obtained by Bristeau at INRIA (see Bristeau, Glowinski and Périaux (1993a,b,c) for further numerical experiments and details).

Scattering by a disk. Before discussing our numerical experiments, let us observe that model (6.234)–(6.236) assumes, implicitly, that its solutions propagates with velocity 1, implying that, here, the wavelength is equal to the period. If $c(> 0)$ is different from 1, we shall rescale x and t, so that $c = 1$. In the following examples, the data are given in the MKSA system before rescaling.

Example 1 (Scattering by a disk) For this problem, B is a disk of radius 0.25 m, $k = 2\pi f$ with $f = 2.4$ GHz, implying that the wavelength is 0.125 m; the disk is illuminated by an incident *planar wave* coming from the left. The artificial boundary is located at a distance of 3λ from B. The *FE triangulation* has 18,816 vertices and 36,970 triangles; the mean length of the edges is $\lambda/15$, the minimal value being $\lambda/28$, while the maximal one is $\lambda/10$.

Fig. 61. Contours of the scattered field (real component).

The value of Δt is $T/35$. To obtain convergence of the iterative method, 74 iterations of algorithm (6.262)–(6.274) were needed (with $\epsilon = 5 \times 10^{-5}$ for the stopping criterion) corresponding to a 3 min computation on a CRAY2. Figure 61 shows the scattered field e^0 (*real component* of the Helmholtz problem solution). For this test problem where the exact solution is known, we have compared on Figures 62 and 63 the computed solution (——) with the exact one ($\cdots\cdots$) on two cross sections (incident direction, opposite to incident direction, respectively). Of course, for this problem, the asymptotic method (just integrating the wave equation from 0 to nT, n 'large') is less CPU time consuming; we have chosen this example just to test the accuracy of the approximations.

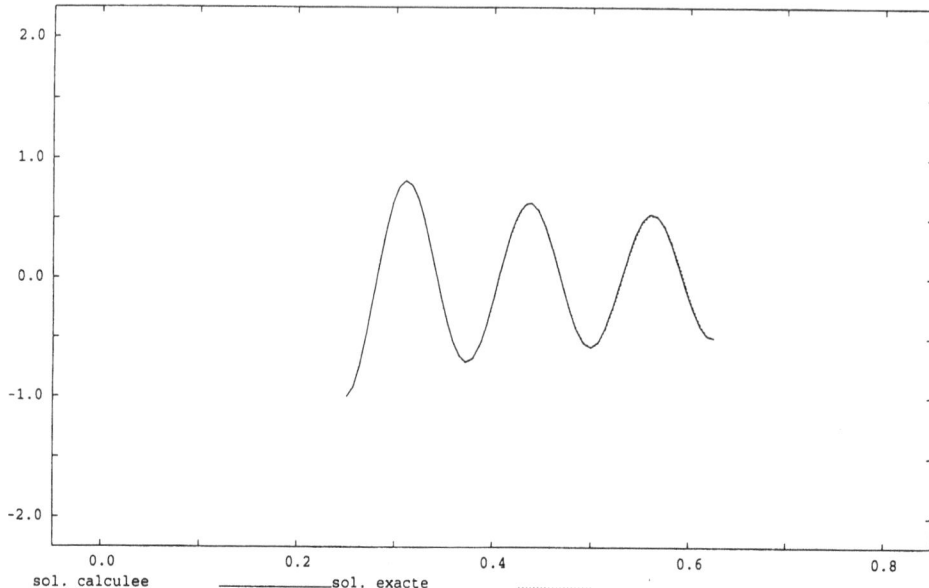

sol. calculee ——————sol. exacte ....

Fig. 62. Comparison between exact ($\cdots\cdots$) and computed (———) scattered fields $e^0(x_1, 0)$ (incident side).

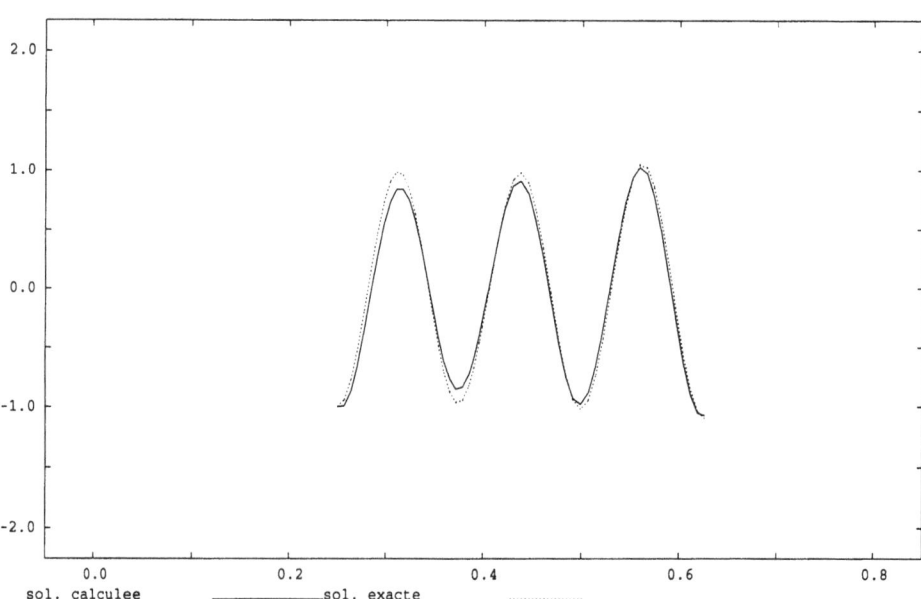

sol. calculee ——————sol. exacte ....

Fig. 63. Comparison between exact ($\cdots\cdots$) and computed (———) scattered fields $e^0(x_1, 0)$ (shadow side).

Remark 6.37 We can substantially increase the accuracy by using on Γ instead of (6.232), *second-order absorbing boundary conditions* like those discussed in Bristeau, Glowinski and Périaux (1993c).

Example 2 (Scattering by semi-open cavities) We have considered two semi-open cavities. We choose $f = 3$ GHz implying that the wavelength is 0.10 m. For the first cavity, the inside dimensions are $4\lambda \times \lambda$ and the thickness of the wall is $\lambda/5$. The FE triangulation has 22,951 vertices and 44,992 triangles. The value of Δt corresponds to 40 time steps per period (i.e. $\Delta t = T/40$). We consider an illuminating monochromatic wave of incidence $\alpha = 30°$, coming from the right. The contours of the scattered fields e^0 (real part) and e^1/k (imaginary part) are shown on Figures 64 and 65, respectively. The convergence is reached with 136 iterations ($\epsilon = 5 \times 10^{-5}$), corresponding to 8 min of CPU time on a CRAY2. Figure 66 shows the convergence of the residuals Re^0_k and Re^1_k associated to the controllability method; these residuals are defined by

$$\text{Re}^0_k = \frac{\|e^0_{k+1} - e^0_k\|_{L^2(\Omega)}}{\|e^0_1 - e^0_0\|_{L^2(\Omega)}}, \quad \text{Re}^1_k = \frac{\|e^1_{k+1} - e^1_k\|_{L^2(\Omega)}}{\|e^1_1 - e^1_0\|_{L^2(\Omega)}}.$$

The asymptotic method gives the same solution, but, for this *nonconvex* obstacle, the convergence is *much slower* (800 iterations, 18 min of CPU time on a CRAY2) than the convergence of the controllability method, as shown on Figure 67.

We have considered a second semi-open cavity for the same frequency and wavelength; the inside dimensions of this larger cavity are $20\lambda \times 5\lambda$, the wall thickness being λ. For this problem where many reflections take place inside the cavity, we need a fine triangulation. The one used here has 208,015 vertices and 412,028 triangles, with $\lambda/30$ as the mean length of the edges inside the cavity ($\lambda/20$ outside). We have taken $\Delta t = T/70$. The test problem corresponds to an illuminating wave of incidence $\alpha = 30°$, coming from the right. The contours of the total field related to e^0 are shown on Figure 68. Figure 69 shows the convergence of the cost function $J(\mathbf{e}_k)$ with $J(\cdot)$ defined by (6.241); we have also shown on Figure 69 the convergence to zero of the two components of this cost function (the one related to e^0_k, and the one related to e^1_k).

For this difficult case the convergence is slower than for the above cavity problem (300 iterations instead of 136).

We have shown on Figure 70 some details of the FE triangulation close to the wall at the entrance of the cavity.

Example 3 (scattering by a two-dimensional aircraft related body) We consider the reflector defined by the cross section of a Dassault Aviation Falcon 50 by its symmetry plane; the shape of the air-intake is given and we have artificially closed it in order to enhance reflections. The plane length is about

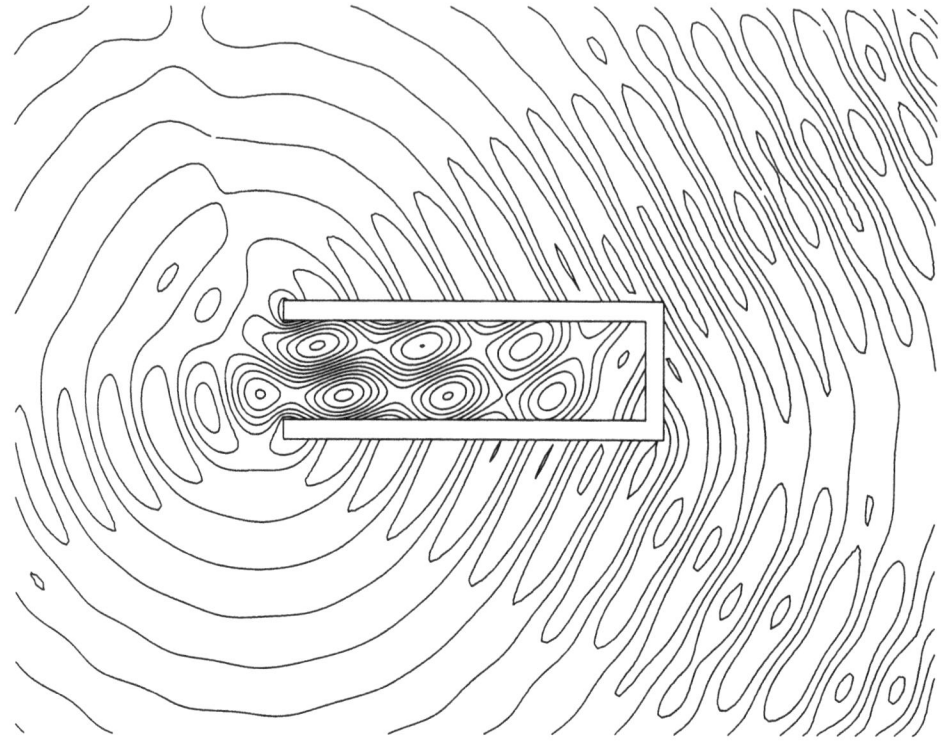

Fig. 64. Contours of the scattered field (real part; $\alpha = 30°$).

18 m, while its height is 6 m. We take $f = 0.6$ GHz, so that $\lambda = 0.5$ m. The FE mesh has 143,850 vertices and 283,873 triangles; Figure 71 shows an enlargement of the mesh near the air intake. We have used $\Delta t = T/40$. We consider an illuminating wave with $\alpha = 0°$ as angle of incidence. The contours of the total field (real part) are presented on Figure 72; we observe the *shadow region* behind the aircraft. The convergence (for $\epsilon = 5 \times 10^{-5}$) is obtained after 260 iterations, corresponding to 90 min of CPU time on a CRAY2; Figure 73 shows the convergence of $J(\mathbf{e}_k)$ to 0 as $k \to +\infty$.

Remark 6.38 For all the test problems discussed above, we have used a direct method based on a *sparse Cholesky solver* to solve the (discrete) elliptic problem encountered at each iteration of the discrete analogue of algorithm (6.262)–(6.274). Despite the respectable size of these systems (up to 2×10^5 unknowns) this part of the algorithm takes no more than a *few percent* of the total computational effort.

Indeed, most of the computational time is spent integrating the forward and backward wave equations; fortunately this is the easiest part to *parallelize* (hopefully in the near future; see Bristeau, Erhel, Glowinski and Périaux (1993)) as it is based on an *explicit time discretization scheme*.

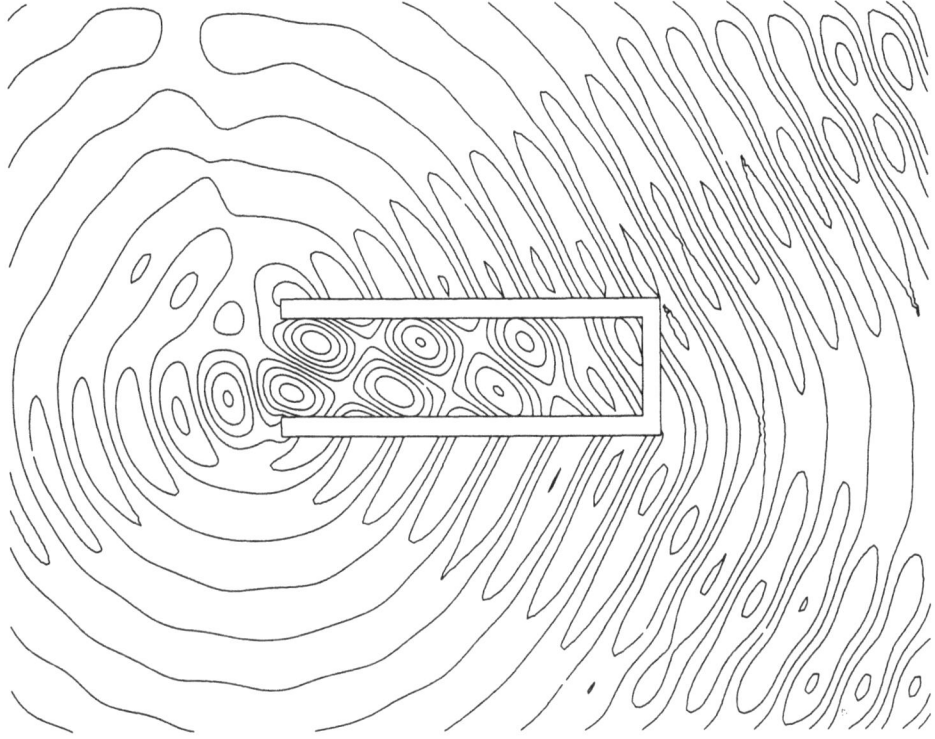

Fig. 65. Contours of the scattered field (imaginary part; $\alpha = 30°$).

6.13.9. Further comments

We have discussed in this section a *controllability* based novel approach to solving the Helmholtz (and two-dimensional harmonic Maxwell) equations for *large wavenumbers* and complicated geometries. The new method so far appears to be more efficient than traditional computational methods which are based on either time asymptotic behaviour or linear algebra algorithms for very large indefinite linear systems.

The new methodology appears to be promising for the *three-dimensional Maxwell equations* and for *heterogeneous media*, including *dissipative* ones. For very large problems, we shall very probably have to combine the above method with *domain decomposition* and/or *fictitious domain methods*, and also to *higher-order* approximations, to reduce the number of grid points.

6.14. Further problems

In this section we have discussed controllability issues concerning wave equations such as

$$u_{tt} - c^2 \Delta u = 0; \qquad (6.275)$$

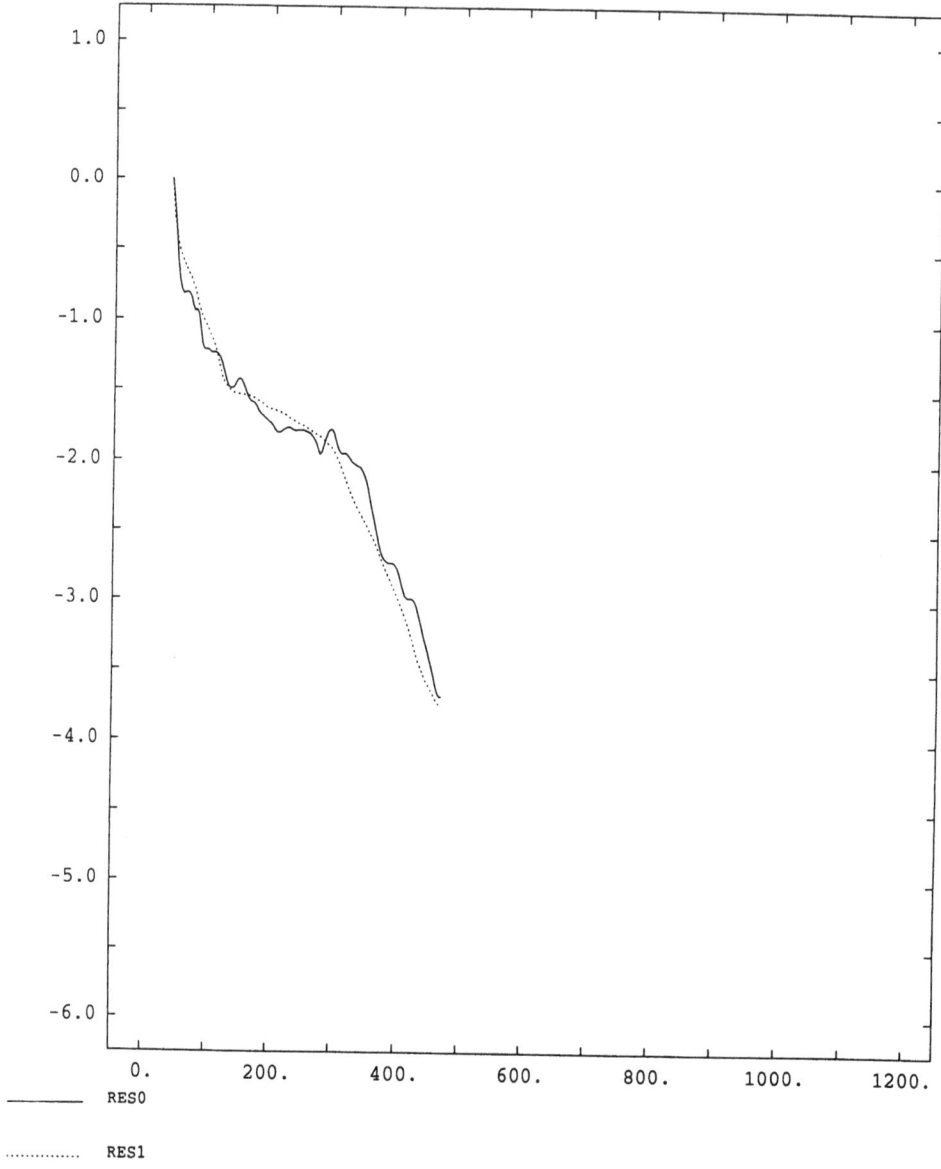

Fig. 66. Convergence of the residual (control solution): ———, residual for y; ······, residual of y_t.

a basic tool for studying exact or approximate controllability for equations such as (3.275) has been the *Hilbert Uniqueness Method* (HUM). Actually, HUM has been applied in Lagnese (1989) to prove the exact boundary controllability of the *Maxwell equations*

$$\varepsilon \frac{\partial \mathbf{E}}{\partial t} - \boldsymbol{\nabla} \times \mathbf{H} = 0, \quad \mu \frac{\partial \mathbf{H}}{\partial t} + \boldsymbol{\nabla} \times \mathbf{E} = 0 \text{ in } Q, \qquad (6.276)$$

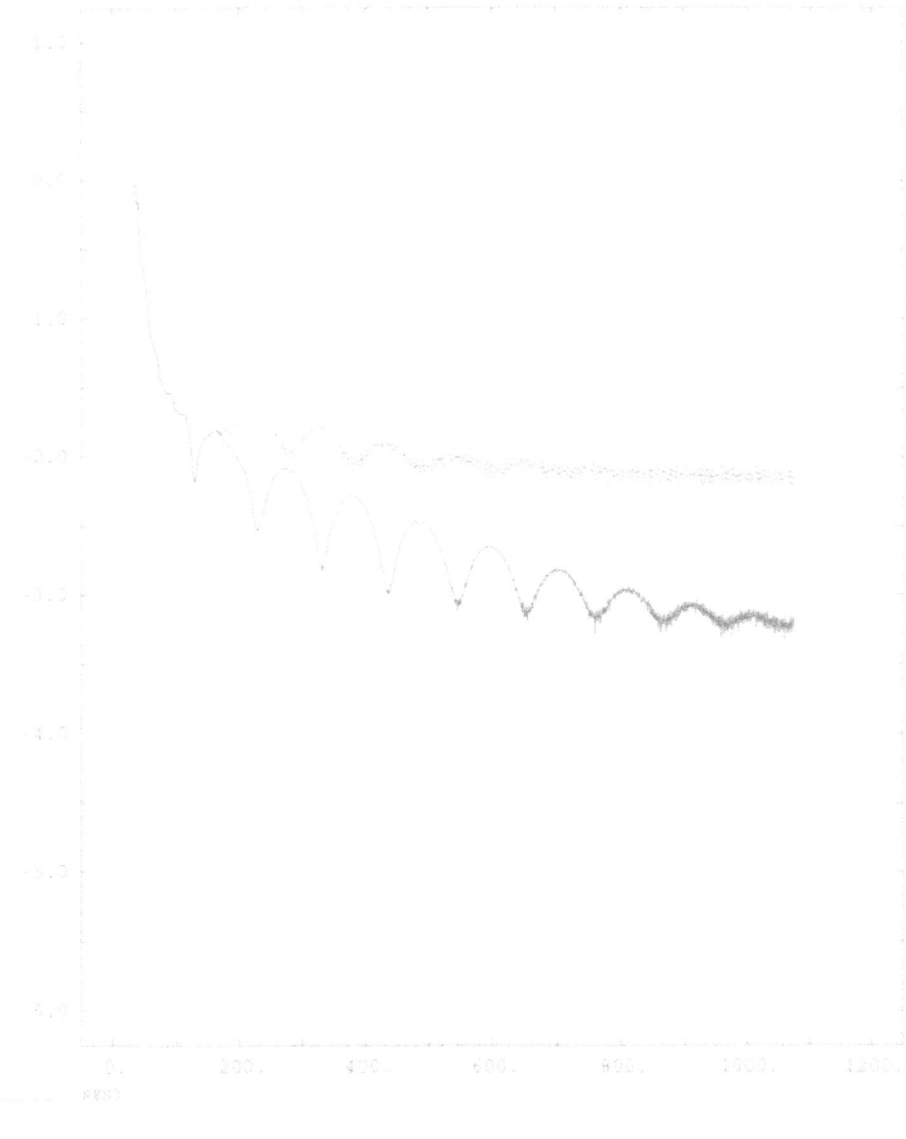

Fig. 67. Convergence of the residual (asymptotic solution): ——, residual for y; $\cdots\cdots$, residual of y_t.

$$\nabla \cdot \mathbf{E} = \nabla \cdot \mathbf{H} = 0 \text{ in } Q \qquad (6.277)$$

(see also Bensoussan (1990)); most computational aspects still have to be explored.

The Hilbert Uniqueness Method has been applied in Lions (1988b) and Lagnese and Lions (1988) to the exact or approximate controllability of systems (mostly from *elasticity*) modelled by *Petrowsky's type equations*.

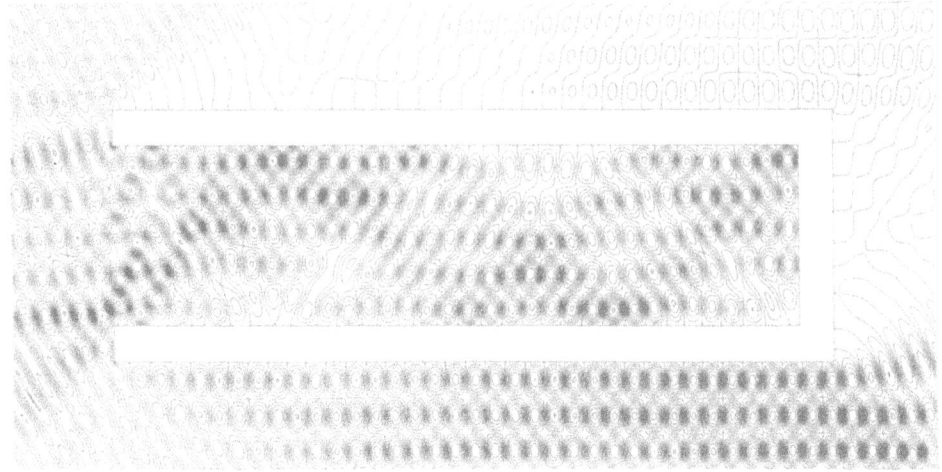

Fig. 68. Contours of the total field ($\alpha = 30°$).

Concerning the numerical application of HUM to the exact controllability of Petrowsky-type equations modelling *elastic shells* vibrations we refer to Marini, Testa and Valente (1994).

Finally, very little is known about the exact or approximate controllability of those (nonlinear) wave (or Petrowsky's type) equations modelling the vibrations of nonlinear systems; we intend, however, to explore the solution of these problems in the near future.

7. COUPLED SYSTEMS

In Sections 1 to 6 we have discussed controllability issues for *diffusion* and *wave* equations, respectively. The control of systems obtained by the coupling of *different types* of equations brings new difficulties which are worth discussing, therefore justifying the present section. The *numerical aspects* will not be addressed here, but in our opinion this Section can be a starting point for investigations in this direction.

In this Section, we shall focus on the controllability of a simplified *Thermoelasticity* system but it is likely that the techniques described here can be applied to systems modelled by more complicated equations.

7.1. A problem from thermoelasticity

Let Ω be a *bounded* domain of $\mathbb{R}^d$, $d \leq 3$, with a smooth boundary Γ. Motivated by applications from *Thermoelasticity* we consider the following

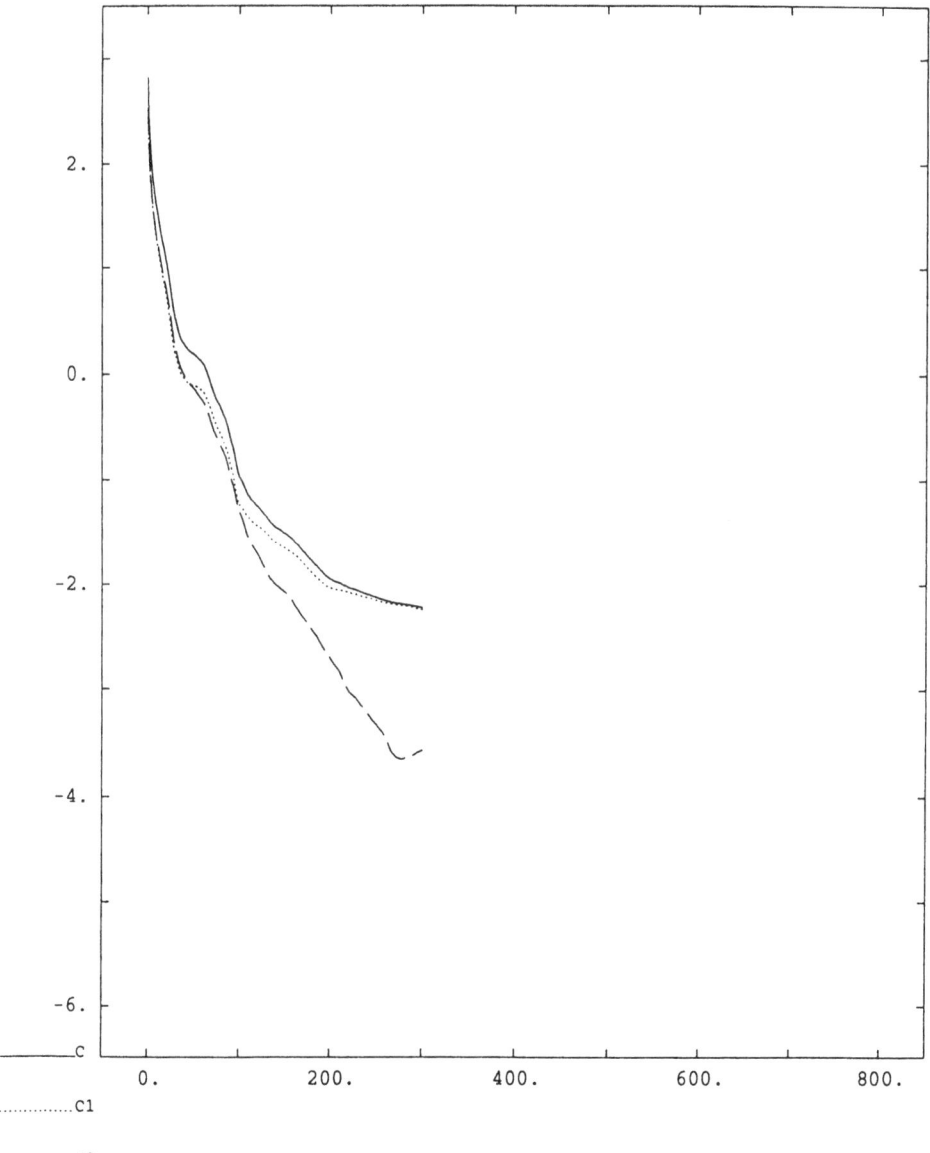

Fig. 69. Convergence of $J(\mathbf{e}_k)$ (——), of the $\mathbf{e}_k^0$ component of $J(\mathbf{e}_k)$ (– – –), and of the $\mathbf{e}_k^1$ component of $J(\mathbf{e}_k)$ (······).

system

$$\frac{\partial^2 \mathbf{y}}{\partial t^2} - \Delta \mathbf{y} + \alpha \nabla \theta = \mathbf{0} \text{ in } Q = \Omega \times (0, T), \qquad (7.1)_1$$

$$\frac{\partial \theta}{\partial t} - \Delta \theta + \alpha \nabla \cdot \frac{\partial \mathbf{y}}{\partial t} = 0 \text{ in } Q, \qquad (7.1)_2$$

Fig. 70. Enlargement of the mesh close to the cavity intake.

where $\mathbf{y} = \{y_i\}_{i=1}^d, \alpha \geq 0$. In (7.1), $\mathbf{y}$ (respectively θ) denotes an *elastic displacement* (respectively a *temperature*) function of x and t. Scaling has been made so that the constants in front of $-\Delta$ are equal to 1 in both equations.

The *initial conditions* are

$$\mathbf{y}(0) = \mathbf{0}, \quad \frac{\partial \mathbf{y}}{\partial t}(0) = \mathbf{0}, \qquad (7.2)_1$$

$$\theta(0) = 0. \qquad (7.2)_2$$

The control is applied on the boundary of Ω, actually on $\Gamma_0 \subset \Gamma$. Also, it is only applied on the component $\mathbf{y}$ of the state vector $\{\mathbf{y}, \theta\}$.

Considering the *boundary conditions*, we shall consider the two following cases:

Case I

$$\mathbf{y} = \begin{cases} \mathbf{v} \text{ on } \Sigma_0 = \Gamma_0 \times (0, T), \mathbf{v} = \{v_i\}_{i=1}^d, \\ \mathbf{0} \text{ on } \Sigma \backslash \Sigma_0, \Sigma = \Gamma \times (0, T) \end{cases} \qquad (7.3)$$

and

$$\theta = \theta_0 \text{ is given on } \Sigma. \qquad (7.4)$$

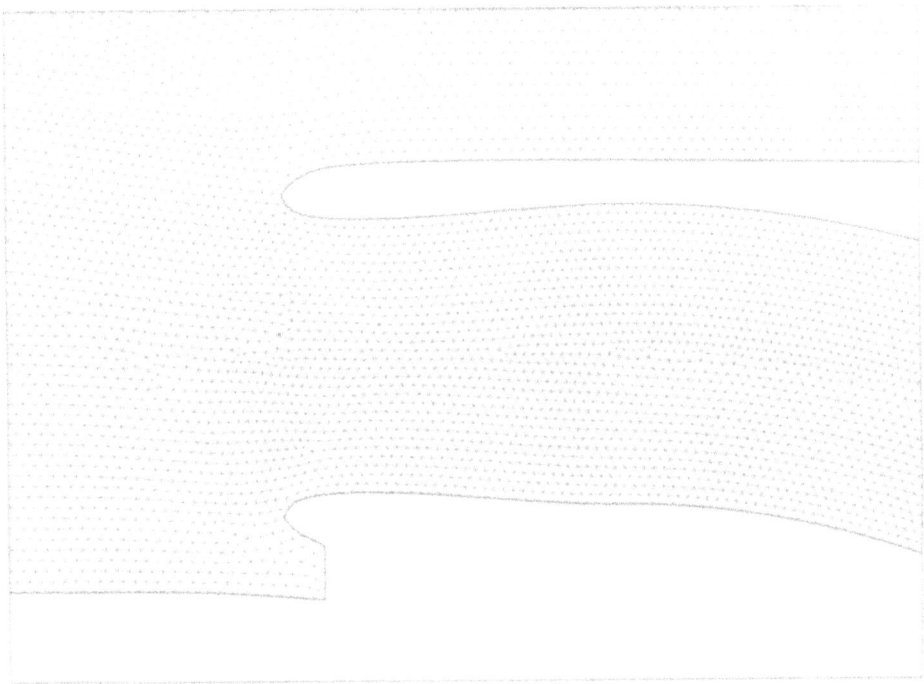

Fig. 71. Enlargement of the mesh close to the aircraft air intake (by courtesy of Dassault Aviation).

Case II

$$\frac{\partial y}{\partial n} = \begin{cases} \mathbf{v} \text{ on } \Sigma_0, \\ \mathbf{0} \text{ on } \Sigma \backslash \Sigma_0 \end{cases} \tag{7.5}$$

with (7.4) unchanged.

Remark 7.1 One can consider a variety of other types of boundary conditions and controls. The corresponding problems can be treated by methods very close to those given below.

Remark 7.2 In order to simplify the proofs and formulae below, we shall take

$$\theta_0 = 0, \tag{7.6}$$

but this is just a technical detail.

In the following sections, we shall study the spaces described by $\mathbf{y}(T)$, $(\partial\mathbf{y}/\partial t)(T)$ and $\theta(T)$; we shall show that under 'reasonable' conditions, one can control $\mathbf{y}(T)$ and $(\partial\mathbf{y}/\partial t)(T)$ but not $\theta(T)$.

Remark 7.3 Controllability for equations (7.1)–(7.4) has been studied in Lions (1988b,Vol. 2) (see also Narukawa (1983)). We follow here a slightly

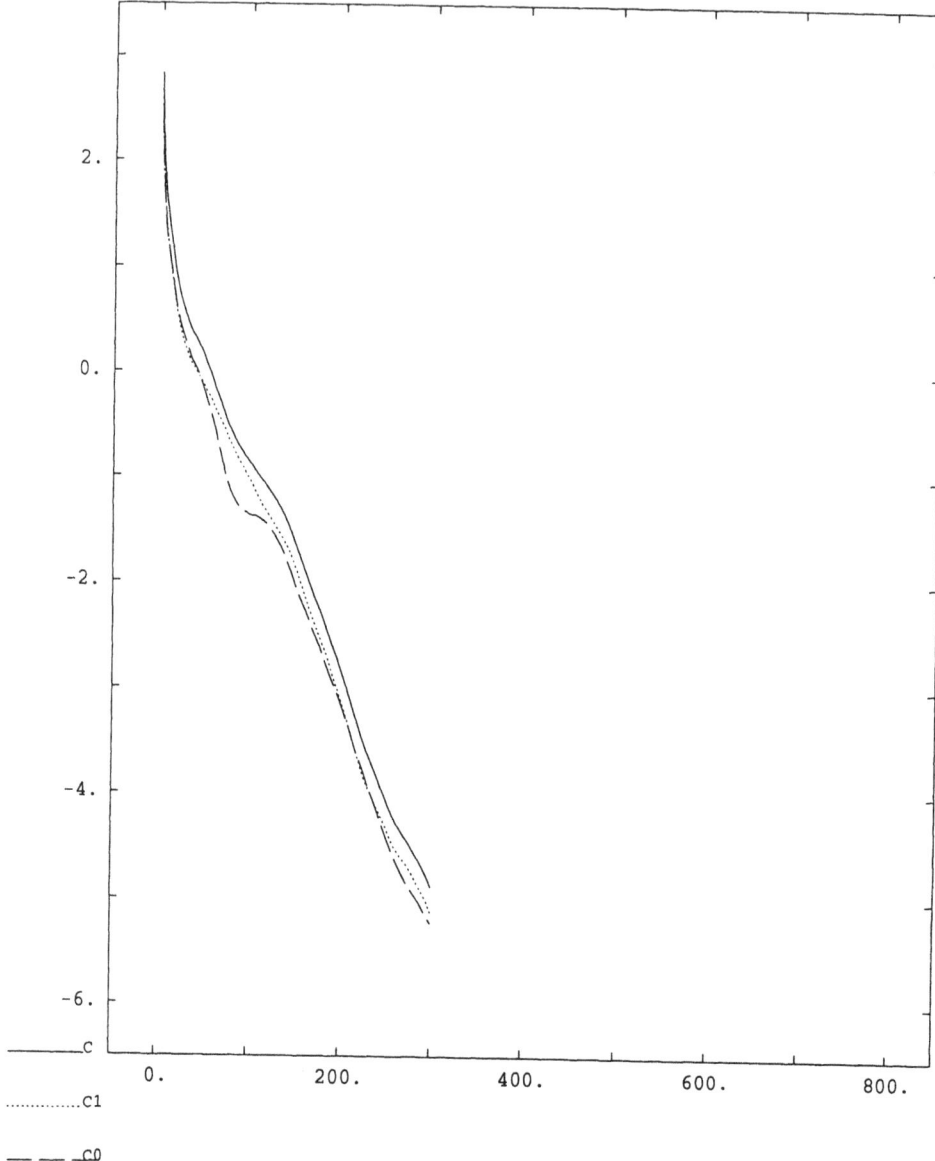

Fig. 72. Convergence of the residuals.

different approach, our goal being to obtain constructive approximation methods.

7.2. The limit cases $\alpha \to 0$ and $\alpha \to +\infty$

In order to obtain a better understanding, the limit cases $\alpha \to 0$ and $\alpha \to +\infty$ are worthwhile looking at. Moreover, they have intrinsic mathematical interest, particularly when $\alpha \to +\infty$.

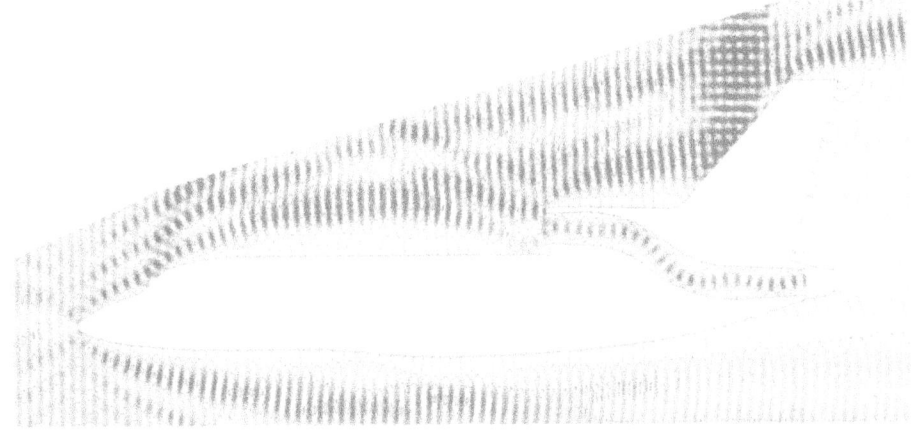

Fig. 73. Contours of the real part of the total field around a Falcon 50 two-dimensional cross section ($\alpha = 0°$) (by courtesy of Dassault Aviation).

7.2.1. *The case* $\alpha \to 0$.

This case is simple. The coupled system (7.1)–(7.4) (or its variant (7.1), (7.2), (7.4), (7.5)) reduces to *uncoupled* wave and heat equations. The control acts only on the **y** components; we then have

$$\frac{\partial^2 \mathbf{y}}{\partial t^2} - \Delta \mathbf{y} = \mathbf{0} \text{ in } Q, \quad \mathbf{y}(0) = \frac{\partial \mathbf{y}}{\partial t}(0) = \mathbf{0}, \ \mathbf{y} = \mathbf{v} \text{ on } \Sigma_0, \ \mathbf{y} = \mathbf{0} \text{ on } \Sigma \backslash \Sigma_0.$$
$$(7.7)$$

Since Δ is a *diagonal* operator we recover cases discussed in Section 6.

Remark 7.4. The *general linear elasticity* system (with Δ replaced by $\lambda \Delta + \mu graddiv$; λ, μ: Lamé coefficients) would lead to similar considerations, with more complicated technical details.

Remark 7.5. Similar considerations apply when (7.3) is replaced by (7.5).

7.2.2. *The case* $\alpha \to +\infty$ *(boundary conditions (7.3).)*

We shall assume (this is *necessary* for what follows) that

$$\int_{\Gamma_0} \mathbf{v} \cdot \mathbf{n} \, d\Gamma = 0.$$
$$(7.8)$$

Then, assuming **v** smooth enough (a condition which does *not* restrict the generality, since we are going to consider *approximate controllability*) and

$$\mathbf{v} = \mathbf{0} \text{ on } \partial \Gamma_0 \times (0, T), \quad \mathbf{v}|_{t=0} = \frac{\partial \mathbf{v}}{\partial t}|_{t=0} = \mathbf{0},$$
$$(7.9)$$

one can construct a function ϕ such that

$$\begin{cases} \phi \text{ is smooth in } \bar{\Omega} \times (0, T), \quad \nabla \cdot \phi = 0 \text{ in } \Omega \times (0, T), \\ \phi = \mathbf{v} \text{ on } \Sigma_0, \quad \phi = \mathbf{0} \text{ on } \Sigma \backslash \Sigma_0. \end{cases}$$
$$(7.10)$$

Then if we introduce $\mathbf{z} = \mathbf{y} - \phi$, we obtain

$$\frac{\partial^2 \mathbf{z}}{\partial t^2} - \Delta\mathbf{z} + \alpha\nabla\theta = -\left(\frac{\partial^2 \phi}{\partial t^2} - \Delta\phi\right) (\equiv \mathbf{f}) \text{ in } Q, \qquad (7.11)_1$$

$$\frac{\partial\theta}{\partial t} - \Delta\theta + \alpha\nabla\cdot\frac{\partial\mathbf{z}}{\partial t} = 0 \text{ in } Q, \qquad (7.11)_2$$

$$\mathbf{z} = \mathbf{0}, \quad \theta = 0 \text{ on } \Sigma, \quad \mathbf{z}(0) = \frac{\partial\mathbf{z}}{\partial t}(0) = \mathbf{0}, \quad \theta(0) = 0. \qquad (7.11)_3$$

We now multiply $(7.11)_1$ (respectively $(7.11)_2$) by $\partial\mathbf{z}/\partial t$ (respectively θ). We obtain with obvious notation $(\|\cdot\| = \|\cdot\|_{L^2(\Omega)})$

$$\frac{1}{2}\frac{d}{dt}\left[\|\frac{\partial\mathbf{z}}{\partial t}\|^2 + \|\nabla\mathbf{z}\|^2 + \|\theta\|^2\right] + \|\nabla\theta\|^2 + \alpha\left[\left(\nabla\theta, \frac{\partial\mathbf{z}}{\partial t}\right) + \left(\nabla\cdot\frac{\partial\mathbf{z}}{\partial t}, \theta\right)\right]$$

$$= \left(\mathbf{f}, \frac{\partial\mathbf{z}}{\partial t}\right). \qquad (7.12)$$

But

$$\left(\nabla\theta, \frac{\partial\mathbf{z}}{\partial t}\right) + \left(\nabla\cdot\frac{\partial\mathbf{z}}{\partial t}, \theta\right) = 0, \qquad (7.13)$$

so that (7.6) leads to *a priori estimates which are independent of* α.

It follows then, that if we denote by $\{\mathbf{z}_\alpha, \theta_\alpha\}$ the solution of (7.11) one has when $\alpha \to +\infty$

$$\begin{cases} \left\{\mathbf{z}_\alpha, \frac{\partial\mathbf{z}_\alpha}{\partial t}\right\} \to \left\{\mathbf{z}, \frac{\partial\mathbf{z}}{\partial t}\right\} \text{ weakly* in } L^\infty(0, T; H_0^1(\Omega) \times L^2(\Omega)), \\ \theta_\alpha \to \theta \text{ weakly in } L^2(0, T; H_0^1(\Omega)) \text{ and weakly * in } L^\infty(0, T; L^2(\Omega)). \end{cases} \qquad (7.14)$$

Returning to the notation $\{\mathbf{y}, \theta\}$, we have that $\mathbf{y}_\alpha \to \mathbf{y}$, where $\mathbf{y}$ is the solution of

$$\begin{cases} \frac{\partial^2 \mathbf{y}}{\partial t^2} - \Delta\mathbf{y} + \nabla p = 0 \text{ in } Q, \quad \nabla\cdot\mathbf{y} = 0 \text{ in } Q, \\ \mathbf{y}(0) = \frac{\partial\mathbf{y}}{\partial t}(0) = \mathbf{0}, \quad \mathbf{y} = \mathbf{v} \text{ on } \Sigma_0, \mathbf{y} = \mathbf{0} \text{ on } \Sigma\backslash\Sigma_0. \end{cases} \qquad (7.15)$$

We clearly see why (7.8) is necessary (from the *divergence theorem*). We observe that the system satisfied by $\{\mathbf{y}, \theta\}$ is again *uncoupled* at the limit when $\alpha \to +\infty$, so that the best thing we can hope is the controllability of $\{\mathbf{y}(T), (\partial\mathbf{y}/\partial t)(T)\}$, but not, of course, the controllability of $\theta(T)$.

Remark 7.6. A systematic study of the controllability of system (7.15) remains to be done (see, however Lions (1990b) for a discussion of the controllability of system (7.15) under strict geometrical conditions).

Remark 7.7. Similar results hold when (7.3) is replaced by (7.5). Then no additional condition such that (7.8) is needed.

7.3. Approximate partial controllability

We now return to the case $0 < \alpha < +\infty$; we assume that

$$
\begin{cases}
\text{when } \mathbf{v} \text{ spans } (L^2(\Sigma_0))^d, \text{ then } \left\{ \mathbf{y}(T; \mathbf{v}), \dfrac{\partial \mathbf{y}}{\partial t}(T; \mathbf{v}) \right\} \\
\text{spans a dense subset of } (L^2(\Omega) \times H^{-1}(\Omega))^d,
\end{cases}
\tag{7.16}
$$

where $\{\mathbf{y}(\mathbf{v}), \theta(\mathbf{v})\}$ is the solution of (7.1)–(7.4).

Remark 7.8. *Sufficient* conditions for (7.16) to be true are given in Chapter 1 of Lions (1988b, Vol. 2); they are of the following type:

(i) Σ_0 is 'sufficiently large',
(ii) $0 < \alpha < \alpha_0$.

Necessary and sufficient conditions for (7.16) to be true do not seem to be known. Interesting results have been obtained by E. Zuazua (1993).

We can then consider the following optimal control problem

$$
\begin{cases}
\inf_{\mathbf{v}} \dfrac{1}{2} \displaystyle\int_{\Sigma_0} |\mathbf{v}|^2 \, d\Sigma, \quad \mathbf{v} \in (L^2(\Sigma_0))^d \text{ such that} \\
\mathbf{y}(T; \mathbf{v}) \in \mathbf{z}^0 + \beta_0 B, \dfrac{\partial \mathbf{y}}{\partial t}(T; \mathbf{v}) \in \mathbf{z}^1 + \beta_1 B_{-1},
\end{cases}
\tag{7.17}
$$

where B (respectively B_{-1}) denotes the unit ball of $(L^2(\Omega))^d$ (respectively of $(H^{-1}(\Omega))^d$).

Problem (7.17) has a unique solution; it can be characterized by a *variational inequality* which can be obtained either directly or by duality methods. Here we use duality, because (among other things) it will be convenient for the next section (where we introduce penalty arguments).

Formulation of a dual problem We follow the same approach as in Section 6.4. We define an operator L from $(L^2(\Sigma_0))^d$ into $(H^{-1}(\Omega))^d \times (L^2(\Omega))^d$ by

$$
L\mathbf{v} = \left\{ -\dfrac{\partial \mathbf{y}}{\partial t}(T; \mathbf{v}), \mathbf{y}(T; \mathbf{v}) \right\}.
\tag{7.18}
$$

We define next F_1 and F_2 by

$$
F_1(\mathbf{v}) = \dfrac{1}{2} \int_{\Sigma_0} |\mathbf{v}|^2 \, d\Sigma,
\tag{7.19$_1$}
$$

$$
F_2(\mathbf{f}^0, \mathbf{f}^1) = \begin{cases} 0 \text{ if } \mathbf{f}^0 \in -\mathbf{z}^1 + \beta_1 B_{-1}, \quad \mathbf{f}^1 \in \mathbf{z}^0 + \beta_0 B, \\ +\infty \text{ otherwise.} \end{cases}
\tag{7.19$_2$}
$$

Problem (7.17) is then equivalent to

$$
\inf_{\mathbf{v} \in (L^2(\Sigma_0))^d} [F_1(\mathbf{v}) + F_2(L\mathbf{v})].
\tag{7.20}
$$

By duality, we obtain

$$\inf_{\mathbf{v}\in(L^2(\Sigma_0))^d}[F_1(\mathbf{v})+F_2(L\mathbf{v})] = -\inf_{\hat{\mathbf{f}}\in(H_0^1(\Omega))^d\times(L^2(\Omega))^d}[F_1^*(L^*\hat{\mathbf{f}})+F_2^*(-\hat{\mathbf{f}})].$$

$$(7.21)$$

The operator L^* is defined as follows. We introduce $\hat{\varphi}$, $\hat{\psi}$ solution of

$$\begin{cases} \dfrac{\partial^2\hat{\varphi}}{\partial t^2} - \Delta\hat{\varphi} + \alpha\boldsymbol{\nabla}\dfrac{\partial\hat{\psi}}{\partial t} = 0 \text{ in } Q, \\[2mm] -\dfrac{\partial\hat{\psi}}{\partial t} - \Delta\hat{\psi} - \alpha\boldsymbol{\nabla}\cdot\hat{\varphi} = 0 \text{ in } Q, \end{cases}$$

$$(7.22)_1$$

$$\hat{\varphi}(T) = \hat{\mathbf{f}}^0 \in (H_0^1(\Omega))^d, \quad \dfrac{\partial\hat{\varphi}}{\partial t}(T) = \hat{\mathbf{f}}^1 \in (L^2(\Omega))^d, \quad \hat{\psi}(T) = 0, \quad (7.22)_2$$

$$\hat{\varphi} = \mathbf{0}, \quad \hat{\phi} = 0 \text{ on } \Sigma. \qquad (7.22)_3$$

Then if $\hat{\mathbf{f}} = \{\hat{\mathbf{f}}^0, \hat{\mathbf{f}}^1\}$, we have

$$L^*\hat{\mathbf{f}} = \dfrac{\partial\hat{\varphi}}{\partial n} \text{ on } \Sigma_0. \qquad (7.23)$$

We obtain thus as dual problem (i.e. for the minimization problem in the right-hand side of (7.21))

$$\inf_{\hat{\mathbf{f}}}\left[\dfrac{1}{2}\int_{\Sigma_0}\left|\dfrac{\partial\hat{\varphi}}{\partial n}\right|^2 d\Sigma + \langle\mathbf{z}^1,\hat{\mathbf{f}}^0\rangle\right.$$

$$\left. - \int_\Omega \mathbf{z}^0\cdot\hat{\mathbf{f}}^1\,dx + \beta_1\|\hat{\mathbf{f}}^0\|_{(H_0^1(\Omega))^d} + \beta_0\|\hat{\mathbf{f}}^1\|_{(L^2(\Omega))^d}\right] \qquad (7.24)$$

Remark 7.9. The same considerations apply to the Neumann controls (i.e. of type (7.5)).

7.4. Approximate controllability via penalty

We consider again (7.1)–(7.4) (with $\theta_0 = 0$) and we introduce (with obvious notation):

$$J_k(\mathbf{v}) = \dfrac{1}{2}\int_{\Sigma_0}|\mathbf{v}|^2\,d\Sigma + \dfrac{k_0}{2}\|\mathbf{y}(T;\mathbf{v})-\mathbf{z}^0\|_{L^2}^2 + \dfrac{k_1}{2}\left\|\dfrac{\partial\mathbf{y}}{\partial t}(T;\mathbf{v})-\mathbf{z}^1\right\|_{H^{-1}}^2. \quad (7.25)$$

In (7.25) we have

$$k = \{k_0, k_1\}, \quad k_i > 0, \quad k_i \text{ 'large' for } i = 0,1. \qquad (7.26)$$

The control problem

$$\inf_{\mathbf{v}\in(L^2(\Sigma_0))^d} J_k(\mathbf{v}) \qquad (7.27)$$

has a *unique* solution, $\mathbf{u}_k$.

Considerations similar to those of Section 6, apply; thus we shall have

$$y(T; \mathbf{u}_k) \in \mathbf{z}^0 + \beta_0 B, \quad \frac{\partial \mathbf{y}}{\partial t}(T; \mathbf{u}_k) \in \mathbf{z}^1 + \beta_1 B_{-1}, \qquad (7.28)$$

for k 'large enough', the 'large enough' not being defined in a constructive way.

In order to obtain estimates on the choice of k, we will now consider the *dual problem* of (7.25). We consider again, therefore, the operator L: $(L^2(\Sigma_0))^d \to (H^{-1}(\Omega))^d \times (L^2(\Omega))^d$, defined by

$$L(v) = \left\{ -\frac{\partial \mathbf{y}}{\partial t}(T; v), \mathbf{y}(T; v) \right\}$$

and we introduce

$$F_3(\mathbf{f}) = \tfrac{1}{2} k_0 \| \mathbf{f}^1 - \mathbf{z}^0 \|_{L^2}^2 + \tfrac{1}{2} k_1 \| \mathbf{f}^0 + \mathbf{z}^1 \|_{H^{-1}}^2. \qquad (7.29)$$

With $F_1(\cdot)$ still defined by $(7.19)_1$, we clearly have

$$\inf_{v \in (L^2(\Sigma_0))^d} J_k(\mathbf{v}) = \inf_{v \in (L^2(\Sigma_0))^d} [F_1(\mathbf{v}) + F_3(L\mathbf{v})]. \qquad (7.30)$$

It follows by duality that

$$\inf_{v \in (L^2(\Sigma_0))^d} J_k(\mathbf{v}) = -\inf_{\mathbf{f}} [F_1^*(L^* \mathbf{f}) + F_3^*(-\mathbf{f})] \qquad (7.31)$$

with $\mathbf{f} = \{\mathbf{f}_0, \mathbf{f}_1\} \in (H_0^1(\Omega))^d \times (L^2(\Omega))^d$ in (7.31).

After some calculations, we obtain

$$\inf_{\mathbf{v}} J_k(\mathbf{v}) = -\inf_{\mathbf{f}} \left[\frac{1}{2} \int_{\Sigma_0} \left| \frac{\partial \varphi}{\partial n} \right|^2 d\Sigma + \frac{1}{2k_0} \| \mathbf{f}^1 \|_{L^2}^2 - \int_{\Omega} \mathbf{z}^0 \cdot \mathbf{f}^1 \, dx \right.$$
$$\left. + \frac{1}{2k_1} \| \mathbf{f}^0 \|_{H_0^1}^2 + \langle \mathbf{z}^0, \mathbf{f}^1 \rangle \right]. \qquad (7.32)$$

The dual problem to the control problem (7.27) is therefore

$$\inf_{\mathbf{f}} \left[\frac{1}{2} \int_{\Sigma_0} \left| \frac{\partial \varphi}{\partial n} \right|^2 d\Sigma + \frac{1}{2k_0} \| \mathbf{f}^1 \|_{L^2}^2 - \int_{\Omega} \mathbf{z}^0 \cdot \mathbf{f}^1 \, dx + \frac{1}{2k_1} \| \mathbf{f}^0 \|_{H_0^1}^2 + \langle \mathbf{z}^1, \mathbf{f}^0 \rangle \right],$$
$$\qquad (7.33)$$

with $\mathbf{f} = \{\mathbf{f}^0, \mathbf{f}^1\} \in (H_0^1(\Omega))^d \times (L^2(\Omega))^d$ in (7.33).

Let us denote by $\mathbf{f}_k$ the solution of (7.33); it is characterized (with obvious notation) by

$$\mathbf{f}_k = \{\mathbf{f}_k^0, \mathbf{f}_k^1\} \in (H_0^1(\Omega))^d \times (L^2(\Omega))^d,$$

$$\int_{\Sigma_0} \frac{\partial \varphi_k}{\partial n} \cdot \frac{\partial \varphi}{\partial n} \, d\Sigma + \frac{1}{k_0} \int_{\Omega} \mathbf{f}_k^1 \cdot \mathbf{f}^1 \, dx + \frac{1}{k_1} \int_{\Omega} \nabla \mathbf{f}_k^0 \cdot \nabla \mathbf{f}^0 \, dx$$

$$= \int_{\Omega} \mathbf{z}^0 \cdot \mathbf{f}^1 \, dx - \langle \mathbf{z}^1, \mathbf{f}^0 \rangle$$

$$\forall \mathbf{f} = \{\mathbf{f}^0, \mathbf{f}^1\} \in (H_0^1(\Omega))^d \times (L^2(\Omega))^d. \qquad (7.34)$$

Similarly, the solution $\mathbf{f}_\beta$ of problem (7.24) is characterized by the following *variational inequality*

$$\mathbf{f}_\beta = \{\mathbf{f}_\beta^0, \mathbf{f}_\beta^1\} \in (H_0^1(\Omega))^d \times (L^2(\Omega))^d$$

$$\int_{\Sigma_0} \frac{\partial \varphi_\beta}{\partial n} \cdot \frac{\partial}{\partial n}(\varphi - \varphi_\beta) \, \mathrm{d}\Sigma + \beta_1(\|\mathbf{f}^0\|_{H_0^1} - \|\mathbf{f}_\beta^0\|_{H_0^1}) + \beta_0(\|\mathbf{f}^1\|_{L^2} - \|\mathbf{f}_\beta^1\|_{L^2})$$

$$\geq \int_\Omega \mathbf{z}^0 \cdot (\mathbf{f}^1 - \mathbf{f}_\beta^1) \, \mathrm{d}x - \langle \mathbf{z}^1, \mathbf{f}^0 - \mathbf{f}_\beta^0 \rangle,$$

$$\forall \mathbf{f} \in (H_0^1(\Omega))^d \times (L^2(\Omega))^d. \tag{7.35}$$

Taking $\mathbf{f} = \mathbf{f}_k$ in (7.34) (respectively $\mathbf{f} = \mathbf{0}$ and $\mathbf{f} = 2\mathbf{f}_\beta$ in (7.35)) we obtain

$$\int_{\Sigma_0} \left| \frac{\partial \varphi_k}{\partial n} \right|^2 \mathrm{d}\Sigma + \frac{1}{k_0} \int_\Omega |\mathbf{f}_k^1|^2 \, \mathrm{d}x + \frac{1}{k_1} \int_\Omega |\nabla \mathbf{f}_k^0|^2 \, \mathrm{d}x$$

$$= \int_\Omega \mathbf{z}^0 \cdot \mathbf{f}_k^1 \, \mathrm{d}x - \langle \mathbf{z}^1, \mathbf{f}_k^0 \rangle, \tag{7.36}$$

$$\int_{\Sigma_0} \left| \frac{\partial \varphi_\beta}{\partial n} \right|^2 \mathrm{d}\Sigma + \beta_0 \|\mathbf{f}_\beta^1\|_{L^2} + \beta_1 \|\mathbf{f}_\beta^0\|_{H_0^1} = \int_\Omega \mathbf{z}^0 \cdot \mathbf{f}_\beta^1 \, \mathrm{d}x - \langle \mathbf{z}^1, \mathbf{f}_\beta^0 \rangle. \tag{7.37}$$

Assuming that problems (7.34), (7.35) have the same solution, it follows from (7.36), (7.37) that

$$\frac{1}{k_0} \|\mathbf{f}_k^1\|_{L^2}^2 + \frac{1}{k_1} \|\mathbf{f}_k^1\|_{H_0^1}^2 = \beta_0 \|\mathbf{f}_k^1\|_{L^2} + \beta_1 \|\mathbf{f}_k^0\|_{H_0^1}, \tag{7.38}$$

which suggests the following simple (may be too simple) rule: adjust k_0, k_1 so that

$$\|\mathbf{f}_k^1\|_{L^2}^2 k_0^{-1} = \beta_0, \quad \|\mathbf{f}_k^0\|_{H_0^1}^2 k_1^{-1} = \beta_1. \tag{7.39}$$

We plan *numerical experiments* to validate (7.39).

REFERENCES

F. Abergel and R. Temam (1990), 'On some control problems in fluid mechanics', *Theoret. Comput. Fluid Dyn.* **1**, 303–326.

G. Auchmuty, E.J. Dean, R. Glowinski and S.C. Zhang (1987), 'Control methods for the numerical computation of periodic solutions of autonomous differential equations', in *Control Problems for Systems Described by Partial Differential Equations and Applications* (I. Lasiecka and R. Triggiani, eds.) Lecture Notes in Control and Information, vol. 97, Springer (Berlin) 64–89.

A. Bamberger (1977), private communication.

C. Bardos, G. Lebeau and J. Rauch (1988), 'Contrôle et stabilisation dans les problèmes hyperboliques', in *Contrôlabilité Exacte, Perturbation et Stabilisation de Systèmes Distribués*, vol. 1 (J.L. Lions (ed.), Masson (Paris) Appendix 2. 492–537.

C. Bardos and J. Rauch (1994), 'Mathematical analysis of control-type numerical algorithms for the solution of the Helmholtz equation'. To appear.

M. Berggren (1992), 'Control and simulation of advection–diffusion systems', MSc Thesis, Department of Mechanical Engineering, University of Houston.

M. Berggren and R. Glowinski (1995), 'Numerical solution of some flow control problems', *Int. J. Fluid Dyn.* To appear.

A. Bensoussan (1990), *Acta Appl. Math.* **20**, 197.

F. Bourquin (1993), 'Approximation theory for the problem of exact controllability of the wave equation with boundary control, in *Mathematical and Numerical Aspects of Wave Propagation* (R. Kleinman, Th. Angell, D. Colton, F. Santosa and I. Stakgold, eds.) SIAM (Philadelphia, PA) 103–112.

F. Brezzi and M. Fortin (1991), *Mixed and Hybrid Finite Element Methods*, Springer (New York).

M.O. Bristeau, J. Erhel, R. Glowinski and J. Periaux (1993), 'A time dependent approach to the solution of the Helmholtz equation at high wave numbers', in *Proc. Sixth SIAM Conf. on Parallel Processing for Scientific Computing*, (R.F. Sincovec, D.E. Keyes, M.R. Leuze, L.R. Petzold and D.A. Reed, eds.) SIAM (Philadelphia, PA).

M.O. Bristeau, R. Glowinski and J. Periaux (1993a), 'Scattering waves using exact controllability methods', *31st AIAA Aerospace Sciences Meeting, Reno, Nevada*, AIAA Paper 93-0460.

M.O. Bristeau, R. Glowinski and J. Periaux (1993b), 'Using exact controllability to solve the Helmholtz equation at high wave numbers', in *Mathematical and Numerical Aspects of Wave propagation* (R. Kleinman, Th. Angell, D. Colton, F. Santosa and I. Stakgold, eds.) SIAM (Philadelphia, PA) 113–127.

M.O. Bristeau, R. Glowinski and J. Periaux (1993c), 'Numerical simulation of high frequency scattering waves using exact controllability methods', in *Nonlinear Hyperbolic Problems: Theoretical, Applied, and Computational Aspects* (A. Donato and F. Oliveri, eds.) Notes in Numerical Fluid Mechanics, vol. 43, Vieweg (Branschweig) 86–108.

J.A. Burns and S. Kang (1991), 'A stabilization problem for Burgers' equation with unbounded control and observation', *International Series of Numerical Mathematics*, vol. 100, Birkhauser (Basel).

J.A. Burns and H. Marrekchi (1993), 'Optimal fixed-finite-dimensional compensator for Burgers' equation with unbounded input/output operators', ICASE Report 93-19, ICASE.

D.M. Buschnell and J.N. Hefner (1990), *Viscous Drag Reduction in Boundary Layers*, American Institute of Aeronautics and Astronautics (Washington, DC).

C. Carthel (1994), 'Numerical methods for the boundary controllability of the heat equation', PhD Thesis, Department of Mathematics, University of Houston.

C. Carthel, R. Glowinski and J.L. Lions (1994), 'On exact and approximate boundary controllabilities for the heat equation: a numerical approach', *J. Opt. Theory Appl.* **82**, 3, 429–484.

A. Chiara (1993), 'Equation des ondes et régularité sur un ouvert lipschitzien', *C.R. Acad. Sci. I* **316**, 33–36.

P.G. Ciarlet (1978), *The Finite Element Method for Elliptic Problems*, North-Holland (Amsterdam).

P.G. Ciarlet (1989), *Introduction to Numerical Linear Algebra and Optimization*, Cambridge University Press (Cambridge).

P.G. Ciarlet (1990a), 'A new class of variational problems arising in the modeling of elastic multi-structures', *Numer. Math.* **57**, 547–560.

P.G. Ciarlet (1990b), *Plates and Junctions in Elastic Multi-Structures: An Asymptotic Analysis*, Masson (Paris).

P.G. Ciarlet (1991), 'Basic error estimates for elliptic problems', in *Handbook of Numerical Analysis*, vol. II (P.G. Ciarlet and J.L. Lions, eds.) North-Holland (Amsterdam) 17–351.

P.G. Ciarlet, H. Le Dret and R. Nzengwa (1989), 'Junctions between three-dimensional and two-dimensional linearly elastic structures', *J. Math. Pures et Appl.* **68**, 261–295.

M.G. Crandall and P.L. Lions (1985), 'Hamilton–Jacobi equations in infinite dimensions, Part I', *J. Funct. Anal.* **62**, 379–396.

M.G. Crandall and P.L. Lions (1986a), 'Hamilton–Jacobi equations in infinite dimensions, Part II', *J. Funct. Anal.* **65**, 368–405.

M.G. Crandall and P.L. Lions (1986b), 'Hamilton–Jacobi equations in infinite dimensions, Part III', *J. Funct. Anal.* **68**, 214–247.

M.G. Crandall and P.L. Lions (1990), 'Hamilton–Jacobi equations in infinite dimensions, Part IV', *J. Funct. Anal.* **90**, 237–283.

M.G. Crandall and P.L. Lions (1991), 'Hamilton–Jacobi equations in infinite dimensions, Part V', *J. Funct. Anal.* **97**, 417–465.

J. Daniel (1970), *The Approximate Minimization of Functionals*, Prentice Hall (Englewood Cliffs, NJ).

E.J. Dean and P. Gubernatis (1991), 'Pointwise control of Burgers' equation – a numerical approach', *Comput. Math. Appl.* **22**, 93–100.

E.J. Dean, R. Glowinski and C.H. Li (1989), 'Supercomputer solution of partial differential equation problems in computational fluid dynamics and in control', *Comput. Phys. Commun.* **53**, 401–439.

J. Dennis and R.B. Schnabel (1983), *Numerical Methods for Unconstrained Optimization and Nonlinear Equations*, Prentice-Hall (Englewood Cliffs, NJ).

J.I. Diaz (1991), 'Sobre la controlabilidad aproximada de problemas no lineales disipativos', *Jornadas Hispano-Francesas Sobre Control de Sistemas Distribuidos*, Universidad de Malaga 41–48.

J.I. Diaz and Fursikov (1994), To appear.

J.D. Downer, K.C. Park and J.C. Chiou (1992), 'Dynamics of flexible beams for multibody systems: a computational procedure', *Comput. Meth. Appl. Mech. Engrg* **96**, 373–408.

T. Dupont, R. Glowinski, W. Kinton and M.F. Wheeler (1992), 'Mixed finite element methods for time-dependent problems: application to control', in *Finite Element in Fluids*, vol. 8 (T. Chung, ed.) Hemisphere (Washington, DC) 137–163.

I. Ekeland and R. Temam (1974), *Analyse Convexe et Problèmes Variationnels*, Dunod (Paris).

B. Engquist, B. Gustafsson and J. Vreeburg (1978), 'Numerical solution of a PDE system describing a catalytic converter', *J. Comput. Phys.* **27**, 295–314.

C. Fabre, J.P. Puel and E. Zuazua (1993), 'Contrôlabilité approchée de l'équation de la chaleur linéaire avec des contrôles de norme L^∞ minimale', *C.R. Acad. Sci. I* **316**, 679–684.

A. Friedman (1988), 'Modeling catalytic converter performance', in *Mathematics in Industrial Problems*, Part 4 (A. Friedman, ed.) Springer (New York) Ch. 7, 70–77.

C.M. Friend (1993), 'Catalysis and surfaces', *Scientific American* (April) 74–79.

H. Fujita and T. Suzuki (1991), 'Evolution problems', in *Handbook of Numerical Analysis* vol. II (P.G. Ciarlet and J.L. Lions, eds.) North-Holland (Amsterdam) 789–928.

V. Fursikov (1992) 'Lagrange principle for problems of optimal control of ill posed or singular distributed systems', *J. Math. Pures et Appl.* **71**(2), 139–194.

D. Gabay (1982), 'Application de la méthode des multiplicateurs aux inéquations variationnelles', in *Méthodes de Lagrangien Augmenté* (M. Fortin and R. Glowinski, eds.) Dunod-Bordas, (Paris).

D. Gabay (1983) 'Application of the method of multipliers to variational inequalities', in *Augmented Lagrangian Methods* (M. Fortin and R. Glowinski, eds.) North-Holland (Amsterdam).

J.A. George (1971), 'Computer implementation of the finite element method', PhD Thesis, Computer Sciences Department, Stanford University.

R. Glowinski (1984), *Numerical Methods for Nonlinear Variational Problems*, Springer (New York).

R. Glowinski (1991), 'Finite element methods for the numerical simulation of incompressible viscous flow. Introduction to the control of the Navier–Stokes equations', in *Lectures in Applied Mathematics*, vol. 28, American Mathematical Society (Providence, RI) 219–301.

R. Glowinski (1992a), 'Ensuring well-posedness by analogy: Stokes problem and boundary control for the wave equation', *J. Comput. Phys.* **103**, 189–221.

R. Glowinski (1992b), 'Boundary controllability problems for the wave and heat equations', in *Boundary Control and Boundary Variation* (J.P. Zolesio, ed.) Lecture Notes in Control and Information Sciences, vol. 178, Springer (Berlin) 221–237.

R. Glowinski and P. Le Tallec (1989), *Augmented Lagrangian and Operator-Splitting Methods in Nonlinear Mechanics*, SIAM (Philadelphia, PA).

R. Glowinski and C.H. Li (1990), 'On the numerical implementation of the Hilbert uniqueness method for the exact boundary controllability of the wave equation', *C.R. Acad. Sci. I* **311**, 135–142.

R. Glowinski and C.H. Li (1991), 'On the exact Neumann boundary controllability of the wave equation', in *Mathematical and Numerical Aspects of Wave Propagation Phenomena* (G. Cohen, L. Halpern and P. Joly, eds.) SIAM (Philadelphia, PA) 15–24.

R. Glowinski, W. Kinton and M.F. Wheeler (1989), 'A mixed finite element formulation for the boundary controllability of the wave equation', *Int. J. Numer. Meth. Engrg* **27**, 623–635.

R. Glowinski, C.H. Li and J.L. Lions (1990), 'A numerical approach to the exact boundary controllability of the wave equation (I) Dirichlet controls: description of the numerical methods', *Japan. J. Appl. Math.* **7**, 1–76.

R. Glowinski, J.L. Lions and R. Tremolières (1976), *Analyse Numérique des Inéquations Variationnelles*, Dunod (Paris).

R. Glowinski, J.L. Lions and R. Tremolières (1981), *Numerical Analysis of Variational Inequalities*, North-Holland (Amsterdam).

G.H. Golub and C. Van Loan (1989), *Matrix Computations*, Johns Hopkins Press (Baltimore, MD).

W. Hackbush (1985), *Multigrids Methods and Applications*, Springer (Berlin).

A. Haraux and S. Jaffard (1991), 'Pointwise and spectral control of plate vibrations', *Revista Matematica Iberoamericana* **7**, 1–24.

J. Henry (1978), 'Contrôle d'un Réacteur Enzymatique à l'Aide de Modèles à Paramètres Distribués. Quelques Problèmes de Contrôlabilité de Systèmes Paraboliques', Thèse d'État, Université P. et M. Curie.

L. Hörmander (1976), *Linear Partial Differential Operators*, Springer (Berlin).

I. Joó (1991), 'Contrôlabilité Exacte et propriétés d'oscillations de l'équation des ondes par analyse non harmonique', *C.R. Acad. Sci.* I **312**, 119–122.

V. Komornik (1994), *Exact Controllability and Stabilization. The Multiplier Method*, Masson (Paris).

O.A. Ladyzenskaya, V.A. Solonnikov and N.N. Ural'ceva (1968), *Linear and Quasilinear Equations of Parabolic Type*, American Mathematical Society (Providence, RI).

J.E. Lagnese (1989), 'Exact boundary controllability of Maxwell's equations in a general region', *SIAM J. Control Optimiz.* **27**, 374–388.

J.E. Lagnese, G. Leugering and G. Schmidt (1992), 'Modelling and controllability of networks of thin beams', in *System Modelling and Optimization*, Lecture Notes in Control and Information, vol. 97, Springer (Berlin), 467–480.

J.E. Lagnese, G. Leugering and G. Schmidt (1994), *Modeling, Analysis and Control of Dynamic Elastic Multi-Link Structures,* Binkhäuser (Boston, Basel, Berlin).

I. Lasiecka (1992), Exponential decay rates for the solutions of Euler Bernoulli equations with boundary dissipation occurring in the moments only. *J. Diff. Equations* **95**, 169–182.

I. Lasiecka and R. Tataru (1994), Uniform boundary stabilization of semilinear wave equations with non linear boundary conditions, *Diff. and Integral Equations*.

T.A. Laursen and J.C. Simo (1993), 'A continuum-based finite element formulation for the implicit solution of multi-body, large deformation frictional contact problems', *Int. J. Numer. Meth. Engrg* **36**, 3451–3486.

P.D. Lax and R.S. Phillips (1989), *Scattering Theory*, Academic (New York).

J.L. Lions (1961), *Equations Différentielles Opérationnelles et Problèmes aux Limites*, Springer (Heidelberg).

J.L. Lions (1968), *Contrôle Optimal des Systèmes Gouvernés par des Equations aux Dérivées Partielles*, Dunod (Paris).

J.L. Lions (1969), *Quelques Méthodes de Résolution des Problèmes aux Limites Non Linéaires*, Dunod (Paris).

J.L. Lions (1983), *Contrôle des Systèmes Distribués Singuliers*, Gauthier-Villars (Paris) (English translation: Gauthier-Villars (Paris) 1985).

J.L. Lions (1986), 'Contrôlabilité exacte des systèmes distribués', *C.R. Acad. Sci.* I **302**, 471–475.

J.L. Lions (1988a), 'Exact controllability, stabilization and perturbation for distributed systems', *SIAM Rev.* **30**, 1–68.

J.L. Lions (1988b), *Contrôlabilité Exacte, Perturbation et Stabilisation des Systèmes Distribués*, vols. 1 and 2, Masson (Paris). (English Translation, Springer 1994).

J.L. Lions (1990), *El Planeta Tierra*, Instituto de España, Espasa Calpe, S.A., Madrid.

J.L. Lions (1990b), 'On some hyperbolic equations with a pressure term', *Proc. CIRM Conf. Trento, Italy, October*.

J.L. Lions (1991a), 'Exact controllability for distributed systems: some trends and some problems', in *Applied and Industrial Mathematics* (R. Spigler, ed.) Kluwer (Dordrecht) 59–84.

J.L. Lions (1991b), 'Approximate controllability for parabolic systems', Harvey Lectures, Haiffa.

J.L. Lions (1993), 'Quelques remarques sur la contrôlabilité en liaison avec des questions d'environnement', in *Les Grands Systèmes des Sciences et de la Technologie* (J. Horowitz and J.L. Lions, eds.), Masson (Paris) 240–264.

J.L. Lions and E. Magenes (1968), *Problèmes aux Limites Non Homogènes*, vol. 1, Dunod (Paris). (English Translation, Springer, 1972).

P.L. Lions and B. Mercier (1979), 'Splitting algorithms for the sum of two nonlinear operators', *SIAM J. Numer. Anal.* **16**, 964–979.

G. Marini, P. Testa and V. Valente (1994), 'Exact controllability of a spherical cap: numerical implementation of HUM', *J. Opt. Theory and Appl.* To appear.

K. McManus, Th. Poinsot and S. Candel (1993), 'Review of active control of combustion instabilities', *Prog. Energy and Combustion Sci.* **19**, 1–29.

Y. Meyer (1989), Private communication.

S. Mizohata (1958), 'Unicité du prolongement des solutions pour quelques opérateurs différentiels paraboliques', *Mem. Coll. Sci. Univ. Kyoto* A **31**, 219–239.

J. Narukawa (1983), 'Boundary value control of thermo-elastic systems', *Hiroshima J. Math.* **13**, 227–272.

J. Nečas (1967), *Les Méthodes Directes en Théorie des Équations Elliptiques*, Masson (Paris).

J. Nocedal (1992), 'Theory of algorithms for unconstrained optimization', *Acta Numerica 1992*, Cambridge University Press (Cambridge) 199–242.

K.C. Park, J.C. Chiou and J.D. Downer (1990), 'Explicit–implicit staggered procedures for multibody dynamics analysis', *J. Guidance Control Dyn.* **13**, 562–570.

D. Peaceman and H. Rachford (1955), 'The numerical solution of parabolic and elliptic differential equations', *J. Soc. Ind. Appl. Math.* **3**, 28–41.

E. Polack (1971), *Computational Methods in Optimization*, Academic (New York).

M.J.D. Powell (1976), 'Some convergence properties of the conjugate gradient method', *Math. Program.* **11**, 42–49.

P.A. Raviart and J.M. Thomas (1988), *Introduction à l'Analyse Numérique des Équations aux Dérivées Partielles*, Masson (Paris).

J.E. Roberts and J.M. Thomas (1991), 'Mixed and hybrid methods', in *Handbook of Numerical Analysis*, vol. II (P.G. Ciarlet and J.L. Lions, eds.) North-Holland (Amsterdam) 523–639.

A.T. Rockafellar (1970), *Convex Analysis*, Princeton University Press (Princeton, NJ).

D.L. Russel (1978), 'Controllability and stabilizability theory for linear partial differential equations. Recent progress and open questions', *SIAM Rev.* **20**, 639–739.

J.M. Samaniego, B. Yip, Th. Poinsot and S. Candel (1993), 'Low frequency combustion instability mechanisms inside dump combustor', *Combustion and Flame* **94**, 363–381.

J. Sanchez Hubert and E. Sanchez Palencia (1989), *Vibrations and Coupling of Continuous Systems (Asymptotic Methods)*, Springer (Berlin).

J.C. Saut and B. Scheurer (1987) 'Unique continuation for some evolution equations', *J. Diff. Eqns* **66**, 118–139.

R.H. Sellin and T. Moses (1989), *Drag Reduction in Fluid Flows*, Ellis Horwood (Chichester).

S.S. Sritharan (1991a), 'Dynamic programming of Navier–Stokes equations', *Syst. Control Lett.* **16**, 299–307.

S.S. Sritharan (1991b), 'An optimal control problem for exterior hydrodynamics', in *Distributed Parameter Control Systems: New Trends and Applications* (G. Chen, E.B. Lee, W. Littman and L. Markus, eds.) Marcel Dekker (New York) 385–417.

A. Taflove (1992), 'Re-inventing electromagnetics, supercomputing solution of Maxwell's equations via direct time integration on space grids', *30th AIAA Aerospace Sciences Meeting, Reno, Nevada*, AIAA Paper 92-0333.

D. Tataru (1994a), Boundary controllability for conservative P.D.E. To appear.

D. Tataru (1994b), Decay rates and attractors for semilinear P.D.E. To appear.

V. Thomée (1990), 'Finite difference methods for linear parabolic equations', in *Handbook of Numerical Analysis*, vol. I (P.G. Ciarlet and J.L. Lions, eds.) North-Holland (Amsterdam) 5–196.

H. Yserentant (1993) 'Old and new convergence proofs for multigrid methods', *Acta Numerica 1993*, Cambridge University Press, (Cambridge) 285–326.

E. Zuazua (1988), 'Contrôlabilité exacte en un temps arbitrairement petit de quelques modèles de plaques', Appendix 1 of *Contrôlabilité Exacte, Pertubations et Stabilisation de Systèmes Distribués*, Vol. I, by J.L. Lions, Masson (Paris), 465–491.

E. Zuazua (1993), 'Contrôlabilité du système de la thermo-élasticité', *C.R. Acad. Sci. Paris* I **317**, 371–376.

ERRATA FOR SECTION 1.10.7

(1) Due to a coding mistake the algorithm used to solve the test problems in Sections 1.10.7.2 and 1.10.7.3 of Part I (see *Acta Numerica 1994*) was not a genuine conjugate gradient algorithm. Indeed, the search direction sequence $\{w^n\}_{n\geq 0}$ was improperly defined leading to a slow convergence; for the 'small' values of k (e.g. 10^2, 10^3) the computed results were essentially correct, but for larger values the slow convergence prevented us from reaching the correct limit since we stopped iterating after a fixed number of iterations (300 or 500, depending of the test problem).

The results obtained with the corrected algorithm are given in Tables 1 to 5 (which replace the coresponding tables of Part I, on pages 345, 356, 357 and 364). Due to the fast convergence properties we have been able to consider much larger values of the penalty parameter k than in Part I (upto 10^{10}). The comments (i) and (ii) in pages 345 and 356 of Part I are still relevant for the values of k used here since the convergence is achieved for a number of iterations which is much smaller than the dimension of the

Table 1. *Summary of numerical results (target defined by (1.461)); $T = 3$, $h = \Delta t = 10^{-2}$.*

b	k	Number of Iterations	$\|u^*\|_{L^2(0,T)}$	$\dfrac{\|y_T^* - y_T\|_{L^2(0,1)}}{\|y_T\|_{L^2(0,1)}}$
$\sqrt{2}/3$	10^2	5	0.921	6.0×10^{-2}
	10^3	7	1.14	2.3×10^{-2}
	10^4	9	1.39	9.1×10^{-3}
	10^5	9	1.66	5.6×10^{-3}
	10^6	12	2.22	4.1×10^{-3}
$1/2$	10^2	5	0.909	5.5×10^{-2}
	10^3	7	1.09	2.1×10^{-2}
	10^4	7	1.30	9.3×10^{-3}
	10^5	9	1.63	5.1×10^{-3}
	10^6	10	2.18	3.2×10^{-3}
$\pi/6$	10^2	5	0.918	5.9×10^{-2}
	10^3	7	1.13	2.3×10^{-2}
	10^4	7	1.37	9.1×10^{-3}
	10^5	9	1.65	5.4×10^{-3}
	10^6	12	2.22	3.9×10^{-3}

Table 2. *Summary of numerical results (target defined by (1.462)); $T = 3$,*
$h = \Delta t = 10^{-2}$.

b	k	Number of Iterations	$\|u^*\|_{L^2(0,T)}$	$\dfrac{\|y_T^* - y_T\|_{L^2(0,1)}}{\|y_T\|_{L^2(0,1)}}$
$\sqrt{2}/3$	10^2	6	1.23	2.2×10^{-1}
	10^3	7	1.96	1.9×10^{-1}
	10^4	10	5.54	1.6×10^{-1}
	10^5	11	11.2	1.4×10^{-1}
	10^6	13	37.1	1.3×10^{-1}
	10^{10}	21	585	6.7×10^{-2}
$1/2$	10^2	5	1.27	1.1×10^{-1}
	10^3	7	1.74	6.3×10^{-2}
	10^4	7	2.42	5.2×10^{-2}
	10^5	9	6.26	3.8×10^{-2}
	10^6	9	12.7	3.0×10^{-2}
	10^{10}	14	66.7	2.5×10^{-2}
$\pi/6$	10^2	5	1.24	1.9×10^{-1}
	10^3	7	1.84	1.6×10^{-1}
	10^4	9	4.84	1.4×10^{-1}
	10^5	12	9.87	1.2×10^{-1}
	10^6	15	29.7	1.1×10^{-1}
	10^{10}	21	534	5.9×10^{-2}

solution of the discrete control problem. However, comment (iii) may apply
to situations where the discrete control problem is really badly conditioned.
The figures corresponding to the new results will be reported elsewhere. For
$k = 10^2, 10^3$ they are practically identical to the corresponding ones in Part
I; for larger values of k we still have a good agreement.

(2) In Section 1.10.7.3 of Part I, the target function $y_T(x) = 27x^2(1 - x)$
was used for the numerical calculations, instead of the function y_T defined
by (1.466), i.e. $y_T(x) = \frac{27}{4}x^2(1 - x)$.

   Table 4 shows the results obtained for $y_T(x) = \frac{27}{4}x^2(1 - x)$ with the
corrected conjugate gradient algorithm; the convergence is quite fast even
for large values of k.

Table 3. *Summary of numerical results (target defined by (1.463)); $T = 3$, $h = \Delta t = 10^{-2}$.*

b	k	Number of Iterations	$\|u^*\|_{L^2(0,T)}$	$\dfrac{\|y_T^* - y_T\|_{L^2(0,1)}}{\|y_T\|_{L^2(0,1)}}$
$\sqrt{2}/3$	10^2	5	0.92	3.47×10^{-1}
	10^3	8	2.3	3.11×10^{-1}
	10^4	10	7.3	2.72×10^{-1}
	10^5	11	16	2.46×10^{-1}
	10^6	13	26	2.4×10^{-1}
	10^{10}	28	2200	2.27×10^{-1}
$1/2$	10^2	5	0.99	3.32×10^{-1}
	10^3	7	2.08	2.47×10^{-1}
	10^4	9	6.3	2.36×10^{-1}
	10^5	9	10	2.27×10^{-1}
	10^6	12	36	2.16×10^{-1}
	10^{10}	15	540	1.82×10^{-1}
$\pi/6$	10^2	5	0.94	3.43×10^{-1}
	10^3	7	2.4	3.02×10^{-1}
	10^4	9	6.9	2.66×10^{-1}
	10^5	10	16	2.40×10^{-1}
	10^6	13	29	2.34×10^{-1}
	10^{10}	28	2200	2.17×10^{-1}

Table 4. *Summary of numerical results (target defined by (1.466)); $T = 3$, $h = \Delta t = 10^{-2}$.*

b	k	Number of Iterations	$\|u^*\|_{L^2(0,T)}$	$\dfrac{\|y_T^* - y_T\|_{L^2(0,1)}}{\|y_T\|_{L^2(0,1)}}$
$\sqrt{2}/3$	10^4	9	10.6	1.5×10^{-1}
	10^5	12	18.4	4.0×10^{-2}
	10^6	15	21.6	1.5×10^{-2}
	10^{10}	16	34.0	8.3×10^{-3}
$1/2$	10^4	7	1.10	3.5×10^{-1}
	10^5	9	1.38	3.5×10^{-1}
	10^6	9	1.92	3.5×10^{-1}
	10^{10}	9	3.28	3.5×10^{-1}
$\pi/6$	10^4	6	10.1	1.8×10^{-1}
	10^5	11	21.1	5.6×10^{-2}
	10^6	12	26.7	2.1×10^{-2}
	10^{10}	15	49.7	1.0×10^{-2}

Table 5. *Summary of numerical results (target defined by (1.467)); $T = 3$, $h = \Delta t = 10^{-2}$.*

b	k	Number of Iterations	$\|u^*\|_{L^2(0,T)}$	$\dfrac{\|y_T^* - y_T\|_{L^2(0,1)}}{\|y_T\|_{L^2(0,1)}}$
$\sqrt{2}/3$	10^3	8	6.53	5.2×10^{-1}
	10^4	11	21.8	3.0×10^{-1}
	10^5	12	41.6	1.6×10^{-1}
	10^6	16	58.8	1.3×10^{-1}
	10^{10}	31	1900	7.1×10^{-2}
$1/2$	10^3	7	2.05	7.1×10^{-1}
	10^4	9	2.68	7.1×10^{-1}
	10^5	9	6.27	7.1×10^{-1}
	10^6	9	12.6	7.1×10^{-1}
	10^{10}	15	66.7	7.1×10^{-1}
$\pi/6$	10^3	8	5.99	6.7×10^{-1}
	10^4	10	27.5	4.2×10^{-1}
	10^5	12	57.6	2.0×10^{-1}
	10^6	16	89.9	1.3×10^{-1}
	10^{10}	16	326	6.2×10^{-2}

Acta Numerica (1995), pp. 335–415

Numerical Solutions to Free Boundary Problems[*]

Thomas Y. Hou

Applied Mathematics,
California Institute of Technology,
Pasadena, CA 91125
E-mail: hou@ama.caltech.edu

Many physically interesting problems involve propagation of free surfaces. Vortex-sheet roll-up in hydrodynamic instability, wave interactions on the ocean's free surface, the solidification problem for crystal growth and Hele–Shaw cells for pattern formation are some of the significant examples. These problems present a great challenge to physicists and applied mathematicians because the underlying problem is very singular. The physical solution is sensitive to small perturbations. Naïve discretisations may lead to numerical instabilities. Other numerical difficulties include singularity formation and possible change of topology in the moving free surfaces, and the severe time-stepping stability constraint due to the stiffness of high-order regularisation effects, such as surface tension.

This paper reviews some of the recent advances in developing stable and efficient numerical algorithms for solving free boundary-value problems arising from fluid dynamics and materials science. In particular, we will consider boundary integral methods and the level-set approach for water waves, general multi-fluid interfaces, Hele–Shaw cells, crystal growth and solidification. We will also consider the stabilising effect of surface tension and curvature regularisation. The issue of numerical stability and convergence will be discussed, and the related theoretical results for the continuum equations will be addressed. This paper is not intended to be a detailed survey and the discussion is limited by both the taste and expertise of the author.

CONTENTS

[*] Research supported in part by an Office of Navel Research Grant N00014-94-1-0310, and a National Science Foundation Grant DMS-9407030.

1. Introduction

Many physically interesting problems involve propagation of free surfaces. Water waves, boundaries between immiscible fluids, vortex sheets, Hele–Shaw-cells, thin-film growth, crystal growth and solidification are some of the better-known examples. These problems present a great challenge to physicists and applied mathematicians because the underlying fluid-dynamic instabilities, such as the Kelvin–Helmholtz and the Rayleigh–Taylor instabilities, produce very rich and complex solution structures. Here we would like to review some of the recent advances in developing efficient and stable numerical approximations for these interfacial flows, and investigate the competing mechanism between fluid-dynamic instabilities and the physical regularising effects such as surface tension. In many applications, surface tension has an important effect on the dynamics of interfaces. It is especially central to understanding such fluid phenomena as pattern formation in Hele–Shaw cells, crystal growth and unstable solidification, the motion of capillary waves on free surfaces, the formation of fluid droplets and noise generation at the ocean surface Prosperetti, Crum and Pumphrey (1989).

We will divide this paper into three parts. The first part is concerned with numerical methods and their stability analysis for locally well-posed interface problems. This includes water waves, multi-fluid interfaces and Hele–Shaw cells with surface tension. The second part is concerned with numerical methods for ill-posed interface problems. This includes vortex sheets and multi-fluid interfaces without surface tension. Typically these problems experience the Kelvin–Helmholtz and/or Rayleigh–Taylor instabilities. The third part is concerned with the level-set approach which uses front capturing techniques. Using this approach, singularity formation and topological changes in the free surfaces can be computed naturally.

1.1. Stable discretisations for locally well-posed interfaces

In this part of the paper, we are concerned with stable numerical methods for water waves and multi-fluid interfaces with surface tension. Accurate simulation of these free surfaces presents a problem of considerable difficulty because the underlying physical problem is very singular and is sensitive to small perturbations. The boundary-integral method has been one of the most common approaches in solving these interfacial problems; see e.g. Pozrikidis (1992). The earliest attempt at using boundary-integral methods can be traced back to Rosenhead (1932) in his study of vortex-sheet roll-up.

In the early Sixties, Birkhoff (1962) extended this method to more general fluid interface problems. The first successful boundary integral method was developed by Longuet-Higgins and Cokelet (1976) to compute plunging breakers. Boundary-integral methods for the exact, time-dependent equations have been developed and used in many other works, including Vinje and Brevig (1981), Baker, Meiron and Orszag (1982), Pullin (1979), Roberts (1983), New, McIver and Peregrine (1985), and Dold (1992). We refer to Schwartz and Fenton (1982) and Yeung (1982) for a review of early works in this area. For small-amplitude surface waves, efficient numerical methods have also been developed based on perturbations about equilibrium in Eulerian variables. An expansion in powers of the surface height is used to calculate Fourier modes. These works include Stiassnie and Shemer (1984), West, Brueckner and Janda (1987), Dommermuth and Yue (1987), Glozman, Agnon and Stiassnie (1993), and Craig and Sulem (1993). The last paper has the advantage that the expansion is uniform in wave number.

The advantage of using boundary-integral methods is that they reduce the two-dimensional problem into a one-dimensional problem involving quantities along the interface only, consequently avoiding the difficulty of differentiating discontinuous fluid quantities across the fluid interface. However, numerical simulations using boundary-integral methods also suffer from sensitivity to numerical instabilities because the underlying problems are very singular (Longuet-Higgins and Cokelet 1976; Roberts 1983; Dold 1992). Straightforward discretisations may lead to numerical instabilities. This includes some of the existing boundary-integral methods. There are two possible sources of numerical instability. First, a certain compatibility is required between the choice of quadrature rule for the singular velocity integral *and* the choice of spatial derivative. This compatibility ensures that a delicate balance of terms at the continuous level is preserved at the discrete level. This balance is crucial for maintaining numerical stability. Second, the periodicity of the numerical solution introduces aliasing errors which affect adversely the balance of terms at the discrete level. Violation of this delicate balance of terms will result in numerical instability.

The key in obtaining stable discretisations is to identify the most singular (or the leading-order) contributions of the method, and to see how various terms balance one another. To this end, we need to study the symbols of discrete singular operators, such as the discrete Hilbert transform and its variants. By studying the leading-order discrete singular operators, we find that a certain amount of Fourier filtering on the interface variables is required for the compatibility of the quadrature and derivative rules, Beale, Hou and Lowengrub (to appear) and Beale *et al.* (1994). The amount of filtering is determined by the quadrature rule in approximating the velocity integral and the derivative rule being used. With this modification, we can prove stability of the boundary-integral method for water waves and multi-fluid interfaces if surface tension effects are included. We also demonstrate that this modification is necessary for stability. Without this modification, the schemes using finite-order derivative operators are numerically unstable; see Subsection 3.2.

Linear analysis has contributed to understanding of numerical instabilities for boundary integral methods. Roberts (1983) showed how to remove a sawtooth instability. Baker and Nachbin (to appear) have performed Fourier analysis near equilibrium for various schemes for a vortex sheet with surface tension, identified sources of instability, and proposed new schemes which are free of linear instabilities. Dold (1992) emphasised the role of time discretisation with respect to instabilities.

While the spatial discretisations are proved to be stable and convergent, stability of the time discretisation is very difficult to obtain in the presence of surface tension. Surface tension introduces a large number of spatial derivatives through local curvature. If an explicit time integration method is used, these high-order derivative terms induce strong stability constraints on the time-step. For example, the time-step stability constraint for the Hele–Shaw flows is given by $\Delta t \leq Ch^3$, where Δt is the time step, and h is the minimum particle spacing. These stability constraints are time dependent, and become more severe by the differential clustering of points along the interface.

In (Hou, Lowengrub and Shelley, 1994) we have successfully removed this stiffness constraint by using an efficient implicit scheme based on a new reformulation of the problem. This reformulation introduces a dynamic change of variables from the (x, y) variables to the arclength metric and tangent angle variables. In this framework, the leading-order singular terms are shown to be linear and have constant coefficients (in space). Thus a Crank–Nicholson-type of discretisation can be used to eliminate the stiffness of the time discretisation. This reformulation greatly improves the stability constraint. For computations of the vortex sheet roll-up in an Euler flow with surface tension using a modest number of points (128), the time step can be chosen 250 times larger than that for an analogous explicit method. Many interfacial problems that were previously unobtainable are now solvable using our method, and new phenomena are discovered.

How to compute beyond the singularity time is a challenging task for the front tracking approach. Here we propose to use curvature regularisation in the boundary-integral formulation to continue beyond the singularity time. This borrows the idea from the level-set approach where curvature regularisation has been used successfully (Osher and Sethian, 1988). Curvature regularisation has an important property of preserving the index of a curve. Consequently, self-crossing of a curve is excluded under this regularisation. Moreover, the curvature regularisation is frame invariant. This is very different from putting an artificial viscosity in the Lagrangian variable. It turns out that the reformulated system of the interface problem can be used most naturally with the curvature regularisation. In motion by mean curvature, the equation for the tangent angle is nonlinear hyperbolic in the absence of curvature regularisation. It is well-known that a smooth initial condition may develop a shock discontinuity at later times. An entropy condition is required to select the unique physical weak solution. The curvature regularisation plays the same role as the viscosity regularisation for hyperbolic conservation laws. Consequently a physical continuation is obtained with curvature regularisation. In practice, we can use high-order Godunov-type methods developed for conservation laws to discretise the equation for the tangent angle (parameterised in arclength variable). This has a similar effect to curvature regularisation. Curvature regularisation can also be used to regularise the ill-posed vortex-sheet problem, providing an attractive alternative to compute vortex-sheet roll-up (Hou and Osher, 1994).

The idea of using curvature regularisation to boundary-integral formulations combines the advantages of both front tracking and front capturing. By applying curvature regularisation to a free surface directly, we do not need to introduce one extra space dimension as in the level-set approach. More accurate numerical methods can be designed since we only deal with the free surface and don't have to differentiate across the free surface. Also, the stiffness can be removed easily using our reformulated system.

1.2. Boundary integral methods for ill-posed interface problems

Methods of boundary-integral type have also been used for the ill-posed cases of fluid-interface motion, including vortex sheets and Rayleigh–Taylor instabilities (Moore, 1981; Anderson, 1985; Baker et al., 1982; Krasny, 1986a,b; Kerr, 1988; Tryggvason, 1988, 1989; Baker and Shelley, 1990; Shelley, 1992). Usually, either a regularisation or filtering of high wavenumbers is required to obtain numerical stability and to maintain an accurate solution. Surface tension and viscosity have been suggested and used as physical regularisations for these ill-posed problems. We refer to (Pullin, 1982; Rangel and Sirignano 1988; Tryggvason and Aref, 1983; Baker and Nachbin, to appear; Hou et al., 1994a; Dai and Shelley, 1993) for numerical study of surface tension regularisation, and Pozrikidis (1992), Tryggvason (1991) for study using viscosity regularisations.

Study of singularity formation in vortex sheets has been an active subject in the past decade. The possibility of a finite-time singularity in vortex sheets was first conjectured by Birkhoff (1962). The first analytical evidence of singularity formation was given by Moore (1979, 1985) in an asymptotic analysis. He predicted that to leading order in the initial amplitude ϵ, the curvature of the vortex sheet blows up at a critical time and the interface forms a branch-point singularity of order 3/2. Using Taylor series in time, Meiron, Baker and Orszag (1982) obtained results in agreement with Moore's. Krasny (1986) performed direct numerical simulations of vortex-sheet motion using the point-vortex approximation and a Fourier filter to control the growth of round-off errors. His results were also consistent with Moore's. Before the singularity time, the numerical solution converged, but convergence was lost after the singularity time. By using an infinite-order approximation combined with Krasny's Fourier filtering, Shelley (1992) has provided strong numerical evidence that the branch singularity of order 3/2 is chosen. Caflisch and Orellana (1989) have found a continuum of explicit solutions to the Birkhoff–Rott equation which display finite-time singularities. However, the physical interpretation of these constructed singularities is not clear since branch-type singularities were built explicitly in their initial data. The selection mechanism of a singularity for the general initial-value problems is not known. Caflisch, Ercolani, Hou and Landis (1993) studied propagation of singularities for the localised Moore's approximations. The 3/2 branch-point singularities were found to be generic. Singularity formation during the Rayleigh–Taylor instability has also been investigated by Baker, Caflisch and Siegel (to appear) using asymptotic and numerical methods.

To compute beyond the singularity time, certain numerical or physical regularisation is required. Moore (1978) has derived an evolution equation for a vortex layer of small thickness. Pullin (1992) has included surface tensions in the evolution equation. Pozrikidis and Higdom (1985) have numerically studied a periodically perturbed layer of constant vorticity. Krasny (Krasny, 1986b, 1987; Nitsche and Krasny, 1994) has used the vortex-blob method to study vortex-sheet roll-up and has obtained a number of interesting results. Baker and Shelley (1990) have considered regularisation of a thin vortex layer. Tryggvason (1989) has considered the vortex-in-cell as a grid-based vortex method, and has used an improved version of the VIC method to study vortex-sheet roll-up. Bell and Marcus (1992) used a second-order projection method for variable density flow to study Rayleigh–Taylor instability. These computational results using different regularisations all produced qualitatively similar results; at least they seem to agree outside the region of vorticity concentration.

The global existence of weak solutions for vortex-sheet initial data is not known in general. Motivated by the numerical studies of vortex sheets in Krasny (1986b, 1987) and Baker and Shelley (1990), DiPerna and Majda (1987a,b), introduced the concept of measure-valued solutions for vortex sheets. These measure-valued solutions may develop regions of vorticity concentration and may have a non-trivial set of defeat measure. If this is the case, the vortex-sheet solution does not satisfy the incompressible, inviscid Euler equations in the weak sense. In the special case of one-signed vorticity, Delort (1991) has recently proved that vortex sheets are global weak solutions of the Euler equations. Using Delort's result, Liu and Xin (1994) have been able to prove that the vortex-blob calculation converges to a weak solution of the Euler equations provided that the initial vorticity is of the same sign. Other theoretical results for vortex-sheet motion assume analytic initial data. Existence and well-posedness of vortex-sheet motion have been established for analytic data for short times; see, for example, Sulem *et al.* (1981), Dochun and Robert (1986, 1988), Caflisch and Orellana (1988) and Ebin (1988).

1.3. Level-set approach

Although it is usually highly desirable to reformulate a problem into boundary-integral equations, there are certain applications for which the boundary-integral method has difficulty handling. For example, in crystal growth and thin-film growth, an initially smooth front can develop cusps and cracklike singularities, and isolated islands of film material can merge (Gray, Chisholm and Kaplan, 1993; Sethian and Strain, 1992; Snyder *et al.*, 1991; Spencer, Voorhees and Davis, 1991). In order to compute up to and continue beyond the singularity time, one has to use local-mesh refinement and local surgery techniques (Unverdi and Tryggvason, 1992). This often introduces some numerical instability and it is done in a somewhat unsatisfactory way. And it becomes increasingly difficult for three-dimensional problems.

The level-set approach developed by Osher and Sethian (1988) provides a powerful numerical method for capturing free surfaces in which topological singularities may form dynamically. The idea is to regard the free surface as a level set of a smooth function defined in one order higher space dimensions than the free surface. Only the information of the zeroth-level set is physically relevant to the free surface we want to compute. Thus we have sufficient freedom in specifying the level-set function away from the zeroth-level set. This freedom makes it possible to select a relatively smooth level-set function at all times. The free surface may form a singularity such as corners or cusps, but the level-set function still remains relatively smooth. Typically, we would like to choose the level-set function to be a signed distance function from the free surface. By viewing the surface as a level set, sharp corners and cusps are handled naturally, and changes of topology in the moving boundary require no additional effort. Furthermore, these methods work in any number of space dimensions.

Another important property of the level-set formulation is that it provides the correct equation of motion for a front propagating with curvature-dependent speed. This equation is of Hamilton–Jacobi type with a right-hand-side that depends on curvature effects. The limit of the right-hand-side as the curvature effect goes to zero satisfies an associated entropy condition. Thus high-order numerical approximations can be devised using techniques developed for the solution of hyperbolic conservation laws (Harten et al., 1987; Colella and Woodward, 1984; Osher and Shu, 1991). Recently, this approach has been applied to computation of minimal surfaces (Chopp, 1993), compressible gas dynamics (Mulder, Osher and Sethian, 1992), crystal growth and dendritic solidification (Sethian and Strain 1992; Osher, private communication), and interaction of incompressible fluid bubbles (Chang et al., 1994; Sussman, Smereka and Osher, 1994). In addition, theoretical analysis of mean curvature flow based on the level-set model presented in Osher and Sethian (1988) has been developed by Evans and Spruck (1991, 1992).

The rest of the paper is organised as follows. In Section 2, we introduce several important examples of free boundary-value problems arising from fluid dynamics and materials science. These include water waves, general two-fluid interfaces, Hele–Shaw cells, a model for crystal growth and unstable solidification. Their boundary-integral formulations will be given. In Section 3, we present a convergent boundary-integral method for water waves. A compatibility condition between the quadrature rule and the discrete derivative is given, and the stability property of the modified boundary-integral method is analysed. Examples of a class of unstable algorithms are given to illustrate why certain Fourier smoothings are necessary for our modified method to be stable. A numerical example of breaking-wave calculation supports the applicability of the method in the fully non-linear regime. In Section 4, we present several numerical methods for calculating vortex sheets and vortex-sheet roll-up. These include Krasny's filtering technique for the point-vortex method, vortex-blob desingularisation for vortex sheets, thin vortex-layer desingularisation and vortex-in-cell method calculations for vortex-sheet roll-up. In Section 5, the stabilising effect of surface tension is considered. We first consider a stable time-continuous discretisation. We then propose an efficient implicit time discretisation that completely removes the stiffness of surface tension by a dynamical reformulation of the interface problems. We also propose a new approach to compute beyond singularity time using our reformulated system together with curvature regularisation. Finally, in Section 6, we consider the level-set approach for computing topological singularities. The basic ideas of level-set approaches are reviewed. Applications to crystal growth and incompressible multi-fluid bubbles are discussed.

2. General two-fluid interfaces

In this section, we consider several examples of interfacial flows arising from fluid mechanics and materials science. They are water waves, stratified two-density interfacial flows, Hele–Shaw flows, crystal growth and solidification.

The stratified interfacial flows have been used as models to understand mixing of fluids, separation of boundary layers, generation of sounds (in bubbly flows) and coherent structures in turbulence models. Theoretical and numerical studies of Hele–Shaw flows and crystal growth have received renewed interest and increasing attention in recent years because of the rich phenomena in the physical solutions and the potential applications in pattern formation and materials science. These interfacial problems have one feature in common. The underlying physical instability generates a rapid growth in the high-frequency components of the solution. Without physical regularisations such as viscosity or surface tension, the problems are ill-posed in the Hadamard sense (except for water waves). It is the competition between the stabilising regularisation effect and the underlying physical instability that generates many fascinating solution structures. Since these problems are highly non-linear and non-local, it is usually difficult to obtain a complete understanding by using only analytical tools. Numerical simulations become essential in our study of these interfacial problems. It is not hard to imagine that this is a very difficult task.

2.1. Water waves without surface tension

Unsteady motion of water waves is one of the most familiar examples of free surfaces in our everyday experience, and it illustrates a rich variety of phenomena in wave motion. One of the spectacular properties of the sea surface is its capacity to turn over on itself and produce breaking waves. Mathematical difficulties in dealing with the exact equations are due to the free boundary, and the inherent non-linear, non-local nature of the problem. The usual linear theory and shallow-water theory for small-amplitude waves have been very useful in studying many important aspects of the wave motion. However, they are valid only where the fluid acceleration is sufficiently small compared to gravity. To obtain a better understanding of the large-amplitude wave interactions such as wave breaking, we need to develop effective numerical methods to compute free surface motion.

There are several different approaches that may be adopted to study free surface motion numerically. We refer to Yeung's (1982) paper for a partial review. The boundary-integral formulation of the equations of water waves leads to a natural approach for computing time-dependent motions. In this approach, the moving interface is tracked explicitly. Only quantities on the interface need be computed. However, high-frequency numerical instabilities are difficult to avoid, because of the non-local and non-linear nature of the problem, and the lack of dissipation.

Consider a two-dimensional incompressible, inviscid and irrotational fluid below a free interface. We parameterise the interface by $\mathbf{x} = (x(\alpha, t), y(\alpha, t))$, where α is a Lagrangian parameter along the interface. The kinematic con-

dition only requires that the normal velocity of the interface be equal to
that of the fluid at the interface. There is no physical constraint on the
tangential velocity of the interface. Tangential motions along the interface
give only changes in frame for its parameterisation and are not physically
specified as they do not affect its shape. Later, in Subsection 5.2, we will
exploit this freedom in choosing a tangential motion to remove the stiffness
of surface tension. Here, we use the usual convention of choosing the tangential velocity to be that of the fluid. Thus the interface is convected by
the fluid velocity $\mathbf{u}$ at the interface:

$$\frac{\partial \mathbf{x}}{\partial t} = \mathbf{u}(\mathbf{x}(\alpha, t), t). \tag{2.1}$$

The fluid velocity $\mathbf{u}$ is determined by the incompressible Euler equations:

$$\rho \left(\frac{\partial \mathbf{u}}{\partial t} + \mathbf{u} \cdot \nabla \mathbf{u} \right) = -\nabla p - \rho g \mathbf{j} \tag{2.2}$$

with the incompressibility constraint $\nabla \cdot \mathbf{u} = 0$, where ρ and p are the fluid
density and pressure, respectively, g is the constant of gravity and $\mathbf{j}$ is a unit
vector in the y direction. In the absence of surface tension, the pressure
is continuous across the interface. Since a vacuum is assumed above the
interface, the pressure is equal to zero at the interface. To simplify the
presentation, we only consider water waves with infinite depth in two space
dimensions. One can easily modify the formulation to accommodate the
bottom geometry if water waves with finite depth are considered, see, for
example, Baker *et al.* (1982). Generalisation to three dimensions can also
be carried out, see, for example Baker (1983) and Kaneda (1990).

Due to irrotationality, we can express velocity in terms of a velocity potential ϕ, that is, $\mathbf{u} = \nabla \phi$. Then incompressibility implies that

$$\Delta \phi = 0$$

in the interior flow region. Furthermore, the momentum equations (2.2) can
be integrated to obtain Bernoulli's equation for the potential:

$$\frac{\partial \phi}{\partial t} + \frac{1}{2} |\nabla \phi|^2 + gy = 0. \tag{2.3}$$

Thus if $\phi(\alpha, 0) = \phi_0(\alpha)$ is given initially along the interface, then we can
evaluate ϕ at later times according to (2.3). To compute the fluid velocity at
the interface, we need to evaluate $\nabla \phi$ at the interface. It is easy to evaluate
the tangential velocity component at the interface. But determining the
normal velocity would require solving the interior problem for ϕ. This could
be a difficult task since the relation between the Dirichlet value of ϕ and its
normal derivative at the interface is non-local. There are several ways to
relate the normal derivative of ϕ to its value at the interface, each involving
a Fredholm integral equation of a different kind.

One way to relate the normal derivative of ϕ to its value at the interface is to use Green's third identity. This is the approach taken by Longuet-Higgins and Cokelet (1976), among others. This corresponds to using a single-layer potential representation. Denote by Γ the interface. Green's third identity gives

$$\int_\Gamma \frac{\partial \phi}{\partial n}(x(\alpha'))G(x(\alpha), x(\alpha'))dx(\alpha')$$

$$= \frac{1}{2}\phi(x(\alpha)) + \int_\Gamma \phi(x(\alpha'))\frac{\partial G}{\partial n}(x(\alpha), x(\alpha'))dx(\alpha')), \quad (2.4)$$

where $\frac{\partial}{\partial n}$ is the exterior normal derivative and G is the Green's function for the Laplace equation. In two dimensions, we have $G(x, x') = \frac{1}{2\pi}\log|x - x'|$. If we prescribe ϕ at the interface, then (2.4) provides an integral relation to determine the normal derivative of ϕ at the interface. Equation (2.4) is a Fredholm integral equation of the first kind for the normal derivative of ϕ. Now the solution procedure is clear. Given $\mathbf{x}(\alpha, t)$ and $\phi(\alpha, t)$ at the interface, we can compute the normal derivative of ϕ by equation (2.4). This determines the normal velocity at the interface. The tangential velocity is given by $\phi_\alpha/|\mathbf{x}_\alpha|$. Then we can update $\mathbf{x}$ in time by (2.1), and update ϕ by Bernoulli's equation.

There are some disadvantages regarding this approach. To solve for $\partial\phi/\partial n$ from equation (2.4), we need to invert a dense N-by-N matrix if we discretise the integral by N grid points. Direct inversion of a $N \times N$ matrix requires $O(N^2)$ storage locations and $O(N^3)$ operation counts. For large N, this becomes prohibitively expensive. We must look for some fast iterative method to approximately invert the dense matrix. However, the matrix associated with a Fredholm integral of the first kind is usually not well conditioned and the number of iterations required increases rapidly with N.

An alternative approach is to use the dipole representation. Following Baker, Meiron and Orszag (1982) and Beale, Hou and Lowengrub (1993a), we express the complex potential by a double-layer representation. Denote by $\mu(\alpha, t)$ the dipole strength and denote the interface position by complex variable $z(\alpha, t) = x(\alpha, t) + iy(\alpha, t)$. We can write the complex potential Φ in the fluid domain in terms of μ

$$\Phi = \frac{1}{2\pi i}\int \frac{1}{z - z(\alpha', t)}\mu(\alpha', t)dz(\alpha'),$$

for z away from the interface. The complex velocity $w = u - iv$ can be obtained by differentiating the complex potential with respect to z and performing integration by parts. We get

$$w = \frac{d\Phi}{dz} = \frac{1}{2\pi i}\int \frac{1}{z - z(\alpha')}\gamma(\alpha')d\alpha',$$

where γ is the non-normalised vortex-sheet strength. It is given as the Lagrangian derivative of the dipole strength, that is, $\gamma = \mu_\alpha$. Using the Plemelj formula, we obtain the limiting velocity from the fluid region on the interface as

$$w(\alpha) = \frac{1}{2\pi i} \int \frac{\gamma(\alpha')}{z(\alpha) - z(\alpha')} d\alpha' + \frac{\gamma(\alpha)}{2z_\alpha(\alpha)}. \tag{2.5}$$

Here the integral is the Cauchy principal-value integral. Denote by ϕ the real potential, $\phi = \mathrm{Re}(\Phi)$. We can determine γ from the condition $\phi_\alpha = \phi_x x_\alpha + \phi_y y_\alpha = \mathrm{Re}(wz_\alpha)$. Using (2.5) we obtain

$$\phi_\alpha = \frac{\gamma}{2} + \mathrm{Re}\left(\frac{z_\alpha(\alpha)}{2\pi i} \int \frac{\gamma(\alpha')}{z(\alpha) - z(\alpha')} d\alpha' \right).$$

This is a Fredholm integral equation of the second kind. The kernel is an adjoint double-layer potential.

It is customary to assume that the interface are periodic in the horizontal direction. Under this assumption, we can express $z(\alpha, t) = \alpha + s(\alpha, t)$, where $s(\alpha, t)$ and $\phi(\alpha, t)$ are periodic in α with period 2π. This also implies that the flow is at rest at infinity. We can sum the singular kernel $1/z$ over periodic intervals to obtain a periodic kernel $\frac{1}{2}\cot(z/2)$ defined over a single period. To summarise, we obtain a system of time-evolution equations for z and ϕ as follows:

$$\bar{z}_t = \frac{1}{4\pi i} \int_{-\pi}^{\pi} \gamma(\alpha')\cot\left(\frac{z(\alpha) - z(\alpha')}{2} \right) d\alpha' + \frac{\gamma(\alpha)}{2z_\alpha(\alpha)}$$

$$\equiv u(\alpha, t) - iv(\alpha, t), \tag{2.6}$$

$$\phi_t = \frac{1}{2}(u^2 + v^2) - g \cdot y, \tag{2.7}$$

$$\phi_\alpha = \frac{\gamma}{2} + \mathrm{Re}\left(\frac{z_\alpha}{4\pi i} \int_{-\pi}^{\pi} \gamma(\alpha')\cot\left(\frac{z(\alpha) - z(\alpha')}{2} \right) d\alpha' \right), \tag{2.8}$$

where $\bar{z}$ is the complex conjugate of z. Equations (2.6)–(2.8) completely determine the motion of the system. We remark that Bernoulli's equation (2.7) is different from (2.3) because of the change to Lagrangian variables.

The advantage of using the dipole representation is that the Fredholm integral of the second kind has a globally convergent Neumann series (Baker, Meiron and Orszag, 1982; and Beale, Hou and Lowengrub, 1993a). This means that the equation for γ may be solved by iteration provided that the interface is reasonably smooth. For example, if γ^j is the jth iterate, then γ^{j+1} is computed by

$$\gamma^{j+1}(x) = 2\phi_\alpha(x) - 2\mathrm{Re}\left(\frac{z_\alpha}{4\pi i} \int_{-\pi}^{\pi} \gamma^j(\alpha')\cot\left(\frac{z(\alpha) - z(\alpha')}{2} \right) d\alpha' \right). \tag{2.9}$$

In general, only several iterations are required for high accuracy if we use the γ from the previous time level as our first iterate. By keeping values of γ at several previous time levels, an extrapolated first iterate can be found which will further reduce the number of iterations to typically 1 or 2. More iterations are required when the solution develops a high-curvature region such as in breaking surface waves. If the interface is not accurately resolved in the high-curvature region, the iteration scheme will eventually stop converging.

2.2. Stratified two-density fluid interfaces

Here we consider the motion of general two-density interfacial flows, that is, an interface separating two inviscid, incompressible and irrotational fluids in the presence of gravity and possibly surface tension in two space dimensions. In the following, the subscript 2 denotes the fluid above the interface and 1 denotes the fluid below the interface. In each fluid, we have Euler's equations

$$\rho_i[\partial_i \mathbf{u}_i + (\mathbf{u}_i \cdot \nabla)\mathbf{u}_i] = -\nabla \mathbf{p}_i - \rho_i g\mathbf{j}, \quad i = 1, 2.$$

The incompressibility and irrotationality constraints imply $\nabla \cdot \mathbf{u}_i = 0$ and $\nabla \times \mathbf{u}_i = 0$. Denote the interface by Γ. At the interface, we impose the Laplace–Young boundary condition which relates the pressure jump to the curvature of the interface, κ, by

$$[p]|_\Gamma = \tau\kappa,$$

where $[p]|_\Gamma$ denotes the jump of pressure across the interface Γ and τ is the surface-tension coefficient. The normal velocity is assumed to be continuous across the interface. As in the case of water waves, these interface problems have vortex-sheet representations. We refer to Birkhoff (1962) and Baker, Meiron and Orszag (1982) for a derivation of the governing equations.

The velocity at the interface is not uniquely defined since the tangential velocity in general has a jump discontinuity across the interface. It is customary to evolve the interface with the average velocity obtained from the limiting velocities above and below the interface. As before, we denote the interface position by complex variable $z(\alpha, t) = x(\alpha, t) + iy(\alpha, t)$, where α is a Lagrangian parameter along the interface. Then, the interface evolves according to

$$\frac{d\bar{z}}{dt} = \frac{1}{2\pi i} \int \frac{\gamma(\alpha', t)}{z(\alpha, t) - z(\alpha', t)} d\alpha', \qquad (2.10)$$

where the integral is understood as the Cauchy principal-value integral and γ is the unnormalised vortex-sheet strength. Equation (2.10) is also called the Birkhoff–Rott equation in the literature. The evolution equation for γ

can be obtained from Bernoulli's equations in both sides of the interface:

$$\frac{d\gamma}{dt} = -2A\left(\mathrm{Re}\left\{\frac{d^2\bar{z}}{dt^2}z_\alpha\right\} + \frac{1}{8}\partial_\alpha\left(\frac{\gamma^2}{|z_\alpha|^2}\right) + gy_\alpha\right) + \tau\kappa_\alpha, \qquad (2.11)$$

where $A = (\rho_1 - \rho_2)/(\rho_1 + \rho_2)$ is the Atwood number. The curvature, κ, is evaluated by

$$\kappa = \frac{x_\alpha y_{\alpha\alpha} - x_{\alpha\alpha} y_\alpha}{(x_\alpha^2 + y_\alpha^2)^{3/2}}.$$

Note that equation (2.11) is actually a Fredholm integral equation of the second kind for $d\gamma/dt$. It has been shown that their Neumann series are globally convergent in the periodic case (Baker, Meiron and Orszag, 1982). The result also holds for unbounded domains (Beale, Hou and Lowengrub, 1993a). Therefore, the integral equation for $d\gamma/dt$ is invertible and can be solved by iteration. Equations (2.10) and (2.11) completely determine the motion of the interface.

2.3. Hele–Shaw flows

A closely related problem is the Hele–Shaw flow which describes the viscosity-dominant creeping flow confined between two closely spaced plates. The case in which one fluid displaces another has been studied extensively. This begins with the theoretical work of Saffman and Taylor (1958). They found exact self-similar fingers for the interface between the two fluids in a channel geometry when surface tension is absent. The subsequent works have mostly focused on the role of surface tension in the selection of finger width (see Pelcé, 1988, for a review). The dynamical behaviour of Hele–Shaw flows has received a lot of interests inspired by the complex patterns formed by an expanding bubble (Paterson, 1981, 1985; Rauseo, Barnes and Maher, 1987). It is believed that surface tension plays an essential role in producing these structures.

Consider an interface Γ that separates two Hele–Shaw fluids of different viscosities and densities. For simplicity, Γ is assumed periodic in the horizontal direction. The fluid below Γ is labelled fluid 1, and that above is labelled 2, and similarly for their respective viscosities and so forth. The velocity in each fluid is given by Darcy's law, together with the incompressibility constraint:

$$(u_j, v_j) = -\frac{b^2}{12\mu_j}\nabla(p_j - \rho_j gy), \qquad \nabla \cdot \mathbf{u}_j = 0.$$

Here b is the gap width of the Hele–Shaw cell, μ_j is the viscosity, p_j is the pressure, ρ_j is the density and gy is the gravitational potential. The boundary conditions are exactly the same as for the two-density fluid interfaces. That is, the normal velocity is continuous across the interface, the jump in the pressure across the interface is proportional to the local mean curvature, and velocity vanishes at infinity. Again, one can derive a boundary-integral formulation for the Hele–Shaw problem using the dipole representation; see, for example, Dai and Shelley (1993). The interface position will satisfy the same Birkhoff–Rott equation as in (2.10). The vortex-sheet strength γ satisfies

$$\gamma(\alpha, t) = -2A_\mu \text{Re} \left\{ \frac{z_\alpha}{2\pi i} \int \frac{\gamma(\alpha', t)}{z(\alpha, t) - z(\alpha', t)} d\alpha' \right\} + \tau \kappa_\alpha - R y_\alpha.$$

Here $A_\mu = (\mu_1 - \mu_2)/(\mu_1 + \mu_2)$ is the Atwood ratio of the viscosities, τ is the non-dimensional surface tension and R is a signed measure of density stratification ($\rho_1 < \rho_2$ implies $R < 0$). As before, the equation for γ is a Fredholm integral of the second kind, and it has a globally convergent Neumann series.

2.4. Crystal growth and dendritic solidification

The last example we consider is concerned with crystal growth in unstable solidification. This problem has attracted considerable interest over the past decade from applied mathematicians, physicists and materials scientists. Here we only consider one particular model for this problem. There are many other interesting free-boundary problems arising from materials science that can be studied by numerical methods similar to those described in this paper. This includes the morphological instability in thin-film growth and the Ostwald ripening problem, see, for example, Gray, Chisholm and Kaplan (1993), Spencer, Voorhees and Davis (1991), Spencer and Meiron (1994), Voorhees *et al.* (1988) and Voorhees (1992).

Consider a container of the liquid phase of the material. Suppose we cool the box smoothly and uniformly below its freezing temperature. If this is done very carefully, the liquid does not freeze. The system is now in a 'metastable' state. A small disturbance, such as dropping a tiny seed of solid phase, will initiate a rapid and unstable process known as *dendritic solidification*. The solid phase will grow from the seed by sending out branching fingers. This growth process is unstable in the sense that small perturbations of the initial data can produce large changes in the solid/liquid boundary. One can model this phenomenon by a moving-boundary problem. The temperature field satisfies a heat equation in each phase, coupled through two boundary conditions on the unknown moving solid/liquid boundary. We refer to Langer (1980), Gurtin (1986) and Caginalp and Fife (1988) for derivations. Extensive asymptotic analysis for the solution has been carried out by several authors in the literature; see Langer (1986), Chadam and Ortoleva (1983), Kessler and Levine (1986), Benamar and Pomeau (1986), and Gaginalp and Fife (1988). It is also possible to reformulate the equations of motion in boundary-integral forms, as is done in Meiron (1986), Strain

(1989), Karma (1986), Kessler and Levine (1986) and Langer (1980). This yields accurate results for smooth boundaries.

The set-up of the problem described below follows the framework of Strain, Sethian and Strain (1992, 1989). Consider a square container, $B = [0,1] \times [0,1]$, filled with the liquid and solid phases of some pure substance. The unknowns are the temperature $u(x,t)$ for $x \in B$, and the solid/liquid boundary $\Gamma(t)$. The temperature field u is taken to satisfy the heat equation in each phase, together with an initial condition in B and boundary conditions on the container walls. Thus we have

$$
\begin{aligned}
u_t &= \Delta u & &\text{in } B \text{ off } \Gamma(t), \\
u(x,t) &= u_0(x) & &\text{in } B \text{ at } t = 0, \\
u(x,t) &= u_B(x) & &\text{for } x \in \partial B.
\end{aligned}
$$

Since the position of the moving boundary $\Gamma(t)$ is unknown, two boundary conditions on $\Gamma(t)$ are required to determine u and $\Gamma(t)$. Let n be the outward normal to the boundary, pointing from solid to liquid. The first boundary condition is the classical Stefan condition:

$$[\partial u/\partial n] = -HV \quad \text{on} \quad \Gamma(t).$$

Here $[\partial u/\partial n]$ is the jump in the normal component of heat flux $\partial u/\partial n$ from solid to liquid across $\Gamma(t)$, V is the normal velocity of $\Gamma(t)$ and H is a constant. The second boundary condition on $\Gamma(t)$ is the classical Gibbs–Thomson relation,

$$u(x,t) = -\epsilon_\kappa(n)\kappa - \epsilon_V(n)V \quad \text{for} \quad x \in \Gamma(t),$$

where κ is the curvature at x on $\Gamma(t)$. Here we model the crystalline anisotropy by assuming that ϵ_κ and ϵ_V depend on the local normal vector n. Now we describe how to put the problem into a boundary-integral form. First we subtract the temperature field due to the initial condition u_0 and the boundary condition u_B. Let $U(x,t)$ be the solution to the heat equation

$$
\begin{aligned}
U_t &= \Delta U & &\text{in } B, \\
U(x,0) &= u_0(x) & &\text{at } t = 0, \\
U(x,t) &= U_B(x,t) & &\text{for } x \in \partial B \text{ and } t > 0.
\end{aligned}
$$

Define $W = u - U$. Then W satisfies

$$
\begin{aligned}
W_t &= \Delta W & &\text{in } B - \Gamma(t), & &(2.12) \\
W(x,0) &= 0 & &\text{at } t = 0, & &(2.13) \\
W(x,t) &= 0 & &\text{for } x \in \partial B, & &(2.14)
\end{aligned}
$$

$$[\partial W/\partial n] = -HV \quad \text{on} \quad \Gamma(t), \qquad (2.15)$$

$$W(x,t) = -\epsilon_\kappa(n)\kappa - \epsilon_V(n)V - U(x,t) \quad \text{for} \quad x \in \Gamma(t). \tag{2.16}$$

The temperature filed U can be obtained easily since there is no free boundary present. For example, we can use any standard numerical discretisation for the initial–boundary-value problem for the heat equation to obtain U. The difficult part is to compute W. We will formulate it as a boundary-integral equation. We use the kernel K of the heat equation to express the solution W to equations (2.12)–(2.16) as a single-layer potential. Given a function V on

$$\Gamma_T = \prod_0^T \Gamma(t) = \{(x,t)|x \in \Gamma(t), 0 \le t \le T\},$$

the single-layer heat potential SV is defined for (x,t) in $B \times [0,T]$ by

$$SV(x,t) = \int_0^t \int_{\Gamma(t')} K(x,x',t-t')V(x',t')dx'dt'.$$

Here the x' integration is over the curves comprising $\Gamma(t')$, and the Green function K of the heat equation in the box $B = [0,1] \times [0,1]$ with Dirichlet boundary conditions on the box walls is given by

$$K(x,x',t') = \sum_{k_1=1}^{\infty} \sum_{k_2=1}^{\infty} e^{-(k_1^2+k_2^2)\pi^2 t}$$
$$\times \sin(k_1\pi x_1)\sin(k_2\pi x_2)\sin(k_1\pi x_1')\sin(k_2\pi x_2'),$$

where $x = (x_1, x_2)$ and $x' = (x_1', x_2')$. The function SV defined above is a continuous function on $B \times [0,T]$, vanishing for $t = 0$ or on ∂B, and satisfying the heat equation every where off Γ_T. Across $\Gamma(t)$, $SV(x,t)$ has a jump in its normal derivative equal to V. Thus, $W(x,t) = H \cdot SV(x,t)$ is the solution to equations (2.12)–(2.16). All that remains is to satisfy the second boundary condition (2.16). This is equivalent to the boundary-integral equation

$$\epsilon_\kappa(n)\kappa + \epsilon_V V + U + H \int_0^t \int_{\Gamma(t')} K(x,x',t-t')V(x',t')dx'dt' = 0, \tag{2.17}$$

for $x \in \Gamma(t)$. Equation (2.17) is an integral equation for the normal velocity, V, of the moving boundary. We note that the velocity V of a point x on $\Gamma(t)$ depends not only on the position of $\Gamma(t)$ but also on its previous history. Thus in order to evaluate $V(x,t)$, we need to store information about the temperature in the previous history of the boundary.

The boundary-integral formulation of the problem requires the evaluation of the single-layer potential on a $M \times M$ grid in B. The computation of $SV(x,t)$ by a quadrature rule at M^2 points at N time steps would require $O(M^3N^2)$ work if there are $O(M)$ points on $\Gamma(t)$ at each time t. For M and N large, this becomes prohibitively expensive (Strain, 1989). A fast

algorithm has been developed by Greengard and Strain (1990) which reduces
the operation count to $O(M^2)$ per time step. This greatly improves the speed
of the calculation. A related, but different, fast method for solidification
has also been proposed and implemented by Spencer and Meiron (1994).
Recently, Osher and his coworker (private communication) have designed an
extremely efficient numerical method based on level-set formulation which
only involves a local heat equation solver without using boundary integral
formulation. The preliminary results seem to be very promising.

2.5. Linear stability around equilibrium solutions

It is instructive to consider the linear stability of equilibrium solutions to
(2.10) and (2.11). The equilibrium solution is the flat sheet: $z(\alpha, t) =
\alpha, \gamma(\alpha, t) = \gamma_0$, where γ_0 is constant. Looking for solutions of (2.10) and
(2.11) of the form $z = \alpha + \epsilon \dot{z}$ and $\gamma = \gamma_0 + \epsilon \dot{\gamma}$, and keeping only the linear
terms in ϵ give the linearised equations

$$
\frac{d\dot{x}}{dt} = -\frac{\gamma_0}{2} H(\dot{y}_\alpha),
$$

$$
\frac{d\dot{y}}{dt} = \frac{1}{2} H(\dot{\gamma}) - \frac{\gamma_0}{2} H(\dot{x}_\alpha),
$$

$$
\frac{d\dot{\gamma}}{dt} = -A\gamma_0 \dot{\gamma}_\alpha + A\gamma_0^2 \dot{x}_{\alpha\alpha} - 2Ag\dot{y}_\alpha + \tau \dot{y}_{\alpha\alpha\alpha},
$$

where $\dot{z} = \dot{x} + i\dot{y}$. Notice that $d\gamma/dt$ is determined explicitly in the third
equation. In the linear level, the integral equation contribution has dropped
out. H is the Hilbert transform defined as

$$
H(f)(\alpha) = \frac{1}{\pi} \int \frac{f(\alpha')d\alpha'}{\alpha - \alpha'}.
$$

It is easy to see that it has the Fourier symbol $\hat{H}(k) = -\mathrm{isgn}(k)$. The
growth rates of the perturbations are determined by the eigenvalues of the
perturbed system which can be calculated explicitly in the Fourier space as
follows:

$$
\lambda(k) = 0, \quad -\frac{A\gamma_0}{2} ik \pm \sqrt{\frac{\gamma_0^2 k^2}{4}(1 - A^2) - Ag|k| - \tau|k|^3}.
$$

The fact that the system admits zero eigenvalue implies that the interface
does not change its shape by translating Lagrangian points along the interface. In the absence of surface tension, that is, $\tau = 0$, the other two
eigenvalues may grow with large k. In fact, if $\gamma_0 \neq 0$, the interface experiences the Kelvin–Helmoltz instability, and the growth rate is proportional
to $O(|k|)$ for large k. On the other hand, if $\gamma_0 = 0$, the stability of the per-
turbation depends on the sign of the Atwood number A. If $A > 0$, then the

interface is stable. If $A < 0$, then the interface experiences the Rayleigh–Taylor instability. The growth rate is proportional to $\sqrt{|k|}$. In general, due to baroclinic generation of vorticity at the interface, the vortex-sheet strength γ can be non-zero in some region even if it is equal to zero initially. Thus the Kelvin–Helmholtz instability is always present for two-density interfaces as long as $A^2 \neq 1$, even if we have lighter fluid on top of a heavy fluid. In the case of positive surface tension, that is, $\tau > 0$, we can see that in the high-wavenumber regime, the surface-tension term dominates. The eigenvalues become imaginary, which can produce oscillations but no growth at the high modes. Thus surface tension is a dispersive regularisation of those instabilities. Of course, for small surface tension, there is still a band of modes below a certain critical wavenumber that can grow exponentially in time.

Similar linearised stability analysis for the Hele–Shaw flows indicates that the surface tension is a dissipative regularisation. For simplicity, we take $A_\mu = 0$. The eigenvalues are as follows:

$$\lambda(k) = 0, \quad -\frac{1}{2}(\tau|k|^3 + R|k|).$$

Therefore, if $R < 0$, there is a band of unstable modes near $k = 0$. This is a Mullins–Sekerka type of instability (1963), driven by the unstable density stratification. At higher wavenumbers, this instability is cut off by the surface-tension term which acts as a third-order dissipation.

3. A convergent boundary-integral method for water waves

The boundary-integral formulation of water waves is naturally suited for numerical computation. There are many ways one can discretise the boundary-integral equations, depending on how we choose to discretise the singular integral and the derivatives. These choices affect critically the accuracy and stability of the numerical method. Straightforward numerical discretisations of (2.6)–(2.8) may lead to rapid growth in the high wavenumbers. In order to avoid numerical instability, a certain compatibility between the choice of quadrature rule for the singular integral *and* that of the discrete derivatives must be satisfied. This compatibility ensures that a delicate balance of terms at the continuous level is preserved at the discrete level. Violation of this compatibility will lead to numerical instability.

We discretise the interval by choosing N equally spaced points $\alpha_j = jh$, where $h = 2\pi/N$. Denote by $z_j(t), \phi_j(t), \gamma_j(t)$ the discrete approximations of $z(\alpha_j, t), \phi(\alpha_j, t), \gamma(\alpha_j, t)$ respectively. To approximate the velocity integral, we use the alternating trapezoidal rule:

$$\int_{-\pi}^{\pi} \gamma(\alpha')\cot\left(\frac{z(\alpha_j) - z(\alpha')}{2}\right) d\alpha' \simeq \sum_{\substack{k=-N/2+1 \\ (k-j)\,\mathrm{odd}}}^{N/2} \gamma_k \cdot \cot\left(\frac{z_j - z_k}{2}\right) 2h, \quad (3.1)$$

The advantage of using this alternating trapezoidal quadrature is that the approximation is spectrally accurate. This and related quadrature rules have been used by several authors in the literature. Baker (1983) used the alternating quadrature rule for a desingularised integrand in water-wave calculations. It gives a quadrature similar to (3.1), but with a different (and desingularised) integrand. Sidi and Israeli (1988) analysed the spectral accuracy of a midpoint-rule approximation for a periodic singular integrand. They realised that the alternating quadrature rule applied to singular, periodic Cauchy kernels such as the integral in (3.1) gives spectral accuracy. Shelley (1992) used scheme (3.1) with Krasny's filtering in the context of studying the vortex-sheet singularity by the point-vortex method. By using the spectral accuracy of the alternating-trapezoidal rule, Hou, Lowengrub, and Krasny (1991) gave a simplified proof of convergence of the point vortex method for vortex sheets (Hou, Lowengrub and Krasny, 1991).

It seems natural to use the alternating-quadrature rule and a finite-order derivative operator (e.g. cubic spline) for the α-derivative. However, as will be seen later, the resulting scheme is numerically unstable at equilibrium; see Baker and Nachbin (to appear), Beale, Hou and Lowengrub (1993b) and Beale et $al.$ (to appear). To see how standard schemes can be modified so that they become numerically stable, we use the discrete Fourier transform. For a discrete function $\{f_j\}$ on the periodic interval, the discrete transform and its inverse are (assuming N is even)

$$\hat{f}_k = \frac{1}{N} \sum_{j=-N/2+1}^{N/2} f_j e^{-ik\alpha_j}, \qquad f_j = \sum_{k=-N/2+1}^{N/2} \hat{f}_k e^{-ik\alpha_j}.$$

We will write a discrete derivative operator in the form

$$D_h^{(\rho)} f_j = \sum_{k=-N/2+1}^{N/2} \rho(kh) ik \hat{f}_k e^{ik\alpha_j}, \quad k = -N/2+1, ..., N/2, \qquad (3.2)$$

where ρ is some non-negative, even function satisfying $\rho(0) = 1, \rho(\pi) = 0$. The choice of $\rho(\xi)$ varies depending on what kind of derivative operator is used. For example, we have $\rho_2(kh) = \sin(kh)/kh$ for the second-order centred differencing; $\rho_c(kh) = 3\sin(kh)/(kh(2 + \cos(kh)))$ for the cubic spline approximation. It is easy to see that the order of accuracy is the order to which $\rho(\xi) \to 1$ as $\xi \to 0$. The spectral derivative without smoothing corresponds to the choice of $\rho \equiv 1$. We denote it by $D_h^{(1)}$. It is well known that the pseudo-spectral methods without smoothing may introduce aliasing errors which could lead to numerical instability (Kreiss and Oliger, 1979; Gottlieb and Orszag, 1977; Tadmor, 1987; and Goodman, Hou and Tadmor, 1994). To suppress aliasing errors, Fourier smoothing is often used. In that case,

ρ should satisfy (i) $\rho \geq 0, \rho(\pm\pi) = 0$; (ii) $\rho(x) = 1$ for $|x| \leq \lambda\pi$ for some $0 < \lambda < 1$; and (iii) ρ is smooth. Condition (ii) ensures spectral accuracy.

Once a derivative operator is chosen, we also use a smoothing based on ρ. For arbitrary periodic function f_j, we define $f_j^{(\rho)}$ by multiplying $\hat{f}_k$ by $\rho(kh)$ and taking the inverse Fourier transform. That is,

$$f_j^{(\rho)} = \sum_{k=-N/2+1}^{N/2} \rho(kh)\hat{f}_k e^{-ik\alpha_j}.$$

Thus, we have $D_h^{(\rho)} f_j = D_h^{(1)} f_j^{(\rho)}$. Similarly, we define $z_j^{(\rho)}$ by applying ρ to the transform of $z_j - \alpha_j$. It is clear that $f^{(\rho)}$ is an rth-order approximation to f if ρ corresponds to the rth-order derivative operator.

Now we can present our numerical algorithm for the water-wave equations (2.6)–(2.8) as follows:

$$\frac{d\bar{z}_j}{dt} = \frac{1}{4\pi i} \sum_{\substack{k=-N/2+1 \\ (k-j)\text{odd}}}^{N/2} \gamma_k \cot\left(\frac{z_j^{(\rho)} - z_k^{(\rho)}}{2}\right) 2h + \frac{\gamma_j}{2(D_h^{(\rho)}z)_j} \equiv u_j - iv_j, \quad (3.3)$$

$$\frac{d\phi_j}{dt} = \frac{1}{2}(u_j^2 + v_j^2) - gy_j, \quad (3.4)$$

$$D_h^{(\rho)}\phi_j = \frac{\gamma_j}{2} + \mathrm{Re}\left(\frac{D_h^{(\rho)}z_j}{4\pi i} \sum_{\substack{k=-N/2+1 \\ (k-j)\text{odd}}}^{N/2} \gamma_k \cot\left(\frac{z_j^{(\rho)} - z_k^{(\rho)}}{2}\right) 2h\right). \quad (3.5)$$

In practice, we solve for γ_j from (3.5) by iteration using (2.9).

Remark. The Fourier smoothing $z^{(\rho)}$ in (3.3) and (3.5) is to balance the high-frequency errors introduced by $D_h^{(\rho)}$. This will become apparent in the discussion of stability below. The choice of such smoothing is sharp. If we use finite-order derivative operators or Fourier smoothing for the spectral derivative, the use of smoothing on z is necessary for stability. We do not need to smooth on γ because γ defined by (3.5) has been smoothed implicitly through $D_h^{(\rho)}$ and $z^{(\rho)}$. We now state the convergence result; see Beale, Hou and Lowengrub (1993) and Beale *et al.* (to appear).

Theorem 1 (Convergence of a Boundary-Integral Method) Assume that $z(\cdot, t), \phi(\cdot, t) \in C^{m+2}[0, 2\pi]$ and $\gamma(\cdot, t) \in C^{m+1}[0, 2\pi]$ for $m \geq 3$, and $|z(\alpha, t) - z(\beta, t)| \geq c|\alpha - \beta|$ for $0 \leq t \leq T$ and $c > 0$. Furthermore, assume that

$$(u_t, v_t) \cdot n - (0, -g) \cdot n \geq c_0 > 0. \quad (3.6)$$

Here (u, v) is the Lagrangian velocity, n is the normal vector to the interface, pointing out of the fluid region, and c_0 is some constant. Then if $D_h^{(\rho)}$

corresponds to an rth-order-derivative approximation, we have for $h \leq h_0(T)$

$$\|z(t) - z(\cdot, t)\|_{l^2} \leq C(T)h^r, \tag{3.7}$$

Similarly, ϕ_j is accurate to order h^r, and γ_j is accurate to order h^{r-1} in the discrete L^2 norm defined by $\|z\|_{l^2}^2 = \sum_{j=1}^N |z_j|^2 h$. If $D_h^{(\rho)}$ corresponds to the spectral approximation, we have the same convergence result as above except for replacing h^r by h^m.

We remark that the sign condition (3.6) is required to prove well-posedness of the water-wave equations (Beale, Hou and Lowengrub, 1993a). It guarantees that the problem is stably stratified. It means that the interface is not accelerating downward, normal to itself, as rapidly as the normal acceleration of gravity. If (3.6) is violated, it would generate the Rayleigh–Taylor instability as if water were above the interface. It can be viewed as a natural generalisation of the criterion of Taylor (1950).

3.1. Discussion of stability analysis

Here we discuss some of the main ingredients in the stability analysis of the scheme given by (3.3)–(3.5). We will mainly focus on the linear stability. Once linear stability is established, non-linear stability can be obtained relatively easily by using the smallness of the error and an induction argument. The reader is referred to Beale, Hou and Lowengrub (to appear) and Beale *et al.* (to appear) for details.

To analyse linear stability, we write equations for the errors $\dot{z}_j(t) \equiv z_j(t) - z(\alpha_j, t)$, and so forth, and try to estimate their growth in time. If we compare the sum in (3.3) for the discrete velocity with the corresponding one for the exact velocity, the terms linear in $\dot{z}_j$, $\dot{\gamma}_j$ are

$$\frac{1}{2\pi i} \sum_{(k-j)\text{odd}} \frac{\dot{\gamma}_k}{z(\alpha_j)^{(\rho)} - z(\alpha_k)^{(\rho)}} 2h$$

$$-\frac{1}{2\pi i} \sum_{(k-j)\text{odd}} \frac{\gamma(\alpha_k)(\dot{z}_j^{(\rho)} - \dot{z}_k^{(\rho)})}{(z(\alpha_j)^{(\rho)} - z(\alpha_k)^{(\rho)})^2} 2h,$$

where we have expanded the periodic sum, with k now unbounded. To identify the most singular terms, we use the Taylor expansion to obtain the most singular symbols

$$\frac{1}{z(\alpha_j) - z(\alpha_k)} = \frac{1}{z_\alpha(\alpha_j)(\alpha_j - \alpha_k)} + f(\alpha_j, \alpha_k),$$

where f is a smooth function. Thus, the most important contribution to the first term is $(2iz_\alpha)^{-1}H_h\dot\gamma_j$, where H_h is the discrete Hilbert transform

$$H_h(\dot\gamma_j) \equiv \frac{1}{\pi} \sum_{(k-j)\text{odd}} \frac{\dot\gamma_k}{\alpha_j - \alpha_k} 2h. \tag{3.8}$$

Similarly, the most important contribution to the second term is $-\gamma(2iz_\alpha^2)^{-1}\Lambda_h(z_j^{(\rho)})$, where Λ_h is defined as follows:

$$\Lambda_h(\dot f_j) \equiv \frac{1}{\pi} \sum_{(k-j)\text{odd}} \frac{\dot f_j - \dot f_k}{(\alpha_j - \alpha_k)^2} 2h. \tag{3.9}$$

Let H and Λ be the corresponding continuous operators for H_h and Λ_h, respectively, that is, with the discrete sums replaced by continuous integrals. At the continuous level, it is easy to show that

$$\Lambda(f) = H(D_\alpha f), \tag{3.10}$$

where D_α is the continuous derivative operator. It turns out that in order to maintain numerical stability of the boundary-integral method, the quadrature rule for the singular integral and the discrete derivative operator $D_h^{(\rho)}$ must satisfy a compatibility condition similar to (3.10). That is, a given quadrature rule, which defines corresponding discrete operators H_h and Λ_h, and a discrete derivative $D_h^{(\rho)}$, must satisfy a compatibility condition similar to (3.10):

$$\Lambda_h(\dot z_i) = H_h(D_h^{(\rho)})(\dot z_i), \tag{3.11}$$

for $\dot z$ satisfying $\hat z_0 = \hat z_{N/2} = 0$. If (3.11) is violated, this will generate a singular operator of the form $(\Lambda_h - H_h(D_h^{(\rho)}))(\dot z)$ in the error equations. This will generate numerical instability; see Subsection 2.6.

For the spectrally accurate alternating-trapezoidal-quadrature rule, the discrete Hilbert transform defined above has properties that are surprisingly similar to those of the continuum counterpart, that is,

$$(\widehat{H_h})_k = -i\,\text{sgn}(k), \quad (\widehat{\Lambda_h})_k = |k|. \tag{3.12}$$

Thus the compatibility condition (3.11) would imply that a spectral derivative operator without smoothing should be used to obtain numerical stability. However, it is well known that aliasing errors can arise from products for spectral methods without smoothing. These aliasing errors will upset the high-mode balance of lower-order terms; see below.

By performing appropriate Fourier smoothing in the approximations of the velocity integral, we can ensure a variant of compatibility condition (3.11) is satisfied, that is,

$$\Lambda_h(\dot z_j^{(\rho)}) = H_h(D_h^{(\rho)})(\dot z_j). \tag{3.13}$$

This can be verified from the spectrum properties of H_h and Λ_h and the definition of the ρ smoothing. This modified compatibility condition is sufficient to ensure stability of our modified boundary-integral method. This explains why we need to smooth z in (3.3) and (3.5) when we approximate the velocity integral. The modified algorithm also allows use of non-spectral derivative operators.

The Fourier smoothing is also needed to eliminate aliasing errors in the discrete integral operator with smooth kernels. Typically, the lower-order terms are of the form $\sum_{(k-j)\text{odd}} f(\alpha_j, \alpha_k)\dot{z}_k 2h$ for a smooth function $f \in C^m$. In the continuous case, we have for any $\dot{z} \in L^2$

$$\int f(\alpha, \alpha')\dot{z}(\alpha')d\alpha' = A_{-m}(\dot{z}),$$

where m is the degree of regularities of f and A_{-m} is a linear bounded operator from H^j to H^{j+m}, H^j being the Sobolev space of function with j derivatives in L^2. This is no longer true at the discrete level due to aliasing errors associated with the alternating-point-quadrature rule. For example, if we let $\dot{z}_j = e^{i(N/2-1)\alpha_j}$ and $f(\alpha, \alpha') = (e^{i2\alpha} - e^{i2\alpha'})/(\alpha - \alpha')$, we can show that

$$\sum_{(k-j)\text{odd}} f(\alpha_j, \alpha_k)\dot{z}_k 2h = -2ie^{i(-N/2+1)\alpha_j},$$

which is of course no smoother than $\dot{z}$. This is a result of aliasing errors at the highest frequencies and is why we must use the Fourier smoothing to eliminate the aliasing errors in the high modes. With the ρ smoothing, we can prove that (Beale, Hou and Lowengrub, to appear)

$$\sum_{(k-j)\text{odd}} f(\alpha_j, \alpha_k)\dot{z}_k^{(\rho)} 2h = A_{-1}(\dot{z}).$$

With these observations, we can derive an error equation for $\dot{z}_j$ that is similar to the continuum counterpart in the linear well-posedness study (Beale, Hou and Lowengrub, 1993a)

$$\frac{d\dot{z}_j}{dt} = z_\alpha^{-1}(I - iH_h)D_h^{(\rho)}\dot{F} + A_0(\dot{z}) + A_{-1}(\dot{\phi}) + O(h^r),$$

where $\dot{F} = \dot{\phi} - u\dot{x} - v\dot{y}$. This also suggests that we should project the error equation onto the local tangential and normal coordinate systems. In these local coordinates, the stability property of the error equations becomes apparent. Let $\dot{z}^N, \dot{z}^T$ be the normal and tangential components of $\dot{z}$ with respect to the underlying curve $z(\alpha)$, where N is the outward normal, and $\hat{\delta} = \dot{z}^T + H_h\dot{z}^N$. We obtain

$$\dot{z}_t^N = \frac{1}{|z_\alpha|}H_h D_h^{(\rho)}\dot{F} + A_{-1}(\dot{\phi}) + A_0(\dot{z}), \tag{3.14}$$

$$\dot{\delta}_t = A_{-1}(\dot{\phi}) + A_0(\dot{z}), \tag{3.15}$$

$$\dot{F}_t = -c(\alpha, t)\dot{z}^N A_{-1}(\dot{z}), \qquad c(\alpha, t) = (u_t, v_t + g) \cdot N, \tag{3.16}$$

where equation (3.16) is obtained by performing error analysis on Bernoulli's equation and using the Euler equations. In this form it is clear that only the normal component of $\dot{z}$ is important. Now it is a trivial matter to establish an energy estimate for the error equations. Note that $H_h D_h^{(\rho)}$ is a positive operator with a symbol $\rho(kh)|k|$. The problem is stable if the sign condition, $c(\alpha, t) > 0$, is satisfied. We refer to Beale, Hou and Lowengrub (to appear) for details.

3.2. Example of unstable schemes

In this section, we present a class of unstable schemes based on an equivalent boundary-integral formulation. We will demonstrate the numerical instability by performing a von Neumann stability analysis around the equilibrium (see also Baker and Nachbin (to appear)). Consider a boundary-integral formulation that uses the dipole strength μ. The vortex-sheet strength is defined in terms of μ by the relationship $\gamma = \mu_\alpha$. The evolution equations for z and ϕ are the same as before. The only difference is in the relation between ϕ and μ.

$$\phi = \frac{\mu}{2} + \text{Re}\left(\frac{1}{4\pi i}\int_{-\pi}^{\pi}\mu(\alpha')z_\alpha(\alpha')\cot\left(\frac{z(\alpha) - z(\alpha')}{2}\right)d\alpha'\right). \tag{3.17}$$

It is clear that equation (2.8) is obtained by differentiating the relation (3.17) with respect to α and integrating by parts. A natural numerical approximation to the above equations is given by

$$\frac{d\bar{z}_j}{dt} = \frac{1}{4\pi i}\sum_{(k-j)\text{odd}}\gamma_k\cot\left(\frac{z_j - z_k}{2}\right)2h + \frac{\gamma_j}{2(D_h^{(\rho)}z_j)} \equiv u_j - iv_j, \tag{3.18}$$

$$\frac{d\phi_j}{dt} = \frac{1}{2}(u_j^2 + v_j^2) - gy_j + \tau\kappa_j, \tag{3.19}$$

$$\phi_j = \frac{\mu_j}{2} + \text{Re}\left(\frac{1}{4\pi i}\sum_{(k-j)\text{odd}}\mu_k D_h^{(\rho)}z_k\cot\left(\frac{z_j - z_k}{2}\right)2h\right), \tag{3.20}$$

$$\gamma_j = D_h^{(\rho)}\mu_j, \tag{3.21}$$

where $D_h^{(\rho)}$ can be any finite-order derivative approximation. As before, we denote its Fourier symbol as $\widehat{D_h^{(\rho)}}_k = ik\rho(kh)$.

Now we perform a linear stability analysis around the equilibrium. Let

$$z = \alpha + \dot{z}; \phi = \dot{\phi}, \mu = 1 + \dot{\mu}.$$

Substituting the above expressions into the discrete equations (3.18)–(3.21), and using (3.12), we obtain to leading order

$$\frac{d}{dt}\bar{z}_j = \frac{1}{2i}H_h(D_h^{(\rho)}\dot{\mu}_j) + D_h^{(\rho)}\dot{\mu}_j/2, \tag{3.22}$$

$$\frac{d}{dt}\dot{\phi}_j = -g\dot{y}_j, \tag{3.23}$$

$$\dot{\phi}_j = \frac{\dot{\mu}_j}{2} + \frac{1}{2}(H_h(D_h^{(\rho)}\dot{y}_j) - \Lambda_h(\dot{y}_j)), \tag{3.24}$$

where H_h and Λ_h are defined as before. This system of equations has constant coefficients and it can be diagonalised in the Fourier space. The eigenvalues of the resulting system in the Fourier space give the growth rates of $\dot{x}, \dot{y}, \dot{\phi}$. They are explicitly given by

$$\lambda_1 = 0; \lambda_2, \lambda_3 = 0.5k^2\rho(1-\rho) \pm 0.5\sqrt{k^4\rho^2(1-\rho)^2 - 16g|k|\rho}.$$

Notice that if $\rho \neq 0, 1$, then $\lambda_2, \lambda_3 \simeq O(k^2)$ for large k. This indicates that the numerical high-mode instability is even stronger than that of Kelvin–Helmholtz! It is clear that the instability is caused by violating the compatibility condition (3.11), that is, $H_h D_h^{(\rho)} - \Lambda_h \neq 0$ (see (3.24)). If the Fourier smoothing is used as in (3.3) and (3.5), then this term vanishes and one can easily see that the modified method is stable. On the other hand, aliasing instabilities cannot be seen from this linear stability analysis around the equilibrium solution because there is no mode mixing for constant-coefficient problems. For computational evidences of numerical instabilities, we refer to Longuet-Higgins and Cokelet (1976); Roberts (1983); Dold (1992); Beale, Hou and Lowengrub (to appear); and Beale *et al.* (to appear).

There are other ways to perform smoothing to partially alleviate the difficulty due to high-mode instability, see Longuet-Higgins and Cokelet (1976), Roberts (1983) and Dold (1992). But they cannot completely eliminate the source of numerical instability since the modified schemes still fail to satisfy the compatibility constraint. So there is still a large number of intermediate to high modes that are numerically unstable. As the number of grid points increases, or as we compute further in time, the numerical scheme will suffer from the high-mode instability. This has been one of the major obstacles in computing free-surface waves.

3.3. A numerical calculation of wave breaking

Here we present a calculation of wave breaking to illustrate how well our modified boundary integral performs in the fully non-linear regime. For a survey of breaking waves; see Peregrine (1983). We use the following initial condition:

$$x(\alpha, 0) = \alpha, y(\alpha, 0) = 0.1\cos(2\pi\alpha), \gamma(\alpha, 0) = 1.0 + 0.1\sin(2\pi\alpha).$$

The gravity coefficient is chosen to be $g = 9.8$. Note that the vortex-sheet strength γ does not have zero mean in this case. This amounts to a convenient choice of frame of reference. Although the derivation of (2.6)–(2.8) was for the special case where ϕ is periodic and γ has zero mean, the formulation is still valid provided only that we apply $D_h^{(\rho)}$ to equation (2.7); only $D_h^{(\rho)}\phi$ is needed in (2.8). The time integration in this numerical example is the fourth-order explicit Adams–Bashforth method. The fourth-order Runge–Kutta method is used to initialise the first three time steps. Also, a fourth-order extrapolation in time is used to obtain a more accurate first iterate in the iterative scheme for γ, as suggested in Baker, Meiron and Orszag (1982). With this improved initial guess, the iteration will converge with an iteration error of order 10^{-10} in two iterations for most time until the wave is close to breaking. In this calculation, we use a 25th-order Fourier smoothing in the spectral derivative with ρ given by

$$\rho(kh) = \exp(-10 \cdot (2|k|/N)^{25}), \quad \text{for} \quad |k| \leq N/2.$$

In Figure 1, we present a series of interface profiles from $t = 0.28$ to $t = 0.5175$. In order to see clearly the time evolution of the water wave, we plot the solution at five different times in a single picture. The first curve from the top is obtained by adding 0.6 to the y coordinate; the successive ones are displaced by multiples of 0.3. Time increases from top to bottom. As we can see from Figure 1b, the interface becomes vertical around $t = 0.32$. After that, the wave turns over. In the mean time, the interface develops large curvature, and requires more-refined numerical resolution. With $N = 256$, we can compute up to $t = 0.5$ with six digits of accuracy in the interface positions. But in order to compute very close to the time of wave breaking, we need to increase our resolution to $N = 512$, or larger. Of course, beyond $t = 0.32$ when the interface becomes vertical, our convergence result will cease to be valid since it violates our condition (3.6) in Theorem 1. But one can see that our numerical calculations remain robust even after condition (3.6) is violated. Without additional filtering, our code can run up to $t \simeq 0.51$. In order to compute all the way up to the time of wave breaking, we need to use Krasny's filtering (see Subsection 4.1) beyond the time of wave turnover ($t \simeq 0.32$) to control the growth of round-off errors due to the Rayleigh–Taylor instability.

In Figure 1c, we plot the enlarged version of the wave fronts from $t = 0.5$ to 0.5175 when the wave is close to breaking. It is evident that the wave will break in finite time. In Figure 1d, we illustrate the number of computational particles near the wave front at the final time of our calculations. We can see that the interface is still well resolved and more particles are clustered near the head of the wave front where the curvature is the

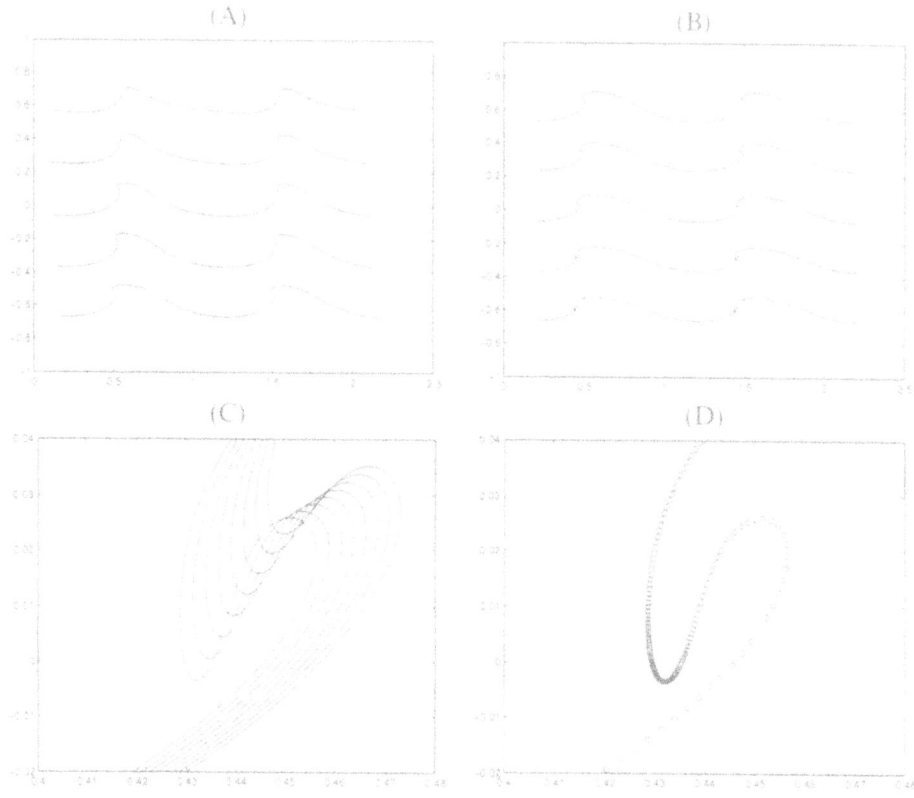

Fig. 1. (A) Water waves at $t = 0.28, 0.32, 0.36, 0.4, 0.44, N = 256, \Delta t = 0.001$. From Beale, Hou and Lowengrub (to appear). (B) Water waves at $t = 0.46, 0.48, 0.5, 0.51, 0.5175, N = 512, \Delta t = 0.00025$. From Beale, Hou and Lowengrub (to appear). (C) Enlarged view of wave fronts at $t = 0.5, 0.5025, ..., 0.5175, N = 512, \Delta t = 0.00025$. From Beale, Hou and Lowengrub (to appear). (D) Enlarged view of wave fronts at $t = 0.5175, N = 512, \Delta t = 0.00025$. From Beale, Hou and Lowengrub (to appear)

largest (about 800 in amplitude). This demonstrates the self-adaptive nature of the boundary-integral method. Details of the calculation and other computational examples can be found in Beale, Hou and Lowengrub (to appear).

4. Numerical computations of vortex-sheet roll-up

The idealisation of a shear layer as a vortex sheet separating two regions of potential flow has often been used as a model to study mixing properties, boundary layers and coherent structures of fluids. A vortex sheet corresponds to the case when the two fluid densities are the same on each side of the interface, but the tangential velocity across the interface has a jump discontinuity. Without physical regularisation such as surface tension or viscosity, the vortex-sheet problem is ill-posed in the Hadamard sense. Small perturbations can lead to exponential growth in high wavenumbers due to the Kelvin–Helmholtz instability. Since the Kelvin–Helmholtz instability is a generic fluid-dynamical instability for multi-fluid interfaces (except for the case of unit Atwood number $A^2 = 1$), the understanding of the numerical and analytical difficulties of the vortex-sheet problem would shed useful light into the general multi-fluid interfaces. In fact, most of the numerical techniques we discuss in this section can be and have been extended to the study of Rayleigh–Taylor instability in multi-fluid interfaces.

Assume the interface is periodic in the horizontal direction, that is, $z(\alpha, t) = \alpha + s(\alpha, t)$ with $s(\alpha, t)$ being periodic in α. The vortex-sheet strength γ is also periodic. For a vortex sheet, the Atwood number is equal to zero. This greatly simplifies the equations of motion for the interface. The governing equations reduce to

$$\frac{d\bar{z}}{dt} = \frac{1}{4\pi i} \int \gamma(\alpha', t) \cot\left(\frac{z(\alpha, t) - z(\alpha', t)}{2}\right) d\alpha',$$

$$\frac{d\gamma}{dt} = 0.$$

This shows that γ is conserved along Lagrangian particle trajectories. If the initial vortex-sheet strength γ_0 is positive, then we can parameterise the interface by its circulation variable Γ. Then the above equations further reduce to a single equation for the interface position z:

$$\frac{d\bar{z}}{dt} = \frac{1}{4\pi i} \int \cot\left(\frac{z(\Gamma, t) - z(\Gamma', t)}{2}\right) d\Gamma', \tag{4.1}$$

Linear stability around equilibrium solution $z = \Gamma$ gives the dispersion relation

$$\lambda^2 = k^2/4, \tag{4.2}$$

so there is one growing eigenmode and one decaying eigenmode. Arbitrarily small perturbation can lead to unbounded exponential growth in high wavenumbers.

4.1. The point-vortex approximation

The point-vortex approximation to (4.1) was first introduced by Rosenhead (1932). The idea is to represent the vortex sheet by a collection of point sources. Let $z_j(t)$ denote the numerical approximation of $z(\Gamma_j, t)$, with $\Gamma_j = jh$, $h = 2\pi/N$. The integral on the right-hand side of (4.1) is approximated by the trapezoidal rule, which omits the infinite contribution due to the self-induced velocity at $\Gamma' = \Gamma$. This gives rise to a system of ordinary differential equations for the particle trajectories

$$\frac{d\bar{z}_j}{dt} = \frac{1}{4\pi i} \sum_{\substack{k=1 \\ k \neq j}}^{N} \cot\left(\frac{z_j - z_k}{2}\right) h \tag{4.3}$$

$$z_j(0) = \Gamma_j + s(\Gamma_j, 0). \tag{4.4}$$

Equations (4.3) have a Hamiltonian structure for the conjugate variables $x_j N^{-1/2}, y_j N^{-1/2}$. The Hamiltonian is given by

$$H_N(t) = \frac{-1}{4\pi N^2} \sum_{j=1}^{N} \sum_{k>j} \ln(\cosh(y_j - y_k) - \cos(x_j - x_k)). \tag{4.5}$$

One immediate consequence from the Hamiltonian is that if the variables $y_j(t)$ remain bounded, then the invariance of $H_N(t)$ implies that the point vortices remain separated (recall that we have assumed one signed vorticity distribution here).

The main issue here is the question of numerical stability of the point-vortex approximation. Using a sinusoidal initial perturbation with a small number of particles, Rosenhead integrated the vortex-sheet equation (4.3) and obtained the expected roll-up of the vortex sheet. These calculations were repeated by Birkhoff (1962). Birkhoff found that the point-vortex approximation did not converge as the number of particles increased. In fact, the increased number of points led to irregular motion of the points and early deterioration of the calculations. Some investigators have tried to use high-order discretisations to resolve this difficulty, and different smoothing techniques have been tried. But irregular motion still persists. We refer to van de Vooren (1980), Higdon and Pozrikidis (1985), Pullin (1982), Moore (1985) and Fink and Soh (1978) for more detailed discussions.

The source of numerical instability was later clarified by Krasny (1986a). Krasny found that there are two types of irregular motion that can occur in the numerical solution of the point-vortex equations for vortex sheets. The first one occurs at smaller times $t > 0$ as the value of N increases. The second type occurs only beyond the vortex-sheet's critical time regardless of the value of N. The second type of irregular motion is due to the loss of regularity of the solution beyond the critical time. Without additional physical or numerical regularisation, the point-vortex method will fail to converge beyond the singularity time. The first type of irregular motion is caused by round-off errors due to the computer's finite-precision arithmetic. Once these round-off error perturbations enter the calculations, they grow according to the equations' dynamics and are subject to the Kelvin–Helmholtz instability. Thus the highest modes will grow the fastest and the growth rate is exponential with increasing wavenumbers. This explains why increasing the number of grid points will lead to rapid growth and early-time irregular motion. In order to control the growth of the round-off-error perturbations, Krasny introduced a non-linear filtering technique (1986a). That is, at every time step, before we evaluate the singular integral, we take the Fourier transform of the particle position z_j. Set to zero those Fourier coefficients that are below a certain cut-off level, say 10^{-13} if a 16-digit arithmetic is used. Then take the inverse Fourier transform.

More specifically, we can consider Krasny's filtering as a projection operator, denoted by P. Given an error tolerance τ, the projection operator P is given by

$$(\widehat{Pf})_k = \begin{cases} \hat{f}_k, & \text{if} \quad |\hat{f}_k| \geq \tau, \\ 0, & \text{otherwise}, \end{cases} \tag{4.6}$$

for any periodic function f. The filter P is non-linear because the wavenumbers at which it is applied depend on the solution. In order for this filtering operator to be effective, we assume that the underlying function, f, has a rapid decay in the Fourier space. Moreover, we want to take the filter level, τ, as small as possible for the sake of accuracy. So it is preferable to perform the numerical calculations in double-precision arithmetic. To illustrate the method, we take the forward Euler discretisation for the point-vortex method as an example. With Krasny's filtering, the numerical method becomes

$$z_j^{n+1} = P\left\{ z_j^n + \Delta t \frac{1}{4\pi i} \sum_{k \neq j} \cot\left(\frac{z_j - z_k}{2}\right) h \right\}. \tag{4.7}$$

The effect of this filtering is dramatic. With this filtering, the first type of irregular motion is eliminated. One can compute up to the time when the curvature singularity is formed. Since the filter level is very small (typically 10^{-13} in a 16-digit arithmetic), it does not affect the accuracy much in the smooth region. Comparison of the filtered calculation and the unfiltered 29-digit calculation shows very good agreement (Krasny, 1986a). Moreover, the filter does not suppress the growth of high-wavenumber modes. With the filtering, the high wavenumber modes can still grow through non-linear interactions. Once the modes grow larger than the filtered level, they are not affected by the filtering.

We include the calculations obtained by Krasny (1986a) for the initial data (note that the period is 1, not 2π):

$$x(\Gamma, 0) = \Gamma + 0.01 \sin 2\pi\Gamma, \quad y(\Gamma, 0) = -0.01 \sin 2\pi\Gamma,$$

which is a small-amplitude perturbation of the equilibrium solution. The filtering technique was used in double precision (16-digit) with $N = 100$. The time step was set to be $\Delta t = 0.01$ for $t \leq 0.25$ and $\Delta t = 0.001$ for $t > 0.25$. A fourth-order Runge–Kutta method was used for time integration. The filter

(A)

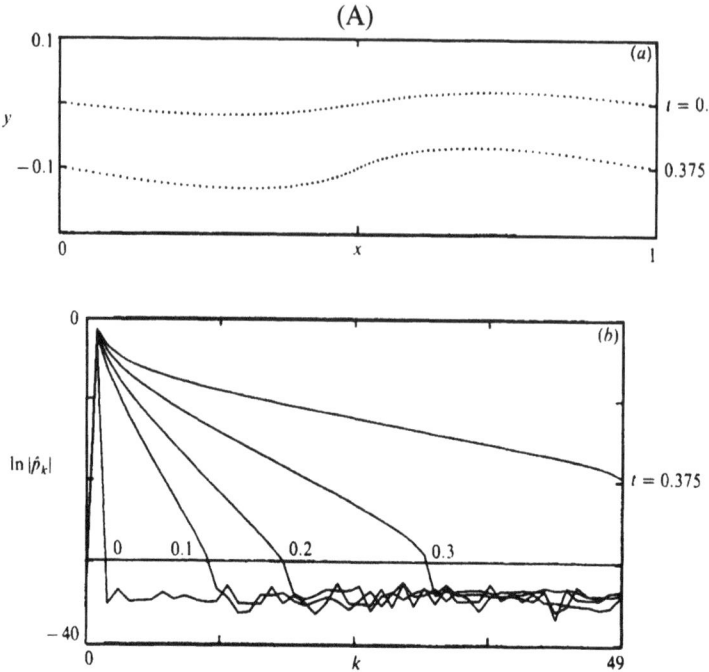

Fig. 2A. Double-precision (16-digit) calculation with filter level set at 10^{-13} (the horizontal line in (b)). This calculation used $N = 100$ and $\Delta t = 0.01$ for $t \le 0.25$ and $\Delta t = 0.001$ for $t > 0.25$: (a) point-vortex positions; (b) log-linear plot of the Fourier coefficients. (2.2) amplitudes versus wavenumber. From Krasny (1986a)

level was set to 10^{-13}. In this case, the filter turned off at $t \simeq 0.35$. The resulting point positions and the Fourier coefficients are plotted in Figure 2A. In this calculation the Hamiltonian was also well conserved. There is no sign of the first type of irregular point motion which had appeared in the unfiltered 16-digit calculation at $t = 0.375$; Using filtering with $N = 200$ in double precision, Krasny can compute up to the singularity time $t = 0.375$; see Figure 2B (d). It is worth noting that without filtering even a 29-digit calculation yielded irregular motion at time $t = 0.375$ (see Fig 2B (c)), not to mention the calculations obtained used single and double precisions (Fig 2B (a) and (b)).

Shelley (1992) used the spectrally accurate alternating quadrature to re-examine the singularity formation in vortex-sheet motion, trying to acquire more precise information on the singularity structure. Shelley's calculations were performed in 30 digits of precision in conjunction with Krasny's filtering technique. The filter level was set to 10^{-25}. This high precision seems to be necessary to discern the asymptotic behaviour of the spectrum. It was found that Moore's asymptotic analysis is valid only at times well before the singularity time. Near the singularity time the form of the singularity

(B)

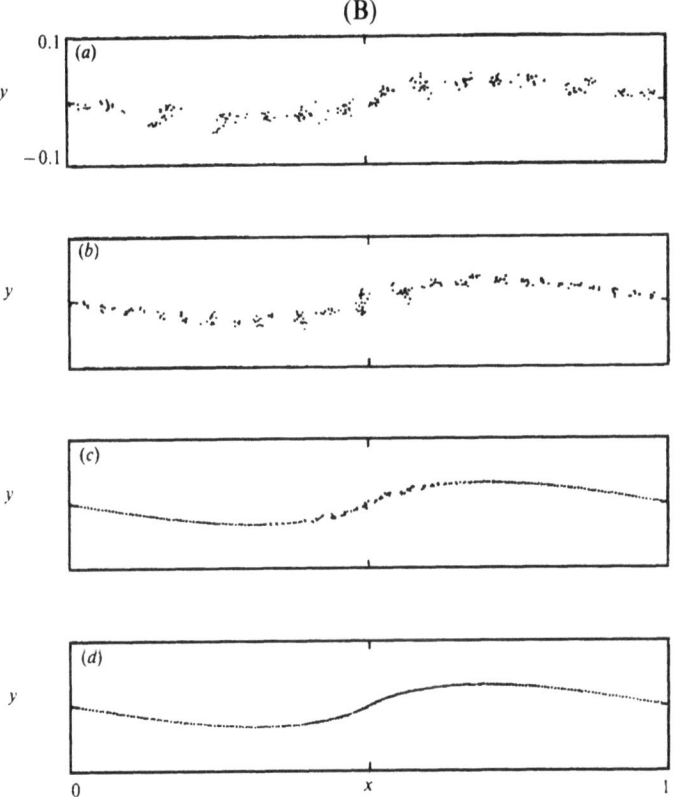

Fig. 2B. Point-vortex positions at $t = 0.375$ for $N = 200$: (a) single precision (7 digit); (b) double precision (16 digit); (c) CDC double precision (29 digit); (d) filtered at level 10^{-13}, double precision (16 digit). From Krasny (1986a).

departs significantly from that predicted by Moore. Moreover, the real and imaginary parts of the solutions behave differently near the singularity time. The form of the singularity also depends upon the amplitude of the initial disturbance.

Convergence of the point-vortex method for vortex sheets was first obtained by Caflisch and Lowengrub (1989) for analytic initial data. A simplified proof was later obtained by Hou, Lowengrub and Krasny (1991), using the spectral accuracy of the alternating quadrature rule. However, these convergence results are for short times, and do not consider the effect of round-off errors. In fact, with a simulated round-off-error term, the convergence result breaks down very quickly as the number of computational particles increases. Recently, Caflisch, Hou and Lowengrub (1994) have been able to prove convergence of the point-vortex method for vortex sheets with Krasny's filtering. The proof is in an analytic function class and uses a discrete form of the Cauchy–Kowalewski theorem. The proof is presented for the case in which the sheet is initially near equilibrium and convergence is obtained nearly up to the singularity time. The analysis in this paper applies directly to other ill-posed problems such as Rayleigh–Taylor unstable interfaces in incompressible, inviscid and irrotational fluids, as well as to Mullins–Sekerka unstable interfaces in Hele–Shaw cells.

4.2. Vortex-blob calculations

In this section, we present a vortex-blob desingularisation to study vortex-sheet roll-up beyond the curvature singularity. The first application of vortex-blob methods for vortex-sheet roll-up was given by Chorin and Bernard (1973) in which they proposed to regularise the point-vortex method by smooth vortex blobs. Subsequently, Anderson (1985) applied vortex-blob methods to study vortex-sheet roll-up using the Bousinesqu approximation, and performed a careful numerical convergence study. In a series of papers (Krasny, 1986b; Krasny, 1987; Nitsche and Krasny, 1994), Krasny has used the vortex-blob method to study vortex-sheet roll-up and has obtained a number of interesting results. Some previous alternative desingularisations for vortex sheets have incorporated a stabilising physical mechanism into the model. Moore (1978) has derived an evolution equation for a vortex layer of small thickness. Pozrikidis and Higdom (1985) have numerically studied a periodically perturbed layer of constant vorticity. Baker and Shelley (1990) have considered regularisation of a thin vortex layer. Pullin (1982) has included surface tension in the evolution equation. We will come back to these other types of regularisation in later sections. Unlike these approaches, the specific form of desingularisation that is used in vortex-blob methods does not correspond precisely to a physical effect. It is a purely numerical regularisation.

The vortex-blob desingularisation for vortex sheets can be described as follows. Let δ be a non-negative real number. We will desingularise the Birkhoff–Rott equations by placing a cut-off in the singular kernel.

$$\frac{\partial x}{\partial t} = \frac{-1}{4\pi} \int_{-\pi}^{\pi} \frac{\sinh(y - y')}{\cosh(y - y') - \cos(x - x') + \delta^2} d\Gamma', \qquad (4.8)$$

$$\frac{\partial y}{\partial t} = \frac{1}{4\pi} \int_{-\pi}^{\pi} \frac{\sin(x - x')}{\cosh(y - y') - \cos(x - x') + \delta^2} d\Gamma', \qquad (4.9)$$

where $x = x(\Gamma, t)$, $x' = x(\Gamma', t)$. When $\delta = 0$, the integral is understood as the Cauchy principal-value integral. In that case, we recover the vortex-sheet equations in the periodic case.

A flat vortex sheet of constant strength is an equilibrium solution of the desingularised equations. It is easy to perform a linear stability analysis around the equilibrium solution. This helps us gain insight into the nature of the desingularisation. The growth rates of the perturbation can be computed by Fourier transform. The dispersion relation is given by Krasny (1986b)

$$\omega^2 = \frac{k(1 - e^{-k \,\cosh^{-1}(1+\delta^2)})e^{-k \,\cosh^{-1}(1+\delta^2)}}{4\delta(2 + \delta^2)^{1/2}}.$$

The positive branch of ω corresponds to the growing perturbations. For a fixed value of $\delta > 0$, there is a wavenumber k_m for which the growth rate $\omega(k_m)$ is maximum. In the limit $k \to \infty$ we have $\omega(k) \to 0$. For example, with $\delta = 0.05$, the maximum growth rate of the eigenvalue is about 3.5. The desingularised equations therefore do not exhibit the severe short wavelength instability of the exact vortex-sheet equations. As $\delta \to 0$ with a fixed wavenumber, we recover the exact dispersion relation $\omega^2 \simeq k^2/4$.

We can apply standard discretisation techniques to solve the initial value problem (4.8)–(4.9). For example, the trapezoidal quadrature of the integrals in (4.8)–(4.9) yields a system of ordinary differential equations in the case of period-one initial condition

$$\frac{dx_j}{dt} = \frac{-1}{2} \sum_{\substack{k=1 \\ k \neq j}}^{N} \frac{\sinh 2\pi(y_j - y_k)}{\cosh 2\pi(y_j - y_k) - \cos 2\pi(x_j - x_k) + \delta^2} h, \quad (4.10)$$

$$\frac{dy_j}{dt} = \frac{1}{2} \sum_{\substack{k=1 \\ k \neq j}}^{N} \frac{\sin 2\pi(x_j - x_k)}{\cosh 2\pi(y_j - y_k) - \cos 2\pi(x_j - x_k) + \delta^2} h. \quad (4.11)$$

If $\delta = 0$, then the above discretisation is the point-vortex approximation of Rosenhead (1932). As for the point-vortex system, for any $\delta \geq 0$, the equations (4.10),(4.11) form a Hamiltonian system for the conjugate variables $x_j \cdot N^{-1/2}$, $y_j \cdot N^{-1/2}$, with the Hamiltonian function given by

$$H_N(t) = \frac{-1}{4\pi N^2} \sum_{j=1}^{N} \sum_{k>j} \ln(\cosh 2\pi(y_j - y_k) - \cos 2\pi(x_j - x_k) + \delta^2). \quad (4.12)$$

Krasny (1986b) used the above vortex-blob method to compute vortex-sheet roll-up using the same initial condition as for the point-vortex method calculation

$$x_j(0) = \Gamma_j + 0.01 \sin 2\pi\Gamma_j, \, y_j(0) = -0.01 \sin 2\pi\Gamma_j. \quad (4.13)$$

The fourth-order Runge–Kutta method was used to perform time integration. There are now three parameters, $\delta, \Delta t$ and h. It turns out that if one simultaneously reduces all three parameters, one does not get a convergent result in general. The strategy, which was first used by Anderson (1985), is to first keep δ fixed, and then reduce Δt and h until we get a convergent solution of the δ equations. By repeating this process for several values of δ, one can extrapolate the limit $\delta \to 0$.

Krasny (1986b) showed that using the δ-desingularisation the vortex sheet

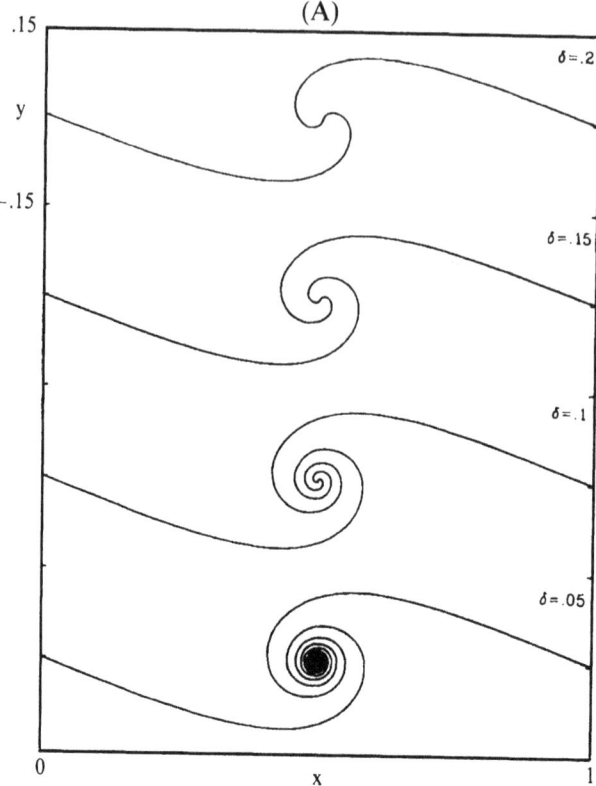

Fig. 3A. Vortex-blob-method solutions of the δ equation at $t = 1$, $N = 400$. From Krasny (1986b).

rolls up into a double-branched spiral past the critical time. The effect of decreasing δ at a fixed time ($t = 1$) beyond the critical time ($t_c = 0.375$) is shown in Figure 3A, which plots the interface for several values of δ between 0.2 and 0.05. These calculations used $N = 400$ and $\Delta t = 0.05$. In the case of $\delta = 0.05$, a smaller time step was used, $\Delta t = 0.01$, and the computation was performed in double-precision arithmetic (16 digits). As δ decreases with $t = 1$ in Figure 3A, more turns appear in the core. For $\delta = 0.05$, the core region is tightly packed. An enlarged view is shown in Figure 3B, which shows that each branch of the spiral contains five complete resolutions. The curve's outer region seems to converge as δ decreases. Some evidence was given in Krasny (1986b).

Krasny also used the vortex-blob method for several other applications, including computing the vortex-sheet roll-up in the Trefftz plane [95], and computing vortex sheet roll-up past a sharp edge to study separation [112]. The calculation of vortex ring formation at the edge of a circular tube in an axisymmetric 3-D vortex-sheet model seems to support the experimental findings (Nitsche and Krasny, 1994). Related numerical studies for axisym-

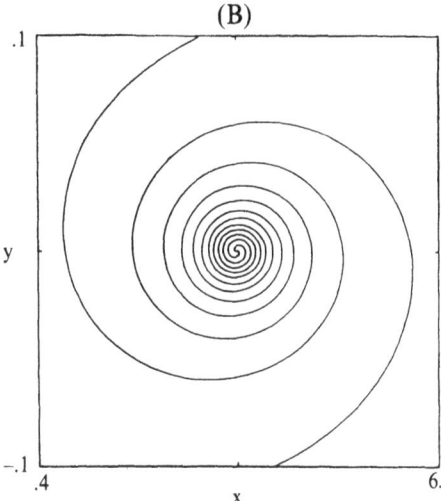

Fig. 3B. An enlarged view of the inner portion of the $\delta = 0.04$ case, $t = 1$, $N = 400$. From Krasny (1986b).

metric vortex-sheet motion have been carried out by Pullin (1979), Caflisch, Li and Shelley (1993), Pugh (1989), Kaneda (1990) and Dahm, Frieler and Tryggvason (1992).

Convergence of vortex-blob methods for vortex sheets has been established by Caflisch and Lowengrub (1989) for short times using analytic data. Following Delort's observation, Liu and Xin (to appear) have been able to prove that Krasny's vortex-blob calculation converges globally to a weak solution of the Euler equations if the initial vorticity does not change sign.

Finally, we remark that it is an easy matter to generalise the vortex-blob method to study Rayleigh–Taylor instability for general two-density interface problems; see, for example, Kerr (1988).

4.3. Thin-vorticity-layer regularisation of vortex sheets

Another approach to compute vortex-sheet motion beyond the singularity time is to study the motion of smoother solutions to the Euler equation. In 1990, Baker and Shelley approximated the vortex sheet by a thin layer of constant and finite vorticity of mean width H. The limiting behaviour of such vortex layers as $H \to 0$ was investigated to determine the possible nature of the vortex sheet past its singularity time. They found that the behaviour of an asymptotically thin vortex layer is given by a vortex sheet whose strength is the local layer width times the vorticity strength.

The problem of vortex layers with constant vorticity, also called vortex patches, is of interest in itself. Accurate and robust numerical methods for vortex-patch problems have been developed by Zabusky et al. (1979, 1983).

They referred to their methods as 'contour dynamics'. It has led to some interesting applications; see, for example, Dritschel (1989) for a review. The mathematical theory of vortex patches has attracted a lot of interest in recent years. The well-known result of Yudovich (1963) provides a theoretical framework for the vortex-patch problem. In particular, it guarantees the global existence of the flow. Yudovich's theory does not preclude the formation of singularities in the boundaries of vortex patches. Majda (1986) proposed the vortex-patch problem, in contour-dynamic form, as a model for the inviscid, incompressible creation of small scales. Motivated by analogy with the stretching of vorticity in three dimensions and by a simple model (1985, 1986), he suggested the possibility of finite-time singularities. In other words, some smooth initial contours might, in finite time, lead to loss of regularity such as infinite length, corners or cusps. This suggestion has been the subject of some debate in the computational literature (Buttke, 1989; Dritschel and McIntyre, 1990). Recently, Chemin (1993) proved that smooth contours stay smooth for all times provided that the initial condition is in $C^{1,\alpha}$ with $\alpha > 0$. A simplified proof was given by Bertozzi and Constantin (1993).

The set-up of the thin layer regularisation is as follows. Consider a periodic vortex layer surrounded by two interfaces. At $t = 0$, these two interfaces are symmetric with respect to the flat interface $y = 0$. The vortex layer is assumed to have mean width H and vorticity $-2U/H$. The lower interface, Γ_1, is parameterised as $z_1(\alpha)$, and the upper interface, Γ_2, as $z_2(\alpha)$, where $z_j(\alpha) = x_j(\alpha) + iy_j(\alpha)$. The thin layer is assumed to be 2π-periodic in the x-direction. It can be shown that the velocity of the fluid at a point $z_j(\alpha, t)$ on the jth interface $(j = 1, 2)$ is given by

$$\frac{\partial \bar{z}_j}{\partial t}(\alpha, t) = \frac{2U}{4\pi Hi} \int_0^{2\pi} (y_1(\alpha, t) - y_j(\alpha'))$$
$$\times \cot\left(\frac{z_j(\alpha, t) - z_1(\alpha', t))}{2}\right) \frac{\partial z_1}{\partial \alpha'}(\alpha', t) d\alpha' \quad (4.14)$$
$$-\frac{2U}{4\pi Hi} \int_0^{2\pi} (y_2(\alpha, t) - y_j(\alpha'))$$
$$\times \cot\left(\frac{z_j(\alpha, t) - z_2(\alpha', t))}{2}\right) \frac{\partial z_2}{\partial \alpha'}(\alpha', t) d\alpha'. \quad (4.15)$$

Note that the motion of vortex layers depends only upon information on the boundaries.

There have been many numerical studies in developing accurate quadrature rules for the boundary integrals in the above contour dynamic equations. Consider the case $j = 1$ as an example. The first integral in (4.14) has a smooth, periodic integrand due to a compensating zero in the function $y_1(\alpha, t) - y_1(\alpha', t)$. Thus the standard trapezoidal rule over equally spaced collocation points gives spectral or infinite-order accuracy. Accuracy is then limited by the approximation to $\partial z_1/\partial \alpha'$ at the collocation points. In Baker and Shelly (1990), derivatives were approximated by periodic, quintic splines with an accuracy of $O(h^6)$. The approximation to the second integral can be derived similarly. It is even simpler in this case since the field point $z_1(\alpha)$ does not sit on the boundary Γ_2.

The numerical study of Baker and Shelley (1990) indicates that the motion of the vortex layer leads to the formation of regions of high curvature, and regions of rapid stretching in the bounding interfaces. To maintain resolution of the interfaces, the mesh was redistributed periodically to resolve the high-curvature regions and collocation points were kept in the regions of rapid stretching. The mesh redistribution was done through a smooth reparameterisation of the interfaces.

The initial conditions for layer interfaces considered in Baker and Shelley (1990) were given by

$$z_1(\alpha, 0) = \alpha - i\frac{H}{2}(1 - a\cos\alpha),\, z_2(\alpha, 0) = \alpha + i\frac{H}{2}(1 - a\cos\alpha),$$

with $a < 1$. The limit of the above initial data as $H \to 0$ corresponds to the vortex-sheet initial data considered by Meiron, Baker and Orszag (1982) in their study of the singularity structure of a vortex sheet. In particular, the vortex sheet acquires a curvature singularity at $\alpha = \pi$.

The evolution of vortex layers with $U = 1/2$ and $a = 1/2$ was calculated for various mean thicknesses, $H = 0.025, 0.05, 0.1$ and 0.2. The case $H = 0.025$ corresponds to an aspect ratio of 250 to 1. This was the smallest value of H that Baker and Shelley (1990) could compute reliably. The critical time of curvature singularity is about $t_c = 1.6$ (Shelley, 1992). Figure 4a shows the location of the layer interfaces with $H = 0.025$ at various times $t = 0, t = 2.0$ and $t = 2.4$. Figure 4b shows several sequences of layer profiles for various thicknesses. Each column gives a sequence of layers at various times with H fixed, and goes as far as the computation is reliable. For a fixed time beyond the critical time, the central region of the layer does not show a converging pattern, but at different times one can observe a similarity in the profiles. This non-uniformity behaviour makes it very difficult to extrapolate the limiting behaviour from the profiles of the layer.

A close examination reveals that the evolution generically occurs in three phases: First, the vorticity advects to the centre (i.e. $\alpha = \pi$), causing a further thickening near the centre. Second, the vorticity in the centre quickly reforms into a roughly elliptical core with trailing arms, which subsequently wrap around the core as it evolves. As the value of H becomes smaller, the vorticity becomes more intense, which leads to faster roll-up. For thinner layers the core structure becomes a smaller fraction of the total layer. The

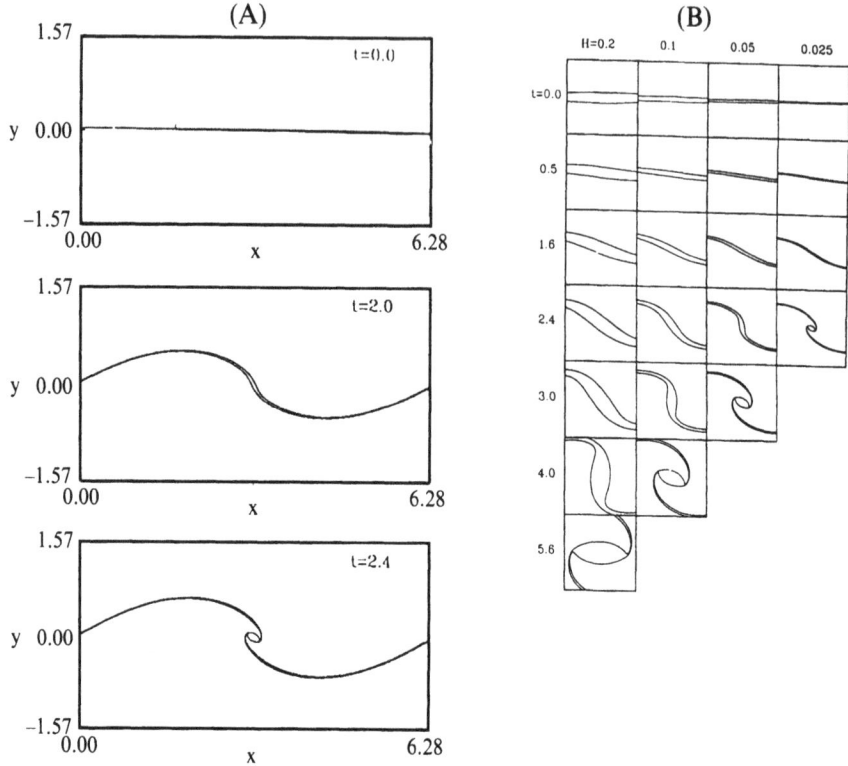

Fig. 4. (A) The location of the layer interfaces for $H = 0.025$ at various times. From Baker and Shelley (1990). (B) Thin-vortex-layer-solutions. The core region of the layer interfaces for various times and thicknesses. From Baker and Shelley (1990).

core seems to collapse to a point with no circulation but infinite vortex-sheet strength. Assuming such a limit exists, it would converge to a weak solution of the Euler equations described by DiPerna and Majda (1987a,b) and Delort (1991). The cores with their trailing arms are very similar to the structures observed by Zabusky *et al.* (1979) in their numerical study of the vortex patches. The simulations also agree qualitatively with the vortex-layer simulations done by Pozrikidis and Higdon (1985).

4.4. Vortex-in-cell method

Here we present the vortex-in-cell (VIC) method for computing vortex sheets by Tryggvason (1989), and compare with the vortex-blob calculations by Krasny (1986b). Usually the VIC method is used only as a device to speed up the calculation of the velocities from the vorticity. However, the grid-particle interpolations usually introduce some numerical smoothing. This smoothing is generally regarded as an unpleasant property of the VIC method because it may suppress the small-scale interactions. On the other hand, one can also regard the VIC method as a grid-based vortex method with the blob size equal to the mesh size. It actually has regularisation properties similar to the vortex-blob method.

We begin with the two-dimensional vortex method for the Euler equation in the vorticity form:

$$\frac{D\omega}{dt} = 0, \quad \nabla^2\psi = -\omega, \quad \mathbf{u} = (\partial_y\psi, -\partial_x\psi), \tag{4.16}$$

where ω is the vorticity, ψ is the stream function and

$$\frac{D}{dt} = \frac{\partial}{\partial t} + \mathbf{u} \cdot \nabla$$

is the material derivative. In vortex methods, the vorticity field is approximated by a collection of discrete point vortices, each with circulation Γ_i. By (4.16), the circulation of each vortex is conserved in time. The Lagrangian particle positions can be found by integrating

$$\frac{d\mathbf{x}_i}{dt} = \mathbf{u}(\mathbf{x}_i, t).$$

To find the velocity from the vorticity, we need to solve the Poisson equation for the stream function. Traditional vortex methods make use of the Biot–Savart kernel, and the fact that the solution can be written as a sum over the singular point sources:

$$\mathbf{u}(\mathbf{x}) = \sum_i \mathbf{K}(\mathbf{x} - \mathbf{x}_i)\Gamma_i,$$

where $\mathbf{K}$ is the Biot–Savart kernel

$$\mathbf{K}(\mathbf{x}) = \frac{1}{2\pi}\frac{(-y, x)}{|\mathbf{x}|^2}, \quad \mathbf{x} = (x, y).$$

Due to the singularity of the Biot–Savart kernel, there has been concern about the possibility of producing unbounded velocity as two neighbouring particles approach each other. To alleviate this difficulty, Chorin (1973) and Chorin and Bernard (1973) introduced a vortex-blob method in which the singular Biot–Savart kernel is replaced by a desingularised kernel, that is,,

$$\mathbf{K}_\delta(\mathbf{x}) = \mathbf{K} * \phi_\delta(\mathbf{x}),$$

where $\phi_\delta(\mathbf{x}) = \phi(\mathbf{x}/\delta)/\delta^2$ is an approximate Delta function, and $\phi(\mathbf{x})$ is its shape function. For example, we can take ϕ to be Gaussian. This modification gives a computationally more stable method than the point-vortex method. The method has been applied and extended to a variety of fluid-dynamical situations (Leonard, 1980). For smooth vorticity fields, it has been proved that the vortex-blob method converges provided that the smoothing blob δ is much larger than the initial grid size h, see, for example, Hald (1979), Beale and Majda (1982), Anderson and Greengard (1985) and Cottet (1988) and the review paper (Hald 1991). For a long time, it has been widely believed that the point-vortex method is numerically unstable without additional regularisation. In Goodman, Hou and Lowengrub (1990), Hou and Lowengrub (1990) and Cottet, Goodman and Hou (1991), we proved a surprising result. The point-vortex method is stable and convergent with second-order accuracy for smooth vorticity fields in two and three space dimensions.

An alternative to the direct summation methods just described are grid-based methods, that work directly with the Poisson equation. The singular point-vortex distribution is approximated by a smoother grid vorticity and the elliptic equation in (4.16) is solved by a fast Poisson solver for difference methods. The grid velocity is obtained by numerical differentiation of ψ over the grid and the velocity of the point vortices is found by interpolating from the grid. Such grid-based methods are generally referred to as vortex-in-cell (or cloud-in-cell) methods, which were first introduced by Christiansen (1973). Since the velocity field is calculated from a smooth grid vorticity, vortex-in-cell methods may be considered as a type of vortex-blob method (Tryggvason, 1989). In 1987, Cottet presented a VIC method for which he was able to show convergence under similar conditions to the blob methods.

In Christiansen's original VIC method, the vorticity of the point vortices is assigned to the corners of the mesh block that each vortex is in by the so-called area–weight rule. This corresponds to giving each vortex an effective area of the order of one mesh block. Thus we may consider the VIC method as a type of vortex-blob method with blob size of the same order as the mesh size. However, since only the nearest four grid nodes are involved, vorticity is not evenly distributed to the nearest four grid nodes. The resulting blob is anisotropic, rather than symmetric, as the blobs in the vortex-blob methods. If the problem being simulated is sensitive to small-scale disturbances, this anisotropy can trigger small-scale Kelvin–Helmholtz instability. This small-scale instability has severely limited previous investigations of the effects of grid refinements (Baker, 1979). Tryggvason (1989) overcame this difficulty by making the blob slightly larger and spreading the vorticity over a larger area on the grid. By doing so, the small-scale anisotropy can be made significantly smaller. The shape function Tryggvason used is the interpolation function suggested by Peskin (1977) in a slightly different context. This conversion of the singular point vortex into a smooth grid vorticity can be viewed as approximating the δ-function by a smoother function. The smoother-shape function at the grid point (i, j) is expressed as a product of two one-dimensional functions:

$$\delta_{ij}(x, y) = \mathrm{d}(x - ih)\mathrm{d}(y - jh)$$

where

$$d(r) = \begin{cases} (1/4h)(1 + \cos(\pi r/2h), & \text{if } |r| < 2h \\ 0, & \text{if } |r| \geq 2h, \end{cases}$$

where h is the mesh size, and the point vortex is located at (x, y). The various aspects of this approximation are discussed in detail by Peskin (1977). Using this approximate Delta function, we approximate the vorticity field at the grid points from the vorticity field at the Lagrangian particle positions $(x_k(t), y_l(t))$ by

$$\omega(ih, jh) = \sum_{k,l} d(ih - x_k(t))d(jh - y_l(t))\omega(x_k(t), y_l(t))h^2.$$

Conversely, we interpolate the velocity field from the grid points to the Lagrangian particle positions by

$$\mathbf{u}(x_i(t), y_j(t)) = \sum_{k,l} d(x_i(t) - kh)d(y_j(t) - lh)\mathbf{u}(kh, lh)h^2.$$

This version of the VIC method has been used successfully by Tryggvason in his numerical study of the Rayleigh–Taylor instability and the vortex-sheet roll-up (Tryggvason, 1988, 1989).

In Figure 5, we present the computations of a vortex-sheet roll-up by Tryggvason (1989) using the VIC method described above. The result is compared with a similar calculation obtained using the vortex-blob method. The VIC simulation in Figure 5a was on a grid with 32 meshes per wavelength; the vortex-blob simulation in Figure 5b used 200 points and $\delta = 0.2$. The vertical dimension of the computational box in the VIC simulation was four times the horizontal one to keep the top and bottom boundaries well away from the interface. The initial conditions were

$$x_i = i/N + 0.05 \sin\left(\frac{2\pi i}{N}\right), \quad y_i = -0.05 \sin\left(\frac{2\pi i}{N}\right),$$

which are the same as those used by Krasny (1986) except that the amplitude was five time larger. This larger amplitude was selected to allow comparisons with runs made by the original four-point VIC code. It was found that the original VIC method was very sensitive to small disturbances from the grid. These disturbances can cause the interface to roll up into more than one vortex.

In Figure 6 we demonstrate the numerical calculations (non-dimensional time equal to 1) obtained using several different methods. Figure 6a and 6b was calculated by the original four-point VIC method. In (a) 16 meshes per wavelength were used, and in (b) 32. Figure 6c and 6d was calculated by the smoother VIC method. In (c) 32 meshes per wavelength were used, and in (d) 64. Figure 6e and 6f was calculated by a vortex-blob method using the modified kernel K_δ. In (e) $\delta = 0.2$, and in (f) $\delta = 0.1$. Both runs

Fig. 5. The roll-up of a vortex sheet. The dimensionless times are 0.0, 0.5, 1.0, 1.5 and 2.0. (A) Calculation with an isotropic vortex-in-cell method. The grid is 32×128. (B) Calculations with a vortex-blob method with 200 points and $\delta = 0.2$. From Tryggvason (1989)

employed a sufficient number of points so that the results were independent of the resolution (in (e) $N = 200$; in (f) $N = 400$). It is evident that the smoother VIC method produces results very similar to those obtained by the vortex-blob method.

We note that several fast algorithms have been developed which give a much faster and accurate evaluation of the particle velocity in the vortex-blob methods. These do not introduce the grid-particle interpolation errors that are present in the VIC method. In (1986), Anderson introduced a fast summation algorithm based on local corrections. It has the advantage of reducing the computational cost of the direct summation without sacrificing the high-order accuracy of the vortex method. The operation count is approximately $O(M \log M) + O(N)$, where M is a constant independent

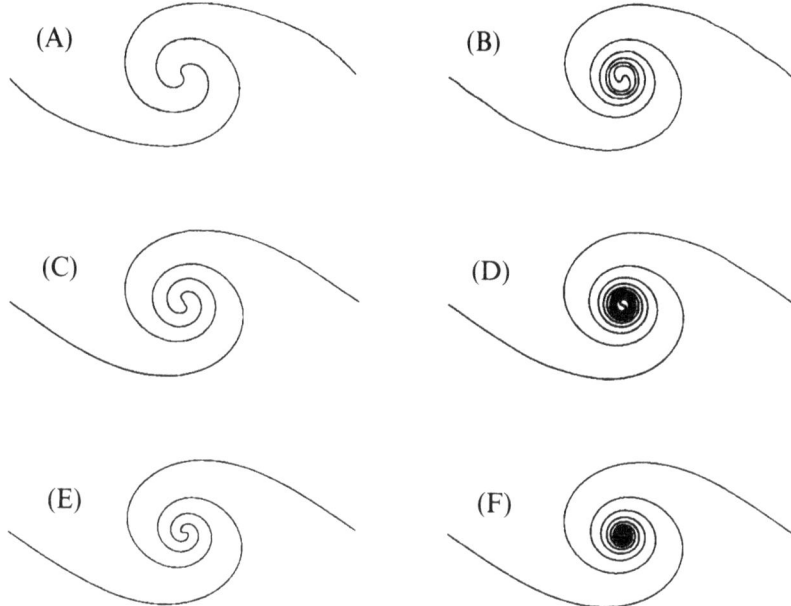

Fig. 6. The large-amplitude stage (at dimensionless time $t = 1.0$) calculated in different ways. (A) Original VIC method on a 16×64 grid. (B) Same as (A) but on a 32×128 grid. (C) Modified (isotropic) VIC method on a 32×128 grid. (D) Same as (C) but on a 64×256 grid. (E) Vortex-blob method 200 points and $\delta = 0.2$. (F) Same as (E) but with 400 points and $\delta = 0.1$. From Tryggvason (1989).

of the number of vortices. The fast multipole summation algorithm developed by Grenngard and Rokhlin (1989) has proved to be very useful. It reduced the operation count from the $O(N^2)$ for direct summation to $O(M \log M) + O(N)$, where M is a constant independent of the number of vortices. A similar fast algorithm has been proposed independently by van Dommelen and Rundensteiner (1989). A somewhat slower, but more flexible version of the fast algorithm based on Taylor expansions has been proposed recently by Draghicescu (1994). A well-vectorised version of the fast multipole algorithm has allowed vortex-method calculations with a large number of vortex particles. For example, in his study of flow past circular cylinders, Koumoutsakos (1993) has used up to $O(10^6)$ vortex particles by efficiently implementing the fast multipole algorithm for vector computer architectures. Fast algorithms have also been used in three-dimensional applications, see, for example, the work of Almgren, Buttke and Colella (1994).

5. Effect of surface tension

The surface tension at an interface between two immiscible fluids arises from the imbalance of their intermolecular cohesive forces. It is one of the most commonly used physical regularisations for interfacial flows. It is believed that surface tension plays a central role in determining the length scales, selection mechanics and large time behaviour of interface dynamics. The understanding of the stabilising effect of surface tension will enhance our understanding of fluid phenomena such as pattern formation in Hele–Shaw cells, the motion of capillary waves on free surfaces, the formation of fluid droplets and the propagation of sound waves in a porous medium.

Surface tension has been used as a physical regularisation for the Kelvin–Helmholtz instability. With surface-tension regularisation, the interface problem is locally well posed. Pullin (1982) was the first to study the stabilising effect of surface tension for vortex sheets. Rangel and Sirignano (1988) also studied the effect of surface tension and density ratio on the nonlinear growth of the Kelvin–Helmholtz instability. Numerical calculations of fluid interfaces with surface tension are more susceptible to numerical instabilities since surface tension introduces high-order spatial derivatives into the governing equations. As in the case of water waves without surface tension, numerical stability requires a certain compatibility between the choice of quadrature rule for the singular integral and the approximation of derivative operators. Violation of this compatibility condition will lead to numerical instability.

5.1. Spatial discretisation

Here we only describe the time-continuous discretisation for general two-density interface problems. Similar discretisation applies to Hele–Shaw flows. Recall that the equations of motion for general two-density fluid interfaces are given by

$$\frac{d\bar{z}}{dt} = \frac{1}{2\pi i} \int \frac{\gamma(\alpha', t)}{z(\alpha, t) - z(\alpha', t)} d\alpha',$$

$$\frac{d\gamma}{dt} = -2A \left(\text{Re} \left\{ \frac{d^2\bar{z}}{dt^2} z_\alpha \right\} + \frac{1}{8} \partial_\alpha \left(\frac{\gamma^2}{|z_\alpha|^2} \right) + g y_\alpha \right) + \tau \kappa_\alpha,$$

Define the derivative operator $D_h^{(\rho)}$ as in the case of water waves; see (3.2). Further, we discretise the singular integral by the alternating-point-trapezoidal rule. The numerical algorithm for which we can prove stability and convergence is given by

$$\frac{d\bar{z}_j}{dt} = \frac{1}{4\pi i} \sum_{(k-j)\text{odd}} \gamma_k^{(\rho)} \cot \left(\frac{z_j^{(\rho)} - z_k^{(\rho)}}{2} \right) 2h; \tag{5.1}$$

$$\frac{d\gamma_j}{dt} = -2A \, \text{Re} \left(\frac{D_h^{(\rho)}}{4\pi i} \sum_{(k-j)\text{odd}} \frac{d\gamma_k}{dt} \cot \left(\frac{z_j^{(\rho)} - z_k^{(\rho)}}{2} \right) 2h \right) \tag{5.2}$$

$$+2A \operatorname{Re}\left(\frac{D_h^{(\rho)}}{4\pi i} \sum_{(k-j)\text{odd}} \gamma_k^{(\rho)} \sec^2\left(\frac{z_j^{(\rho)} - z_k^{(\rho)}}{2}\right) \cdot ((z_j)_t - (z_k)_t)2h\right)$$

(5.3)

$$-\frac{A}{4}D_h^{(\rho)}\left(\frac{\gamma_j^2}{|D_h^{(\rho)}z_j|^2}\right) - 2AgD_h^{(\rho)}y_j + \tau D_h^{(\rho)}\kappa_j,$$

(5.4)

where

$$\kappa_j = \left(D_h^{(\rho)}x_j^{(q)}(D_h^{(\rho)})^2 y_j - D_h^{(\rho)}y_j^{(q)}(D_h^{(\rho)})^2 x_j\right) / ((D_h^{(\rho)}x_j^{(q)})^2 + (D_h^{(\rho)}y_j^{(q)})^{3/2}$$

(5.5)

and $\hat{x}_k^\rho = \rho(kh)\hat{x}_k$ and $\hat{x}_k^q = q(kh)\hat{x}_k$, where $q(x) = \frac{d}{dx}(x\rho(x))$.

The use of x^q, y^q in the curvature computation is to balance the aliasing errors in the high modes due to the non-linearity of the curvature term. Its use is determined by the discrete product rule

$$D_h^{(\rho)}(f\dot{z}) = fD_h^{(\rho)}\dot{z} + (D_h^{(\rho)}f)\dot{z}^q + hA_0(\dot{z})$$

(5.6)

for any smooth function f. To illustrate the algorithm for a practical example, we take the second-order finite-difference derivative operator as an example. Note that $\rho(x) = \sin(x)/x$ and $q(x) = \cos(x)$ if $D_h^{(\rho)}$ corresponds to the second-order centred difference derivative. It is easy to see that

$$f_j^q = \frac{1}{2}(f_{j+1} + f_{j-1}),$$

(5.7)

$$D_h^{(\rho)}f_j^q = \frac{f_{j+2} - f_{j-2}}{4h} = D_{2h}^{(\rho)}f_j,$$

(5.8)

$$(D_h^{(\rho)})^2 f_j = \frac{f_{j+2} - 2f_j + f_{j-2}}{h^2}.$$

(5.9)

Thus, equations (5.8)–(5.9) imply that we should simply use every other grid point when discretising the curvature.

In the presence of surface tension, a higher-order norm is used to estimate the growth rate of the errors. This requires a better control of aliasing errors introduced in the approximation of the singular integrals. For finite-order derivative approximations, $\rho'(\pm\pi) \neq 0$ in general, and so the natural filtering associated with $D_h^{(\rho)}$ is not strong enough to control the aliasing errors. We will need to apply an additional filtering to achieve this result. In equations (5.2)–(5.4), we need to replace $D_h^{(\rho)}$ by $\tilde{D}_h^{(\rho)}$, where

$$\tilde{D}_h^{(\rho)}x_j = D_h^{(\rho)}x_j^s \quad \text{and} \quad \hat{x}_k^s = s(kh)\hat{x}_k$$

(5.10)

where s satisfies

$$|s(kh) - 1| \leq C(kh)^r, s(kh) > 0 \quad \text{and} \quad s(\pm\pi) = 0.$$

(5.11)

The evaluation of the curvature remains unchanged. It is computed exactly the same as in equation (5.5). With these modifications, we can prove the convergence of the algorithm defined by (5.1)–(5.5) (Beale, Hou and Lowengrub, to appear).

Theorem 2 Convergence with Surface Tension Assume that $z(\cdot, t)$, $\phi(\cdot, t) \in C^{m+3}[0, 2\pi]$ and $\gamma(\cdot, t) \in C^{m+2}[0, 2\pi]$ for $t \leq T$ and $m \geq 4$. If D_h corresponds to an rth-order derivative operator, with $r \geq 4$ and $h \leq h_0(T)$, then

$$\|z.(t) - z(\cdot, t)\|_{H_h^1} \leq C(T)h^r, \tag{5.12}$$

$$\|\gamma.(t) - \gamma(\cdot, t)\|_{H_h^{1/2}} \leq C(T)h^{r-1}, \tag{5.13}$$

where

$$\|\phi\|^2_{H_h^{1/2}} = \sum_{|k| \leq N/2} (1 + |k|\rho(kh))|\hat{\phi}_k|^2 \quad \text{and} \quad \|\phi\|^2_{H_h^1} = \|\phi\|^2_{l^2} + \|D_h\phi\|^2_{l^2}. \tag{5.14}$$

If $D_h^{(\rho)}$ corresponds to a spectral derivative approximation, the result is the same with r replaced by m.

As we can see, there are many choices of quadrature rule and derivative rule. Also, it is not clear which term needs to be smoothed, and which term need not be smoothed. Our analysis indicates that the combination of these choices must satisfy certain compatibility conditions in order to be stable. These compatibility conditions can be determined by performing linear stability analysis around the arbitrary smooth solution of the interface. In principle, such analysis is non-trivial and could be very messy. By studying the leading-order linear singular operators and projecting them into the appropriate local coordinates, a simplified system can be derived from which stability of the numerical method becomes apparent. We note that with surface-tension regularisation, the interface is locally well-posed (Craig, 1985; Beale, Hou and Lowengrub, 1993a). The sign of gravity plays no role. The convergence result holds even if the fluid is unstably stratified.

The proof of Theorem 2 relies on an estimate of the linearised error in the curvature. Let

$$\dot{\kappa}_j = \kappa_j - \kappa(\alpha_j). \tag{5.15}$$

Using the discrete product rule and the fact that $(fg)^q = gf^q + hA_0(f)$ for any smooth g, we can show that the linear part of $\dot{\kappa}_j$ is given by

$$\dot{\kappa}^i_j = \frac{1}{s_\alpha(\alpha_j)} D_h^{(\rho)} \left[\frac{1}{s_\alpha(\alpha_j)} D_h^{(\rho)} \dot{x}^N_j \right] + A_0(\dot{x}^T) + A_0(\dot{x}^N). \tag{5.16}$$

We refer the reader to Beale, Hou and Lowengrub (to appear) for details.

Performing a linear stability analysis similar to that for water waves, we obtain

$$\frac{d\dot{\phi}}{dt} = -\frac{\gamma\sigma^2}{2}D_h^{(\rho)}\dot{\psi} + A_0(\dot{\phi} + \dot{\psi}) + A_{-1}(\dot{\Gamma}), \tag{5.17}$$

$$\frac{d\dot{\psi}}{dt} = -2\sigma D_h^{(\rho)}\dot{\Gamma} - \frac{\gamma\sigma^2}{2}D_h^{(\rho)}\dot{\phi} - A\gamma\sigma^2 D_h^{(\rho)}\dot{\phi} + A_0(\dot{\phi} + \dot{\psi}) + A_{-1}(\dot{\Gamma}), \tag{5.18}$$

$$\frac{d\dot{\Gamma}}{dt} = \tau\sigma D_h^{(\rho)}\left(\sigma\left(\frac{\gamma^2\sigma}{4\tau} - \Lambda_h\right)\dot{\psi}\right) + A_0(\dot{\phi} + \dot{\psi} + \dot{\Gamma}), \tag{5.19}$$

where $\sigma = 1/s_\alpha$ and $D_h^{(\rho)}$ and Λ_h are defined as in (3.2) and (3.9). Here $\dot{\phi}$ denotes a variant of $D_h^{(\rho)}\dot{x}^T$, $\dot{\psi}$ denotes a variant of $\Lambda_h\dot{x}^N$ and $\dot{\Gamma}$ denotes a variant of $\dot{\gamma}$. Recall that the operator Λ_h is positive, that is, $\widehat{D_{hk}^{(\rho)}} = |k|\rho(kh)$. In deriving equations (5.17)–(5.19), we have changed variables so that the coupling between $\dot{\phi}$ and $\dot{\psi}$ changes from elliptic to hyperbolic. By doing this, we have successfully put the term responsible for the Kelvin–Helmholtz instability to the third equation. It appears as $\gamma^2\sigma/4\tau$ that is added to $-\Lambda_h$. The Λ_h term represents the dispersive effect of surface tension. We can see that the dispersive regularisation dominates the destabilising term for those wavenumbers k satisfying $|k| > \max_\alpha\{\gamma^2\sigma/4\tau\}$. This observation leads to our energy estimate and convergence proof. The details are given in Beale, Hou and Lowengrub (to appear); see also Beale, Hou and Lowengrub (to appear).

We remark that the numerical approximations discussed for two-density interfaces can easily be generalised for Hele–Shaw flows and other multi-fluid interfaces.

5.2. Removing the stiffness of surface tension for interfacial flows

It turns out that it is difficult to obtain a stable and efficient time integration scheme for fluid interfaces with surface tension. If an explicit time discretisation is used, there is a severe time-step stability constraint. This constraint arises because of the presence of surface tension, and is a major obstacle to performing high-resolution, long-time numerical simulations. In this section, we present a new approach that successfully removes the high-order time-step constraint induced by surface tension. This approach was developed in detail, and demonstrated through numerical simulations, in Hou, Lowengrub, and Shelley (1994a). Using our method, it is possible to perform accurate and large time integration of fluid interfaces with surface tension. Many previously untenable problems now become possible using our approach. The application of these methods has led to the discovery of interesting new phenomena. For example, numerical calculations of the vortex-sheet roll-up with surface tension, using up to 8192 points, reveal the late time self-intersection of the interface, which creates trapped bubbles of fluid. This is very interesting. A collision of interfaces is a singularity in the evolution, and is of a type that has not been observed previously for such flows in 2-D. We will describe this further below. Our methods have also been applied recently to problems of topological singularity formation in Hele–Shaw flows (Goldstein, Pesci and Shelley 1993) and to studies of the effect of anisotropy in quasi-static solidification (Almgren, Dai and Hakim, 1993).

By stiffness, we mean the presence of strict time-step stability constraints. The stiffness is introduced by the curvature term in the Laplace–Young boundary condition. For incompressible fluid interfaces, it is especially difficult to remove the stiffness of surface tension because the stiffness enters non-linearly and non-locally. Straightforward implicit discretisation would not work since it could be as expensive to solve for the implicit solution. By performing the frozen coefficient Fourier analysis of the interface equations, we can derive the dynamic stability constraint

$$\Delta t < C \cdot (\bar{s}_\alpha h)^{3/2}/\tau, \tag{5.20}$$

where $\bar{s}_\alpha = \min_\alpha s_\alpha$. Therefore, the stability constraint is determined by the minimum grid spacing in arclength ($\Delta s \approx h s_\alpha$), which is strongly time dependent. Our experience is that the Lagrangian motion of the points can lead to 'point clustering' and hence to very stiff systems, even for flows in which the interface is smooth and the surface tension is small. For example, in previous calculations of the motion of vortex sheets with surface tension, a fourth-order (explicit) Runge–Kutta method was used to advance the system. An adaptive time-stepping strategy was used to satisfy the stability constraint. With $N = 256$, the time step had become as small as 10^{-6}, and soon thereafter the computation became too expensive to continue. For Hele–Shaw flows the situation is even worse. A similar analysis gives the constraint

$$\Delta t < C \cdot (\bar{s}_\alpha h)^3 /\tau. \tag{5.21}$$

Our approach relies on two key observations. The first is to introduce a new set of variables for which curvature can be evaluated 'linearly' through these new variables. The second observation is to factor out the leading-order singular linear operators from the non-linear and non-local system. This gives rise to a much simplified leading-order system to which standard implicit methods such as the Crank–Nicholson scheme can be trivially applied.

Our new set of dynamical variables consists of the tangent angle, θ, and arclength metric, s_α, of the interface. This is strongly motivated by the formula $\kappa = \theta_s = \theta_\alpha/s_\alpha$. Furthermore, we would like to impose α as an arclength variable. This is equivalent to imposing s_α is a function of time alone. By doing this, we need to reparameterise the interface dynamically, which amounts to a change of frame in time by introducing a particular tangential velocity T.

Given an equation of motion of a free interface,

$$(x(\alpha, t), y(\alpha, t))_t = Un + T\hat{s},$$

where n and $\hat{s}$ are the unit local normal and tangential vectors, respectively and U and T are the local normal and tangential components of the interface velocity. Define the tangent angle θ and the arclength metric s_α as $\tan\theta = y_\alpha/x_\alpha$, $s_\alpha = \sqrt{x_\alpha^2 + y_\alpha^2}$. It is easy to derive an equivalent equation of motion for θ and s_α

$$(s_\alpha)_t = T_\alpha - \theta_\alpha U, \tag{5.22}$$

$$\theta_t = \left(\frac{1}{s_\alpha}\right)(U_\alpha + \theta_\alpha T). \tag{5.23}$$

For most interface problems of practical interest, the motion of the interface is determined only by the normal velocity. The tangential velocity would determine the frame or parameterisation of the interface, but it does not affect the shape of the interface. We will exploit this degree of freedom in choosing T to derive a simpler evolution equation for s_α and θ. Ideally, we would like to choose a frame such that the moving particles $\{(x(\alpha_j, t), y(\alpha_j, t))\}_j$ are equally spaced at all times if they are so initially. This corresponds to imposing s_α to be independent of α, varying with time only. To achieve this, we choose the tangential velocity T such that

$$T_\alpha - \theta_\alpha U = \frac{1}{2\pi}\int_0^{2\pi}(T_\alpha - \theta_\alpha U)\mathrm{d}\alpha.$$

Since T is periodic with respect to α, we get

$$T(\alpha, t) = T(0, t) + \int_0^\alpha \theta_{\alpha'} U \mathrm{d}\alpha' - \frac{\alpha}{2\pi}\int_0^{2\pi}\theta_{\alpha'}U\mathrm{d}\alpha'. \tag{5.24}$$

This expresses T entirely in terms of θ and U. The spatial constant $T(0, t)$ just gives an overall temporal shift in frame. With this choice of T, the evolution equations for θ and s_α reduce to

$$(s_\alpha)_t = -\frac{1}{2\pi}\int_0^{2\pi}\theta_\alpha U\mathrm{d}\alpha. \tag{5.25}$$

$$\theta_t = \left(\frac{1}{s_\alpha}\right)(U_\alpha + \theta_\alpha T). \tag{5.26}$$

This system should be solved, together with the evolution equations governing other dynamical variables, such as vortex-sheet strength, velocity potential, etc. This is a complete reformulation of the evolution problem. Once we obtain s_α and θ in time, we can recover the interface position (x, y)

by an integration (up to a constant of integration). This formulation of plane curve motion is not new. See, for example, Strain (1989) in the context of unstable solidification.

Next we would like to factor out the leading-order linear singular operators in the evolution equations. To illustrate the idea, we take vortex sheets with surface tension as an example. It is important to note that the time-step stability constraint is a result of high-order derivative and singular operators. They enter only at small spatial scales or high-frequency components of the solution. Thus it is essential to single out the leading singular operators and high-order derivative terms, and treat these terms implicitly. Although the Birkhkoff–Rott equation is highly non-linear and non-local, its leading-order approximation at small scales is extremely simple. It can be expressed in terms of the Hilbert transform. With some manipulation, we find that for a vortex-sheet flow with surface tension, the normal velocity U behaves at small scales as

$$U \sim \frac{1}{2} \left(\frac{2\pi}{L} \right) \mathcal{H}[\gamma],$$

while for γ,

$$\gamma_t \sim \tau \kappa_\alpha.$$

It is worth noting that the Hilbert transform is diagonalisable under the Fourier transform. Now we can recompose the equations of motion to a form suitable for applying implicit time-integration methods. We will separate the leading-order singular operators from the smoother and lower-order operators. The leading-order terms dominate at small scales, and will be treated implicitly. The smoother and lower-order terms are non-linear and non-local. We will treat them explicitly. There is no stiffness in the equation for s_α since only the space-averaged quantity enters the equation. The stiffness of the system is in the coupling of the θ and γ equations. The recomposed system for θ and γ is given by

$$\theta_t = \left(\frac{1}{2s_\alpha^2} \right) \mathcal{H}[\gamma_\alpha] + P, \tag{5.27}$$

$$\gamma_t = \left(\frac{\tau}{2s_\alpha} \right) \theta_{\alpha\alpha} + Q. \tag{5.28}$$

The first term in each equation is the leading-order term, dominant at small scales. P and Q represent the smoother and lower-order terms. They are obtained by subtracting off the leading-order terms from the right-hand sides of the θ and γ equations respectively. We term this form of the equations of evolution *Small Scale Decomposition* (SSD). It is the leading terms that introduce the stiffness into the system. These leading terms diagonalise under the Fourier transform, and can be treated implicitly very easily. A very similar, and simpler, recomposition can be found for Hele–Shaw flows.

A stable second-order integration can be obtained by discretising the leading-order stiff terms implicitly using a Crank–Nicholson discretisation, and leap-frogging on the non-linear terms. To simplify the notation, we denote $2\pi s_\alpha$ by $L(t)$. It is the total arclength of the interface. In Fourier space, this gives

$$\frac{\hat{\theta}^{n+1} - \hat{\theta}^{n-1}}{2\Delta t} = \frac{|k|}{4} \left(\left(\frac{2\pi}{L^{n+1}} \right)^2 \hat{\gamma}^{n+1} + \left(\frac{2\pi}{L^{n-1}} \right)^2 \hat{\gamma}^{n-1} \right) + \hat{P}^n(k), \quad (5.29)$$

$$\frac{\hat{\gamma}^{n+1} - \hat{\gamma}^{n-1}}{2\Delta t} = -\frac{\tau}{2} k^2 \left(\frac{2\pi}{L^{n+1}} \hat{\theta}^{n+1} + \frac{2\pi}{L^{n-1}} \hat{\theta}^{n-1} \right) + \hat{Q}^n(k). \quad (5.30)$$

Given L^{n+1}, $\hat{\theta}^{n+1}(k)$ and $\hat{\gamma}^{n+1}(k)$ can be found explicitly by inverting a 2×2 matrix. By using an explicit method to integrate L (a non-stiff ODE), L^{n+1} is found before updating θ and γ. At most, a first-order CFL condition must be satisfied because of transport terms in the θ and γ evolutions. And indeed, numerical simulations of the fully non-linear flow show no high-order time-step constraint from the surface tension, but do reveal a first-order CFL constraint. Further details on implementation are found in Hou, Lowengrub and Shelley (1994). Recently, we have been able to prove convergence of the above reformulated boundary-integral method for general two-density interface problems with surface tension (Ciniceros and Hou, to appear). This includes Hele–Shaw flows and water waves.

A fourth-order implicit discretisation

We can also design a fourth-order implicit multistep discretisation in time. Motivated by the work of Ascher, Ruuth and Wetton (to appear), we propose the following fourth-order implicit discretisation in time:

$$(25/12)\hat{\theta}^{n+1} - 4\hat{\theta}^n + 3\hat{\theta}^{n-1} - (4/3)\hat{\theta}^{n-2} + (1/4)\hat{\theta}^{n-3}$$

$$= \Delta t \left(\frac{|k|}{2} \left(\frac{2\pi}{L^{n+1}} \right)^2 \hat{\gamma}^{n+1} + 4\hat{P}^n(k) \right.$$

$$\left. - 6\,\hat{P}^{n-1}(k) + 4\hat{P}^{n-2}(k) - \hat{P}^{n-3}(k) \right)$$

$$(25/12)\hat{\gamma}^{n+1} - 4\hat{\gamma}^n + 3\hat{\gamma}^{n-1} - (4/3)\hat{\gamma}^{n-2} + (1/4)\hat{\gamma}^{n-3}$$

$$= \Delta t \left(-\tau k^2 \frac{2\pi}{L^{n+1}} \hat{\theta}^{n+1} + 4\hat{Q}^n(k) \right.$$

$$\left. - 6\,\hat{Q}^{n-1}(k) + 4\hat{Q}^{n-2}(k) - \hat{Q}^{n-3}(k) \right).$$

We have tested this fourth-order version of implicit discretisation. It indeed gave a fourth-order convergence with a CFL stability constraint which is about half of that in the second-order Crank–Nicholson discretisation

(Hou, Lowengrub and Shelley, to appear). The improved order of accuracy
in time is very important for large time integration of free interfaces. And it
is especially useful in our study of the formation of topological singularities;
see; Hou, Lowengrub and Shelley (to appear).

We would like to emphasise that the equal arclength frame we described
above is one convenient choice. This choice leads to a constant coefficient
system to the leading order which makes the inversion explicit by using the
fast Fourier transform. But there are other situations where we may want to
choose a non-equal arclength frame that is adapted to the local property of
the interface. For example, we may want to cluster computational particles
near a singular region. This can be carried out in a similar way. In this
case, the implicit solutions become a variable coefficient problem, which can
be solved by some iterative methods such as the preconditioned conjugate
gradient method. We refer to Hou, Lowengrub and Shelley (1994; to appear)
for more discussions of the formulation and implementation issues.

5.3. Numerical examples

In this section, we present some very interesting numerical simulations that
serve to demonstrate the utility of the SSD. The numerical methods are
based on the Crank–Nicholson discretisation discussed above. This yields
a stable, second-order in time, infinite-order in space discretisation. We
are interested in understanding the competing effects of surface tension and
the Kelvin–Helmholtz instability on the motion of a vortex sheet. In our
calculation, $\tau = 0.005$, with the initial condition

$$x_0(\alpha) = \alpha + 0.01 \sin 2\pi\alpha, \, y_0(\alpha) = -0.01 \sin 2\pi\alpha, \, \gamma_0(\alpha) = 1. \qquad (5.31)$$

This initial data was used by Krasny (1986a,b) in the absence of surface
tension to study singularity formation through the Kelvin–Helmholtz instability. In the case of zero surface tension, a curvature singularity was shown
to occur at the centre ($\alpha = 1/2$) at $t \approx 0.375$. With $\tau = 0.005$, the linear
dispersion analysis gives approximately 16 linearly growing modes above
$k = 0$. Modes higher than 16 are all linearly stable and are dispersively
regularised by surface tension.

Figure 7 shows a sequence of interface positions, starting from the initial condition (Figure 7A). At early times, the interface steepens and behaves similarly to the zero-surface-tension case. However, it passes smoothly
through the $\tau = 0$ singularity time, and becomes vertical at the centre at
$t \approx 0.45$. At about this time, dispersive waves are generated at the centre
and propagate outwards. By $t = 0.6$ (Figure 7B) the interface has rolled
over and has begun to roll-up into a spiral. However, at later times (see
Figure 7C and D), sections of interface within the inner turns of the spiral
appear to be attracted towards one another, and in the process appear to

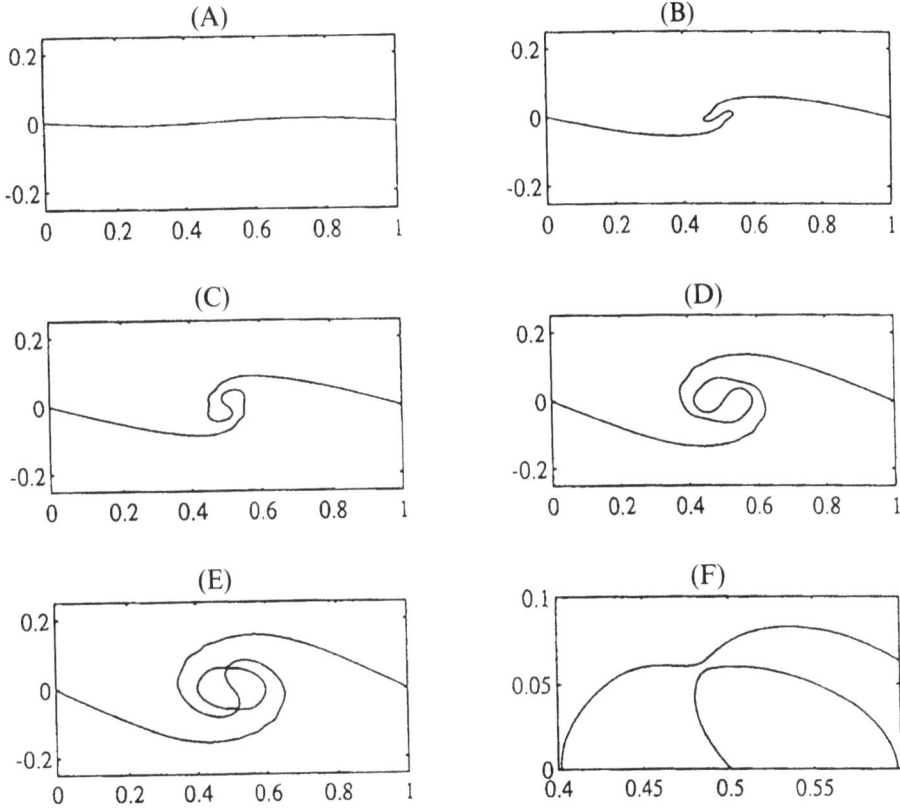

Fig. 7. Point-vortex-method solution of vortex sheets with surface tension, $\tau = 0.005, N = 1024, \Delta t = 1.25 \times 10^{-4}$. (A) $t = 0$; (B) $t = 0.6$; (C) $t = 0.8$; (D) $t = 1.2$; (E) $t = 1.4$; (F) close-up of top pinching region, $t = 1.4$. From Hou, Lowengrub and Shelley (1994)

pinch off interior 'bubbles' of fluid (see Figure 7E). A close-up of the pinch region is shown in Figure 7F.

These sections of interface in the pinching region appear to collide at a finite time. Figure 8 shows the minimum distance between the two sections of interface in the pinching region as a function of time for several spatial and temporal resolutions. Figure 8B shows that the total energy is conserved up to 6-digit accuracy very close to the time of pinching for $N \leq 2048$. This figure suggests strongly that the pinching occurs at a finite time. Moreover, the width apparently vanishes with infinite slope, which indicates that the pinching rate intensifies as the width narrows.

We remark that this apparent singularity is of a completely different type from that of the $\tau = 0$ singularity. This singularity is a *topological singularity*. Beyond the pinching singularity, change in the topology of the flow may occur. More fundamentally, with $\tau = 0$ the singularity occurs

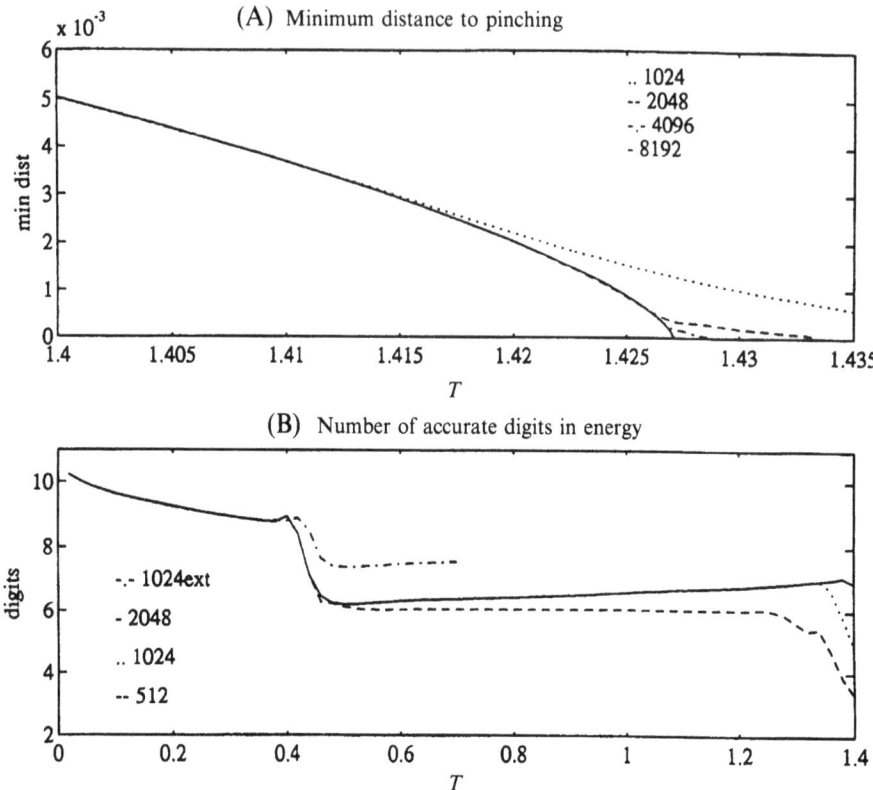

Fig. 8. Pinching in, and accuracy in the energy of, the inertial vortex sheet: (A) minimum width of the top pinching region, $\tau = 0.005$, $N = 1024$ with $\Delta t = 2.5 \times 10^{-4}$, $N = 2048$ with $\Delta t = 1.25 \times 10^{-4}$, $N = 4096$ with $\Delta t = 6.25 \times 10^{-5}$, $N = 8192$ with $\Delta t = 3.125 \times 10^{-5}$; (B) number of accurate digits in the energy, τ 0.005, $N = 512, N = 1024, N = 2048$ with $\Delta t = 1.25 \times 10^{-4}$, and $N = 1024$ with $\Delta t = 3.125 \times 10^{-5}$(1024ext). From Hou, Lowengrub and Shelley (1994).

through a rapid *compression* of vorticity along the sheet at a single isolated point (Moore, 1979; Krasny, 1986; Baker and Shelley, 1990; Shelley, 1992). Here the singularity occurs through a rapid *production* of vorticity that is associated with the surface tension. And indeed, the maximum vortex-sheet strength appears to diverge at the singularity time, unlike the $\tau = 0$ case. Further details, physical interpretation and modelling are given in Hou, Lowengrub and Shelley (1994) and Hou, Lowengrub and Shelley (to appear).

The second example we consider is the expanding bubble. This is a calculation of a gas bubble expanding into a Hele–Shaw fluid; see Figure 9. The dynamics of expanding bubbles in the radial geometry have attracted a great deal of attention due to the formation of striking patterns observed in experiments. In our set-up, the viscosity inside the bubble is set to zero, but the viscosity outside the bubble is equal to 1. Therefore, there is a viscosity contrast $A_\mu = 1$. Thus the γ equation now reads

$$\gamma = -2A_\mu s_\alpha U^T + \tau \kappa_\alpha,$$

where U^T is the tangential velocity component of the interface velocity (u, v). In this case, γ is defined implicitly through an integral relation. An iterative scheme is needed to solve for γ. The initial condition is given by

$$(x_0(\alpha), y_0(\alpha)) = r(\alpha)(\cos \alpha, \sin \alpha),$$
$$\text{with } r(\alpha) = 1 + 0.1 \sin 2\alpha + 0.1 \cos 3\alpha.$$

See the innermost curve in Figure 9. This choice of initial condition is to avoid particular symmetry in the initial interface. The value of surface tension is $\tau = 0.001$, the time step is $\Delta t = 0.001$ and $N = 4096$. Figure 9 shows the expansion of this bubble from $t = 0$ to $t = 20$, printed at unit intervals of time. We can see that the interface develops oscillations in the moving front, and subsequently produces many fingers and pedals as time evolves. These petals expand outwards and eventually tip-split into two petals. This process repeats itself. In performing this calculation, we have done a careful resolution study. This calculation agrees very well with lower-resolution calculations. It is quite remarkable that we can now use such a large time step $\Delta t = 0.001$ for $N = 4096$. Without the new formulation, the time step would have been at least one thousand times smaller to achieve stability for an explicit method.

5.4. A note on computing beyond the singularity time

Topological changes or formation of singularities in free interfaces occur in many physical applications. For example , in crystal growth and thin-film growth, an initial smooth front can develop cusps and crack-like singularities, and isolated islands of film material can merge (Gray, Chisholm and Kaplan, 1993; Sethian, 1985; Snyder *et al.*, 1991; Spencer, Vorhees and Davis, 1991). Computing beyond the singularity time using front tracking methods is usually very difficult and complicated. Local grid surgery is required to reconnect the Lagrangian particles near the singularity region; see, for example, Unverdi and Tryggvason (1992). Also, how we reconnect the interfaces may affect the solution at later times. So it would be highly desirable to develop a more systematic framework for boundary-integral methods to compute beyond the (topological) singularity time. In the next section, we will discuss front capturing methods based on the level-set approach. But here we would like to exploit further what we can do within the framework of boundary-integral methods.

Here we propose a new approach to continue our boundary-integral calculation beyond the topological singularity. This borrows ideas from the

Fig. 9. An expanding Hele–Shaw bubble. $N = 4096, \Delta t = 0.001, \tau = 0.001$, $t = 0, 1, 2, \ldots, 20$. From Hou, Lowengrub and Shelley (1994).

level-set approach (see the next section). The idea is to use curvature regularisation, as has been successfully used for the level-set approach. By curvature regularisation, we mean that we add to the normal velocity component a term proportional to the local mean curvature, that is, $U' = U + \epsilon\kappa$. Here U is the normal component of the interface velocity, κ is the (mean) curvature and U' is the regularised normal velocity. We are interested in studying the limiting solution as $\epsilon \to 0^+$, beyond the singularity time.

To illustrate the idea, we take the example of motion by mean curvature. The normal velocity is given by $U = 1 + \epsilon\kappa$, where κ is local curvature. In the limit of $\epsilon = 0$, the equal arclength frame would choose a tangential velocity $T = \theta$. With this choice of T, our reformulated system of equations for θ and s_α becomes a variant of the viscous Burgers equation:

$$(s_\alpha)_t = 1 - \epsilon(\theta_\alpha)^2/s_\alpha, \tag{5.32}$$

$$\theta_t = \frac{\theta\theta_\alpha}{s_\alpha} + \frac{\epsilon}{s_\alpha}\left(\frac{\theta_\alpha}{s_\alpha}\right)_\alpha. \tag{5.33}$$

In the limit of $\epsilon = 0$, the θ equation becomes the inviscid Burgers equation, and s_α becomes constant (positive) in space. Now it is clear that a cusp or topological singularity corresponds to a shock discontinuity in the tangent angle. It is well known that an entropy condition is required to select the physical weak solution beyond the time a shock discontinuity is formed. For positive ϵ, the curvature regularisation plays exactly the same role as the viscosity regularisation. Thus using an upwinding scheme or high-order Godonov scheme for computing θ would give the correct continuation beyond the singularity. By applying curvature regularisation to a free surface directly, we do not need to introduce one extra space dimension as in the level-set approach. More accurate numerical methods can be designed since we only deal with the free surface and don't have to differentiate across the free surface. Also, the stiffness can be removed easily using our reformulated system.

We have used our formulation to reproduce some of the calculations presented in Osher and Sethian (1988) using the level-set formulation. We obtained the same results for computations of the cusp and corner singularities. In Figure 10, we plot the evolution of a sinusoidal initial condition propagating with unit normal velocity. The initial condition is given by $x(\alpha, 0) = \alpha, y(\alpha, 0) = -0.05 \sin(2\pi\alpha)$. $N = 128$, and an upwinding scheme was used to integrate the θ equation in time. Since the curve propagates into itself with unit normal velocity, a corner singularity is formed at later times. It is clear that applying an upwinding scheme to our reformulated system produced the entropy-satisfying continuation beyond the singularity time.

Merging of interfaces can also be handled similarly. We can determine accurately the time of merging by monitoring the minimum distance between the two interfaces, as we did for the vortex-sheet calculation. At the time of merging, we need to reparameterise the merged interface. This can be done by combining the original parameterisation of the two interfaces. For the merged interface, there is a jump discontinuity for θ at the point of contact. This will generate a cusp or corner singularity after the merging of the two interfaces. But using the curvature regularisation described above, the reformulated method can capture the cusp or corner singularities with no additional effort. And the entropy condition is satisfied automatically. Apparently, this idea can be applied to water waves, interaction of fluid bubbles and droplet formation. Detailed description and computational results will be presented elsewhere (Hou and Osher, 1994). Generalisation of this idea to three space dimensional problems is our active on-going research.

We would like to emphasise that curvature regularisation is a geometric (or topological) regularisation. It is frame-independent, and consequently it is an intrinsic regularisation. It has an important property of preserving the index of a curve. As a consequence, a curve cannot cross itself under the curvature regularisation. Of course, if we use curvature regularisation in the

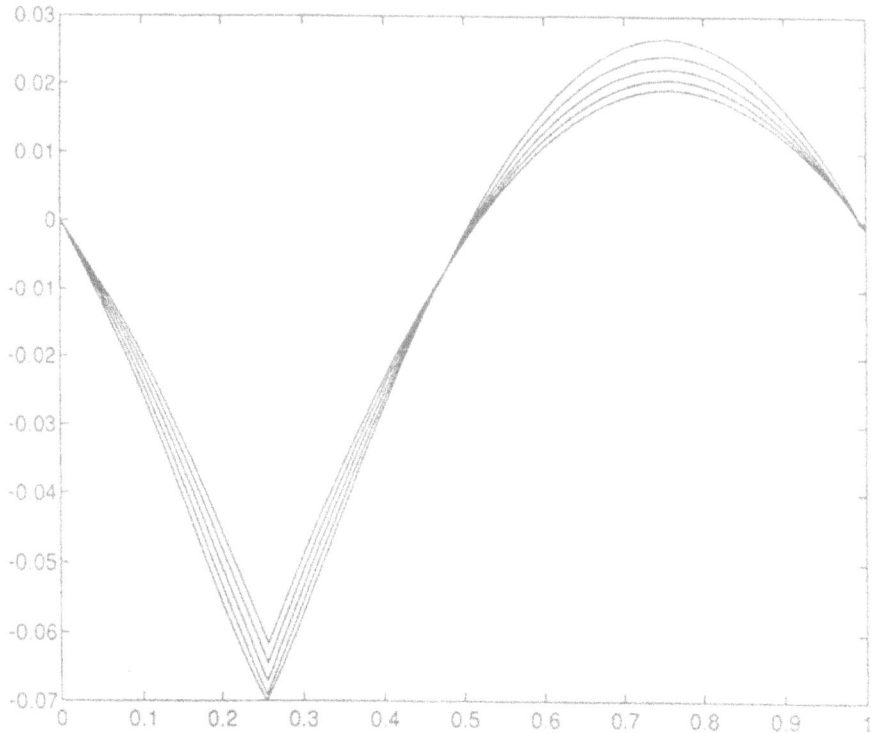

Fig. 10. Motion by mean curvature using reformulated front tracking method with curvature regularisation. The normal velocity $U = 1$. The initial condition is $x = \alpha, y = 0.05\sin(2\pi\alpha)$. $N = 128$. The upwinding scheme was used to integrate in time. From Hou and Osher (1994)

original Lagrangian frame, the differential point clustering of particles will result in a very stiff system to solve. So it is essential to apply curvature regularisation to our reformulated system in which an equal arclength frame is imposed dynamically. The curvature regularisation can also be used to regularise ill-posed problems. Using the point-vortex method approximation with curvature regularisation, we can compute beyond the Kelvin–Helmholtz singularity, and obtain a roll-up solution of vortex sheets. But it is more effective if a small vortex blob of the order of the mesh size is used. In Figure 11A, we present our vortex-sheet calculation using the point-vortex method and the curvature regularisation. The same initial condition as Krasny's was used. The curvature regularisation coefficient is 0.01. The solution is plotted at $t = 1.24$ with $N = 256$. This clearly gives a vortex-sheet roll-up solution. But it seems to require more resolutions to compute further in time. In Figure 11B, we present the same vortex-sheet calculation using a small blob. The blob size is equal to 0.01. The vortex-sheet positions at three different times are shown in Fig. 11B, with $N = 512$. The solutions

(A)

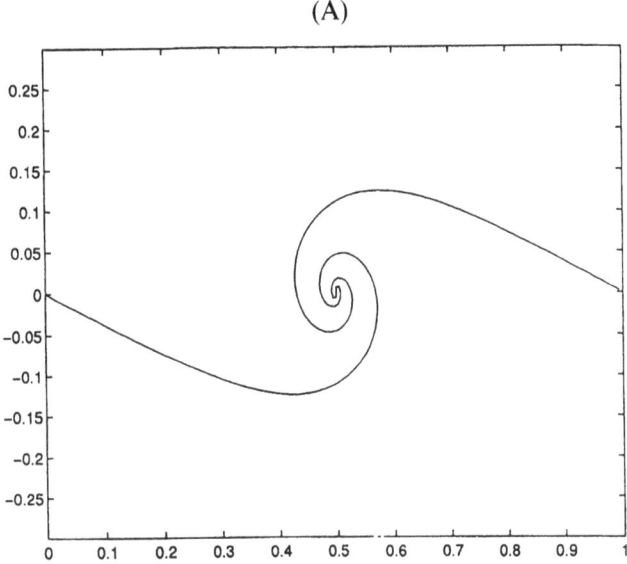

Fig. 11A. Vortex-sheet roll-up calculation using the point-vortex method and the curvature regularisation. The same initial condition as Krasny's was used. The curvature regularisation coefficient is 0.01. The solution is plotted at $t = 1.24$. $N = 256$. From Hou and Osher (1994).

are very similar to those obtained by Krasny using vortex-blob calculations with larger blobs. By using the curvature regularisation, it is possible to study the limiting solution as the regularisation parameters tend to zero simultaneously with the mesh size.

Curvature regularisation introduces a *dissipative* regularisation for the θ equation with respect to the *arclength variable*. It is important that such dissipative regularisation is with respect to the arclength variable. If we naïvely add a dissipative regularisation in the original Lagrangian frame α, the result is quite different. Such Lagrangian regularisation would allow interface self-crossing, producing a non-physical continuation beyond the Kelvin–Helmholtz singularity time (Hou and Osher, 1994).

6. The Level-Set Approach

The level-set approach is an effective front capturing method for computing free surfaces. It was originally introduced by Osher and Sethian in 1988. The basic idea is to consider the free surface as a zeroth-level set of some smooth function which is defined in one higher space dimension than the free surface. So advancing the free surface is reduced to advancing the level-set function. Since only the zeroth-level set is physically relevant to the free surface, there is a lot of freedom in advancing the level-set function away from the zeroth-level set. Such freedom can be exploited to design a

(B)

Fig. 11B. Vortex-sheet roll-up calculation using the curvature regularisation with a small vortex blob. The same initial condition as Krasny's was used. The curvature regularisation coefficient and the blob size are equal to 0.01. $N = 512$. The solutions are plotted at $t = 1.0, 1.2, 1.4$. The frame dimension of each plot is $0 \leq x \leq 1, -4 \leq y \leq 0.4$. From Hou and Osher (1994).

smooth level-set function throughout the numerical computation. Thus, a free surface may develop a topological singularity such as a cusp, a corner or merging of two surfaces; the level-set function remains relatively smooth (the level-set function is Lipschitz continuous at the singularity of the interface). Moreover, the level-set function satisfies a Hamilton–Jacobi-type equation, and curvature regularisation corresponds to an entropy condition. Therefore, high-order Godonov methods developed for hyperbolic conservation laws can be used to compute the level-set function. Unlike the front tracking approach, no special effort is required at the interface singularity. The interface is recovered at the end of the computation by locating the zeroth-level set. Generalisation to three space dimensional problems requires no additional effort.

In this section, we describe the level-set algorithm for propagating a curve or union of curves $\Gamma(t)$. We assume that the motions of these curves are completely determined by the normal velocity, V. Let B be a fixed domain which contains the union of curves in all times of interest. The main idea is to construct a function $\phi(x, t)$ defined on B, such that the level set $\{\phi = 0\}$ corresponds to the moving curves $\Gamma(t)$, that is,

$$\Gamma(t) = \{x : \phi(x, t) = 0\}.$$

We now derive a partial differential equation for ϕ, which holds on $B \times [0, T]$. First, we need to construct a smooth extension of the normal velocity, V, of the curves to the entire domain of B such that

$$F(x, t) = V(x, t) \quad \text{for} \quad x \in \Gamma(t).$$

Now consider the motion of an arbitrary level set $\{\phi(\mathbf{x}, t) = C\}$. We will follow the derivation of Osher and Sethian (1988). Let $\mathbf{x}(\alpha, t)$ be the Lagrangian trajectory of this level-set. This implies that

$$\phi(\mathbf{x}(\alpha, t), t) = C.$$

Differentiating the above relation with respect to time, we get

$$\phi_t + \frac{\partial \mathbf{x}}{\partial t} \cdot \nabla \phi = 0.$$

Note that $\nabla \phi$ is normal to the level set $\{\phi(\mathbf{x}, t) = C\}$, and $\frac{\partial \mathbf{x}}{\partial t} \cdot n = F$, where $n = \nabla \phi / |\nabla \phi|$ is the unit normal vector to the level set $\phi = C$. This consideration implies that the evolution equation for the level-set function ϕ is given by

$$\phi_t + F|\nabla \phi| = 0, \tag{6.1}$$
$$\phi(x, 0) = \text{given}. \tag{6.2}$$

Equation (6.1) yields the motion of $\Gamma(t)$ with normal velocity V on the level set $\phi = 0$. We refer to equation (6.1) as the level-set 'Hamilton–Jacobi' formulation.

One essential property of the level-set function is that it always remains a function, even if the free surface (corresponding to $\phi = 0$) changes topology, breaks, merges or forms sharp corners. Parameterisations of the boundary become multivalued or singular in these cases. Furthermore, since the level-set formulation is completely Eulerian, finite-difference approximations over a fixed grid may be used to discretise the equation in space and time. Thus, there is no need to explicitly track the free surface during a numerical calculation. The free surface is recovered only at the end of the computation.

To illustrate, suppose we wish to follow an initial curve $\Gamma(t = 0)$ propagating with normal velocity $V = 1 - \epsilon \kappa$, where κ is the local curvature of the boundary. The curvature of the level curve passing through a point (x, y, t) is given by

$$\kappa = \nabla \cdot \left(\frac{\nabla \phi}{|\nabla \phi|} \right) = -\frac{\phi_y^2 \phi_{xx} - 2\phi_x \phi_y \phi_{xy} + \phi_x^2 \phi_{yy}}{(\phi_x^2 + \phi_y^2)^{3/2}}.$$

The minus sign occurs because we have initialised the surface so that $\nabla\phi$ points inwards and we want κ to be positive for a circle. The smooth extension of V to F is straightforward, and equation (6.1) becomes

$$\phi_t + (\phi_x^2 + \phi_y^2)^{1/2} = \epsilon\nabla \cdot \left(\frac{\nabla\phi}{|\nabla\phi|}\right)$$

$$\phi(x, y, t = 0) = \pm\text{distance from } (x, y) \text{ to } \Gamma(t = 0).$$

As shown in Sethian (1985), for $\epsilon > 0$, the parabolic right-hand side diffuses sharp gradients and forces ϕ to stay smooth for all time. This is not true for $\epsilon = 0$ and $F = 1$. A corner singularity must develop in time.

Thus the goal is to produce approximations to the spatial derivative that (1) do not smooth sharp corners artificially and (2) pick up the correct entropy solution when singularities develop. The schemes are motivated by the fact (Osher and Sethian 1988) that the entropy condition for propagating boundaries is identical to the one for hyperbolic conservation laws, where stable, consistent, entropy-satisfying algorithms have a rich history.

In discretising the term $F|\nabla\phi|$, we decompose F into two components:

$$F = F_A + F_G.$$

Here, F_A is an advection term containing that part of the velocity that is independent of the moving boundary, and F_G contains those terms that depend on the geometric properties of the boundary, such as the curvature and normal. We begin by splitting the influence of F, and rewrite the equation for ϕ as

$$\phi_t = -(F_A|\nabla\phi| + F_G|\nabla\phi|).$$

In two space dimensions, one can easily devise an iterative type of scheme based on dimension-by-dimension splitting (Osher and Sethian, 1988; Osher and Shu, 1991):

$$\phi_{ij}^{n+1} = \phi_{ij}^n - F_A\Delta t(\max(D_x^-\phi_{ij}, 0))^2(\min(D_x^+\phi_{ij}, 0))^2$$
$$+(\max(D_y^-\phi_{ij}, 0))^2(\min(D_y^+\phi_{ij}, 0))^2 - \Delta t F_G|\nabla\phi|.$$

Here we have not approximated the final term $F_G|\nabla\phi|$; one may use a straightforward centred difference approximation to this term. This is the first-order multi-dimensional algorithm described in Osher and Sethian (1988). High-order schemes have also been derived, see Osher and Shu (1991). In Figure 12 we show this technique applied to the case of a star propagating outwards with speed $F = 1$, $\Delta t = 0.01$, and a mesh size of 50 points in each direction in a box. The cusp singularities were captured properly. The curve became circular as it evolved (Osher and Sethian, 1988).

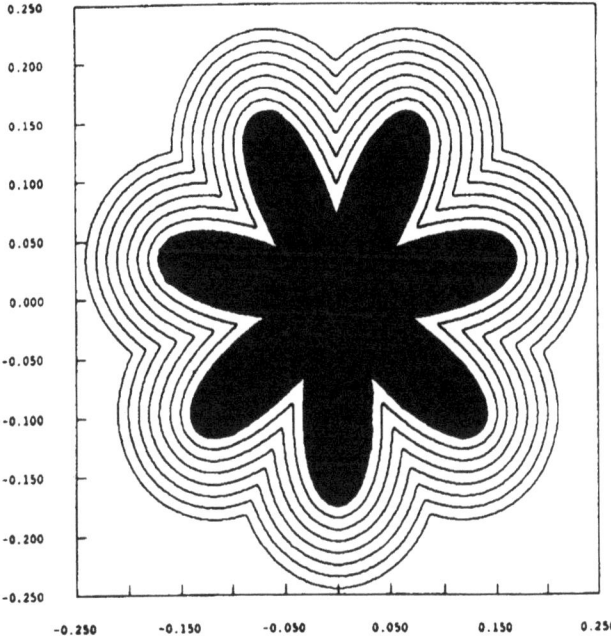

Fig. 12. Expanding star, $F(\kappa) = 1$, $t = 0.0, 0.7(0.01)$. $N = 50$. From Osher and Sethian (1988).

6.1. Crystal growth and solidification

We have described the boundary-integral formulation in Subsection 2.4. Here we describe how to extend the velocity V to a globally defined speed function F. Such an extension is needed to use the level-set formulation. What we will describe below is contained in the paper by Sethian and Strain (1992).

The most natural extension makes direct use of the integral equation

$$\epsilon_\kappa \kappa + \epsilon_V V + U + H \int_0^t \int_{\Gamma(t')} K(x, x', t - t')V(x', t')\mathrm{d}x'\mathrm{d}t' = 0, \quad (6.3)$$

for $x \in \Gamma(t)$. Each term in (6.3) can be evaluated anywhere in B, once V is known on $\Gamma(t')$ for $0 \le t' \le t$ and ϕ is known on B. Thus, given the set $\Gamma(t)$, plus all its previous positions and velocities for $0 \le t' \le t$, one could first solve an integral equation to find the velocity V for all points on $\Gamma(t)$ and then find $F(x, t)$ by solving the equation

$$\epsilon_\kappa \kappa(x, t) + \epsilon_V F(x, t)$$
$$+ U(x, t) + H \int_0^t \int_{\Gamma(t')} K(x, x', t - t')V(x', t')\mathrm{d}x'\mathrm{d}t' = 0,$$

for F throughout B. The curvature away from $\Gamma(t)$ is evaluated by

$$\kappa = \nabla \cdot \left(\frac{\nabla \phi}{|\nabla \phi|} \right) = \nabla \cdot n, \quad n = \frac{\nabla \phi}{|\nabla \phi|}.$$

These expressions make sense everywhere in B. This defines the extension of V away from $\Gamma(t)$.

Furthermore, it was observed by Greengard and Strain (1990) that one can decompose the single-layer potential into a history part $S_\delta V$ and a local part $S_L V$ as follows:

$$
\begin{aligned}
SV(x,t) &= \int_0^{t-\delta} \int_{\Gamma(t')} K(x, x', t - t') V(x', t') \, dx' dt' \\
&\quad + \int_{t-\delta}^{t} \int_{\Gamma(t')} K(x, x', t - t') V(x', t') \, dx' dt' \\
&\equiv S_\delta V + S_L V.
\end{aligned}
$$

Here δ is a small regularisation parameter. Heuristically, we try to separate the local part, which is causing the jump in the normal derivative of the potential, from the history part, which is smooth and independent of current velocity. It was shown (Sethian and Strain, 1992; Greengard and Strain, 1990) that the local part $S_L V$ can be approximated by

$$S_L V(x,t) = \sqrt{\delta/\pi} V(x, T) + O(\delta^{3/2}),$$

at point x on $\Gamma(t)$. The history part $S_\delta V$ depends only on values of V at times t' bounded away from the current time, $t' \leq t - \delta$. This is a smooth function. A fast summation method has been developed to evaluate the history part efficiently, requiring only $O(M^2)$ calculations per time step. Finite-difference approximations can also be used to obtain a fast evaluation of the history part; see Brattkus and Meiron (1992). Now we can define the extended velocity F explicitly through the history part of the single-layer potential:

$$F = \frac{-1}{\epsilon_V(n) + H\sqrt{\delta/\pi}} [\epsilon_\kappa \kappa + U + H S_L V].$$

We have reduced the equation of motion, with an $O(\delta^{3/2})$ error, to a pair of equations on fixed domain B:

$$\phi_t + F|\nabla \phi| = 0,$$

$$F = \frac{-1}{\epsilon_V(n) + H\sqrt{\delta/\pi}} [\epsilon_\kappa \kappa + U + H S_L V].$$

Numerical approximation of these coupled equations gives rise to a robust algorithm which can handle topological singularities, cusps, and corners.

In Figure 13, we plot a sequence of fingered growth under mesh refinement

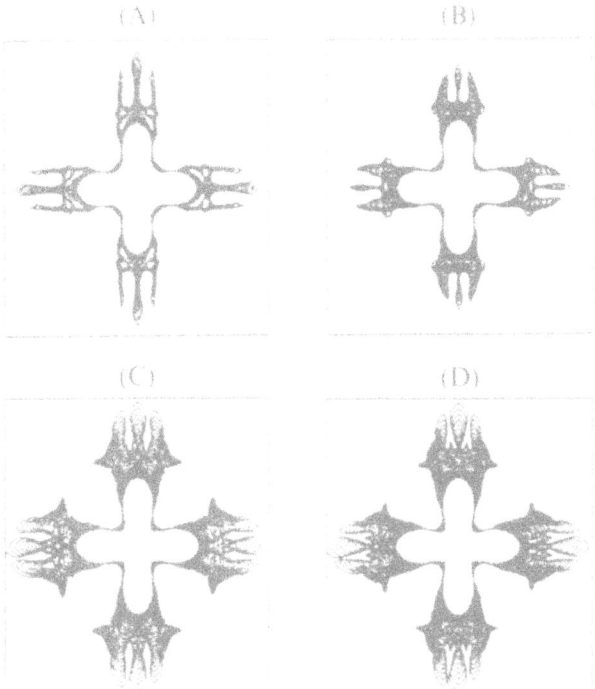

Fig. 13. Fingered crystal: effect of refining both grid size and time step, $H = 1, A = 0, \epsilon_\kappa = 0.001, \epsilon_V = 0.001, k_A = 0$: (A) 32×32 mesh, $\Delta t = 0.005$; (B) 48×48 mesh, $\Delta t = 0.005$; (C) 96×96 mesh, $\Delta t = 0.001\,25$; (D) 128×128 mesh, $\Delta t = 0.00125$. From Sethian and Strain (1992).

(Sethian and Strain, 1992). Here the physical parameters were specified as follows: $\epsilon_V = 0.001, \epsilon_\kappa = 0.001, H = 1$. There was no anisotropy in the coefficient, and the constant undercooling was set to be -1. In Figure 13A, a 32×32 grid was used with $\Delta t = 0.005$. In Figure 13B, a 48×48 grid was used with $\Delta t = 0.005$. In Figure 13C, a 96×96 grid was used with $\Delta t = 0.001\,25$. In Figure 13D, a 128×128 grid was used with $\Delta t = 0.001\,25$. On the coarsest mesh (32×32), only the gross features of the fingering and tip-splitting process are seen. As the numerical parameters are refined, the basic pattern emerges. It is clear that the resulting shapes are qualitatively the same, and there is little qualitative difference between Figure 13c and Figure 13d. We refer to Sethian and Strain (1992) for more details.

One disadvantage of this approach is that computing the normal velocity at each time step requires solving the boundary-integral problem. So it is not a completely Eulerian formulation. Since the boundary-integral problem is history dependent and the integration is non-local in space, it is usually very expensive. Even using the fast algorithm for heat potentials developed by Greengard and Strain (1990), numerical calculations by this approach are still slower compared with finite-difference approximations for the heat equation. Recently, Osher and his co-workers (private communication) have developed a completely Eulerian level-set formulation to compute solidification problems. The method is in principle as fast as standard finite-difference methods for the heat equation. The preliminary results seem to be very encouraging.

Numerical study of unstable solidification has been a very active research area in the past decade. Other numerical studies of solidification problems include the works of Meiron (1986), Kessler and Levine (1986), Langer (1980), Karma (1986), Voorhees *et al.* (1988), Almgren (1993), Bratkkus and Meiron (1992), Greenbaum *et al.* (1993).

6.2. Level-set formulation for incompressible-fluid surfaces

We have described a number of boundary-integral methods for computing fluid interfaces in previous sections. We can see that they are very effective as long as the interface stays smooth. However, when the interface develops pinching singularity, as seen in Subsection 5.2, corners and topological changes, boundary-integral methods are difficult to compute beyond time singularities. Here we would like to present a level-set formulation for incompressible-fluid interfaces with discontinuous densities and viscosities. Detailed derivation can be found in Chang *et al.* (to appear). Here we just present the result in our reformulation.

The equations governing the motion of an unsteady, viscous, incompressible, immiscible two-fluid system are the Navier–Stokes equations. In conservation form, the equations are

$$\rho(\mathbf{u}_t + \nabla \cdot (\mathbf{uu})) = -\nabla p + \rho \mathbf{g} + \nabla \cdot (2\mu \mathbf{D}),$$

where $\mathbf{u}$ is velocity and ρ and μ are discontinuous density and viscosity fields respectively. $\mathbf{D}$ is the rate-of-deformation tensor whose components are $D_{ij} = \frac{1}{2}(u_{i,j} + u_{j,i})$. The density and viscosity are purely convected by the fluid velocity:

$$\frac{\partial}{\partial t}(\rho) + \nabla \cdot (\mathbf{u}\rho) = 0,$$

$$\frac{\partial}{\partial t}(\mu) + \nabla \cdot (\mathbf{u}\mu) = 0.$$

These equations are coupled to the incompressibility condition

$$\nabla \cdot \mathbf{u} = 0.$$

Denote the stress tensor by $\sigma(\mathbf{x})$, which is given by

$$\sigma(\mathbf{x}) = -p\mathbf{I} + 2\mu \mathbf{D},$$

where $\mathbf{I}$ is the identity matrix, $\mathbf{D}$ is the deformation tensor and p is the pressure. We let Γ denote the fluid interface. The effect of surface tension is to balance the jump of the normal stress along the fluid interface. This gives rise to a free-boundary condition for the discontinuity of the normal stress across Γ

$$[\sigma_{ij}n_j] \,|_\Gamma = \tau\kappa n_i, \tag{6.4}$$

where $[p]$ denotes the jump of p across the interface, κ is the curvature of Γ, τ is the surface-tension coefficient and $\mathbf{n}$ is a unit outward normal vector along Γ. Note that in the case of inviscid flows, the above jump condition is reduced to

$$[p] \,|_\Gamma = \tau\kappa. \tag{6.5}$$

In this case, the effect of surface tension is to introduce a discontinuity in pressure across the interface proportional to the (mean) curvature.

Our level-set formulation is based on the following observation. The effect of surface tension can be expressed in terms of a singular source function that is defined by our level-set function. This is similar in spirit to Peskin's formulation for the immersed boundary-value problem for blood flows through a heart valve (Peskin, 1977); see also Unverdi and Tryggvason (1992). Let us denote by ϕ the level-set function. The fluid interface Γ corresponds to the zero-level set of ϕ. In Chang et $al.$ (to appear), we derived a completely Eulerian level-set formulation for multi-fluid interface problems with surface tension. The evolution equations are given by

$$\rho(\mathbf{u}_t + \nabla\cdot\mathbf{u}\mathbf{u}) = -\nabla p + \rho\mathbf{g} + \nabla\cdot(2\mu\mathbf{D}) + \tau\kappa(\phi)\nabla\phi\delta(\phi), \tag{6.6}$$

$$\frac{\partial}{\partial t}\phi + \mathbf{u}\cdot\nabla\phi = 0, \tag{6.7}$$

where $\delta(\phi)$ is a one-dimensional Dirac Delta function and ϕ is chosen in such way that $\nabla\phi$ is in the outward normal direction when evaluated on Γ. The curvature $\kappa(\phi)$ can be expressed by ϕ and its derivatives

$$\kappa(\phi) = -\frac{\phi_y^2\phi_{xx} - 2\phi_x\phi_y\phi_{xy} + \phi_x^2\phi_{yy}}{(\phi_x^2 + \phi_y^2)^{\frac{3}{2}}}. \tag{6.8}$$

Our level-set formulation was partially motivated by the work of Unverdi and Tryggvason (1992). The work of Unverdi and Tryggvason was formulated as a vortex-in-cell method. The free surface is tracked explicitly by following the Lagrangian markers of the free surface. A fixed underlying grid is used to invert the Poisson equation. The interface velocity is obtained by grid/particle interpolation, in the same way as we described in the previous section on the VIC method. But in this semi-Lagrangian formulation, the coupling between the Delta function source term and the momentum equations is non-$local$. If $\mathbf{x}(s,t)$ is a parameterisation of the fluid interface Γ, with s being the *arclength* variable and t being the time variable, and $\delta(\mathbf{x})$ is the two-dimensional Dirac Delta function, then the momentum equations become

$$\rho(\mathbf{u}_t + \nabla \cdot (\mathbf{uu})) = -\nabla p + \rho \mathbf{g} + \nabla \cdot (2\mu \mathbf{D})$$
$$+ \int_\Gamma \tau \kappa(\mathbf{x}(s,t)) \delta(\mathbf{x} - \mathbf{x}(s,t)) \mathbf{n} ds. \qquad (6.9)$$

Note that the singular source term is a 2-D Delta function and it is non-local. This is in contrast with the local and 1-D Delta function source term. In fact, equation (6.9) was not derived explicitly in Unverdi and Tryggvason (1992). It was based on our derivation of the level-set reformulation that we gave an independent derivation of (6.9). Also, consistency of equation (6.9) with the original interface problem requires s to be an arclength variable. Since $\mathbf{x}(s,t)$ is advected by the fluid velocity, s will not remain as an arclength variable even if it is so chosen initially. Therefore, if a Lagrangian variable α is used to parameterise the interface, that is, $\mathbf{x}(\alpha, t)$, then a factor $|\mathbf{x}_\alpha|$ should be added on to the integration with respect to α. This point was not clearly stated before, and it caused some confusion in the literature.

After the completion of our work on the level-set formulation, the work of Brackbill, Kothe and Zemach (1992) was brought to our attention. They have derived a continuum method for modelling surface tension for multi-fluid flows which is almost the same as our formulation if we replace the Dirac Delta function by a regularised one. Brackbill *et al.* used a 'colour' function to describe the smoothed interface. The colour function changes continuously in the transition region of finite thickness. Brackhill *et al.*'s derivation was based on a physical argument.

We now describe how to discretise the level-set formulation. Assume that we have chosen the initial level-set function such that $\phi < 0$ defines region 1 of the fluid, and $\phi > 0$ defines region 2. Further, we assume that ρ_1 and ρ_2 are the constant densities in region 1 and region 2, respectively, and μ_1 and μ_2 are the constant viscosities in region 1 and region 2 respectively. Then we have $\rho = \rho_1 + (\rho_2 - \rho_1)H(\phi)$, where H is the Heaviside function that satisfies $H(x) = 1$ for $x > 0$ and $H(x) = 0$ for $x < 0$. Similarly, we have $\mu = \mu_1 + (\mu_2 - \mu_1)H(\phi)$. In numerical computations, we approximate H by a regularised Heaviside function, and approximate δ by a regularised Delta function, just as in Peskin (1977). The regularised Delta function $\delta_\epsilon(x)$ has support in $\{|x| \leq \epsilon\}$. Typically, we choose $\epsilon = 4h$ in our calculations.

The evolution equations can be solved by a projection method. In its most basic form, the projection method requires the solution of advection–diffusion equations, which are then projected onto the space of divergence-free vector fields. The projection uses the Hodge decomposition which states that any vector $\mathbf{V}$ can be uniquely decomposed into a divergence-free field $\mathbf{V}_d$ and a gradient field ∇p, that is, $\mathbf{V} = \mathbf{V}_d + \nabla p$. Moreover $\mathbf{V}_d$ is orthogonal to the gradient field. For more detailed descriptions of projection methods and their applications, we refer to Chorin (1968) and Bell, Colella and Glaz (1989) and the review paper by Gresho and Sani (1987). For our problem, density is not a constant in the entire domain. A modification of the standard projection method is required. A second-order projection method for variable density has been introduced by Bell and Marcus (1992), and it has been applied successfully to a number of interesting multi-fluid interface problems.

From the evolution equations (6.7), we have $\mathbf{u}_t = L\mathbf{u} - \nabla p/\rho$. As in Bell and Marcus (1992), we introduce a density-weighted inner product such that we can decompose $\mathbf{V}$ into $\mathbf{V}_d$ and $\nabla p/\rho$. In the density-weighted norm, $\mathbf{V}_d$ is orthogonal to ∇p. Given a vector $\mathbf{V}$, we define our projection operator as $\mathbf{P}_d(\mathbf{V}) = \mathbf{V}_d$. Since the Hodge decomposition is unique and $\mathbf{u}_t$ is divergence free, we have $\mathbf{u}_t = \mathbf{P}_d(L\mathbf{u})$. In order to compute the projection, we take the divergence of both sides of the equation $\mathbf{V} = \mathbf{V}_d + \nabla p/\rho$ to obtain

$$\nabla \cdot \left(\frac{1}{\rho}\nabla p\right) = \nabla \cdot \mathbf{V}.$$

The orthogonality condition implies the boundary condition $\partial p/\partial n = 0$ on the boundary. Another way to compute the projection is to take the curl of both sides of the equation $\mathbf{V} = \mathbf{V}_d + \nabla p/\rho$. This also gives a variable elliptic problem for p with a different boundary condition. We refer to Bell and Marcus (1992) for more detailed discussions.

The convection terms can be approximated by high-order ENO schemes Harten *et al.* (1987) or by other high-order Godunov schemes. Apparently, our level-set formulation works for both two-dimensional and three-dimensional problems. There are no additional complications to extend the method to three-dimensional problems.

To obtain an effective method, it is important to keep the level-set function as smooth as possible at all times. For this reason, it is desirable to keep the level-set function as a signed distance function from the moving surface. This also ensures that the regularised surface has a finite thickness of order ϵ for all time. However, even if we initialise the level-set function ϕ as a signed distance from the free surface, the level-set function in general will not remain a distance function at later times. In Sussman, Smereka and Osher (to appear), an iterative procedure was proposed to reinitialise the level-set function at each time step so that the reinitialised level-set function remains a distance function from the front. Specifically, given a level-set function, ϕ_0, at time t, solve for the steady-state solution of the equation

$$\frac{\partial}{\partial t}\phi = \text{sgn}(\phi_0)(1 - |\nabla\phi|),$$
$$\phi(\mathbf{x}, 0) = \phi_0(\mathbf{x}),$$

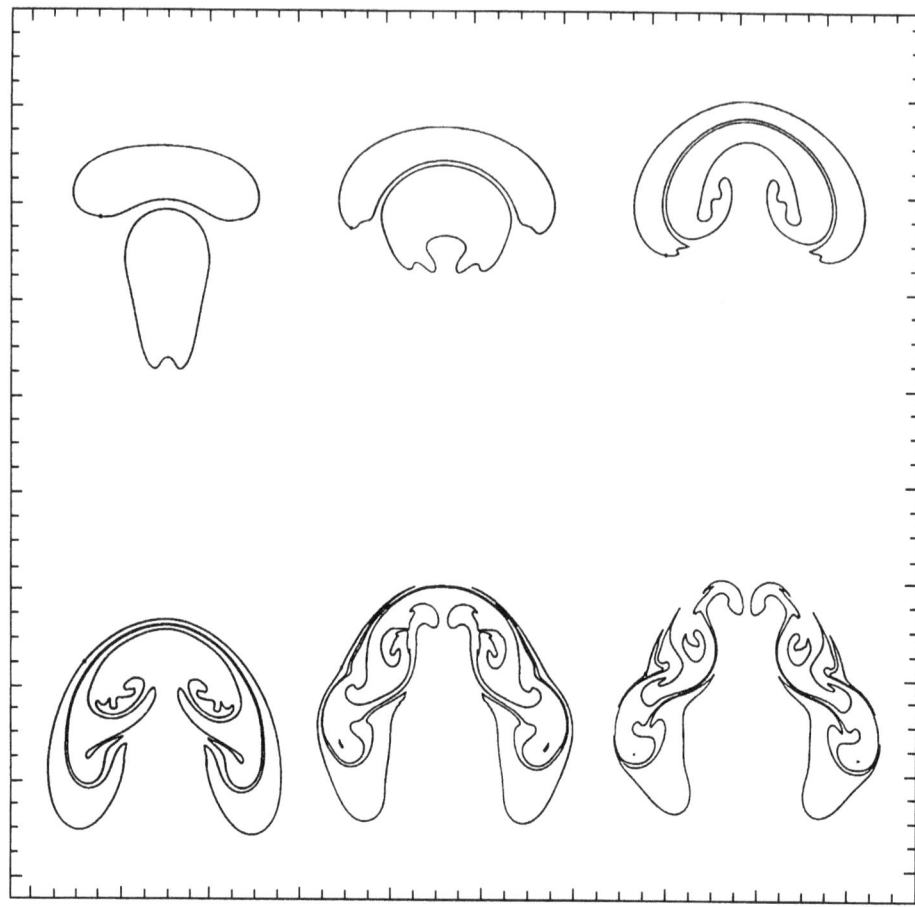

Fig. 14. Fourth-order difference approximations for two-fluid bubbles with different densities. The density ratio is 1:60:3600, with the bottom bubble being the lightest. The viscosity is equal to 0.000 125 in all fluids. $t = 0.1, 0.15, 0.2$ for the first row ($N = 256$), and $t = 0.275, 0.325, 0.35$ for the second row ($N = 512$). From Chang *et al.* (to appear).

where sgn is the sign function. The solution ϕ has the same zero-level set as ϕ_0, and satisfies $|\nabla\phi| = 1$, and so is a distance function for the front. It was found in Sussman, Smereka and Osher (to appear) that such reinitialisation is crucial in maintaining the accuracy of large-time integrations, especially when the density ratio between the two fluids is large. In Sussman, Smereka and Osher, the motion of bubbles in water and falling water drops in air were studied numerically using our level-set formulation, together with the reinitialisation procedure described above. The density ratio is 1 to 1000. The numerical results were in good agreement with some experimental results.

In Figure 14, we illustrate the method by considering the interaction of two fluid bubbles with different densities using 256×256 grid points. The density for the bubble on the top is 60, the density for the bubble on the bottom is 1 and the background density is 3600. The initial interfaces of the bubbles are elliptical in shape. A Bousinesqu approximation was used in these calculations. We assume that viscosities are the same in all fluids and are equal to 0.000 125. The problem is set up in such a way that both bubbles rise in time and the bottom bubble rises the fastest. As the bottom bubble rises in time, we see that the top portions of the bubble interfaces are almost in contact. But they cannot merge into a single bubble in this case because the densities are different for these two bubbles. In this calculation, we labelled the two interfaces with two different level-set values. That is, Γ_1 corresponds to $\phi = c_1$ and Γ_2 corresponds to $\phi = c_2$, with $c_1 \neq c_2$. In the mean time, the bubble in the bottom develops a roll-up. We plot the solutions at $t = 0.1, 0.15, 0.2, 0.275, 0.325, 0.35$. We increase our numerical resolutions to 512×512 for times larger than $t = 0.2$. Part of the interface that has rolled up pinches off before $t = 0.275$; two smaller bubbles are detached from the bottom bubble, and have their own dynamics. As the region between the top portions of two bubbles becomes thinner and thinner in time, they eventually pinch off at $t = 0.325$ and $t = 0.35$ respectively. In the process, many small-scale structures are produced due to the unstable stratification of the fluids.

REFERENCES

A. Almgren, T. Buttke and P. Colella (1994), 'A fast adaptive vortex method in 3 dimensions', *J. Comp. Phys.*, **113**, 117–200.

R. Almgren (1993), 'Variational algorithms and pattern-formation in dendritic solidification', *J. Comp. Phys.*, **106**, 337–354.

R. Almgren, W.-S. Dai and V. Hakim (1993), 'Scaling behaviour in anisotropic Hele–Shaw flow', *Phys. Rev. Lett.*, **71**, 3461–3464.

C. Anderson (1985), 'A vortex method for flows with slight density variations', *J. Comp. Phys.*, **61**, 417–444.

C. Anderson (1986), 'A method of local corrections for computing the velocity due to a distribution of virtex blobs', *J. Comp. Phys.*, **62**, 111–123.

C. Anderson and C. Greengard (1985), 'On vortex methods', *SIAM J. Numer. Anal.*, **22**, 413–440.

U. Ascher, S Ruuth and B. Wetton, 'Implicit–explicit methods for time-dependent PDE's', to appear in *SIAM J. Numer. Anal.*

G. Baker (1979), 'The 'cloud-in-cell' technique applied to the roll-up of vortex sheets', *J. Comp. Phys.*, **31**, 76–95.

G. Baker (1983), Generalized vortex methods for free-surface flows, in *Waves on Fluid Interfaces*, (R. E. Meyer, ed.), Academic Press pp. 53–81.

G. Baker, R. Caflisch and M. Siegal, 'Singularity formation during the Rayleigh–Taylor instability', to appear *J. Fluid Mech.*

G. Baker, D. Meiron and S. Orszag (1982), 'Generalized vortex methods for free-surface flow problems', *J. Fluid Mech.*, **123**, 477–501.

G. R. Baker and M. J. Shelley (1990), 'On the connection between thin vortex layers and vortex sheets', *J. Fluid Mech.*, **215**, 161–194.

G. Baker and A. Nachbin, 'Stable methods for vortex sheet motion in the presence of surface tension', to appear in *J. Comp. Phys.*

J. T. Beale and A. Majda (1982), 'Vortex methods, I: Convergence in three dimensions', *Math. Comp.*, **32**, 1–27.

J. T. Beale, T. Y. Hou and J. S. Lowengrub (1993a), 'Growth rates for the linear motion of fluid interfaces far from equilibrium', *Comm. Pure Appl. Math.*, **46**, 1269–1301.

J. T. Beale, T. Y. Hou and J. S. Lowengrub, 'Convergence of a boundary integral method for water waves', to appear in *SIAM J. Num. Anal.*

J. T. Beale, T. Y. Hou and J. S. Lowengrub (1993b), On the well-posedness of two fluid interfacial flows with surface tension, *Singularities in Fluids, Plasmas and Optics*, (eds. R. C. Caflisch and G. C. Papanicolaou), Kluwer Academic, London, pp. 11–38.

J. T. Beale, T. Y. Hou, J. Lowengrub and M. Shelley, 'Spatial and temperal stability issues for interfacial flows with surface tension', to appear in *J. Math. Computer Modelling*, Vol. 20, No. 10/11, pp. 1–27, 1994.

J. T. Beale, T. Y. Hou and J. S. Lowengrub, Two Fuid Flows with Surface Tension, Part 1: Growth Rates of the Linearized Motion Far From Equilibrium; Part 2: Convergence of Boundary Integral Methods, in preparation, Applied Math., Caltech.

J. B. Bell, P. Colella and H. M. Glaz (1989), 'A 2nd order projection method for the incompressible Navier–Stokes equations', *J. Comp. Phys.*, **85**, 257–283.

J. B. Bell and D. L. Marcus (1992), 'A second-order projection method for variable-density flows', *J. Comp. Phys.*, **101**, 334–348

M. Benamar and Y. Pomeau (1986), 'Theory of dendritic growth in a weakly undercooled melt', *Europhys. Lett.*, **2**, 307–314.

A. L. Bertozzi and P. Constantin (1993), 'Global regularity for vortex patches, *Comm. Math. Phys.*', **152**, 19–28.

G. Birkhoff (1962), 'Helmholtz and Taylor instability', in *Proc. Symp. Appl. Math.*, **13**, 55–76, Vol. XIII, Publ. American Math. Society, Providence, R.I.

J. U. Brackbill, D. B. Kothe and C. Zemach (1992), 'A continuum method modeling surface tension', *J. Comp. Phys.*, **100**, 335–354.

K. Brattkus and D. I. Meiron (1992), 'Numerical simulation of unsteady crystal growth', *SIAM J. Appl. Math.*, **52**, 1303–1320.

T. F. Buttke (1989), 'The observation of singularities in the boundary of patches of constant vorticity', *Phys. Fluids A*, **1**, 1283–1285.

R. Caflisch and O. Orellana (1988), 'Long time existence for a slightly perturbed vortex sheet', *Comm. Pure Appl. Math.*, **39**, 807–838.

R. Caflisch and O. Orellana (1989), 'Singularity solutions and ill-posedness for the evolution of vortex sheets', *SIAM J. Math. Anal.*, **20**, 293–307.

R. Caflisch and J. Lowengrub (1989), 'Convergence of the vortex method for vortex sheets', *SIAM J. Numer. Anal.*, **26**, 1060–1080.

R. Caflisch, X. Li and M. Shelley (1993), 'The collapse of an axisymmetrical, swirling vortex sheet', *Nonlinearity*, **6**, 843–867

R. Caflisch, N. Ercolani, T. Y. Hou and Y. Landis (1993), 'Multi-valued solutions and branch point singularities for nonlinear and elliptic systems', *Comm Pure Appl Math.*, **46**, 453–499.

R. Caflisch, T. Y. Hou and J. Lowengrub (1994), Almost Optimal Convergence of the Point Vortex Method for Vortex Sheets Using Numerical Filtering, preprint, Dept of Math., Univ. of Minnesota, submitted to *Math. Comput.*

G. Gaginalp and P. C. Fife (1988), 'Dynamics of layered interfaces arising from phase boundaries', *SIAM J. Appl. Math.*, **48**, 506–518.

J. Chadam and P. Ortoleva (1983), 'The stabilizing effect of surface tension on the development of the free-boundary in a planar, one-dimensional, Cauchy–Stefan problem', *IMA J. Appl. Math.*, **30**, 57–66.

Y. C. Chang, T. Y. Hou, B. Merriman and S. Osher, 'Eularian capturing methods based on a level set formulation for incompressible fluid interfaces', to appear in *J. Comp. Phys.*

H. Ceniceros and T. Y. Hou, Convergence of a Non-stiff Boundary Integral Method for Interfacial Flows with Surface Tension, preprint, Applied Math., Caltech.

J.-Y. Chemin (1993), 'Presistency of geometric structures in bidimensional incompressible fluids', *Ann. Sci. EC*, **26**, 517–542.

D. L. Chopp (1993), 'Computing minimal surfaces via level set curvature flow,' *J. Comp. Phys.*, **106**, 77–91.

A. Chorin (1968), 'Numerical solution of the Navier–Stokes equations', *Math. Comput.*, **22**, 745–762.

A. Chorin (1973), 'Numerical study of slightly viscous flow', *J. Fluid Mech.*, **57**, 785–796.

A. Chorin and P. S. Bernard (1973), 'Discretization of a vortex sheet, with an example of roll-up', *J. Comp. Phys.*, **13**, 423–429.

J. P. Christiansen (1973), 'Numerical simulation of hydrodynamics by the method of point vortices', *J. Comp. Phys.*, **13**, 363–379.

P. Constantin, P. D. Lax and A. Majda (1985), 'A simple one dimensional model for the three dimensional vorticity equations', *Comm. Pure Appl. Math.*, **38**, 715–724.

P. Colella and P. R. Woodward (1984), 'The piecewise parabolic method (PPM) for gas dynamical simulations', *J. Comp. Phys.*, **54**, 174–201.

G. H. Cottet (1988), 'On the convergence of vortex methods in two and three dimensions', *Ann. Insti. H. Poincaré*, **5**, 641–672.

G. H. Cottet, J. Goodman and T. Y. Hou (1991), 'Convergence of the grid free point vortex method for the 3-D Euler equations', *SIAM J. Numer. Anal.*, **28**, 291–307.

G. H. Cottet (1987), 'Convergence of a vortex in cell method for the two-dimensional Euler equations', *Math. Comp.*, **49**, 407–425.

W. Craig (1985), 'An existence theory for water waves and the Bousinesq and the Kortewegde–de Vries scaling limits', *Comm. Partial Diff. Eqns.*, **10**, 787–1003.

W. Craig and C. Sulem (1993), 'Numerical simulation of Gravity Waves, *J. Comp. Phys.*', **108**, 78–83.

W. J. A. Dahm, C. E. Frieler and G. Tryggvason (1992), 'Vortex structure and dynamics in near field of a coaxial jet', *J. Fluid Mech.*, **241**, 371–402.

W.-S. Dai and M. J. Shelley (1993), 'A numerical study of the effect of surface tension and noise on an expanding Hele–Shaw bubble', *Phys. Fluids A*, **5(9)**, 2131–2146.

J.-M. Delort (1991), 'Existence de nappes de tourbillon en dimension deux', *J. Amer. Math. Soc.*, **4**, 553–586.

R. DiPerna and A. Majda (1987a), 'Concentration in regularizations for 2-D incompressible flow', *Comm. Pure Appl. Math.*, **40**, 301–345.

R. DiPerna and A. Majda (1987b), 'Oscillations and concentrations in weak solutions of the incompressible fluid equations', *Comm. Math. Phys.*, **108**, 667–689.

J. W. Dold (1992), 'An efficient surface-integral algorithm applied to unsteady gravity waves', *J. Comp. Phys.*, **103**, 90–115.

D. G. Dommermuth and D. K. P. Yue (1987), 'A high order spectral method for the study of nonlinear gravity waves'. *J. Fluid Mech.*, **184**, 267–288.

C. I. Draghicescu (1994), 'An efficient implementation of particle methods for the incompressible Euler equations', *SIAM J. Numer. Anal.*, **31**, 1090–1108.

D. G. Dritschel (1989), 'Contour dynamics and contour surgery – Numerical algorithms for extended, high-resolution modelling of vortex dynamics in two-dimensional, inviscid, incompressible flows', *Comp. Phys. Review*, **10**, 77–146.

D. G. Dritschel and M. E. McIntyre (1990), 'Does contour dynamics go singular?' *Phys. Fluids A*, **2**, 748–753.

J. Duchon and R. Robert (1986), 'Solution globales avec nape tourbillionaire pour les equations d's Euler dans le plan', *C. R. Acad. Sc. Paris*, **302**, Series I. 5, 183–186.

J. Duchon and R. Robert (1988), 'Global vortex sheet solutions of Euler equations in the plane', *J. Diff. Eqn.*, **73**, 215–224.

D. Ebin (1988), 'Ill-posedness of the Rayleigh–Taylor and Helmholtz problems for incompressible fluids', *Comm. PDE*, **13**, 1265–1295.

L. C. Evans and J. Spruck (1991), 'Motion of level sets by mean-curvature, I', *J. Diff. Geom.*, **33**, 635–681.

L. C. Evans and J. Spruck (1992), 'Motion of level sets by mean-curvature, II', *Tran. Amer. Math. Soc.*, **330**, 321–332.

J. D. Fenton and M. M. Rienecker (1982), 'A Fourier method for solving nonlinear water-wave problems: Application to solitary-wave interactions', *J. Fluid Mech.*, **118**, 411–443.

P. T. Fink and W. K. Soh (1978), 'A new approach to roll-up calculations of vortex sheets', *Proc. Roy. Soc. London A.*, **362**, 195–209.

M. Glozman, Y. Agnon and M. Stiassnie (1993), 'High-order formulation of the water-wave problem', *Physica D*, **66**, 347–367.

R. Goldstein, A. Pesci and M. Shelley (1993), 'Topology transitions and singularities in viscous flows', *Phys. Rev. Lett.*, **70**, 3043–3046.

J. Goodman, T. Y. Hou and J. Lowengrub (1990), 'The convergence of the point vortex method for the 2-D Euler equations', *Comm. Pure and Appl. Math.*, **43**, 415–430.

J. Goodman, T. Y. Hou and E. Tadmor (1994), 'On the stability of the unsmoothed Fourier method for hyperbolic equations', *Numer. Math.*, **67**, 93–129.

L. J. Gray, M. F. Chisholm and T. Kaplan (1993), Morphological stability of thin films, in *Boundary Element Technology VIII*, (H. Pina and C. A. Brebbia, eds.), Southampton, UK, Computational Mechanics Publications, pp. 181–190.

A. Greenbaum, L. Greengard and G. B. McFadden (1993), 'Laplace's equation and the Dirichlet–Neumann map in multiply connected domains', *J. Comp. Phys.*, **105**, 267–278.

L. Greengard and V. Rokhlin (1987), 'A fast algorithm for particle summations', *J. Comp. Phys.*, **73**, 325–348.

L. Greengard and J. Strain (1990), 'A fast algorithm for the evaluation of heat potentials', *Comm. Pure Appl. Math.*, **43**, 949–963.

P. M. Gresho and R. L. Sani (1987), 'On pressure boundary conditions for the incompressible Navier–Stokes equations', *Internat. J. Numer. Methods Fluids*, **7**, 1111–1145.

D. Gottlieb and S. A. Orszag (1977), *Numerical Analysis of Spectral Methods: Theory and Applications*, CBMS–NSF Regional Conference Series in Applied Mathematics **26**, SIAM Publ., Philadelphia.

M. E. Gurtin (1986), 'On the 2-phase stefan problem with interfacial energy and entropy', *Arch. Ration. Mech. Anal.*, **96**, 199–241.

O. Hald (1979), 'Convergence of vortex methods II', *SIAM J. Numer. Anal.*, **16**, 726–755.

O. Hald (1991), Convergence of vortex methods, in *Vortex Methods and Vortex Motion*, (Gustafson and Sethian, eds.), SIAM publications, Philadelphia, pp. 33–58.

A. Harten, B. Engquist, S. Osher and S. R. Chakravarthy (1987), 'Uniformly high order accurate essentially non-oscillatory schemes, III', *J. Comp. Phys.*, **71**, 231–303.

J. J. L. Higdon and C. Pozrikidis (1985), 'The self-induced motion of vortex sheets', *J. Fluid Mech.*, **150**, 203–231.

T. Y. Hou and J. Lowengrub (1990), 'Convergence of a point vortex method for 3-D Euler equations', *Comm. Pure Appl. Math.*, **43**, 965–981.

T. Y. Hou and S. Osher (1994), *Computing Topological Singularities Using a Tracking Method: Geometric regularization*, preprint, Applied Math., Caltech.

T. Y. Hou, J. S. Lowengrub and M. J. Shelley (1994), 'Removing the stiffness from interfacial flows with surface tension', *J. Comp. Phys.*, **114**, 312–338.

T. Y. Hou, J. S. Lowengrub and M. J. Shelley, *The Roll-up and Self-intersection of Vortex Sheets Under Surface Tension*, preprint, Courant Institute.

T. Y. Hou, J. S. Lowengrub and R. Krasny (1991), 'Convergence of a point vortex method for vortex sheets', *SIAM J. Numer. Anal.*, **28**, 308–320.

Y. Kaneda (1990), 'A representation of the motion of a vortex sheet in a three dimensional flow', *Phys. Fluids A*, **2**, 458–461.

A. Karma (1986), 'Wavelength selection in dendritic solidification', *Phys. Rev. Lett.*, **57**, 858–861.

R. M. Kerr (1988), 'Simulation of Rayleigh–Taylor flows using vortex blobs', *J. Comput. Phys.*, **76**, 48–84.

D. A. Kessler and H. Levine (1986), 'Stability of dendritic crystals', *Phys. Rev. Lett.*, **57**, 3069–3072.

P. Koumoutsakos (1993), Direct Numerical Simulations of Unsteady Separated Flows Using Vortex Methods, Ph.D. thesis, Graduate School of Aeronautics, Caltech.

R. Krasny (1986a), 'A study of singularity formation in a vortex sheet by the point vortex approximation', *J. Fluid Mech.*, **167**, 65–93.

R. Krasny (1986b), 'Desingularization of periodic vortex sheet roll-up', *J. Comp. Phys.*, **65**, 292–313.

R. Krasny (1987), 'Computation of vortex sheet roll-up in the Trefftz plane', *J. Fluid Mech.*, **184**, 123–155.

H.-O. Kreiss and J. Oliger (1979), 'Stability of the Fourier method', *SIAM J. Num. Anal.*, **16**, 421–433.

J. S. Langer (1980), 'Instabilities and pattern formation in crystal growth', *Rev. Modern Phys.*, **52**, 1–28.

J. S. Langer (1986), 'Existence of needle crystals in local models of solidification', *Phys. Rev. A*, **33**, 435–441.

A. Leonard (1980), 'Vortex methods for flow simulation', *J. Comp. Phys.*, **37**, 289–335.

J. G. Liu and Z. P. Xin, *Convergence of Vortex Methods for Weak Solutions to the 2-D Euler Equations with Vortex Sheet Data*, preprint, Courant Institute.

M. S. Longuet-Higgins and E. D. Cokelet (1976), 'The deformation of steep surface waves on water, I. A numerical method of computation', *Proc. Roy. Soc. London A*, **350**, 1–26.

A. Majda (1986), 'Vorticity and the mathematical theory of incompressible fluid flow', *Comm. Pure Appl. Math.*, **39**, 5187–5220.

D. I. Meiron (1986), 'Boundary Integral formulation of the two-dimensional symmetric model of dendritic growth', *Physica-D*, **23**, 329–339.

D. I. Meiron, G. R. Baker and S. A. Orszag (1982), 'Analytic structure of vortex sheet dynamics, 1. Kelvin–Helmholtz instability', *J. Fluid Mech.*, **114**, 283–298.

W. Mulder, S. Osher and J. Sethian (1992), 'Computing interface motion in compressible gas dynamics', *J. Comp. Phys.*, **100**, 209–228.

D. W. Moore (1979), 'The spontaneous appearance of a singularity in the shape of an evolving vortex sheet', *Proc. R. Soc. Lond. Ser. A*, **365**, 105–119.

D. W. Moore (1981), 'On the point vortex method'. *SIAM J. Sci. Stat.*, **2**, 65–84.

D. W. Moore (1978), 'The equations of motion of a vortex layer of small thickness', *Stud. Appl. Math.*, **58**, 119–140.

D. W. Moore (1985), Numerical and analytical aspects of Helmholtz instability. In Theoretical and Applied Mechanics, *Proc. XVI IUTAM* (eds., Niodsen and Olhoff), pp. 263.

W. Mullins and R. Sekerka (1963), 'Morphological stability of a particle growing by diffusion or heat flow', *J. Appl. Phys.*, **34**, 323–329.

A. L. New, P. McIver and D. H. Peregrine (1985), 'Computations of overturning waves', *J. Fluid Mech.*, **150**, 233–251.

M. Nitsche and R. Krasny (1994), *A Numerical Study of Vortex Ring Formation at the Edge of a Circular Tube*, preprint, Dept. of Math., Univ. of Michigan.

S. Osher and J. Sethian (1988), 'Fronts propagating with curvature-dependent speed: Algorithms based on Hamilton–Jacobi formulations', *J. Comp. Phys.*, **79**, 12–49.

S. Osher and C. W. Shu (1991), 'High order essentially nonoscillatory schemes for Hamilton–Jacobi equations', *J. Comp. Phys.*, **28**, 907–922.

S. Osher, private communication.

L. Paterson (1981), 'Radial fingering in a Hele–Shaw cell', *J. Fluid Mech.*, **113**, 513–519.

L. Paterson (1985), 'Fingering with miscible fluids in a Hele–Shaw cell', *Phys Fluids.*, **28**, 26–30.

P. Pelcé (1988), *Dynamics of Curved Fronts*, Academic Press, Inc., San Diego.

D. H. Peregrine (1983), 'Breaking waves on beaches', *Ann. Rev. Fluid Mech.*, **15**, 149–178.

C. Peskin (1977), 'Numerical analysis of blood flow in the heart', *J. Comp. Phys.*, **25**, 220–252.

C. Pozrikidis (1992), *Boundary Integral and Singularity Methods for Linearized Viscous Flow*, Cambridge University Press.

C. Pozrikidis and J. J. L. Higdon (1985), 'Nonlinear Kelvin–Helmholtz instability of a finite vortex layer', *J. Fluid Mech.*, **157**, 225–263.

A. Prosperetti, L. A. Crum and H. C. Pumphrey (1989), 'The underwater noise of rain', *J. Geophys. Res.*, **94** , 3255–3259.

D. A. Pugh (1989), Development of Vortex Sheets in Boussinesq Flows – Formation of Singularities, Ph.D. Thesis, Imperial College, London

D. I. Pullin (1979), 'Vortex ring formation at tube and orifice openings', *Phys. Fluids*, **22**, 401–403.

D. I. Pullin (1982), 'Numerical studies of surface tension effects in nonlinear Kelvin–Helmholtz and Rayleigh–Taylor instability', *J. Fluid Mech.*, **119**, 507–532.

R. Rangel and W. Sirignano (1988), 'Nonlinear growth of the Kelvin–Helmholtz instability: effect of surface tension and density ratio', *Phys. Fluids*, **31**, 1845–1855.

S. N. Rauseo, P. D. Barnes, Jr and J. V. Maher (1987), 'Development of radial fingering patterns', *Phys. Rev. A*, **35**, 1245–1251.

A. J. Roberts (1983), 'A stable and accurate numerical method to calculate the motion of a sharp interface between fluids', *IMA J. Appl. Math.*, **31**, 13–35.

L. Rosenhead (1932), 'The point vortex approximation of a vortex sheet', *Proc. Roy. Soc. London Ser. A*, **134**, 170–192.

P. G. Saffman and G. I. Taylor (1958), 'The penetration of a fluid into a porous medium of Hele–Shaw cell containing a more viscous fluid', *Proc. Roy. Soc. London Ser. A*, **245**, 312–329.

L. W. Schwartz and J. D. Fenton (1982), 'Strongly nonlinear waves', *Ann. Rev. Fl. Mech.*, **14**, 39–60.

M. Shelley (1992), 'A study of singularity formation in vortex sheet motion by a spectrally accurate vortex method', *J. Fluid Mech.*, **244**, 493–526.

J. A. Sethian (1985), 'Curvature and the evolution of fronts', *Comm. Math. Phys.*, **101**, 487–499.

J. Sethian and J. Strain (1992), 'Crystal growth and dendritic solidification', *J. Comp. Phys.*, **98**, 231-253.

A. Sidi and M. Israeli (1988), 'Quadrature methods for periodic singular and weakly singular Fredholm integral equations', *J. Sci. Comp.*, **3**, 201–231.

C. W. Snyder, B. G. Orr, D. Kessler and L. M. Sander (1991), 'Effect of strain on surface morphology in highly strained InGaAs', *Phys. Rev. Lett.*, **66**, 3032–3035.

B. J. Spencer, P. W. Voorhees and S. H. Davis (1991), 'Morphological instability in epitaxially dislocation-free solid films', *Phys. Rev. Lett.*, **67**, 3696–3999.

B. J. Spencer and D. I. Meiron (1994), 'Nonlinear Evolution of the stress-driven morphological instability in a 2-dimensional semi-infinite solid', *Act. Met. Mat.*, **42**, 3629–3641.

M. Stiassnie and L. Shemer (1984), 'On modifications of the Zakharov equation for surface gravity waves', *J. Fluid Mech.*, **143**, 47–67.

J. Strain (1989), 'A boundary integral approach to unstable solidification', *J. Comp. Phys.*, **85**, 342–389.

P. Sulem, C. Sulem, C. Bardos and U. Frisch (1981), 'Finite time analyticity for the two and three dimensional Kelvin–Helmholtz instability', *Comm. Math. Phys.*, **80**, 485–516.

M. Sussman, P. Smereka and S. Osher, 'A level set approach for computing solutions to incompressible 2-phase flow', to appear in *J. Comp. Phys.*

E. Tadmor (1987), 'Stability analysis of finite difference, pseudospectral and Fourier–Galerkin approximations for time dependent problems', *SIAM Review*, **29**, 525–555.

G. I. Taylor (1950), 'The instability of liquid surfaces when accelerated in a direction perpendicular to their planes', *Proc. Roy. Soc. London Ser. A.*, **201**, 192–196.

G. Tryggvason and H. Aref (1983), 'Numerical experiments on Hele–Shaw flow with a sharp interface', *J. Fluid Mech.*, **136**, 1–30.

G. Tryggvason (1988), 'Numerical simulations of the Rayleigh–Taylor instabilities', *J. Comp. Phys.*, **75**, 253–282.

G. Tryggvason (1989), 'Simulations of vortex sheet roll-up by vortex methods', *J. Comp. Phys.*, **80**, 1–16.

G. Tryggvason , W. J. A. Dahm and K. Sbeih (1991), 'Fine structure of vortex sheet roll-up by viscous and inviscid simulation', *J. Fluid Eng.*, **113**, 31–36.

S. O. Unverdi and G. Tryggvason (1992), 'A front-tracking method for viscous, incompressible, multi-fluid flows', *J. Comp. Phys.*, **100**, 25–37.

A. van de Vooren (1980), 'A numerical investigation of the rolling-up of vortex sheets', *Proc. Roy. Soc. London Ser. A*, **373**, 67–91.

L. van Dommelen and E. A. Rundensteiner (1989), 'Fast, adaptive summation of point forces in the two-dimensional Poisson equation', *J. Comp. Phys.*, **83**, 126–147.

T. Vinje and P. Brevig (1981), 'Numerical simulation of breaking waves', *Adv. Water Resources*, **4**, 77–82.

P. W. Voorhees, G. B. McFadden, R. F. Boisvert and D. I. Meiron (1988), 'Numerical simulation of morphological development during Ostwald ripening', *Acta Metall.*, **36**, 207–222.

P. W. Voorhees (1992), 'Ostwald ripening of 2-phase mixtures', *Annual review of Material Sci.*, **22**, 197–215.

B. J. West, K. A. Brueckner and R. S. Janda (1987), 'A new numerical method for surface hydrodynamics', *J. Geophys. Res. C*, **92**, 11803–24.

R. W. Yeung (1982), 'Numerical methods in free-surface flows', *Ann. Rev. Fl. Mech.*, **14**, 395–442.

V. I. Yudovich (1963), 'Non-stationary flow of an ideal incompressible liquid', *Zh. Vych. Mat.*, **3**, 1032–1066 (in Russian)

N. J. Zabusky, M. H. Hughes and K. V. Robert (1979), 'Contour dynamics for the Euler equations in two dimensions', *J. Comp. Phys.*, **30**, 96–106.

N. J. Zabunksy and E. A. Overman (1983), 'Regularization of contour dynamics algorithms , 1 Tangential regularization', *J. Comp. Phys.*, **52**, 351–373.

Acta Numerica (1995), *pp.* 417–457

Particle Methods for the Boltzmann Equation

Helmut Neunzert and Jens Struckmeier

Department of Mathematics
University of Kaiserslautern
Germany
E-mail: neunzert@mathematik.uni-kl.de
and
struckm@mathematik.uni-kl.de

CONTENTS

1. Introduction

In the following chapters we will discuss particle methods for the numerical simulation of rarefied gas flows.

We will mainly treat a billiard game, that is, our particles will be hard spheres. But we will also touch upon cases where particles have internal energies due to rotation or vibration, which they exchange in a collision, and we will talk about chemical reactions happening during a collision.

Due to the limited size of this paper, we are only able to mention the principles of these real-gas effects. On the other hand, the general concepts of particle methods to be presented may be used for other kinds of kinetic equations, such as the semiconductor device simulation. We leave this part of the research to subsequent papers.

Finally, this paper is written by mathematicians. Missing physical intuition needed to 'simulate the game of nature' (Bird, 1989), we have to describe rarefied gas flows by a kinetic equation – this is the modelling part – and then we have to solve this equation numerically.

In a first – a modelling – part we will describe how to get the 'correct' kinetic equation. In a second part we shall describe our basic ideas for solving these equations. They lead to particle methods or – as we sometimes prefer to call them in order to stress the principal similarity to finite differences or finite elements – finite point set methods (FPM).

In Section 4 we shall talk about the practical aspects of a realization of particle methods and the rôle of random numbers and give a comparison between existing codes. In the last part we shall touch on several techniques to improve particle codes, to accelerate the algorithm and to use particle methods on massively parallel machines.

Finally we present some numerical results obtained with particle codes.

2. Collision Integrals for the Boltzmann Equation

Our mathematical model will be a kinetic equation describing the time evolution of a density in position–velocity space

$$t \to f(t, x, v), \quad x \in \Omega, v \in \mathbb{R}^3,$$

which may depend on internal energies too. A kinetic equation has the form

$$\frac{\partial f}{\partial t} + v \cdot \frac{\partial f}{\partial x} + E \cdot \frac{\partial f}{\partial v} = \hat{I}(f),$$

where E is an exterior or self-consistent force field and $\hat{I}(f)$ denotes the collision term.

A prototypical kinetic equation is the Boltzmann equation stated in 1874 by Ludwig Boltzmann. The equation describes the microscopic behaviour of a dilute gas undergoing binary collisions. For the rest of the paper we assume that the force field E vanishes. Hence the main aspect in the modelling part is the derivation of the collision integral.

2.1. Collision Integral

Bobylev (1993) gives a systematic derivation of $\hat{I}(f)$ from several quite simple postulates. We shall shortly review these results since they seem to offer a new approach for collision modelling – the classical approach due to Boltzmann or improved versions of it as given by Cercignani (1989) are well known.

(a)   We take into account only binary collisions – hence $\hat{I}$ is a quadratic, time-independent operator

$$\hat{I}(f)(v) = I(f, f)(v) = \int\limits_{\mathbb{R}^3} \int\limits_{\mathbb{R}^3} K(v \mid v_1, v_2) f(v_1) f(v_2) \, dv_1 \, dv_2.$$

Remark 1   This assumption fails if one has to consider recombination in chemical reactions, where a third collision partner is needed as an energy source.

(b) The collision operator $\hat{I}$ is invariant under translation in the velocity space: if

$$f_a(v) := f(v + a),$$

then

$$\hat{I}(f_a) = \hat{I}(f)_a$$

Remark 2 This assumption is not true for semiconductor devices.

From (a) and (b) one gets

$$K(v \mid v_1, v_2) = Q(v_1 - v, v_2 - v)$$

and

$$\hat{I}(f)(v) = \int_{\mathbf{R}^3} \int_{\mathbf{R}^3} Q(u_1, u_2) f(v + u_1) f(v + u_2)\, du_1\, du_2.$$

(c) The collision operator $\hat{I}$ is invariant under rotations in v-space. Then

$$Q(u_1, u_2) = \tilde{Q}(|u_1|, |u_2|, < u_1, u_2 >).$$

(d) The collision operator $\hat{I}$ can be decomposed into a gain and a loss term

$$Q = Q^+ - Q^- \text{ with } Q^\pm \geq 0$$

and $\hat{I}^-(f) = 0$ if $f = 0$: nothing can be lost, if there is nothing. Then

$$Q^-(u_1, u_2) = \frac{1}{2} \left[g(|u_1|)\delta(u_1) + g(|u_2|)\delta(u_1) \right],$$

where $g(|u|)$ is an arbitrary function.
With q defined by $Q^+(u_1, u_2) = 2^3 q(2u_1, 2u_2)$ we get

$$\hat{I}(f)(v) \quad = \quad \int_{\mathbf{R}^3} \int_{\mathbf{R}^3} q(u - u', u + u') f(v') f(w')\, du'\, dw$$

$$- f(v) \int_{\mathbf{R}^3} g(|u|) f(w)\, dw,$$

where $u = v - w$, $v' = v + \frac{1}{2}(u' - u)$, $w' = w - \frac{1}{2}(u' - u)$.

(e) We have conservation of mass (or particles):

$$\int_{\mathbf{R}^3} \hat{I}(f)(v)\, dv = 0.$$

Then

$$g(|u|) = \int_{\mathbf{R}^3} q(u' - u, u' + u)\, du'$$

and with $p(u' \mid u) = q(u - u', u + u')$ (transition probability) we get

$$\hat{I}(f)(v) = \int_{\mathbb{R}^3} \int_{\mathbb{R}^3} [p(u' \mid u)f(v')f(w') - p(u \mid u')f(v)f(w)] \, du' \, dw.$$

(f) Microreversibility means $p(-u' \mid -u) = p(u \mid u')$. If we include the symmetry (c), we get

$$p(u \mid u') = \tilde{p}\left(|u|, |u'|, < u, u' >\right)$$

and

$$p(u' \mid u) = p(u \mid u').$$

From (f) we get the H-theorem

$$\int_{\mathbb{R}^3} \hat{I}(f)(v)\ln f(v) \, dv \le 0.$$

Remark 3 Assumption (b) implies also conservation of momentum

$$\int_{\mathbb{R}^3} v\hat{I}(f)(v) \, dv = 0.$$

(g) Conservation of energy implies $p(u \mid u') = 0$ if $|u| \ne |u'|$. Then

$$p(u \mid u') = 2\delta\left((|u'|)^2 - |u|^2\right) \sigma\left(|u|, \frac{< u, u' >}{|u|^2}\right).$$

With $u' = |u'| \cdot \eta = |u| \cdot \eta$, $u = v - w$ we get finally

$$\hat{I}(f)(v) = \int_{\mathbb{R}^3} \int_{S^2} |u|\sigma\left(|u|, \frac{< u, \eta >}{|u|}\right) [f(v')f(w') - f(v)f(w)] \, dw \, d\omega,$$

where $\sigma(|u|, \cos\theta)$ is now the only undetermined function, the differential cross section.

The differential cross section σ is now to be chosen in such a way that we are able to reproduce measurements. These measurements are mainly those on transport coefficients – for example, the dependence of the kinematic viscosity on temperature. The simplest idea for σ is given by considering a billiard gas (in the phenomenological derivation)

$$\sigma(|u|, \cos\theta) = d \cdot \cos\theta,$$

where d is a constant connected with the diameter of the molecules. But this gives wrong macroscopic laws; for example, the viscosity η does not depend on T as experiments tell us, which is reflected in the Sutherland formula:

$$\frac{\eta}{\sqrt{T}} \sim \frac{T}{T + T_s}.$$

A better agreement can be achieved by changing σ a bit, using the so-called variable hard sphere (VHS) model

$$\sigma(|u|, \cos\theta) = d\left(1 + \frac{\alpha}{|u|^2}\right)\cos\theta.$$

In this model, the diameter 'shrinks' if the relative velocity $|u| = |v - w|$ is larger; this is not microscopically realistic, but reasonable in the sense of modelling.

2.2. Real-Gas Effects

If one wants to include real-gas effects such as inelastic scattering or chemical reactions, the model gets much more complicated. We will sketch the approach to these phenomena.

Assume we have a mixture of molecules A_2 and the corresponding atoms A. There are essentially (neglecting ionization) five kinds of collision processes that we have to take into account:

$$
\begin{array}{rcll}
A + A & \rightleftharpoons & A + A, & (i) \\
A + A_2 & \rightleftharpoons & A + A_2, & (ii) \\
A_2 + A_2 & \rightleftharpoons & A_2 + A_2, & (iii) \\
A + A_2 & \rightleftharpoons & A + A + A, & (iv) \\
A_2 + A_2 & \rightleftharpoons & A + A + A_2. & (v)
\end{array}
$$

The equations (i)–(iii) describe scattering processes, where (i) corresponds to the classical Boltzmann case. The possibility of dissociation and recombination is stated in (iv) and (v). Note that in the case of recombinations we have to consider triple collisions in order to fulfil energy and momentum conservation.

In 1960 Ludwig and Heil formulated a system of generalized Boltzmann equations describing the aforementioned collision processes. Following Kuščer (1991) we reformulate these equations in terms of differential cross sections.

Let $f(v, t)$ and $g(v, \epsilon, t)$ be the distribution functions for the components A and A_2 of the mixture, where ϵ represents the internal energy of the molecule. Then the Boltzmann (or 'Ludwig–Heil') equations for f (and g, respectively) have collision terms representing these 5 collision processes. The differential cross sections depend on the total energy E of the process (instead of $|u|$) and on internal energies. As an example we show just one expression occurring in the description of the dissociation of molecules:

$$\sum_{\epsilon_2,\epsilon',\epsilon'_1} \int \frac{m^2|u'|^2}{16E_{tr}^2} \sigma_{dm}\left(E,\eta',\epsilon',\epsilon'_1 \to E_{tr},\omega,\epsilon_2\right)$$

$$\times \left[g'g'_1 - \left(\frac{2h}{m}\right)^3 ff_1g_2\right] dv_1\, dv_2\, d\omega(\eta')$$

and it would take some time to explain all the terms here. Recombination is shown in $\left(\frac{2h}{m}\right)^3 ff_1g_2$ and it is still an open question whether recombination plays a significant rôle in real applications.

The undetermined part is again σ; for the nonreactive part, where the molecules and atoms are just scattered, one uses a generalization of the so-called Larssen–Borgnakke model (Borgnakke–Larssen, 1975) that consists essentially in dividing the differential cross section into three parts and performing 'detailed balance'. For collisions among diatomic molecules the model is as follows:

$$\sigma_{sm}(E,\eta\cdot\eta',\epsilon',V_i,\epsilon'_1,V_j \to \epsilon,V_k,\epsilon_1,V_\ell) = (1-a-b)\sigma_{sm,el} + a\sigma_{sm,ve} + b\sigma_{sm,in}$$

with

$$\sigma_{sm,el} = \frac{1}{4\pi}\sigma_{sm}^{tot}(E)\cdot\delta(\epsilon-\epsilon')\delta(\epsilon_1-\epsilon'_1)\delta_{ik}\delta_{j\ell},$$

$$\sigma_{sm,ve} = \frac{3}{2\pi E^3}\sigma_{sm}^{tot}(E)\cdot(E-\epsilon-\epsilon_1)\delta_{ik}\delta_{j\ell},$$

$$\sigma_{sm,in} = C(E)\cdot(E-\epsilon-\epsilon_1-V_k-V_\ell)\sigma_{sm}^{tot}(E),$$

where ϵ is continuous rotational energy, V_i is discrete vibrational energy with level index i and σ_{sm}^{tot} is total scattering cross section. Note that σ_{sm}^{tot} depends on the collision energy E as in the VHS model.

In the generalized Larssen–Borgnakke model three kinds of scattering are considered:

(i)   completely elastic ($\sigma_{sm,el}$),
(ii) vibrationally elastic but maximally inelastic with respect to rotation ($\sigma_{sm,ve}$),
(iii) completely inelastic ($\sigma_{sm,in}$).

The explicit form of the factor $C(E)$ (depending on the vibrational model) is somewhat lengthy and therefore not quoted here. The parameters a and b are chosen to reproduce measured transport coefficients.

For the dissociation reaction we assume (since we do not have enough measurements) for the differential cross section a uniform probability distribution over the energy shell in phase space. This concept is widely used in high-energy physics and often successful in describing decay processes. The differential cross sections for the dissociation reactions (iv) and (v) are the following:

$$\sigma_{da}(E', \eta', \epsilon' \to E^{tr}, \omega) = \frac{1}{4\pi^2} \sigma_{da}^{tot}(E', \epsilon'),$$

$$\sigma_{dm}(E', \eta', \epsilon', V_i', \epsilon_1', V_j' \to \epsilon, V_k, \epsilon_{tr}, \omega)$$
$$= C_{vib}(E)(E - \epsilon - V_k)^2 \sigma_{dm}^{tot}(E', \epsilon', V_i', \epsilon_1', V_j')$$

with ('threshold cross section')

$$\sigma_{da,dm}^{tot} = \sigma^{(n)} \frac{(E' - E_B)^n}{E_{tr}'} \cdot \Theta(E' - E_B),$$

where E_B is the binding energy of the molecule and Θ is the Heaviside function. The parameters $\sigma^{(n)}$ and n have to be chosen to reproduce the measured 'rate coefficient' in equilibrium. This means that averaging of $|u| \cdot \sigma_{da(m)}^{tot}$ over Maxwell–Boltzmann distributions should lead to a form of the rate coefficient similar to the well known 'Arrhenius law':

$$K(T) = AT^s \exp\left(\frac{-E_B}{K_B T}\right),$$

where K_B is Boltzmann's constant and T is temperature. The modelling becomes complicated, but is still possible to handle. We finally want to mention that – besides recombination – ionization, radiative energy transfer and soforth are not yet included and much work remains to be done. We refer the reader to Kuščer (1991) and Bärwinkel and Wolters (1975).

3. Particle Methods for the Boltzmann Equation

There are two aspects of particle methods for the Boltzmann equation: the first one is the theoretical derivation of a particle method; the second the practical aspects of implementing such a simulation scheme.

In this chapter we will discuss the first aspect starting with the definition of particle approximations. The fundamental part in the time evolution of particles is the collision integral; hence we first consider in Subsection 3.2 the homogeneous Boltzmann equation. Finally we explain how to derive particle methods for the full inhomogeneous equation.

3.1. Particle Approximations

A particle is characterized by its position x, velocity v and mass (or charge) α. In order to simplify the notation we put $p = (x, v)$. A particle ensemble (or finite point set) is given by

$$\omega_N = \{(\alpha_1, p_1), \ldots, (\alpha_N, p_N)\}$$

or – in another notation – by

$$\delta_{\omega_N} = \sum_{i=1}^{N} \alpha_i \delta_{p_i}.$$

We consider sequences of particle ensembles

$$\omega_N^N = \left\{ (\alpha_1^N, p_1^N), \ldots, (\alpha_N^N, p_N^N) \right\}$$

or

$$\delta_{\omega_N^N} = \sum_{i=1}^{N} \alpha_i^N \delta_{p_i^N}.$$

Often p_i^N are taken from a sequence of $p_1, p_2, \ldots$, that is, more and more particles are brought into the game; then

$$\left\{ p_1^N, \ldots, p_N^N \right\} = \left\{ p_1, \ldots, p_N \right\}.$$

One can in general not expect as good results for sequences of velocities as for sequences of ensembles.

Now, for a given density $f \in \mathcal{L}_+^1(\mathbb{R}^3)$ we say that '$\delta_{\omega_N^N}$ converges to f' if

$$\lim_{N \to \infty} \sum_{i=1}^{N} \alpha_i^N \varphi(p_i^N) = \int f \cdot \varphi \, dv \, dx \qquad \text{for all } \varphi \in C^b(\mathbb{R}^3 \times \mathbb{R}^3).$$

This means that the discrete measure $\delta_{\omega_N^N}$ weak* converges to $f \, dv \, dx$.

Remark 4

(a) We may interpret this as an integration rule, where we integrate the function φ with respect to the measure $f dv dx$. Knots and weights depend on f, not on φ. Estimates should distinguish between a distance between ω_N^N and f and a smoothness property of φ.

(b) We should be aware that if f does not have a bounded support, we are not able to include unbounded φ such as $|v|^2$ or $|v|^2 v$ etc. So we do not get the convergence of moments we need for physical reasons (as temperature or heat transfer). This is a serious problem, which we see also numerically, if we compute the heat transfer. Some improvements in this direction may be found in Struckmeier (1994).

We would like to measure the distance between ω_N^N and f. This might be done by any distance in measure spaces (such as the Prohorov metric or bounded Lipschitz distance), but also – since the limit $f \, dv \, dx$ is absolutely continuous with respect to the Lebesgue measure – with the help of the 'discrepancy'. Consider an axis-parallel 'rectangle' $R \subset \mathbb{R}^3 \times \mathbb{R}^3$ and the mass of w_N^N in R:

$$\sum_{i=1}^{N} \alpha_i^N \mathcal{X}_R(\boldsymbol{p}_i^N) \qquad \text{with } \mathcal{X}_R(P) := \begin{cases} 1 & \text{if} \qquad\qquad P \in R, \\ 0 & \text{otherwise.} \end{cases}$$

Compare it with the mass in R as given by f, that is, $\int_R f \, dv$. The largest possible deviation, that is,

$$\sup_R \left| \sum_{i=1}^{N} \alpha_i^N \mathcal{X}_R(\boldsymbol{p}_i^N) - \int_R f \, dv \, dx \right| =: D\left(w_N^N, f\right)$$

is called 'discrepancy'. It is a distance between w_N^N and f and we have

$$\delta_{w_N^N} \to f \quad \text{iff} \quad D\left(w_N^N, f\right) \to 0.$$

There are other similar definitions of discrepancy using the class of convex sets and so forth instead of rectangles – but this does not change the situation.

There are two consequences of our definition – at least for equal weights $\alpha_i^N = \frac{M}{N}$:

(a) The Koksma–Hlawka inequality:

$$\left| \int \varphi f \, dv \, dx - \frac{M}{N} \sum_{j=1}^{N} \varphi(v_j^N) \right| \le \mathrm{Var}[\varphi] \cdot D(w_N^N, f).$$

We see that in fact $\delta_{w_N^N} \to f$ if $D(w_N^N, f) \to 0$ and that it goes linear with D. The variation of φ, which we denote by $\mathrm{Var}[\varphi]$, is for one-dimensional v the usual total variation and might be substituted by $\int |\varphi'(v)| \, dv$, if φ is differentiable. In dimension 3 or higher it is the so-called 'Hardy–Krause' variation, a quite lengthy concept based on the Vitali variation. One realizes that the estimate separates the distance D from the properties of the test function. For f we assume nothing more than that it is a density.

(b) We are now able to discuss an optimal speed of convergence: How fast does $D(w_N^N, f)$ converge to zero? Clearly, the speed depends on the definition of D and we get mainly relative information. For $f = \mathcal{X}_{[0,1]^k}(v)$, the uniform distribution in the unit cube, there are very strong number-theoretic results:

With $D(w_N^N) = D(w_N^N, \mathcal{X}_{[0,1]^k})$ one gets that there exist constants C_k, C_k' with

$$D(w_N^N) \le C_k \frac{\ln N^{k-1}}{N} \quad \text{for some } w_N^N$$

and

$$D(w_N^N) > C_k' \frac{\ln N^{\frac{k-1}{2}}}{N} \quad \text{for all } w_N^N.$$

Since one can construct sequences of w_N^N, which have a convergence rate given by $\ln N^{k-1}/N$, one may say that this is 'almost optimal' today and not much can be gained in principle. The convergence is slow, but faster than $N^{-\frac{1}{2}}$, which would be the rate for random numbers. And it grows relatively slowly with the dimension k – this is the reason why particle methods are useful for higher dimensions! We shall see that for us k will be typically $2 \times 3 + 2 = 8$. We shall come back to the question of how to construct this optimal convergence order in Section 4.

Remark 5

(a) Do we gain much by using weighted particles? We have more free parameters, but realize: we want to improve $D(w_N^N, f)$, not $|\int f\varphi \, dv - \sum \alpha_i^N \varphi(v_i^N)|$, for a concrete φ! The only answer that is yet known is for a very simple case: Take $k = 1$ and $f = \mathcal{X}_{[0,1]}$. Then the best we can get without weights is $\frac{1}{N}$, and with weights $\frac{1}{N+1}$ – but only if $\sum_{i=1}^N \alpha_i^N = \frac{N}{N+1}$. The order of convergence is not changed in this case.

(b) If we construct w_N^N using a sequence $(p_j)_{j \in N}$, by just adding a new particle when moving from N to $N + 1$, we loose a bit of convergence speed: Now

$$\mathcal{O}\left(\frac{\ln N^k}{N}\right)$$

is the optimal order we can achieve.

3.2. The Homogeneous Boltzmann Equation

The spatially homogeneous Boltzmann equation is given by

$$f_t(t, x, v) = \hat{I}^+(f) - f \int \int k f(t, w) \, d\omega(\eta) \, dw.$$

We have to discretize this equation with respect to t, putting $f_j(v) = f(j\Delta t, v)$, and we may do that either by just a simple Euler step

$$f_{j+1} = \left(1 - \Delta t \int k f_j \, d\omega(\eta) \, dw\right) f_j + \Delta t \int k f_j(v') f_j(w') \, d\omega(\eta) \, dw \quad (3.1)$$

or by integrating

$$\frac{\partial f}{\partial t} = \hat{I}^+(f_j) - f \int k f_j \, d\omega(\eta) \, dw \quad (3.2)$$

over $j\Delta t \le t \le (j+1)\Delta t$ with f_j as initial value.

For the first idea we have as a price to pay a severe restriction on Δt – but we pay it, since the second idea is computationally very expensive without a restriction on Δt. There has been no investigation yet as to whether it might be occasionally cheaper to combine both methods.

Anyhow, we proceed with the simple explicit discretization and use a weak formulation, which we get by multiplying both sides by a bounded continuous test function $\varphi \in C^b$ and integrating over v (in principle, we should realize that $f(t, \cdot)$ is a density of a measure and measures are quite natural mathematical objects for dealing with mass or charge distributions etc.; we could derive a measure formulation of any kinetic equation, which would be a natural starting point for our particle approximations, but the weak formulation is equivalent to a measure formulation). We get using $dv' \, dw' = dv \, dw$, $|v' - w'| = |v - w|$ and $v = v' - \eta < v' - w', \eta >$

$$\int \varphi(v) f_{j+1}(v) \, dv = \int_{\mathbb{R}^3} \int_{\mathbb{R}^3} (K_{v,w}\varphi) f_j(v) f_j(w) \, dv \, dw \qquad (3.3)$$

with

$$K_{v,w}\varphi = \left(1 - \Delta t \int k(|v - w|, \theta) \, d\omega(\eta)\right) \varphi(v) + \Delta t \int k(|v - w|, \theta) \varphi(v') \, d\omega(\eta).$$

Equation (3.1) is equivalent to (3.3), if we use $\int f_j(v) \, dv = 1$, which is guaranteed by the conservation of mass. The 'transition kernel' $K_{v,w}\varphi$ is here independent of f_j – this would be different for (3.2).

We need to transform $K_{v,w}\varphi$ into a form like

$$K_{v,w}\varphi = \int \varphi(\psi(v, w, x)) \chi(x) \, dx \qquad (3.4)$$

with an auxiliary k-dimensional variable x, since we then get

$$\int \varphi(v) f_{j+1}(v) \, dv = \int \int \int \varphi(\psi(v, w, x)) f_j(v) f_j(w) \chi(x) \, dx \, dv \, dw$$

and we shall see that a point approximation of the $(6+k)$-dimensional density $f_j(v) f_j(v) \chi(x)$ leads immediately to an approximation of f_{j+1}. Assuming that we have such an approximation for f_j, we have to construct one for $f_j(v) f_j(w) \chi(x)$ and get the approximation for the time evolution $j \to j+1$.

The representation (3.4) is due to Babovsky (1989):

Let B be a ball in $\mathbb{R}^2$ of area 1 (radius $\frac{1}{\sqrt{\pi}}$); then we can construct a function $\phi_{v,w} : B \to S_+^2$ such that

$$\psi(v, w, x) = T_{v,w}(\phi_{v,w}(x))$$

and χ is the characteristic function of B; here $T_{v,w}(\eta)$ is just v', that is, $T_{v,w}(\eta) = v - \eta < v - w, \eta >$. So $\phi_{v,w}(x)$ is nothing but another representation of the 'impact parameter η'. But more is hidden: the formulation includes at the end 'dummy collisions', that is, collisions without effect – a useful strategy (as we shall see) originally used by Nanbu (1980), Neunzert, Gropengiesser and Struckmeier (1991) and Ivanov and Rogasinsky (1988).

We shall give the construction of ϕ, since it is the basis of our simulation code: We fix v, w and take $v - w$ as polar axis in a polar coordinate system (α, β) for η, where α is the angle between η and $v - w$, that is, θ. We get

$$k(\theta)\,\mathrm{d}\eta = k(\alpha)\sin\alpha\,\mathrm{d}\alpha\,\mathrm{d}\beta.$$

Choose a function $r(\alpha)$ such that

$$r(\alpha)\frac{\mathrm{d}r}{\mathrm{d}\alpha} = \Delta t \cdot k(\alpha)\sin\alpha.$$

Since $\eta \in S_+^2$, that is, $0 \le \alpha \le \frac{\pi}{2}$, the right-hand side is positive for $\alpha > 0$ and $r(\alpha)$ is invertible with inverse $\alpha(r)$. The maximal value of $r^2(\alpha)$ is

$$r^2(\tfrac{\pi}{2}) = 2\Delta t \int_0^{\frac{\pi}{2}} k(\alpha)\sin\alpha\,\mathrm{d}\alpha.$$

Now the restriction for the Euler scheme comes into play. We have to guarantee nonnegativity of f_{j+1} if f_j is nonnegative; this is achieved by

$$1 - \Delta t \int_{S_2^+} k(|v - w|, \theta)\,\mathrm{d}\omega(\eta) \ge 0 \qquad \text{for all } v, w,$$

that is,

$$\Delta t \int_0^{2\pi}\int_0^{\frac{\pi}{2}} k(\alpha)\sin\alpha\,\mathrm{d}\alpha\,\mathrm{d}\beta \le 1$$

or

$$2\Delta t \int_0^{\frac{\pi}{2}} k(\alpha)\sin\alpha\,\mathrm{d}\alpha = r^2\left(\frac{\pi}{2}\right) \le \frac{1}{\pi}.$$

This is a serious restriction on Δt! With $r_{\max} = r(\frac{\pi}{2})$ we get

$$\begin{aligned}
\Delta t \int k(\theta)\varphi(v')\,\mathrm{d}\omega(\eta) &= \Delta t \int \varphi(T_{v,w}(\eta))k(\theta)\,\mathrm{d}\omega(\eta) \\
&= \int_0^{2\pi}\int_0^{r_{\max}} \varphi\left(T_{v,w}(\alpha(r), \beta)\right) r\,\mathrm{d}r\,\mathrm{d}\beta \\
&= \int_{B_{r_{\max}}} \varphi\left(T_{v,w}(\phi_{v,w}(x))\right)\,\mathrm{d}^2 x,
\end{aligned}$$

if $\phi_{v,w}(x)$ is just the mapping $x \sim (r, \beta) \to (\alpha(r), \beta)$ $((r, \beta)$ are the polar coordinates of the point x in the ball $B_{r_{\max}}$ with radius $r_{\max}$).

We have defined $\phi_{v,w}(x)$ for $x \in B_{r_{\max}} \subset B$; this describes the case when 'real' collisions happen – v' is different from v. The other part – corresponding to $(1 - \Delta t \int k \, d\omega)\varphi(v)$ – reflects the probability that no collision occurs and so we define $\phi_{v,w}(x)$ as follows.

If $x = (r \cos \beta, r \sin \beta) \notin B_{r_{\max}}$, then

$$\phi_{v,w}(x) := \left(\frac{\pi}{2}, \beta\right).$$

If $\alpha = \frac{\pi}{2}$, $v - w$ is orthogonal to η and $v' = v$! Therefore, if x is in the annulus $r_{\max} \leq r \leq \frac{1}{\sqrt{\pi}}$ we have dummy collisions.

$\phi_{v,w}$ is now defined for all $x \in B$ and since

$$\left(1 - \Delta t \int k(\theta) \, d\omega(\eta)\right) \varphi(v) = \int_{r_{\max} \leq \frac{1}{\sqrt{\pi}}} \varphi\left(T_{v,w}(\phi_{v,w}(x))\right) \, dx,$$

it does what it should do:

$$K_{v,w}\varphi = \int_B \varphi\left(\psi(v, w, x)\right) \, dx$$

with $\chi(x) = 1$ for all $x \in B$.

What we have to solve numerically is

$$\int_{\mathbb{R}^3} \varphi(v) f_{j+1}(v) \, dv = \int_{\mathbb{R}^3} \int_{\mathbb{R}^3} \int_B \varphi\left(\psi(v, w, x)\right) f_j(v) f_j(w) \, dx \, dv \, dw. \quad (3.5)$$

Assume that we have an approximation $\left\{v_1^N(j), \ldots, v_N^N(j)\right\}$ of f_j and we want to construct an approximation of f_{j+1}.

The right-hand side of (3.5) tells us what we have to do: The measure over which we integrate is

$$f_j(v) f_j(w) \mathcal{X}_B(x) \, dv \, dw \, dx,$$

where $\mathcal{X}_B$ is the characteristic function of B. We need therefore a 'finite point set' that approximates $f_j(v) f_j(w) \mathcal{X}_B(x)$, which is an 8-dimensional density of total 'mass' $1/M^2$.

If we construct a set $\left\{\left(v_1^N(*), w_1^N(*), x_1^N\right), \ldots, \left(v_N^N(*), w_N^N(*), x_N^N\right)\right\}$ (with weights M/N) approximating this density, then

$$\frac{M}{N} \sum_{i=1}^N \varphi\left(\psi\left(v_i^N(*), w_i^N(*), x_i^N\right)\right)$$

approximates $\int \varphi(v) f_{j+1}(v) dv$ and

$$v_i^N(j+1) = \psi\left(v_i^N(*), w_i^N(*), x_i^N\right)$$

is an approximation of f_{j+1}!

This gives the simulation procedure and a convergence criterion:

Given an approximation $\left\{v_i^N(j), \ldots, v_N^N(j)\right\}$ of f_j, construct from that an approximation

$$\left\{\left(v_1^N(*), w_1^N(*), x_1^N\right), \ldots, \left(v_N^N(*), w_N^N(*), x_N^N\right)\right\}$$

of $f_j(v) f_j(w) \mathcal{X}_B(x)$. Then

$$v_i^N(j+1) = \psi(v_i^N(*), w_i^N(*), x_i^N), \quad i = 1, \ldots, N,$$

approximates f_{j+1}.

The main question remains: how do we get $(v_i^N(*), w_i^N(*))$? We have only $v_i^N(j)$, $i = 1, \ldots, N$, but we have a lot of freedom – the only theoretical condition is the convergence condition. We will come back to this question in Subsection 4.1.

Practically speaking, we have more conditions – it is necessary to maintain all conservation properties (mass, momentum, energy) even for the evolution in the simulation process, which means for equal weights

$$\sum_{i=1}^{N} v_i^N(j) = \sum_{i=1}^{N} v_i^N(j+1)$$

and

$$\sum_{i=1}^{N} \|v_i^N(j)\|^2 = \sum_{i=1}^{N} \|v_i^N(j+1)\|^2.$$

All practical computations show the importance of the numerical conservation of these quantities (see also Greengard and Reyna (1992)).

3.3. Particle Methods for Inhomogeneous Problems

In the previous subsection we derived a particle method for the spatially homogeneous Boltzmann equation. If we solve an inhomogeneous problem we have to take into account the spatial location of a particle.

Concerning the discretization of the inhomogeneous equation we may use

$$\bar{f}((j+1)\Delta, x, v) = f(j\Delta t, x - \Delta t v, v),$$

$$\frac{\partial \bar{f}}{\partial t} = \hat{I}(\bar{f}),$$

that is, there is a decoupling of the free flow of particles and the collisions among them.

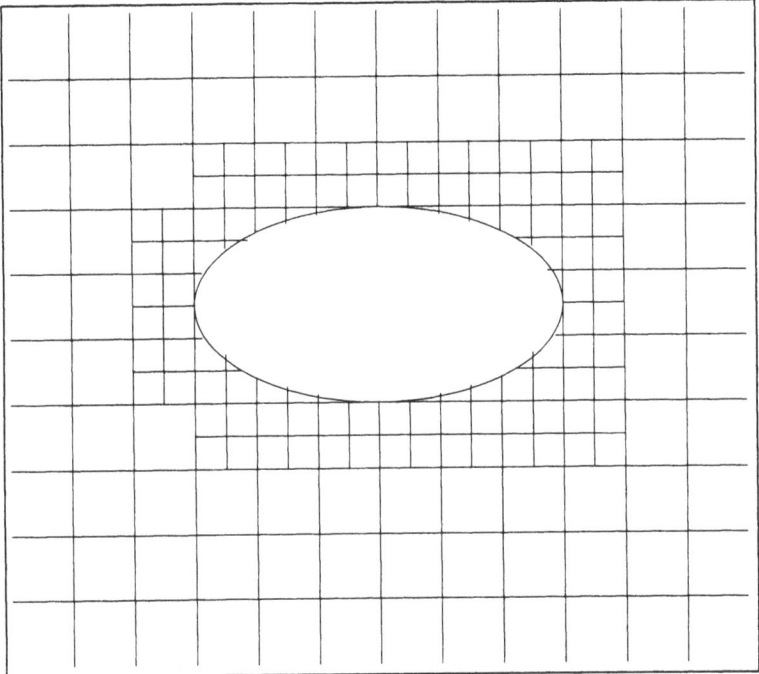

Fig. 1. Adaptive regular grid structure

Given an approximation of $f(j\Delta t, x, v)$ by a finite point set we have no problems with the first equation: we just move the particles over the time increment Δt with the particle velocity and no spatial discretization is required.

The second equation is much more complicated. Remember that $\bar{f}$ depends on x, which for finite point sets is the x-coordinate of the particle, but the collision operator is local in space.

The easiest way to get rid of the difficulties – this approach is used by nearly all methods – is

(a) to introduce a spatial cell structure, like that shown in Figure 1,

(b) to substitute $\bar{f}$ by a step function

$$\bar{f}(t, x, v) = f_{C_i}(t, v) \quad \text{if } x \in C_i$$

(c) and to consider in each cell the homogeneous Boltzmann equation and use the algorithm presented in Subsection 3.2.

Hence, only particles that are located in the same cell can form a collision pair.

Several important remarks have to be made here:

(a) One problem with this approach is that in every time step the particles have to be resampled, after the free flow, from the cell structure. If the cell structure is like in classical FEM given by triangles or tetrahedral cells this procedure requires an enormous computational effort. Hence one uses regular meshes like the one shown in Figure 1.

(b) The size of a cell has to be smaller than the local mean free path, the appropiate resolution scale. Now the local mean free path depends strongly on the local macroscopic gas density. This quantity may vary by orders of magnitude in different regions: in front of an obstacle the density may increase by a factor of 10 whereas on the lee side the density may decrease by the same factor. There are two ways to overcome this problem. The first is to use an adaptive grid structure, like in Figure 1 the second is to use different particle weights in different regions (this is discussed in Subsection 5.1).

(c) In the first part of the simulation process, the free flow of the particles, some particles may hit the spatial boundary, leave the domain or enter it. One has to take care of the corresponding boundary conditions:
• particles may leave the computational domain (absorption),
• particles may be re-emitted at a physical boundary (gas–surface interaction),
• particles may be reflected because of symmetry,
• particles may enter the spatial domain at parts of the boundary.
The gas–surface-interaction part is the most important phenomenon. The boundary condition is defined by a scattering kernel describing the velocity (respectively internal energy) change of particles hitting the surface. For monatomic gases the boundary condition is

$$|(v, n)| f(t, x, v) = \int\limits_{(v', n) < 0} R(v' \longrightarrow v; t, x) |(v', n)| f(t, x, v') \, dv'$$

for all times $t \in \mathbb{R}_+$, x on the spatial boundary and $(v, n) > 0$.

The classical model (for monatomic gases) is the diffuse reflection with complete thermal accommodation. Several other models, such the as Maxwell model (Cercignani, 1989), Cercignani–Lampis model (Lord, 1991) or Nocilla model (Nocilla, 1961), exist in the literature.

Different boundary models lead to different aerodynamic characteristics, whence a concrete knowledge of the real interaction law is fundamental for the description of rarefied gas flows.

4. Practical Aspects of Particle Methods

In Section 3 a description of the main idea was given. But still particle methods have enormous demands of computational time and storage. Therefore many minor tricks are needed to improve the efficiency and reliability of the method. These tricks are the treasure different groups accumulate during the development and the use of this code. In the following chapters we describe some of the details of our code.

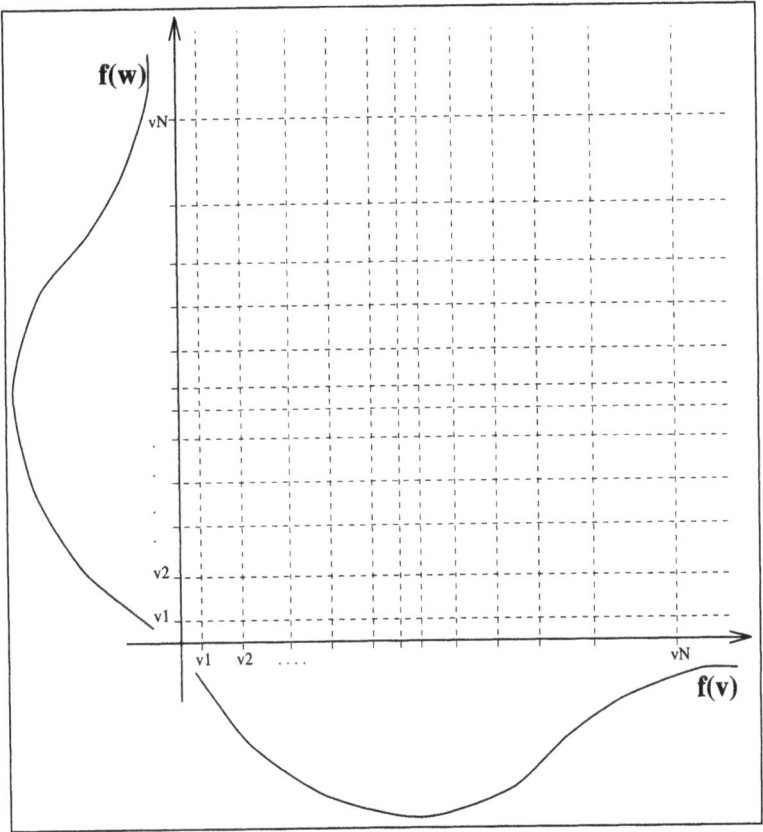

Fig. 2. Approximation of the product $f(0,v) \cdot f(0,w)$

4.1. Collision Selection and Conservation Quantities

The main part of the particle method given in the last chapter is the collision procedure which may be described as follows: Given N particles (of equal weights) at $v_1, \ldots, v_N$ (we omit the indices which are not necessary), determine N pairs $\{(v_1^*, w_1^*), \ldots (v_N^*, w_N^*)\}$ and 'impact parameters' $x_1^*, \ldots, x_N^*$ appropriately and get the new velocities by

$$\psi\left(v_1^*, w_1^*, x_1^*\right), \ldots, \psi\left(v_N^*, w_N^*, x_N^*\right).$$

There is no theoretical 'necessity' to form the pairs out of the set of particles already given – but it is quite natural. Then we have N^2 candidates for those pairs:

$$\left(v_i, v_j\right), \quad 1 \leq i, j \leq N.$$

Figure 2 gives a 1-dimensional impression. How do we select N pairs out of N^2 possible ones in order to get an approximation of $f_j(v) f_j(w)$? Denote the selected pairs by $\left(v_1, v_{j(1)}\right), \ldots, \left(v_N, v_{j(N)}\right)$.

If we have the pair $(v_i, v_{j(i)})$, we find an impact parameter x_i. $\{x_1, \ldots, x_N\}$

must approximate $\mathcal{X}_B$, that is, the uniform distribution in a ball, and they may do that independently of $\left(v_i, v_{j(i)}\right)$. This defines the new velocity $\psi\left(v_i, v_{j(i)}, x_i\right)$. So, where do we put our cross in the ith column of the (v_i, v_j)-diagram, that is, what is $j(i)$?

The first idea due to Nanbu (1980) was a stochastic one:

Select a random number r_i from a uniform distribution in $[0, 1]$ and put $j(i) = [Nr_i] + 1$; then $j(i) \in \{1, \ldots, N\}$, but it might happen that two different i get the same partner $j(i)$. We distribute the crosses randomly in each column. We need to show that, for fixed velocities $\tilde{v}$, $\tilde{w}$ and $R_{\tilde{v}} \times R_{\tilde{w}} = \{(v, w) \mid v \le \tilde{v}, w \le \tilde{w}\}$,

$$\frac{1}{N} \sum \mathcal{X}_{R_{\tilde{v}} \times R_{\tilde{w}}} \left(v_i, v_{j(i)}\right) \to \int_{v \le \tilde{v}} f(v) \, dv \int_{w \le \tilde{w}} f(w) \, dw.$$

Using the central limit theorem, Babovsky (1989) showed that this is true for almost all sequences $(r_i)_{i \in N}$, that is, the procedure converges with probability 1.

In principle we are through – but only in principle: There are many necessary and possible improvements.

For example, in the Nanbu procedure described above, there is no conservation of total momentum or energy – this is true only 'on average'. The practical consequences were such that Nanbu's method could not compete with the so-called Direct Simulation Monte Carlo (DSMC) of Bird (1976), which we shall describe in Subsection 4.3.

Babovsky gave an improvement that does not have this drawback.

Assume that $N = 2n$. Then divide the set $\{v_1, \ldots, v_N\}$ randomly into two subsets $\{v_1^1, \ldots, v_n^1\}$ and $\{v_1^2, \ldots, v_n^2\}$, each containing half of the particles. Now choose a permutation π of $\{1, \ldots, n\}$ at random (i.e. each permutation has the same probability) and consider $\left(v_i^1, v_{\pi(i)}^2\right)$ as well as $\left(v_{\pi(i)}^2, v_i^1\right)$ as pairs: we make our crosses symmetric with respect to the main diagonal. Finally, we choose the same impact parameter x_i for both pairs and get two new velocities

$$\psi\left(v_i^1, v_{\pi(i)}^2, x_i\right) \text{ and } \psi\left(v_{\pi(i)}^2, v_i^1, x_i\right).$$

This procedure keeps the idea of a binary collision and it conserves energy and momentum, since they are conserved 'pairwise'

$$v_i^1 + v_{\pi(i)}^2 = \psi\left(v_i^1, v_{\pi(i)}^2, x_i\right) + \psi\left(v_{\pi(i)}^2, v_i^1, x_i\right)$$

and similarly for the energy $\|v_i^1\|^2 + \|v_{\pi(i)}^2\|^2$.

So, symmetry guarantees these conservation laws – but only for equally weighted particles. Babovsky has also shown convergence in probability for this procedure.

If we have different weights for approximating different species in a mixture with great differences in the concentrations, then, in one cell, one might have particles with different weights α_i.

Conservation of momentum and energy in a cell would mean

$$\sum_{j=1}^{k} \int f^j(t, v) v \, dv \;=\; \text{constant},$$

$$\sum_{j=1}^{k} \int f^j(t, v) \|v\|^2 \, dv \;=\; \text{constant},$$

where $f^j(t, v)$ designates the distribution of the jth species, which is assumed to have a total mass M_j.

Approximating f^j by $M_j \alpha_j \sum_{i=1}^{N_j} \delta_{v_i^j}$, where α_j is the weight of the jth species, we would get the discrete conservation of total momentum C_M and total energy C_E

$$\sum_{j=1}^{k} M_j \alpha_j \sum_{i=1}^{N_j} v_i^j \;=\; C_M, \tag{4.1}$$

$$\sum_{j=1}^{k} M_j \alpha_j \sum_{i=1}^{N_j} \|v_i^j\|^2 \;=\; C_E. \tag{4.2}$$

If we would now consider binary collisions and try to conserve momentum and energy 'individually' in each of these binary collisions, we would fail if two particles representing different species were involved:

$$\alpha_j M_j v^j + \alpha_i M_i v^i = \alpha_j M_j v^{j'} + \alpha_i M_i v^{i'}$$

and

$$\alpha_j M_j \|v^j\|^2 + \alpha_i M_i \|v^i\|^2 = \alpha_j M_j \|v^{j'}\|^2 + \alpha_i M_i \|v^{i'}\|^2$$

are solvable only if $\alpha_i = \alpha_j$ or if no collision happens.

However, it is possible to conserve momentum and energy with weighted particles for the particle ensemble $\{(\alpha_1^1, v_1^1), \ldots, (\alpha_k, v_{N_k}^k)\}$ – not 'pairwise', but by choosing the collision parameters x_i^k such that equations (4.1) and (4.2) are fulfilled for the post-collision velocities [27].

4.2. Random Numbers and the Generation of Random Variates

We discuss now the question of how much stochasticity is necessary in a particle code. What we need is to have

(a)   a good approximation of the initial value $f_0(v)$ by a particle set;

(b)   a selection of N pairs $(v_1, w_1), \ldots, (v_N, w_N)$ out of N^2 candidates (v_i, v_j) such that they are a good approximation of $f(v)f(w)$, if $v_1, \ldots, v_N$ is a good approximation of f;

(c) N 2-dimensional points $x_1, \ldots, x_N$ approximating $\mathcal{X}_B(x)$;

(d)   for the case where there are stochastic boundary conditions (such as diffuse reflection etc.) an approximation of the distribution of the fluxes leaving the boundary.

One may use random number generators for all purposes, that is, one takes a 1D random number generator (for a uniform distribution in [0,1]), uses sections of length k to get k-dimensional points, which should be uniformly distributed in $[0, 1]^k$ and transforms them to get a sample distributed with the given density f – this is what we have to do for (a) and for (c). How we use random number generators in (b) was described in the previous pages.

The main question is: do we need the 'random property' of these generators and how should we define this property?

We give one version of stochasticity for uniformly distributed random numbers on $[0, 1]$:

If one has to construct a set of N points $x_1, \ldots, x_N$ approximating $\mathcal{X}_{[0,1]}(x)$ in an optimal way, the best solution is simply the set

$$\left\{ \frac{1}{2N}, \frac{3}{2N}, \ldots, \frac{2N-1}{2N} \right\}.$$

The discrepancy is $\frac{1}{N}$ and this is optimal, but certainly not very random. We can see that by constructing 2D points from it, for example,

$$\left(\frac{1}{2N}, \frac{3}{2N} \right), \left(\frac{3}{2N}, \frac{5}{2N} \right), \ldots, \left(\frac{2N-3}{2N}, \frac{2N-1}{2N} \right);$$

all points are near the diagonal of the unit square $[0, 1]^2$ and therefore certainly not a good approximation of $\mathcal{X}_{[0,1]^2}$.

The best discrepancy we can get (for the 1D and 2D sets) is now of order $\ln N/N$. The points $x_1, \ldots, x_N$ seem now to be more stochastic – let's call them stochastic of order 1. We may realize that it is only pseudo random by looking at sections of length k: $(x_1, \ldots, x_k), (x_2, \ldots, x_{k+1}), \ldots$ and considering them as points in $[0, 1]^k$. If they are still good approximations of $\mathcal{X}_{[0,1]^k}$, we say they have stochasticity of order $k - 1$. A real random number generator should have stochasticity of order ∞ – if we use it for Monte Carlo methods in reactor physics, we need a stochasticity of very high order (the dimension is proportional to the number of collisions a neutron has with a nucleus).

In starting a simulation we should check how much stochasticity is needed – and only then can we decide how to generate our particles. For problems (a) and (c), we need just 3D or 2D approximations of $f(v)$ or $\mathcal{X}_B(x)$. We might do this by using sections of length 3 or 2 – or by other constructions, called low-discrepancy methods, which we shall describe now.

For (b) we need – for our permutation method – a stochastic separation of a $2n$-set into two n-sets and then a sequence $r_1, \ldots r_N$ of $[0,1]$ random numbers. Since the convergence proof shows convergence in probability, we need the independence of $r_1, \ldots r_N$, in our language stochasticity of order $N - 1$. But this is only due to our method of selecting N pairs out of N^2. We could do that completely deterministically but haven't done it yet. Why not select just one cross pattern $(i, j(i))$, which represents a uniform distribution of the crosses (one may play with introducing an index discrepancy just defined on $\{1, \ldots, N\}^2$ and find an optimal $j(i)$), and apply it in each collision process? It would fulfil our convergence condition but would presumably insert a small but systematic error, which may accumulate during the evolution. This is the only risk in using as little stochasticity as possible: The fluctuations get smaller, but might be 'one-sided' and do not average out in the evolution.

Such a problem may occur in treating boundary conditions (Missmahl 1990), where one gets one-sided errors, which lead to a 'numerical cooling'. Changing the deterministic procedure just a bit, one may get rid of the effect – but this has to be done carefully.

But for (a) and (c) one may easily use low-discrepancy methods extensively described by H. Niederreiter (1992).

We want to construct point sequences (not ensemble sequences) $x_1, x_2, \ldots$ such that $\omega_N = \{x_1, \ldots, x_N\}$ has a low discrepancy against $\mathcal{X}_{[0,1]^k}$, that is,

$$D(\omega_N, \mathcal{X}_{[0,1]^k}) = D(\omega_N) = \mathcal{O}\left(\frac{\ln N^k}{N}\right)$$

(remember: sequences of ensembles could have $\mathcal{O}\left(\frac{\ln N^{k-1}}{N}\right)$).

For $k = 1$, we get as optimal order $\dfrac{\ln N}{N}$, not $\dfrac{1}{2N}$ as for $\left\{\dfrac{1}{2N}, \ldots, \right.$ $\left. \dfrac{2N-1}{2N}\right\}$, which is an ensemble sequence.

The starting point is an old idea by van der Corput, defining x_i as follows: Take the dual representation of $i = \ell_1 + \ell_2 2^1 + \cdots + \ell_m 2^{m-1}$, $\ell_k = 0, 1$, and put

$$x_i = \phi_2(i) := \ell_1 2^{-1} + \ell_2 2^{-2} + \cdots + \ell_m 2^{-m} \in [0, 1].$$

For it

$$D(\omega_N) = \frac{2}{3\ln 2} \frac{\ln N}{N} + \mathcal{O}\left(\frac{1}{N}\right),$$

so it has optimal order. We can change the basis 2 and use any p-adic representation of i as well; this was done by Hammersley and is denoted by $\phi_p(i)$.

To get k-dimensional sequences, Halton proposed taking numbers $p_1, \ldots, p_k$ relatively prime and constructing

$$x_i = (\phi_{p_1}(i), \ldots, \phi_{p_k}(i)), \quad i \in \mathbf{N}.$$

Here again $D(\omega_N) = \mathcal{O}\left(\frac{\ln N^k}{N}\right)$, that is, is optimal.

Please realize that we do not construct k-dimensional points by using sections of 1D sequences. Therefore we need only stochasticity 0.

There are several other methods of constructing k-dimensional low-discrepancy sequences, mainly by Faure (1982), Sobol (1969) and Niederreiter (1992). They differ in the $\mathcal{O}$-constants, which depend on the dimension k – and they may have especially low discrepancy for certain N. Since our k is never higher than 10, we do not care about it too much. There are many tests on the behaviour of different LD-sequences by G. Pagès (1992).

From a practical point of view it is fundamental to have fast algorithms for generating x_i – the algorithms should not be slower than the linear congruential methods used in normal random number generators.

A fast algorithm for a special class of low-discrepancy sequences can be found in Struckmeier (1993). It uses the p-adic Neumann–Kakutani transformations $T_p: [0, 1] \to [0, 1]$, which might be written as $T_p(x) = x \oplus \frac{1}{p}$ with a 'left addition $\oplus$' or as

$$T_p(x) = x + b_j^p$$

with

$$b_j^p = \frac{1}{p^j}(p + 1 - p^j)i$$

and

$$j = j(x) = \left[-\frac{\ln(1-x)}{\ln p}\right] + 1.$$

Now x_i defined by $x_i = T_p(x_{i-1})$, $x_0 \in [0, 1]$ arbitrary, is a low-discrepancy sequence, called a generalized Halton sequence, and has the same optimal behaviour.

The algorithm is clear: one generates $b_j^p \ \forall j \in \mathbf{N}$ and then iterates as follows.

Given x_n, we compute $j(x_n)$ and then $x_{n+1} = x_n + b_{j(x_n)}^p$ (in practice it is sufficient to compute only the first 32 points of b_j^p). In k dimension, we use relatively prime numbers $p_1, \ldots, p_k$ and define the mth component, x_i^m, of x_i by

$$x_i^m = T_{p_m}(x_{i-1}^m), \ 1 \le m \le k.$$

Table 1. *CPU time in seconds to generate 10^6 numbers on $[0,1]$*

Hardware	g.H. (b=2)	LC (F77)	rand() (UNIX)
IBM 6000/530	1.9	2.8	1.6
HP 9000/835 SRX	4.8	25.8	12.9
HP 9000/710	1.0	3.1	2.0
nCUBE 2S 1 node	6.3	5.4	-

Table 2. *Discrepancy and variation of different sequences*

Sequence	D_N	V_M	D_N	V_M	D_N	V_M
Optimal	$1.72 \cdot 10^{-2}$		$5.15 \cdot 10^{-3}$		$2.89 \cdot 10^{-3}$	
rand()	$1.30 \cdot 10^{-1}$	$1.6 \cdot 10^{-3}$	$7.76 \cdot 10^{-2}$	$6.7 \cdot 10^{-4}$	$6.40 \cdot 10^{-2}$	$3.0 \cdot 10^{-4}$
g.H. (b=2)	$3.97 \cdot 10^{-2}$	$7.1 \cdot 10^{-5}$	$1.25 \cdot 10^{-2}$	$5.7 \cdot 10^{-6}$	$9.71 \cdot 10^{-3}$	$8.1 \cdot 10^{-7}$
g.H. (b=3)	$3.50 \cdot 10^{-2}$	$6.1 \cdot 10^{-5}$	$1.64 \cdot 10^{-2}$	$8.1 \cdot 10^{-6}$	$8.99 \cdot 10^{-3}$	$3.6 \cdot 10^{-6}$
g.H. (b=5)	$3.43 \cdot 10^{-2}$	$6.1 \cdot 10^{-5}$	$1.57 \cdot 10^{-2}$	$1.1 \cdot 10^{-5}$	$9.63 \cdot 10^{-3}$	$2.3 \cdot 10^{-6}$
	$N = 29$	$M = 20$	$N = 97$	$M = 20$	$N = 173$	$M = 20$

This method works quite well in low dimensions, but not for very high dimensions k: then p_k becomes very large and T_p produces worse results for very large p (the $\mathcal{O}$-constant depends on p and tends to ∞ exponentially fast).

Here are some of the numerical results given in Struckmeier (1993): First the time needed to generate 10^6 numbers on different machines is given in Table 1. Then some discrepancies averaged over samples of size M – we average the discrepancy and compute the variation V_M – are given in Table 2.

Further numerical examples are given in Subsection 5.5.

Up to now, all the effort has been put into the generation of uniformly distributed sequences on $[0,1]^k$. But the densities in rarefied gas dynamics, which we want to approximate, are never constant; typical densities are, for example, Maxwellians. Therefore we have to transform uniformly distributed sequences into f-distributed ones, where f is a given density. This is easy for Maxwellians: the densities factorize, so the problem may be reduced to 1D problems. The 1D case is simple – especially since one can use the so-called Box–Muller algorithm.

If the k-dimensional density does not factorize, the problem is more complicated. Hlawka and Mück (1972) have constructed a transformation T whose inverse transforms uniformly distributed point sets into f-distributed ones. The transformation $T = (T_1, \ldots, T_k)$, which has to be inverted, has a diagonal structure

$$T_j(x_1, \ldots, x_k) = T_j(x_1, \ldots, x_j), \ j = 1, \ldots, k.$$

This can be used for a numerical inversion – an extensive study on the optimal numerical method was done by M. Hack (1993). The estimates for the discrepancy are worse in this case (Hlawka and Mück, 1972)

$$D\left(T^{-1}\omega_N^N, f\right) \leq C \cdot D\left(\omega_N^N\right)^{\frac{1}{k}},$$

but the computations show much better behaviour. Fortunately, the problems we have treated until now have not called for the construction of point sets with low discrepancy against an arbitrary f (the simulation algorithm did it).

4.3. Bird's DSMC Method

We shall now describe the DSMC version, originally developed by G. Bird, and compare it with our method.

One main difference is that the original DSMC method does not consider dummy collisions, that is, one checks whether a pair really performs a collision (i.e. if $x \in B_{r_{\max}}$). If so then we call it a 'collision pair'.

To decide whether a given pair (v_i, v_j) is a collision pair (cp), one uses an acceptance–rejection method with a parameter $V_{\max}$, which is supposed to be the maximum relative speed of all particles

$$V_{\max} = \max\left\{\|v_i - v_j\| \mid 1 \leq i, j \leq N\right\}.$$

Then a pair is a cp if a $[0,1]$-uniformly distributed random number r is larger than

$$\frac{\|v_i - v_j\|}{V_{\max}}.$$

In this case an impact parameter is chosen and a collision is performed. The computation of $V_{\max}$ requires N^2 operations; therefore one starts with a guess V of $V_{\max}$ and updates it if one finds a larger $\|v_i - v_j\|$. We get the following procedure:

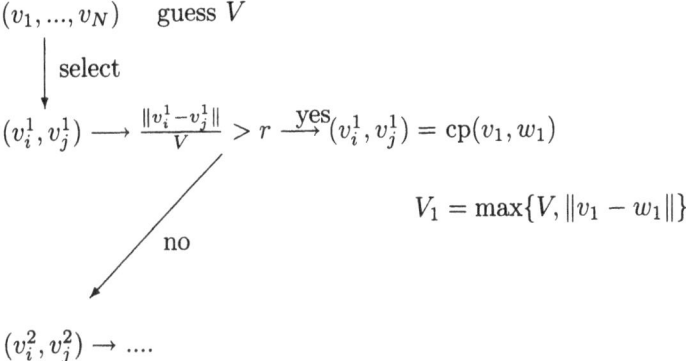

If $cp(v_1, w_1)$ is selected, we determine a time increment

$$\Delta\tau_1 = \frac{C}{N\|v_1 - w_1\|} \quad \text{(C is a gas-dependent constant).}$$

We substitute (v_i', v_j') for (v_i, v_j), that is, we update our particle ensemble after $\Delta\tau_1$, and we repeat the process until we reach Δt, that is, until

$$\Delta\tau_1 + \cdots + \Delta\tau_k \geq \Delta t.$$

(For our space-independent problem, Δt has lost its original meaning: our time step is $\Delta\tau$ and it is chosen such that only one collision happens during this interval; in this case, the time discretization is coupled with N – the time step tends to zero as N goes to ∞. In the finite point set method, N may go to ∞ without Δt tending to 0. In a space-dependent problem, Δt keeps its importance: we move the particles in space over Δt.)

For the correct procedure (with the real $V_{\max}$), Wagner (1992) has shown convergence as a stochastic process, that is, in probability. In practice, the results are sensitive to wrong initial guesses of $V_{\max}$.

The 'no time counter' version of Bird, mainly used today for computational reasons, seems similar: instead of changing time steps $\Delta\tau_i$ choose one fixed $\Delta\tau$, which is supposed to be the average time, in which one collision happens

$$\Delta\tau = \frac{C}{N \cdot V}$$

(i.e. V instead of $\|v - w\|$). V is updated at the end of Δt, not after $\Delta\tau$. The algorithm works quite well, again up to a sensitivity with respect to V.

To compare shortly our finite point set method with permutations, we have just

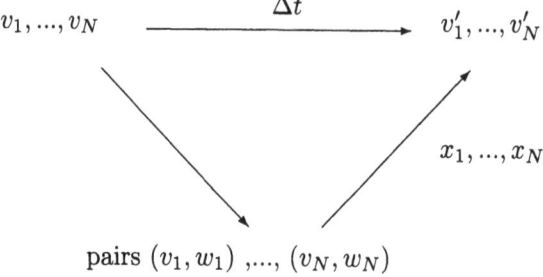

Δt is restricted by

$$1 - \Delta t \int_{s_2^+} k(\|v - w\|, \theta) \, d\omega(\eta) \geq 0$$

for all (possible) v, w!

Finally, we may also do updating during the collision process: We perform each collision immediately, that is, substitute (v_i', v_j') for (v_i, v_j) after $\Delta t/N$. The difference is that we keep N collisions (including the dummy ones) and work with the small timestep $\Delta t/N$, but do not need a guess of V_{max}.

The differences in computing time are less than 10 per cent, the results are demonstrated by the examples given in Subsection 5.5.

5. Some Ideas on How to Improve and Extend the Code

In this last section we shall report on some ideas on how to improve the code, to accelerate the algorithm and to extend it to more realistic situations. These topics will be:

5.1 'Different weights for particles in different regions'.
   This is different from 'different weights for different species' and does not create the same problem of conserving energy and momentum when particles of different weights collide. There is a detailed study of it by Schreiner (1991).

5.2 'The use of symmetry in particle codes'.
   If point sets are considered in a physical way – as representations of real particle sets – it is not easy to take advantage of geometrical symmetries of the problem (and the solution). To do that we have to exploit the idea of approximation by discrete measure; for example, if the density has cylindrical symmetry, depending only on $x_1, v_1, \|\tilde{x}\|$, $\|\tilde{v}\|$ and $< \tilde{x}, \tilde{v} >$ (where $\tilde{x} = (x_2, x_3)$, $\tilde{v} = (v_2, v_3)$), then our measures will be measures in this 5D (instead of 6D) space. One can save a lot of computing time, as is shown by Struckmeier and Steiner (1993).

5.3 'Matching' of kinetic equations with diffusion or aerodynamic limits.

This must be promising: each kinetic equation has some singular limits ('Diffusion approximation', Euler or Navier–Stokes equation etc.), which hold at least in some parts of position space. Solving these simpler equations in these parts and matching the solutions with those of the kinetic equations, which one gets in the 'kinetic rest' of the domain, poses a new problem in domain decomposition. There have been attempts in this direction – see, for example, Illner and Neunzert (1993), Bourgat et al. (1994) and Klar (1994).

5.4 'Efficiency on massively parallel systems'.

During the past few years several authors have investigated the performance of a particle method on massively parallel systems (see e.g. Barteland Plimpton (1992), Dagum (1991), Struckmeier and Pfreundt (1993) and Wong and Long (1992)). In this section we will follow the approach given in Struckmeier and Pfreundt (1993).

5.1. Spatially Weighted Particles

'Different weights in different regions, but equal weights in each cell' is an easily solvable weighting problem. In Schreiner (1991) the author describes how to find an appropriate particle mass in each cell in position space and how to change our particles (by splitting them or pasting them: Splipa) so that each particle has this desired mass.

Clearly, this desired mass m^* has to be small if the density in a cell is small (e.g. behind a space vehicle) – and it has to be large if the density is high (in the bow shock). In this way one may control the number of particles in each cell. During the free flow, particles of different masses may enter the same cell – but since we want to perform collisions only with particles of the same mass, we have to homogenize them. We allow only integer values for particle masses and we assume that m^* is always of the form 2^j; therefore homogenization might be done by splitting particles of mass 2^{j+k} into 2^k particles of mass m^* or by pasting minor particles together (by first splitting them into particles of minimal mass and then unifying them two at a time again and again until they have grown enough). The only problem here is that one should do this in such a way that mass, momentum and energy are conserved in each Splipa procedure; in particular the velocities after pasting have to be chosen carefully and there are only some signs to be chosen freely.

One might save time by these ideas. For a 2D problem (flow around an ellipse), Schreiner (1991) used 25 or 64 particles per cell in the beginning; the simulation without any weighting is then called A25 or A64, and with 3 or 4 different weights we call it B25-3 or B25-4, respectively (see Table 3).

Table 3. *CPU times and number of particles in the stationary state*

	CPU	Partnr
A64	44'41"	706,000
A25	24'52"	275,000
B25-3	29'13"	334,000
B25-4	26'17"	248,000

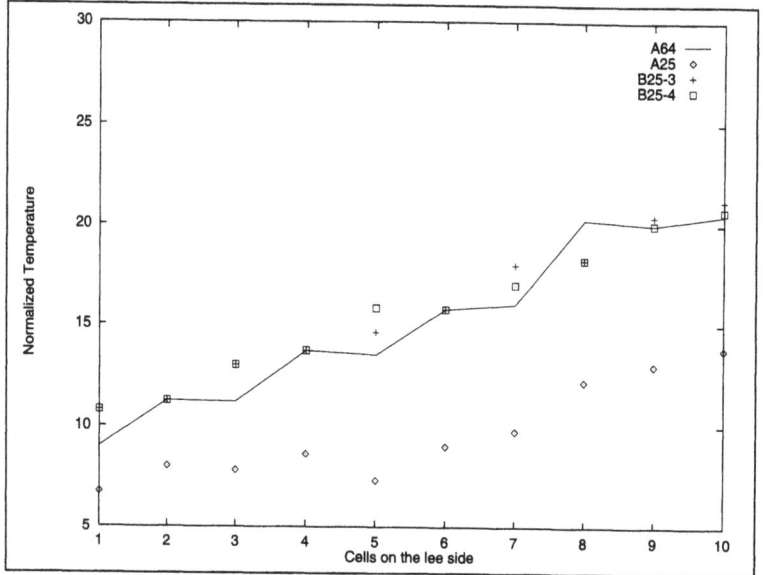

Fig. 3. Temperature along line in flow direction

The results differ – at least the temperature (it is a second moment) shows big changes from A25 to B25-3 behind the vehicle (see Figure 3). So, it is cheap and rewarding to use this weighting. But we want to recall that weights for different species, where homogenization is not possible, create much bigger problems.

5.2. Simulations with Axisymmetric Geometry

Symmetry reduces dimension in any numerical method, but normally not for particle methods. The reason is the usual one: particles are considered as physical quantities, not as approximations of densities.

Assume that we have cylindrical symmetry, that is, the boundary has a rotational symmetry with respect to the x-axis. Introducing cylindrical coordinates means to substitute (x, r, φ) for (x, y, z) and (v_x, v_r, v_φ) for (v_x, v_y, v_z).

Since v_r, v_φ depend on φ, we get a more complicated free-streaming term and have to transform $\hat{I}(f)$ to cylindrical coordinates (which was done in Niclot (1987)). A new collision strategy has to be defined – this way seems to be too elaborate. We may use (x, r, φ) together with (v_x, v_y, v_z) – things will not fit completely, but some important aspects remain unchanged. We get

$$\frac{\partial F}{\partial t} + v_x \frac{\partial F}{\partial x} + (\cos \varphi v_y + \sin \varphi v_z)\frac{\partial F}{\partial r} + \frac{-\sin \varphi v_y + \cos \varphi v_z}{r}\frac{\partial F}{\partial \varphi} = \hat{I}(F).$$

$\hat{I}(F)$ is not changed here. Free streaming means solving $\dot{x} = v_x$, $\dot{r} = (\cos \varphi v_y + \sin \varphi v_z)$, $\dot{\varphi} = (-\sin \varphi v_y + \cos \varphi v_z)/r$ with initial values (x_0, r_0, φ_0). The solution is

$$T_x(t, x_0, r_0, \varphi_0, v) = x_0 + tv_x,$$

$$T_r(t, x_0, r_0, \varphi_0, v) = \left(r_0^2 + 2tr_0(\cos \varphi_0 v_y + \sin \varphi_0 v_z) + t^2(v_y^2 + v_z^2)\right)^{1/2},$$

$$T_\varphi(t, x_0, r_0, \varphi_0, v) = \arctan\left(\frac{r_0 \sin \varphi_0 + tv_z}{r_0 \cos \varphi_0 + tv_y}\right).$$

For $\hat{I}(f) = 0$ we get $F(t, x, r, \varphi, v) = F_0(T(-t, x, r, \varphi), v)$.

Now we define $G = r^{-1}F$ and consider the corresponding equation. If, for example, F is a uniform distribution in position space with respect to the Lebesgue measure (in polar coordinates $r dr d\varphi dx$), then G can be regarded as a uniform distribution with respect to the 'cartesian' measure $dr\, d\varphi\, dx$, since $G\, r\, dr\, d\varphi\, dx = F\, dr\, d\varphi\, dx$.

To be more flexible, we consider

$$G(t, x, r, \varphi, v) = R(r)F(t, x, r, \varphi, v).$$

The equation for G is similar to that for F, but it has on the left-hand side an additional term $-(\cos \varphi v_y + \sin \varphi v_z)\partial_r(\ln R)g$ and instead of $\hat{I}(F)$ we have $R^{-1}\hat{I}(G)$. This additional term changes the solution of the free-streaming part into

$$G(t, x, r, \varphi, v) = \frac{R(T_r(-t))}{R(r)}G_0(T(-t), v)$$

and the factor $R(T_r(-t))/R(r)$ may be handled as a weight: a particle, moving from $P_i = (x_i, r_i, \varphi_i, v_i)$ to $P_i(\Delta t) = (T(\Delta t, P_i), v_i)$ changes its weight in proportion to $R(r_i)/R(r_i(\Delta t))$. For the natural choice $R(r) = r^{-1}$ the particles become heavier in moving away from the axis – the number of particles in a ring of thickness Δr remains unchanged (since the mass in a ring $(i-1)\Delta r \le r \le i\Delta r$ grows linearly with i, the weight of a particle has to grow linearly with i too in order to keep the particle numbers constant).

But now we have particles of different weights in the same cells – something we wanted to avoid. Even in the beginning, $R(r) = r^{-1}$ would give

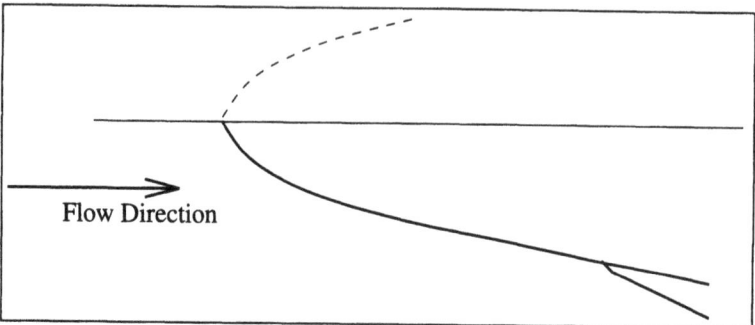

Fig. 4. Geometry of the hyperboloid flare

different weights. Therefore R is chosen as a step function approximating r^{-1}; but still differently weighted particles may enter a ring. Besides homogenization described under (b) one may follow a general idea by Bird: if the weight changes by a factor α less than one, just keep the particle with its old weight but with a survival probability of α. If α is larger than 1, say $\alpha = m + \alpha'$, $m \in \mathbf{N}$, $0 \leq \alpha' < 1$, create m new particles of the same weight and one other with probability α'. Again such a strategy does not work if we have different species of gas, but is successful here. No rigorous proof is yet available.

This reduces the computational costs drastically. Struckmeier and Steiner (1993) have done a study for HERMES with a flap at the leading edge (see Figure 4). Some results are shown in Table 4.

5.3. Domain Decomposition Techniques

We believe that the most promising prospect practically as well as theoretically is to use kinetic equations only where one is forced to use them – and to use the appropriate limits wherever it is possible. This idea materializes in two questions:

(a)   The 'where' problem: What are the regions where the diffusion limit or the Euler equation is valid, but in their complements the kinetic equations are necessary.

(b)   The other problem is the 'how' problem: how do we patch or match the solution of the kinetic equation with those of the limits.

Kinetic equations deal with position-velocity densities and the limits with macroscopic quantities, which can be interpreted as some moments of the kinetic density: What kind of boundary conditions for the two types of equations are the 'correct' ones? (Assuming the kinetic solution everywhere is the truth, which boundary conditions at the transition give a 'combined solution' as near as possible to the truth?) Until now, only the continuity of the macroscopic quantities across the transition boundary has been tried

Table 4. *Numerical parameters and global surface quantities*

Altitude[km]	Gas	$T_\infty[\mathrm{K}]$	Ma	$T_w[\mathrm{K}]$	$\lambda_\infty[\mathrm{m}]$
120	N_2	368	20	1400	2.69
110	N_2	247	23	1400	0.60
100	N_2	194	25	1400	0.137

Altitude[km]	Partnr	Cellnr	Part/Cell	Timesteps	CPU[h]
120	570,000	11,264	64	1000	1.5
110	925,000	11,264	64	1000	2.5
100	2,000,000	40,960	36	1000	4.0

Altitude[km]	$C_{d,0°}$	$C_{l,0°}$	$(L/D)_{0°}$	$C_{h,0°}$	$C_{m,0°}$
120	2.191	.890	.406	.868	.882
110	1.688	1.048	.621	.539	.641
100	1.360	1.170	.860	.313	.490

Altitude[km]	$C_{d,12°}$	$C_{l,12°}$	$(L/D)_{12°}$	$C_{h,12°}$	$C_{m,12°}$
120	2.304	.941	.408	.901	.974
110	1.785	1.109	.621	.557	.727
100	1.461	1.246	.853	.325	.584

Coefficient	Drag	Lift	Lift/Drag	Heat	Pitching

to be realized; details are described in Lukshin, Neunzert and Struckmeier (1992).

Since we focus on collisions, we just want to stress one comparing the simulation of collisions with the solution of an Euler equation (we choose Euler since it is – as a singular limit – much better understood than Navier–Stokes). The Boltzmann equation is solved by a particle code that moves the particles in a free flow over Δt and then treats the collisions at the end of the time step. The Euler equation can be solved by a very similar procedure: move particles in a free flow over Δt, but then redistribute them according to a Maxwellian distribution whose moments are given by particles at the end of a time step. The Euler equation gives a time evolution that is a free flow with a constraint: stay on the manifold given by $\{f : \hat{I}(f) = 0\}$. The ordinary free flow, starting at this manifold, moves away – so we have to project back onto it; this is the redistribution (see figure 5). So, the difference

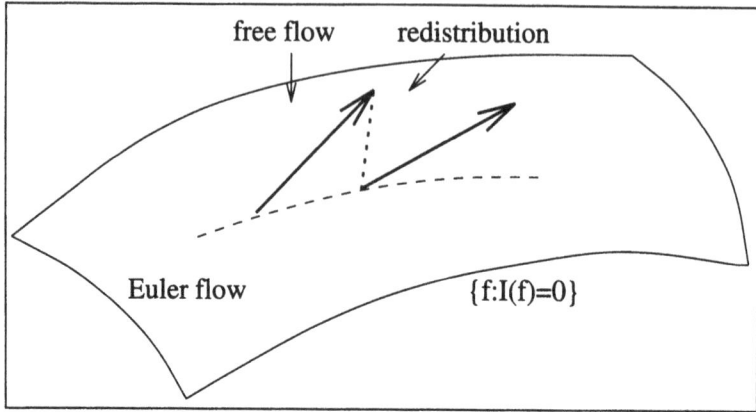

Fig. 5. Redistribution of the kinetic density

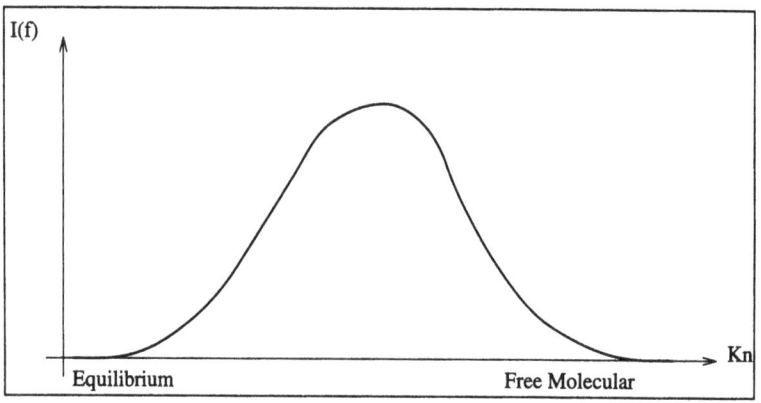

Fig. 6. Influence of $\hat{I}(f)$ in dependence on the Knudsen number

between the Boltzmann and Euler equations is the difference between the collision procedure and the projection. The projection is numerically much cheaper – so do projection whenever it is possible and collisions when it is necessary. What we try to use here is the fact that $\hat{I}(f)$ becomes small when f becomes very rarefied – and, when f becomes very dense and near to a Maxwellian, frequent collisions create an equilibrium distribution f for which $\hat{I}(f) = 0$. The denser f is the more expensive the collision procedure becomes – but at the same time, the smaller $\hat{I}(f)$ becomes (see Figure 6). To avoid this effect, we may use these projections. The key words here are 'kinetic schemes' and they may be converted into particle schemes for the Euler equations – see Schreiner (1994).

Using kinetic schemes, matching of the two codes is a minor problem: just do projection or collisions cellwise, but otherwise move freely without caring where you are.

5.4. Particle Methods on Parallel Computers

Still, realistic problems need enormous computational effort. Therefore it is reasonable to investigate the performance of Boltzmann simulation codes on massively parallel systems.

A parallelization of the code refers mainly to the grid structure on the spatial domain; the cells are, for example, cubes with a length smaller than the mean free path of the unperturbed gas. The collision process in a cell is independent of those in other cells – and it is the most time consuming part. One parallelizes the code by assigning a certain number of cells to each processor. In general one would have much more cells than processors, whence it is necessary to include a communication procedures.

In the first part of the time-iteration process particles may leave cells and enter others – if these cells belong to different processors, this means communication between processors.

The partition of cells has to be done such that this communication, that is, the number of particles crossing processor boundaries, is minimized. But a static partition, fixed at the beginning of the computation according to a priori information about the flow fields (most of the particles move essentially with the stream velocity), does not produce a good load balance of the processors – particle numbers per processor change and result in a very insufficient load balance; this reduces the speed-up factor as compared to single processors.

To get an adaptive procedure, we put cells lying in a row with respect to the main stream velocity together to form 'spatial sticks'. Several spatial sticks are assigned to processors – and the adaption consists simply of exchanging sticks from the minimally to the maximally loaded processors. This exchange creates an iteration procedure until we get near to the partition when the numbers of particles in the processor domains are near the average number. The procedure creates partitions where the local character of the stick–processor assignment is destroyed (see Table 5).

One may suspect that this gives rise to a high communication time; that this is not the case is shown in Figure 7.

The speed-up factor is constant near 30 (here the factor 32 is – using 32 processors – optimal) if one compares different Knudsen numbers, that is, different densities and therefore a different collision frequency (see Figure 8). A comparison of CPU times on the nCUBE2s with a VP100 shows that the higher peak performance of the vector machine does not lead to lower CPU time (Table 6).

5.5. Numerical Examples

First we come back to the comparison of the finite point set method and the DSMC method of Bird (see Subsection 4.3).

Table 5. *Final state of the adaptive processor partition (zy-plane)*

6	27	25	4	5	8	2	11	18	4	3	18
23	19	14	25	15	2	17	8	32	27	2	5
18	27	14	22	3	29	4	8	24	29	16	22
30	4	4	24	21	22	14	16	10	9	11	15
6	30	20	10	2	24	3	23	8	21	29	26
15	26	9	9	13	16	8	27	1	7	30	3
21	12	22	5	7	30	30	6	23	26	6	2
23	31	23	18	13	7	22	10	12	3	29	27
32	29	10	13	31	23	21	14	18	26	13	14
26	9	7	2	10	9	24	29	1	16	17	18
8	21	24	30	19	7	15	30	12	13	6	12
11	3	24	30	14	12	25	4	25	22	31	6
21	16	18	1	29	15	1	20	13	4	28	26
28	25	29	3	2	5	19	20	23	2	28	32
1	5	18	29	1	11	26	20	32	21	28	22
7	15	28	20	28	17	17	14	6	5	28	25
25	24	5	23	16	8	17	10	21	9	17	4
13	12	20	14	6	19	16	14	30	20	31	17
2	11	11	9	8	1	29	25	5	9	30	30
10	19	4	9	29	3	7	25	18	1	10	30
12	32	22	17	1	20	15	5	22	19	30	30
28	26	28	10	12	11	11	15	15	6	27	19
3	8	21	27	27	16	24	19	23	20	12	30
7	16	13	7	17	26	11	19	32	13	27	24

We consider the flow around a hyperboloid flare (see Figure 4) at high Mach number and an altitude around 100 km. Here one may use the axisymmetric version of the particle codes described in Subsection 5.2. We calculate 'global' quantities acting on the body such as the drag, lift or heat-transfer coefficient.

The main task is to investigate the sensitivity of the different approaches to the number of particles used in the simulation.

Looking at Figures 9 and 10, we realize that, except for the DSMC time-counter version, both methods, DSMC as well as the finite point set method, show nearly the same behaviour with respect to the particle number. A lot of more results can be found in Struckmeier and Steiner (1993).

The high degree of modelling necessary to describe real-gas effects requires a validation of the models used in a particle method. The following example has become a 'classic' test-case, mainly due to the establishment of the European Hypersonic Data Base (EHDB).

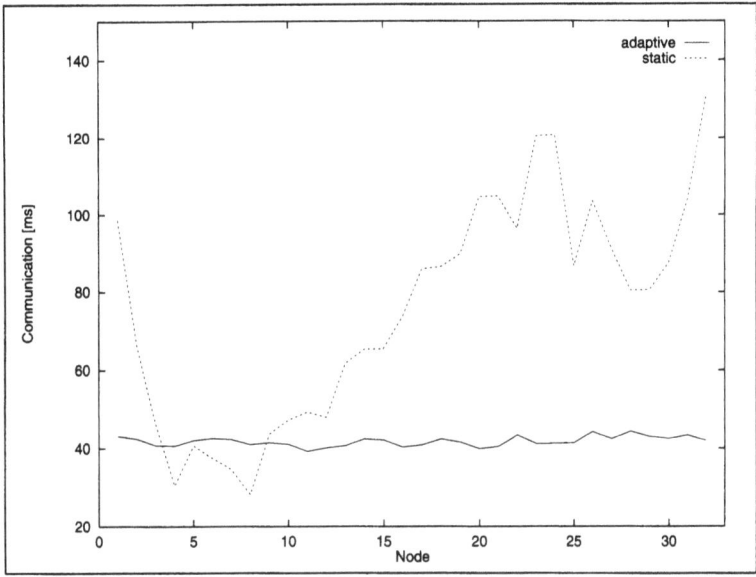

Fig. 7. Communication time vs. nodes at Kn = 0.5 and 45° angle of attack

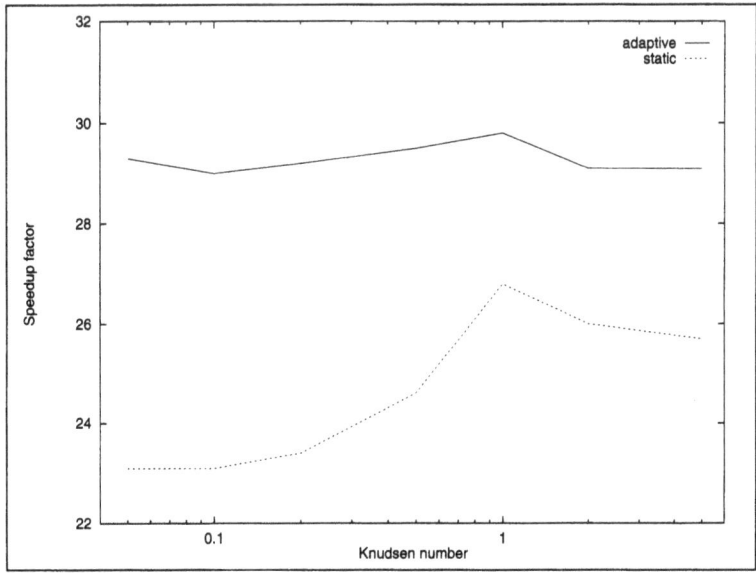

Fig. 8. Speed-up factor vs. Knudsen number at 45° angle of attack

We consider the flow of nitrogen gas around a 3D delta-wing at a high Mach number. The measured quantities are global surface quantities like the drag or the heat-transfer coefficient. Figure 11 shows the drag coefficient versus the Knudsen number at Mach 20.2 and 45° angle of attack, Figure 12 the heat-transfer coefficient.

Table 6. *CPU times for a 3D computation with Kn = 0.5 and 45° angle of attack*

	MFLOP	CPU[s]	ratio
nCUBE2s/8	35	579	3.7
nCUBE2s/16	70	297	1.9
nCUBE2s/32	140	156	1.0
Fujitsu VP100	285	1075	6.9

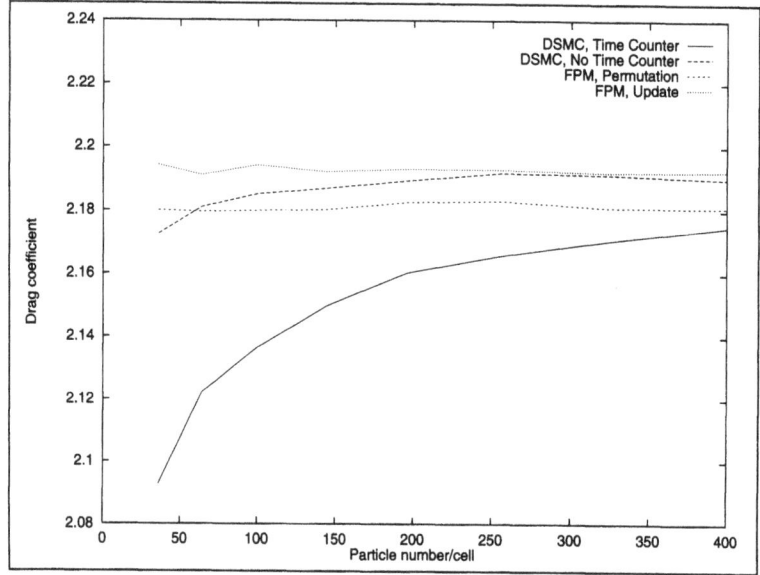

Fig. 9. Drag coefficient versus particle number

The agreement between numerical results and measurements is quite convincing; hence, the model for the exchange of different energy types like rotational and vibrational energies (in this case the Larsen–Borgnakke model) is accurate enough to reproduce the physical situation.

As a last example (see Figure 13) we compare the local pressure distribution along the surface line calculated by a particle method with the prediction given by the modified Newton theory. We consider again the hyperboloid flare (Figure 4) with flap angle of 0° respectively 12° at Mach 25 and an altitude of 100 km. The given altitude corresponds to the 'small' Knudsen number of $9 \cdot 10^{-3}$; hence, one may expect that the modified Newton theory gives reasonably accurate results.

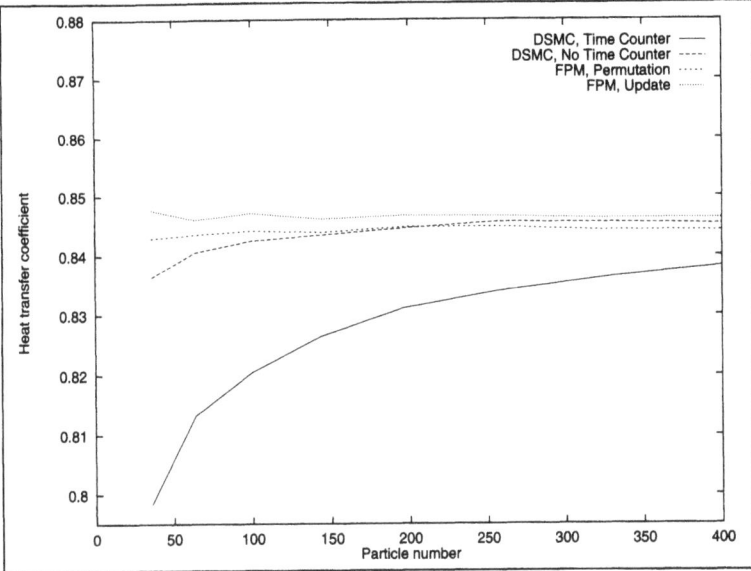

Fig. 10. Heat-transfer coefficient versus particle number

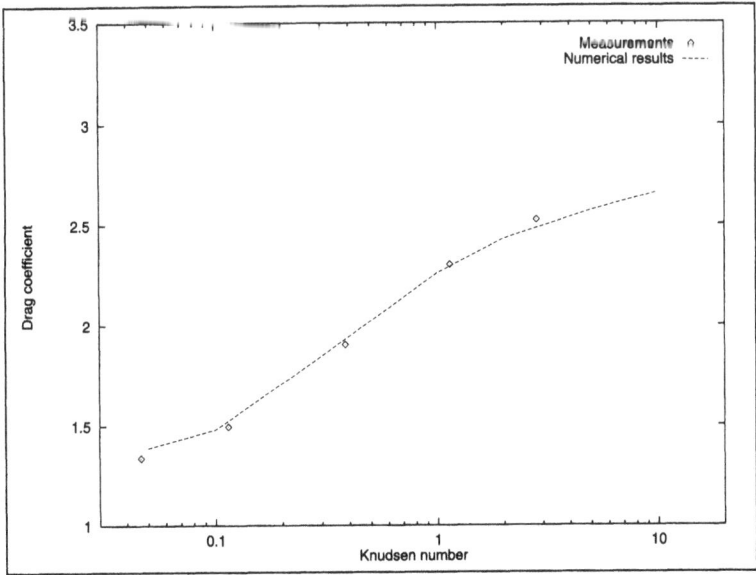

Fig. 11. Drag coefficient versus Knudsen number

6. Final Remarks

We believe that particle methods have become a reliable instrument in rarefied gas dynamics. Using massively parallel systems one can treat realistic problems with a reasonable effort. However, some physical effects like ionization or recombination are still neglected (or handled in an unreliable way); therefore further improvements are needed. The most promising ansatz is

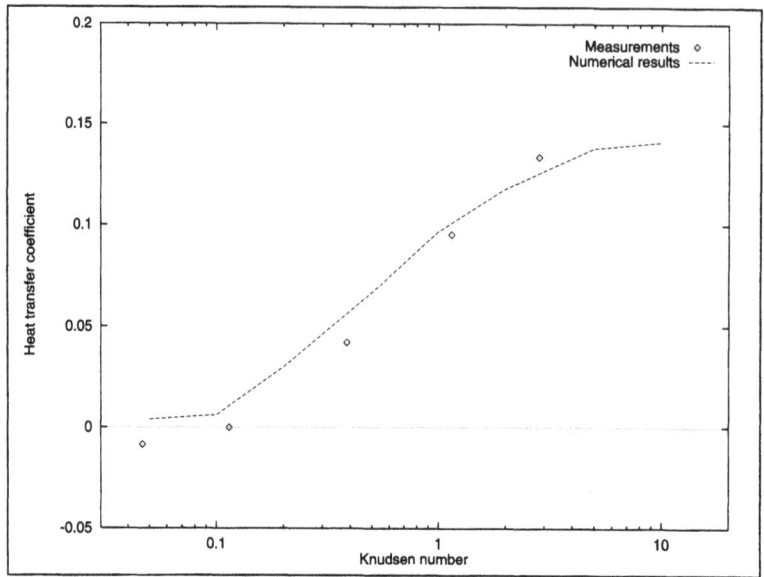

Fig. 12. Heat-transfer coefficient versus Knudsen number

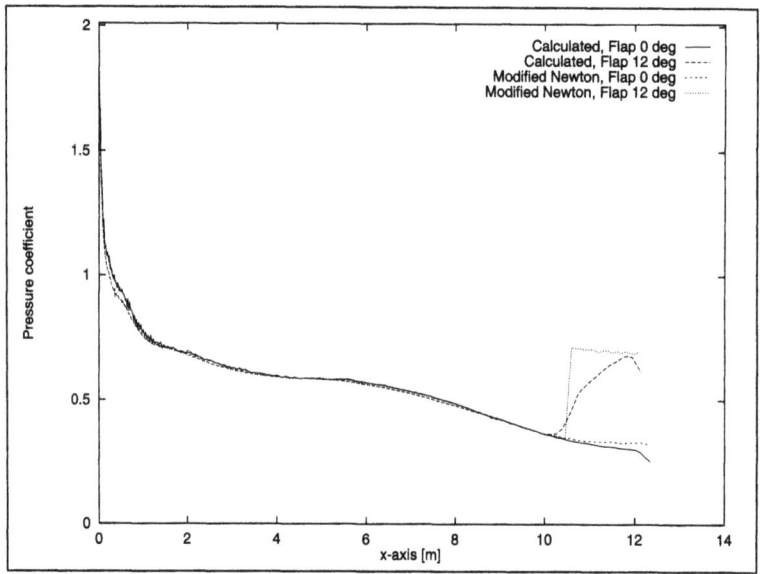

Fig. 13. Local pressure coefficient along the surface line

the combination of asymptotic analysis with numerical methods (as in many fields!); we will solve the simpler 'singular limit equations' whenever it is possible – and the more complicated kinetic equations when it is necessary.

Why are particle methods not just finite difference methods or something similar? Kinetic equations are high-dimensional. Approximations of densities by discrete measures are more robust with respect to dimensions. This might be the reason that competing methods until now have failed in treating 3D problems. Why are there still some stochastic elements and why is there still a lot of 'Monte Carlo'? The explanation for this may be given by the theory of information-based complexity; there are many hints that this theory provides ideas to answer the question of why Monte Carlo – correctly applied – is advantageous.

There are still gaps between existence theory and numerics; but the theory cannot provide us with uniqueness which is what we need to bridge the gap. There is still a long and exciting way to go.

REFERENCES

H. Babovsky (1989), 'Convergence proof for Nanbu's Boltzmann simulation scheme', *Europ. J. of Mech. B/Fluids* **1**, 41–55.

T. J. Bartel and S. J. Plimpton (1992), DSMC Simulation of Rarefied Gas Dynamics on a Large Hypercube Supercomputer, Paper–92–2860, Am. Inst. Aero. Astro., Washington.

K. Bärwinkel and H. Wolters (1975), Reaktionskinetik und Transport in relaxierenden Gasen, Report BMF-FB W W 75–28, Dornier, Friedrichshafen.

G. A. Bird (1976), *Molecular Gas Dynamics*, Clarendon Press, Oxford.

G. A. Bird (1989), 'Perception of Numerical Methods in Rarefied Gas Dynamics', *Progr. Astro. and Aero.* **118**, 211–226.

A. V. Bobylev (1993), 'The Boltzmann Equation and the Group Transformations', *Math. Models & Methods in Appl. Sciences* **3**, No. 4, 443–476.

C. Borgnakke and P. S. Larssen (1975), 'Statistical collision model for Monte Carlo simulation of polyatomic gas mixtures', *J. Comput. Phys.* **18**, 405–420.

J. F. Bourgat, P. LeTallec, B. Perthame and Y. Qiu (1994), 'Coupling Boltzmann and Euler Equations without Overlapping', *Contemporary Mathematics* **157**, 377–398.

C. Cercignani (1989), *The Boltzmann Equation and Its Applications*, Spinger, Berlin.

L. Dagum (1991), Three Dimensional Direct Particle Simulation on the Connection Machine, Paper–91–1365, Am. Inst. Aero. Astro., Washington.

H. Faure (1982), 'Discrepance des suites associées à une système de numération (en dimension s)', *Acta Arithmeticae* **41**, 337–351.

C. Greengard and L. G. Reyna (1992), 'Conservation of expected momentum and energy in Monte Carlo particle simulation', *Phys. Fluids A* **4**, 849–852.

M. Hack (1993), 'Construction of Particlesets to Simulate Rarefied Gases', Report No. 89, Lab. Technomathematics, University of Kaiserslautern.

J. H. Halton (1960), 'On the efficiency of certain quasi-random sequences of points in evaluating multi-dimensional integrals', *Numer. Math.* **2**, 84–90.

J. M. Hammersley (1960), 'Monte Carlo methods for solving multivariate problems', *Ann. New York Acad. Sci.* **86**, 844–874.

E. Hlawka and R. Mück (1972), A transformation of equidistributed sequences, Appl. of Number Theory to Numerical Analysis, ed. S. K. Zaremba, Academic Press, New York, 371–388

R. Illner and H. Neunzert (1993), 'Domain Decomposition: Linking of Aerodynamic and Kinetic Descriptions', Report No. 90, Lab. Technomathematics, University of Kaiserslautern.

M. S. Ivanov and S. V. Rogasinsky (1988), 'Analysis of numerical techniques of the direct simulation Monte Carlo method in the rarefied gas dynamics', *Sov. J. Numer. Anal. Math. Modelling* **3**, No. 6, 453–465.

A. Klar (1994), 'Domain Decomposition for Kinetic and Aerodynamic Equations', PhD thesis, Dept. Math., University of Kaiserslautern.

I. Kuščer (1991), 'Dissociation and Recombination in an Inhomogeneous Gas', *Physica A* **176**, 542–556.

G. Lord (1991), 'Some extensions to the Cercignani–Lampis gas–surface scattering kernel', *Phys. Fluids A* **4**, 706–710.

G. Ludwig and M. Heil (1960), 'Boundary-layer theory with dissociation and ionization', in *Advances of Applied Mechanics* , vol. 6, Academic Press, New York.

A. Lukshin, H. Neunzert and J. Struckmeier (1992), 'Interim Report for the Project DPH 6473/91: Coupling of Navier–Stokes and Boltzmann Regions', Internal Report, Dept. Math., University of Kaiserslautern.

G. Missmahl (1990), 'Randwertprobleme bei der Boltzmann–Simulation', Diploma thesis, Dept. Math., University of Kaiserslautern.

K. Nanbu (1980), 'Direct Simulation Scheme Derived from the Boltzmann Equation', *J. Phys. Japan* **49**, 2042–2049.

H. Neunzert, F. Gropengiesser and J. Struckmeier (1991), 'Computational methods for the Boltzmann equation', in *Venice 1989: The State of Art in Appl. and Ind. Math.* (ed.), R. Spigler, Kluwer, Dordrecht, 111–140.

H. Neunzert, K. Steiner and J. Wick (1993), 'Entwicklung und Validierung eines Partikelverfahrens zur Berechnung von Strömungen um Raumfahrzeuge im Bereich verdünnter ionisierter Gase', Report DFG-FB Ne 269/8–1, Lab. Technomathematics, University of Kaiserslautern.

B. Niclot (1987), The Two Particle Boltzmann Collision Operator in Axisymmetric Geometry, Report No. 164, Centre de Mathématiques Appliquées, Ecole Polytechnique, Palaiseau.

H. Niederreiter (1992), *Random Number Generation and Quasi–Monte Carlo Methods*, SIAM, Philadelphia.

S. Nocilla (1961), 'On the Interactions between Stream and Body in Free-molecule Flow', *Proc. 2nd Int. Symp. on Rarefied Gas Mechanics*, (L. Talbot, ed.), Academic Press, New York.

G. Pagés (1992), 'Van der Corput sequences, Kakutani transform and one-dimensional numerical integration', *J. Comp. & Appl. Math.* **44**, 21–39.

K. Sobol (1969), *Multidimensional quadrature formulae and Haar functions*, Nauka, Moscow.

M. Schreiner (1991), 'Weighted Particles in the Finite Pointset Method', Report No. 62, Lab. Technomathematics, University of Kaiserslautern.

W. Schreiner (1994), Partikelverfahren für kinetische Schemata zu den Eulergleichungen, PhD thesis, Dept. Math., University of Kaiserslautern.

J. Struckmeier (1993), 'Fast Generation of Low-Discrepancy Sequences', Report No. 93, Lab. Technomathematics, University of Kaiserslautern, to appear in *J. of Comp. & Appl. Math.*

J. Struckmeier (1994), 'Die Methode der finiten Punktmengen: Neue Ideen und Anregungen', PhD thesis, Dept. Math., University of Kaiserslautern.

J. Struckmeier and F. J. Pfreundt (1993), 'On the efficiency of simulation methods for the Boltzmann equation on parallel computers', *Parallel Computing* **19**, 103–119.

J. Struckmeier and K. Steiner (1993), 'A Comparison of Simulation Methods for the Boltzmann Equation', Report No. 91, Lab. Technomathematics, University of Kaiserslautern, submitted to *J. Comp. Physics.*

J. G. Van der Corput (1935), 'Verteilungsfunktionen I,II', *Nederl. Akad. Wetensch. Proc. Ser. B* **38**, 813–821; 1058–1066.

W. Wagner (1992), 'A convergence proof for Bird's direct simulation Monte Carlo method for the Boltzmann equation', *J. Stat. Phys.* **66**, 689–722.

B. C. Wong and L. N. Long (1992), Direct Simulation Monte Carlo (DSMC) on the Connection Machine, Paper–92–0564, Am. Inst. Aero. Astro., Washington.

Acta Numerica (1995), pp. 459–491

The New qd Algorithms

Beresford N. Parlett *

Department of Mathematics

University of California

Berkeley CA 94720,

USA

E-mail: parlett@math.berkeley.edu

CONTENTS

1. Introduction

Let us think about ways to find both eigenvalues and eigenvectors of tridiagonal matrices. An important special case is the computation of singular values and singular vectors of bidiagonal matrices. The discussion is addressed both to specialists in matrix computation and to other scientists whose main interests lie elsewhere. The reason for hoping to communicate with two such diverse sets of readers at the same time is that the content of the survey, though of recent origin, is quite elementary and does not demand familiarity with much beyond triangular factorization and the

* The author is grateful for partial support under Contract ONR N00014-90-J-1372.

Gram–Schmidt process for orthogonalizing a set of vectors. For some readers the survey will cover familiar territory but from a novel perspective. The justification for presenting these ideas is that they lead to new variations of current methods that run a lot faster while achieving greater accuracy.

Tridiagonal matrices have received a great deal of attention since the 1950s. The (i, j) entries of these matrices vanish if $|i - j| > 1$ and the interest in them stems from the fact that they include the most narrowly banded representations of a matrix that can be obtained by a finite number of similarity transformations using rational expressions in the matrix entries together with square roots. This statement needs a little justification because the rational canonical form (RCF) is, by construction, *the* matrix with fewest nonzero entries that can be achieved by rational operations on the data. For an $n \times n$ matrix the RCF has only n parameters while a tridiagonal form has $3n - 2$, but $2n - 1$ when normalized, and so RCF seems preferable.

The fact is that, apart from theorists doing exact computations, the RCF is not used in eigenvalue computations. The main reason is that the representation is too condensed for standard floating-point computation. The coefficients of the characteristic polynomial have to be known to many, many more decimal places than do the matrix entries in order to determine the eigenvalues to the same accuracy.

Beyond that one might add that the RCF is a Hessenberg matrix (entry (i, j) vanishes if $i - j > 1$) and, by design, there are no further useful similarity transformations that can be applied to it. In contrast to the RCF there are infinitely many tridiagonal matrices in a similarity class and so there is hope of computing a sequence of them that converges to bidiagonal or diagonal form. This is where our interest lies.

It should be mentioned that the tridiagonal form is probably also too condensed for the most difficult cases, (see Parlett (1992)), but it is rich enough to suffice for many applications and we shall stay with it here.

Our topic, the new qd algorithms, will be developed as the consequence of two ideas.

The first concerns the representation of tridiagonal matrices and we mention it briefly here. In the eigenvalue context there is no loss of generality in supposing that our tridiagonals are normalized so that all entries in positions $(i, i + 1)$ are either 0 or 1. Zeros here make calculations easier so we may assume that all these entries are 1. Such tridiagonals are denoted by J. Most, but not all, such J permit triangular factorization

$$J = LU,$$

where the precise forms L and U are shown at the beginning of the next section. It is clear that in the $n \times n$ case L and U together are defined by $2n - 1$

parameters; exactly the same degree of freedom as in J. Section 3 argues that the pair L, U is preferable to J itself. Consequently all transformations on tridiagonals should be re-examined in this representation.

The second idea relates to the LR algorithm (LR) discovered by H. Rutishauser in 1957, (see Rutishauser (1958)), and presented in Section 5 here. When LR is rewritten in the L, U representation one obtains the (progressive) qd algorithm. The letters q and d are lower case because they do not stand for matrices and the matrix-computation community tries to reserve capital letters for matrices. Thus q here has nothing to do with the Q of the QR factorization.

In the new representation the LR algorithm spends most of its time computing the triangular factorization not of a single matrix, but of a product, namely UL. It is not well enough appreciated that finding the LU factorization of any product BC is equivalent to applying a generalized Gram–Schmidt process to the rows of B and the columns of C so that $B = \hat{L}P^*$, $C = Q\hat{U}$, and P^*Q is diagonal. When this Gram–Schmidt process is applied to UL in an efficient manner one obtains a little-known variant of qd, called the *differential* qd algorithm (dqd), that requires a little more arithmetic effort than qd itself. Rutishauser discovered dqd, (see Rutishauser (1958)) near the end of his life and never mentioned the shifted version dqds that K. V. Fernando and I discovered, (see Fernando and Parlett (1994)), independently of Rutishauser's work, in 1991, while trying to improve on the Demmel and Kahan QR algorithm (Demmel and Kahan, 1990) for computing singular values of bidiagonals. The connection of dqd with the generalized Gram–Schmidt process on bidiagonals is new and constitutes the second idea that underpins this survey.

The presentation here runs completely counter to history. The paper by Fernando and Parlett (1994), develops dqds in the singular-value context, gives historical comments and shows the connections with continued fractions. However, none of that is necessary and it is simplicity we pursue here.

This survey develops several qd algorithms (six in all) in a matrix context in the most elementary way. It is not difficult to see several directions in which these ideas may be generalized or modified to good effect.

The differential forms of qd algorithms are the right ones for parallel computation.

A sceptic might say that since no one uses LR algorithms there is no point in finding fancy versions of them. In the general case there is still much work to be done in finding clever shift strategies that approach an eigenvalue in a stable way. However, in the symmetric case even the current simple shift strategies achieve high relative accuracy in all eigenvalues and are between 2 and 10 times faster than QR; see Fernando and Parlett (1994). Yet the most powerful argument in favor of qd algorithms may turn out to be the efficient computation of accurate eigenvectors.

The general plan of this survey is conveyed adequately by the table of contents.

2. Bidiagonals versus Tridiagonals

Bidiagonal matrices of a special form will play a leading role in this essay. Whenever possible 6×6 matrices will be used to illustrate the general pattern.

$$
L = \begin{bmatrix}
1 \\
l_1 & 1 \\
 & l_2 & 1 \\
 & & l_3 & 1 \\
 & & & l_4 & 1 \\
 & & & & l_5 & 1
\end{bmatrix}, \quad
U = \begin{bmatrix}
u_1 & 1 \\
 & u_2 & 1 \\
 & & u_3 & 1 \\
 & & & u_4 & 1 \\
 & & & & u_5 & 1 \\
 & & & & & u_6
\end{bmatrix}.
$$

To save space these matrices may be written as

$$
L = bidiag \begin{pmatrix} 1 & 1 & 1 & 1 & 1 & 1 \\ l_1 & l_2 & l_3 & l_4 & l_5 & \end{pmatrix},
$$

$$
U = bidiag \begin{pmatrix} & 1 & 1 & 1 & 1 & 1 \\ u_1 & u_2 & u_3 & u_4 & u_5 & u_6 \end{pmatrix}.
$$

The pair L, U determine two triangular matrices; first

$$
J = LU = \begin{bmatrix}
u_1 & 1 \\
l_1 u_1 & l_1 + u_2 & 1 \\
 & l_2 u_2 & l_2 + u_3 & 1 \\
 & & l_3 u_3 & l_3 + u_4 & 1 \\
 & & & l_4 u_4 & l_4 + u_5 & 1 \\
 & & & & l_5 u_5 & l_5 + u_6
\end{bmatrix},
$$

which may be written

$$
J = tridiag \begin{pmatrix} & 1 & & 1 & \bullet & 1 & \\ u_1 & l_1 + u_2 & & \bullet & \bullet & l_5 + u_6 \\ l_1 u_1 & & l_2 u_2 & & \bullet & l_5 u_5 & \end{pmatrix},
$$

and second

$$
J' = UL = \begin{bmatrix}
u_1 + l_1 & 1 \\
u_2 l_1 & u_2 + l_2 & 1 \\
 & u_3 l_2 & u_3 + l_3 & 1 \\
 & & u_4 l_3 & u_4 + l_4 & 1 \\
 & & & u_5 l_4 & u_5 + l_5 & 1 \\
 & & & & u_6 l_5 & u_6
\end{bmatrix},
$$

which may be written

$$J' = tridiag \begin{pmatrix} & 1 & & 1 & \bullet & & 1 & \\ u_1 + l_1 & & u_2 + l_2 & & \bullet & \bullet & & u_6 \\ & u_2 l_1 & & u_3 l_2 & & \bullet & & u_6 l_5 \end{pmatrix}.$$

Note that both tridiagonals have their superdiagonal entries, that is, entries $(j, j+1)$, equal to 1. Also note that $J' = L^{-1}JL$. The reader should note the pattern of indices in J and J' because frequent reference will be made to them throughout the survey.

The attractive feature here is that because the 1's need not be represented explicitly the factored form of J requires no more storage than J itself; there are $2n - 1$ parameters for $n \times n$ matrices in each case.

Advantages of the factored form

1   L, U determines the entries of J to greater than working-precision accuracy because the addition and multiplication of l's and u's is implicit. Thus J_{ii} is given by $l_{i-1} + u_i$ implicitly but by $fl(l_{i-1} + u_i)$ explicitly.

2   The mapping $L, U \rightarrow J$ is naturally parallel; for example, $l * u$ gives the off-diagonal entries. In contrast the mapping $J \rightarrow L, U$, that is, Gaussian elimination, is intrinsically sequential.

3   Singularity of J is detectable by inspection when L and U are given, but only by calculation from J.

4   Solution of $Jx = b$ takes half the time when L and U are available.

Disadvantages of the factored form

The mapping $J \rightarrow L, U$ is not everywhere defined. Even when the factorization exists it can happen that $\|L\|$ and $\|U\|$ greatly exceed $\|J\|$. This is very bad for applying the LR algorithm but harmless when eigenvectors are to be calculated. So we should be careful to consider the goal before stigmatizing a process as unstable. Moreover in the eigenvalue context we are free to replace J by $J - \sigma I = LU$ for some suitably chosen shift σ that gives acceptable L and U.

Frequently we make the *nonzero assumption*: $l_j u_j \neq 0$, $j = 1, \ldots, n - 1$. If $u_n = 0$ then a glance at $J' = UL$ shows that its last row is zero. However, it is not valid to simply discard l_{n-1} along with u_n. In other words we do not have a factored form of the leading $(n-1) \times (n-1)$ principal submatrix of J' unless l_{n-1} is negligible compared to u_{n-1}. Similarly if u_1 is zero we must not ignore l_1. In fact any zero values among $\{l_j, u_j; \; j = 1, \ldots, n-1\}$ are readily exploited.

Splitting: If $l_j = 0$, $j < n$, then the spectrum of J is the union of the spectra of two smaller tridiagonals given in factored form by $\{l_i, u_i; \; i = 1, j\}$ and $\{l_i, u_i; \; i = j+1, n\}$. Here $l_n = 0$.

Singularity: If $u_j = 0$, $j \leq n$, then zero is an eigenvalue. However, some computation is necessary to deflate this eigenvalue and obtain L and U factors of an $(n-1) \times (n-1)$ matrix. One pass of the qd algorithm (given later) will suffice in exact arithmetic.

Any tridiagonal matrix that does not split, or its transpose, is diagonally similar to a form with 1's above the diagonal, that is, a J matrix. So for eigenvalue hunting there is no loss of generality in using this normalization. However, if the normalization is not convenient then there is an alternative factorization of tridiagonals that was considered by H. Rutishauser; see Rutishauser (1990). Let

$$N = bidiag \begin{pmatrix} 1 & & 1 & & 1 & & 1 & & 1 & & 1 \\ & 1 & & 1 & & 1 & & 1 & & 1 & \end{pmatrix},$$

$$B = bidiag \begin{pmatrix} & e_1 & & e_2 & & e_3 & & e_4 & & e_5 & \\ q_1 & & q_2 & & q_3 & & q_4 & & q_5 & & q_6 \end{pmatrix}.$$

Then

$$NB = tridiag \begin{pmatrix} & e_1 & & e_2 & & \bullet & & e_5 & \\ q_1 & & e_1 + q_2 & & \bullet & & \bullet & & e_5 + q_6 \\ & q_1 & & q_2 & & \bullet & & q_5 & \end{pmatrix}$$

and

$$BN = tridiag \begin{pmatrix} & e_1 & & e_2 & & \bullet & & e_5 & \\ q_1 + e_1 & & q_2 + e_2 & & \bullet & & \bullet & & q_6 \\ & q_2 & & q_3 & & \bullet & & q_6 & \end{pmatrix}.$$

Note that, after we identify $e_i = l_i$, $q_i = u_i$, for all i, $J = DNBD^{-1}$ and $J' = DBND^{-1}$ with $D = diag(1, e_1, e_1 e_2, \ldots, e_1 \cdots e_5)$.

Since all the various qd algorithms relate naturally to J matrices we will stay with the L, U factors rather than the N, B representation.

In the past most attention has been paid to the *positive case*: $l_i > 0$, $i = 1, \ldots, n-1$, $u_j > 0$, $j = 1, \ldots, n$. Note in passing the following standard results.

Lemma 1 If $l_i u_i > 0$, $i = 1, \ldots, n-1$, then J is symmetrizable by a diagonal similarity and the number of positive (negative) u_i is the number of positive (negative) eigenvalues.

Lemma 2 If $l_i u_{i+1} > 0$, $i = 1, \ldots, n-1$, then J' is symmetrizable by a diagonal similarity and the number of positive (negative) u_i is the number of positive (negative) eigenvalues.

For a real symmetric or complex Hermitian matrix a preliminary reduction to tridiagonal form has proved to be a stable and efficient step in computing the eigenvalues. In the general case preliminary reduction to tridiagonal form has been less successful. Stability is not guaranteed for any current methods; see Parlett (1992). Sometimes users are lucky but as larger and larger matrices are tried unsatisfactory experiences are more frequent. It may well be that the tridiagonal form is too compact for the difficult cases. The attentive reader will note in the following pages that some of the algorithms can be extended to fatter forms, such as block tridiagonal, but such ideas will not be pursued here.

3. Stationary qd Algorithms

Triangular factors change in a complicated way under translation. Given L and U of the form given in Section 2 the task here is to compute $\bar{L}$ and $\bar{U}$ so that

$$J - \sigma I = LU - \sigma I = \bar{L}\bar{U}$$

for a given suitable shift σ. Equating entries on each side shows that

$$
\begin{aligned}
l_i + u_{i+1} - \sigma &= \bar{l}_i + \bar{u}_{i+1}, \quad i = 0, \ldots, n-1, \; l_0 = 0, \\
l_i u_i &= \bar{l}_i \bar{u}_i, \quad i = 1, \ldots, n-1.
\end{aligned}
$$

These relations yield the so called stationary qd algorithm:

$stqd(\sigma)$: $\bar{u}_1 = u_1 - \sigma$;
        for $i = 1, n-1$
          $\bar{l}_i = l_i u_i / \bar{u}_i$
          $\bar{u}_{i+1} = l_i + u_{i+1} - \sigma - \bar{l}_i$
        end for.

Naturally it fails if $\bar{u}_i = 0$ for some $i < n$.

At this point the sceptical reader might object that $stqd(\sigma)$ is exactly the same algorithm that would be obtained by forming $J - \sigma I$ and performing Gaussian elimination. Indeed if $stqd(\sigma)$ is executed with the operations proceeding from left to right, for example,

$$\bar{u}_{i+1} = fl(fl(fl(l_i + u_{i+1}) - \sigma) - \bar{l}_i),$$

then the two procedures are not just mathematically equivalent but also computationally identical. However, it is not necessary to follow this left-to-right ordering. For example, one could write

$$\bar{u}_{i+1} = (l_i - \bar{l}_i - \sigma) + u_{i+1}$$

and, if the compiler respects parentheses, then stqd(σ) will quite often produce different output than Gaussian elimination on $J - \sigma I$. If l_i and $\bar{l}_i$ are much larger than u_{i+1} then the second form is more accurate than the first.

The preceding thoughts lead to an alternative algorithm, easily missed, for $\bar{L}$ and $\bar{U}$. It involves more arithmetic effort and an auxiliary storage cell but has some striking advantages in accuracy for finite-precision arithmetic.

To derive the algorithm define variables $\{t_i\}$ by

$$t_{i+1} \equiv \bar{u}_{i+1} - u_{i+1} = l_i - \bar{l}_i - \sigma.$$

Observe that

$$
\begin{aligned}
t_{i+1} &= l_i - l_i u_i / \bar{u}_i - \sigma \\
&= l_i(\bar{u}_i - u_i)/\bar{u}_i - \sigma \\
&= t_i l_i / \bar{u}_i - \sigma.
\end{aligned}
$$

For reasons that are not clear Rutishauser called the associated algorithm the *differential* form of stqd. We call it dstqd.

dstqd(σ): $t_1 = -\sigma$;
        for $i = 1, n - 1$
           $\bar{u}_i = u_i + t_i$
           $\bar{l}_i = u_i(l_i/\bar{u}_i)$
           $t_{i+1} = t_i(l_i/\bar{u}_i) - \sigma$
        end for
        $\bar{u}_n = u_n + t_n$.

In practice the t-values may be written over each other in a single variable t. If the common subexpression $l_i/\bar{u}_i$ is recognized then only one division is needed. Thus dstqd exchanges a subtraction for a multiplication so the extra cost is not excessive.

At first sight stqd may not seem relevant to the eigenvalue problem but if λ is a very accurate approximation to an eigenvalue then dstqd(λ) is needed to approximate the associated eigenvector; see Section 12. In this application huge values among the $\{\bar{u}_i\}$ are to be expected and do not have deleterious effect on the computed eigenvector. In fact dstqd(σ) may be used to find eigenvalues too by extracting a good approximate eigenvalue from the t-values for a shift.

4. Progressive qd Algorithms

This section seeks the triangular factorization of $J' - \sigma I$, not $J - \sigma I$:

$$J' - \sigma I = UL - \sigma I = \hat{L}\hat{U}$$

for a suitable shift σ. Equating entries on each side of the defining equation gives the so-called rhombus rules of H. Rutishauser (see Rutishauser (1954), in German and, in English, Henrici (1958)):

$$u_{i+1} + l_{i+1} - \sigma = \hat{l}_i + \hat{u}_{i+1} \quad \text{and} \quad l_i u_{i+1} = \hat{l}_i \hat{u}_i.$$

These relations give the so-called progressive qd algorithm with shift which we call qds(σ).

qds(σ): $\hat{u}_1 = u_1 + l_1 - \sigma$;
　　　for $i = 1, n - 1$
　　　　　$\hat{l}_i = l_i u_{i+1}/\hat{u}_i$
　　　　　$\hat{u}_{i+1} = u_{i+1} + l_{i+1} - \sigma - \hat{l}_i$
　　　end for.

The algorithm qds fails when $\hat{u}_i = 0$ for some $i < n$. In contrast to the stationary algorithm the mapping $\sigma, L, U \rightarrow \hat{L}, \hat{U}$ is nontrivial even when $\sigma = 0$. When $\sigma = 0$ we write simply qd, not qds.

At this point the skeptical reader might object that qds(σ) is exactly the same algorithm that would be obtained by forming $J' - \sigma I$ and performing Gaussian elimination. Indeed if the operations are done proceeding from left to right, for example,

$$\hat{u}_{i+1} = fl(fl(fl(u_{i+1} + l_{i+1}) - \sigma) - \hat{l}_i),$$

then the two procedures are not just mathematically equivalent but also computationally identical. However it is not necessary to follow this ordering. For example, one could write

$$\hat{u}_{i+1} = (l_{i+1} - \hat{l}_i - \sigma) + u_{i+1}$$

and, if the compiler respects parentheses, then the output will quite often be different.

There is an alternative implementation of qds that is easy to miss. In fact Rutishauser never wrote it down. The new version is slightly slower than qds but has compensating advantages. Here we derive it simply as a clever observation leaving to later sections the task of making it independent of qds.

As suggested in an earlier paragraph we might define an auxiliary variable

$$d_{i+1} \equiv u_{i+1} - \hat{l}_i - \sigma \ (= \hat{u}_{i+1} - l_{i+1}).$$

Observe that

$$
\begin{aligned}
d_{i+1} &= u_{i+1} - (l_i u_{i+1}/\hat{u}_i) - \sigma \\
&= (u_{i+1}(\hat{u}_i - l_i)/\hat{u}_i) - \sigma \\
&= (u_{i+1}d_i/\hat{u}_i) - \sigma.
\end{aligned}
$$

Rutishauser seems to have discovered the unshifted version two or three years before he died, perhaps 15 years after discovering qd, but he did not make much use of it; see Rutishauser (1990). He called it the *differential qd* algorithm (dqd for short) and so we call the new shifted algorithm dqds (*differential* qd with *shifts*).

dqds(σ): $d_1 = u_1 - \sigma$;
$\qquad$ **for** $i = 1, n - 1$
$\qquad\qquad \hat{u}_i = d_i + l_i$
$\qquad\qquad \hat{l}_i = l_i(u_{i+1}/\hat{u}_i)$
$\qquad\qquad d_{i+1} = d_i(u_{i+1}/\hat{u}_i) - \sigma$
$\qquad$ **end for**
$\qquad\qquad \hat{u}_n = d_n.$

By definition, dqd = dqds(0). In the positive case dqd requires *no subtractions* and enjoys very high relative stability; see Section 8. In practice each d_{i+1} may be written over its predecessor in a single variable d. Looking ahead we mention that the quantity $\min |d_j|$ gives useful information on the eigenvalue nearest 0.

5. The LR Algorithm for J Matrices

As mentioned in Section 1 our exposition reverses the historical process. Rutishauser discovered LR by interpreting qd in terms of bidiagonal matrices, a brilliant and fruitful insight. This is worth explaining. By definition, the LR transform of J is J' and of J' is the matrix J'' defined in two steps by

$$
J' = \hat{L}\hat{U}, \quad J'' = \hat{U}\hat{L}.
$$

Now qd applied to L and U yields $\hat{L}$ and $\hat{U}$ and so defines J'' implicitly. There is no need to form J' or J''.

When shifts are employed the situation is a little more complicated. It is necessary to look at two successive steps with shifts σ_1 and σ_2.
$\qquad$ In shifted LR

$$
\begin{aligned}
J_1 - \sigma_1 I &= L_1 U_1, \\
J_2 &= U_1 L_1 + \sigma_1 I, \\
J_2 - \sigma_2 I &= L_2 U_2, \\
J_3 &= U_2 L_2 + \sigma_2 I.
\end{aligned}
$$

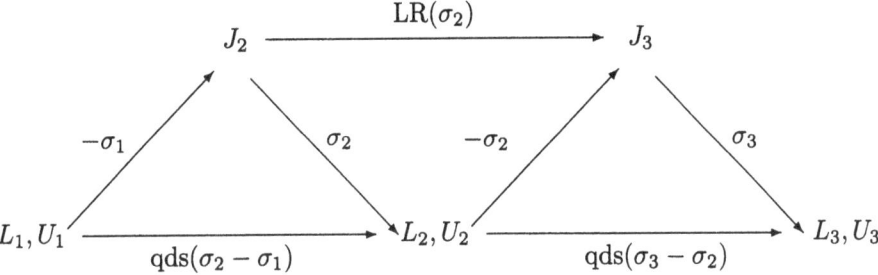

Fig. 1. Relation of LR to qds

In other words, the shifts are restored so that J_1, J_2, J_3 are similar. Note that

$$J_2 = U_1 L_1 + \sigma_1 I = L_1^{-1}(J_1 - \sigma_1 I)L_1 + \sigma_1 I$$
$$= L_1^{-1} J_1 L_1.$$

However, if J_2 is not to be formed one cannot *explicitly* add σ_1 back to the diagonal. On the other hand

$$J_2 - \sigma_2 I = U_1 L_1 - (\sigma_2 - \sigma_1 I) = L_2 U_2.$$

Thus to find L_2 and U_2 from L_1 and U_1 it is only necessary to apply qds$(\sigma_2 - \sigma_1)$. In other words to get qds equivalent to LR with shifts $\{\sigma_i\}_{i=1}^\infty$ it is necessary to use the differences $(\sigma_i - \sigma_{i-1})$ with qds. In LR the shifts should converge to an eigenvalue of the original J or J'. In qds the shifts should converge to 0 and $u_n \to 0$, $l_{n-1} \to 0$ too and all shifts must be accumulated. It is worth recording the relationship in a diagram in Figure 1.

In practice the LR algorithm avoids explicit calculation of the L's and U's and the transformation $J_i \to J_{i+1}$ is effected via a sequence of elementary similarity transformations. As implemented in the late 1950s and early 1960s, LR proved insufficiently reliable and was displaced by the QR algorithm in the mid 1960s. In one important class of applications both LR and qds were accurate and efficient: the positive case $l_i > 0$, $u_i > 0$ for all i.

For comparison purposes we present two implementations of LR: the explicit and the implicit shift versions. Let

$$J = tridiag \begin{pmatrix} & 1 & & 1 & \bullet & & 1 & \\ \alpha_1 & & \alpha_2 & & \bullet & \bullet & & \alpha_n \\ & \beta_1 & & \beta_2 & \bullet & & \beta_{n-1} & \end{pmatrix},$$

let $J - \sigma I = LU$, $\hat{J} = UL + \sigma I$ and write $\hat{J}$ in the same notation as J. The matrix L may be written as a product of lower-triangular plane transformers N_i. So

$$U = N_{n-1}^{-1} \cdots N_1^{-1}(J - \sigma I), \quad \hat{J} = U N_1 \cdots N_{n-1} + \sigma I$$

and the active part of N_i is $\begin{pmatrix} 1 & 0 \\ e_i & 1 \end{pmatrix}$, where the multiplier e_i is in position $(i+1, i)$. The following diagram shows a typical stage.

$$\begin{pmatrix} 1 & 0 \\ -e_j & 1 \end{pmatrix} \begin{pmatrix} d_j e_{j-1} & d_j & 1 \\ \beta_j e_{j-1} & \beta_j & \alpha_{j+1} - \sigma \end{pmatrix} = \begin{pmatrix} \hat{\beta}_{j-1} & d_j & 1 \\ 0 & 0 & d_{j+1} \end{pmatrix},$$

$$\begin{pmatrix} d_j & 1 \\ 0 & d_{j+1} \\ 0 & \beta_{j+1} \end{pmatrix} \begin{pmatrix} 1 & 0 \\ e_j & 1 \end{pmatrix} = \begin{pmatrix} d_j + e_j & 1 \\ e_j d_{j+1} & d_{j+1} \\ e_j \beta_{j+1} & \beta_{j+1} \end{pmatrix}.$$

LR (explicit shift σ): $d_1 = \alpha_1 - \sigma$;
$\qquad\qquad$ for $i = 1, n - 1$
$\qquad\qquad\qquad$ if $d_i = 0$ then exit (fail)
$\qquad\qquad\qquad$ $e_i = \beta_i / d_i$
$\qquad\qquad\qquad$ $\hat{\alpha}_i = d_i + (e_i + \sigma)$
$\qquad\qquad\qquad$ $d_{i+1} = \alpha_{i+1} - (e_i + \sigma)$
$\qquad\qquad\qquad$ $\hat{\beta}_i = d_{i+1} * e_i$
$\qquad\qquad$ end for
$\qquad\qquad$ $\alpha_n = d_n + \sigma$.

In practice we write d and e for d_i and e_i.

To derive the implicit shift version we note first that

$$\hat{J} = N_{n-1}^{-1} \cdots N_1^{-1} J N_1 \cdots N_{n-1}$$

and, of more importance, that the 2×2 submatrix

$$\begin{pmatrix} \hat{\beta}_{j-1} & d_j \\ \beta_j e_{j-1} & \beta_j \end{pmatrix}$$

has rank one in exact arithmetic. Thus the multiplier e_j could be computed from $e_j = \beta_j e_{j-1} / \hat{\beta}_{j-1}$ and then the shift σ disappears from the inner loop. The initial value $e_1 = \beta_1 / (\alpha_1 - \sigma)$ is the only occasion on which σ appears. Indeed if $\alpha_1 \gg \sigma$ then much of the information in σ is irretrievably lost.

The algorithm is sometimes described as chasing the bulge $\beta_j e_{j-1}$ in position $(j+1, j-1)$ down the matrix and off the bottom as $j = 2, 3, \dots, n-1$ and n. We write δ instead of d to emphasize that these quantities differ from the corresponding ones in the explicit shift algorithm.

LR (implicit shift σ): $\delta = \alpha_1$
$\qquad\qquad$ if $\delta = \sigma$ then exit (fail)
$\qquad\qquad$ $e = \beta_1 / (\delta - \sigma)$
$\qquad\qquad$ for $i = 1, n - 1$
$\qquad\qquad\qquad$ $\hat{\alpha}_i = \delta + e$
$\qquad\qquad\qquad$ $\hat{\beta}_i = \beta_i - e * (\hat{\alpha}_i - \alpha_{i+1})$
$\qquad\qquad\qquad$ $\delta = \alpha_{i+1} - e$

$$\textbf{if } \hat{\beta}_i = 0 \textbf{ then} \text{ exit (fail)}$$
$$e = e * \beta_{i+1}/\hat{\beta}_i$$
$$\textbf{end for}$$
$$\hat{\alpha}_n = \delta.$$

The attraction of this algorithm is that it employs nothing but explicit similarities on J.

6. Gram–Schmidt Factors

This section shows that dqd could have been discovered independently of qd.

To most people the Gram–Schmidt process is the standard way of producing an orthonormal set of vectors $q_1, q_2, \ldots, q_k$ from a linearly independent set $f_1, f_2, \ldots, f_k$. The defining property is that $span(q_1, q_2, \ldots, q_j) = span(f_1, f_2, \ldots, f_j)$, for each $j = 1, 2, \ldots, k$. The matrix formulation of this process is the QR factorization: $F = QR$, where $F = [f_1, f_2, \ldots, f_k]$, $Q = [q_1, q_2, \ldots, q_k]$ and R is $k \times k$ and upper triangular.

The generalization of this process to a pair of vector sets $\{f_1, f_2, \ldots, f_k\}$ and $\{g_1, g_2, \ldots, g_k\}$ is so natural that there can be little objection to keeping the name Gram–Schmidt; the context determines immediately whether one or two sets of vectors are involved. Denote by F^* the conjugate transpose of F.

Theorem 1 Let F and G be complex $n \times k$ matrices, $n \geq k$, such that G^*F permits triangular factorization:

$$G^*F = \tilde{L}\tilde{D}\tilde{R},$$

where $\tilde{L}$ and $\tilde{R}$ are unit triangular (left and right), respectively, and $\tilde{D}$ is diagonal. Then there there exist unique $n \times k$ matrices $\tilde{Q}$ and $\tilde{P}$ such that

$$F = \tilde{Q}\tilde{R}, \quad G = \tilde{P}\tilde{L}^*, \quad \tilde{P}^*\tilde{Q} = \tilde{D}.$$

Remark. When $G = F$ the traditional QR factorization is recovered with an unconventional normalization: generally $Q = \tilde{Q}\tilde{D}^{-1/2}$. F^*F permits triangular factorization when, and only when, the leading $k - 1$ columns of F are linearly independent.

Remark. In practice, when $n = k$ and $\tilde{D}$ is invertible one can omit Q and write $F = \tilde{P}^{-1}(\tilde{D}\tilde{R})$, $G = \tilde{P}\tilde{L}^*$ and still call it *the* Gram–Schmidt factorization. The important feature is the uniqueness of $\tilde{Q}$ and $\tilde{P}$. The columns of $\tilde{Q}$ and rows of $\tilde{P}^*$ form a pair of dual bases for the space of n-vectors (columns) and its dual (the row n-vectors). There is no notion of orthogonality or inner product here; $p_i^*q_j = 0$ says that p_i^* annihilates q_j, $i \neq j$.

We omit the proof of the theorem to save space.

The Gram–Schmidt factorization leads directly to the differential qd algorithms. Let us show how.

Corollary 1 Let bidiagonal matrices L and U be given as in Section 2. If UL permits factorization

$$UL = \hat{L}\hat{D}\hat{R} = \hat{L}\hat{U},$$

where $\hat{L}$ and $\hat{R}$ are unit bidiagonal, then there exist unique matrices $\tilde{P}$ and $\tilde{Q}$ such that

$$U = \hat{L}\tilde{P}^*, \quad L = \tilde{Q}\hat{R}, \quad \tilde{P}^*\tilde{Q} = \hat{D}.$$

Remark. In words, apply Gram–Schmidt to the columns of L and the rows of U, in the natural order, to obtain $\hat{U}$ and $\hat{R}$. Then note that $\hat{U} = \hat{D}\hat{R}$

Note that if $u_i = 0$, $i < n$, then UL does not permit triangular factorization. However, the theorem allows $u_n = 0$. When $u_n \neq 0$ then U is invertible and so is $\hat{D}$. In this case we can rewrite the factorization as

$$KL = \hat{D}\hat{R} = \hat{U}, \quad UK^{-1} = \hat{L}, \quad K = \hat{D}\tilde{Q}^{-1}.$$

The matrix K is hidden when we just write $UL = \hat{L}\hat{U}$. However, the identification of K with the Gram–Schmidt process goes only half way in the derivation of the dqd algorithm. The nature of the Gram–Schmidt process shows that $\tilde{P}$ and $\tilde{Q}$ are upper Hessenberg matrices. Fortunately $\tilde{Q}$ and $\tilde{P}$ are *special* Hessenberg matrices that depend on only $2n$ parameters, not $n(n-1)/2$. We are going to show that they may be written as the product of $(n-1)$ simple matrices that are non-orthogonal analogues of plane rotations. That means that L may be changed into $\hat{U}$ and U into $\hat{L}$ by a sequence of simple transformations and neither K, $\tilde{Q}$ nor $\tilde{P}$ need appear explicitly.

Definition 1 A *plane transformer*, in plane (i, j), $i \neq j$, is an identity matrix except for the entries (i, i), (i, j), (j, i) and (j, j). The 2×2 submatrix they define must be invertible.

Let us describe the first minor step in mapping $L \to \hat{U}$, $U \to \hat{L}$. We seek an invertible 2×2 matrix such that

$$\begin{pmatrix} x & z \\ -y & w \end{pmatrix}\begin{pmatrix} 1 & 0 \\ l_1 & 1 \end{pmatrix} = \begin{pmatrix} \hat{u}_1 & 1 \\ 0 & 1 \end{pmatrix},$$

$$\begin{pmatrix} u_1 & 1 \\ 0 & u_2 \end{pmatrix}\begin{pmatrix} w & -z \\ y & x \end{pmatrix} = \begin{pmatrix} 1 & 0 \\ \hat{l}_1 & * \end{pmatrix} \cdot det,$$

where $det = xw + yz$ and $*$ may be anything. A glance at the last column of the top equation shows that $z = w = 1$. The 0 in the $(2, 1)$ entry on the right shows that $y = l_1$ and the 0 in the $(1, 2)$ entry in the second equation shows that $x = u_1$. Thus $det = wx + yz = u_1 + l_1 \equiv \hat{u}_1$. ¿From the $(2, 1)$ entry of the second equation we learn that

$$u_2 l_1 = u_2 y = \hat{l}_1 det = \hat{l}_1 \hat{u}_1.$$

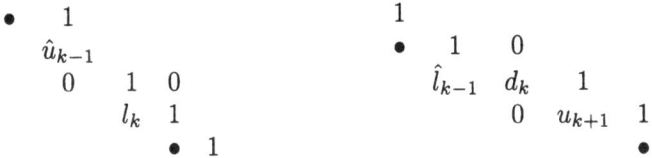

Fig. 2. Active entries

Finally, and of most interest,

$$* = u_2 x / \det = u_2 u_1 / \hat{u}_1.$$

This is the intermediate quantity d_2 in dqd and we see it here as something that gets carried to the next minor step. If we write $d_1 = u_1$ we obtain the start of the inner loop of dqd:

$$\begin{aligned} \hat{u}_1 &= d_1 + l_1, \\ \hat{l}_1 &= l_1(u_2/\hat{u}_1), \\ d_2 &= d_1(u_2/\hat{u}_1). \end{aligned}$$

The typical minor step is similar. It is instructive to look at the matrices part way through the transformation $L \to \hat{U}$, $U \to \hat{L}$ as shown in Figure 2. At minor step k the plane transformed is $(k, k+1)$ and the active part of the plane transformer is $\begin{pmatrix} d_k & 1 \\ -l_k & 1 \end{pmatrix}$ on the left and $\begin{pmatrix} 1 & -1 \\ l_k & d_k \end{pmatrix}$ on the right, with $\det = \hat{u}_{k+1}$. Finally at the end of minor step $(n-1)$ the trailing 2×2 submatrices are

$$\begin{pmatrix} \hat{u}_{n-1} & 1 \\ 0 & 1 \end{pmatrix} \quad \text{and} \quad \begin{pmatrix} 1 & 0 \\ \hat{l}_{n-1} & d_n \end{pmatrix} = \begin{pmatrix} 1 & 0 \\ \hat{l}_{n-1} & 1 \end{pmatrix} \begin{pmatrix} 1 & 0 \\ 0 & d_n \end{pmatrix}.$$

If $d_n \neq 0$ we simply multiply row n on the left by d_n and divide column n on the right by d_n as a final similarity transformation. When $d_n = 0$ the matrices $\hat{L}$ and $\hat{R}$ remain invertible. Thus $\hat{u}_n = d_n \hat{r}_{nn} = 0 \cdot 1 = 0$.

So we have derived the dqd algorithm without reference to qd. Of more significance is the fact that the quantities d_i, $i = 1, \ldots, n$, provide useful information about UL that qd does not reveal and so dqd facilitates the choice of shift.

7. The Meaning of d_i

Theorem 2 Consider L and U as described in Section 2. If U is invertible then the quantities d_i, $i = 1, \ldots, n$, generated by the dqd algorithm applied to L and U satisfy

$$d_i^{-1} = [(UL)^{-1}]_{ii}, \quad i = 1, \ldots, n.$$

Proof. The algorithm may be considered as transforming L to $\hat{U}$ by premultiplications and U to $\hat{L}$ by inverse multiplications on the right as described in the previous section. At the end of the $(k-1)$th plane transformation the situation is as indicated below:

$$G_{k-1}L = \begin{bmatrix} \bullet & \bullet & & & \\ & \hat{u}_{k-1} & 1 & & \\ & 0 & 1 & 0 & \\ & & l_k & 1 & \\ & & & \bullet & \bullet \end{bmatrix}, \quad \begin{bmatrix} \bullet & & & & \\ \bullet & 1 & 0 & & \\ \hat{l}_{k-1} & d_k & 1 & & \\ & 0 & u_{k+1} & 1 & \\ & & \bullet & \bullet & \end{bmatrix} = UG_{k-1}^{-1},$$

where

$$G_{k-1} = \Phi_{k-1}\Phi_{k-2}\cdots\Phi_1, \quad \Phi_i \text{ transforms plane } (i, i+1), \quad G_0 = I_n.$$

The striking fact is that row k of $G_{k-1}L$ and column k of UG_{k-1}^{-1} are singletons. If e_j denotes column j of I_n then

$$e_k^t G_{k-1}L = e_k^t, \quad e_k d_k = UG_{k-1}^{-1}e_k, \quad 1 \le k \le n.$$

Rearranging these equations yields, for $k = 1, 2, \ldots, n$,

$$\begin{aligned} d_k^{-1} &= (e_k^t G_{k-1})(G_{k-1}^{-1}e_k d_k^{-1}) \\ &= (e_k^t L^{-1})(U^{-1}e_k) = [(UL)^{-1}]_{kk}. \end{aligned}$$

$\square$

In the positive case ($l_i > 0$, $u_i > 0$), UL is diagonally similar to a symmetric positive-definite matrix.

Corollary 2 In the positive case,

$$\left(\sum_{i=1}^n d_i^{-1}\right)^{-1} < \lambda_{\min}(UL) \le \min_i d_i.$$

Proof. For any matrix M that is diagonally similar to a positive-definite symmetric matrix

$$\max_j m_{jj} \le \lambda_{\max}[M] < \text{trace}[M].$$

Take $M = (UL)^{-1}$. $\square$

Even in the general case, as $u_n \to 0$, $\min_i |d_i|$ becomes an increasingly accurate approximation to $|\lambda_{\min}|$.

8. Incorporation of Shifts

The algorithms and theorems presented so far serve only as background. LR, QR and qd algorithms are only as good as their shift strategies. In practice one uses qds and dqds, the shifted versions of qd and dqd.

The derivation of dqds(σ) in terms of a Gram–Schmidt process is not obvi-ous. Formally we write $UL - \sigma I = (U - \sigma L^{-1})L = \hat{L}\hat{U}$ and apply the Gram–Schmidt process to the columns of L and the rows of $U - \sigma L^{-1}$ to obtain

$$L = G\hat{R}, \quad U - \sigma L^{-1} = \hat{L}F, \quad FG = \hat{D}.$$

Eliminating G yields

$$L = (G\hat{D}^{-1})(\hat{D}\hat{R}) = F^{-1}\hat{U}, \quad U - \sigma L^{-1} = \hat{L}F.$$

At first sight the new term $-\sigma L^{-1}$ appears to spoil the derivation of F as a product of plane transformers. However, it is not necessary to know all the terms of L^{-1} but only the $(i + 1, i)$ entries immediately below the main diagonal. The change from the unshifted case is small. The active parts of the two transformations are given by

$$\begin{pmatrix} d_i & 1 \\ -l_i & 1 \end{pmatrix}\begin{pmatrix} 1 & 0 \\ l_i & 1 \end{pmatrix} = \begin{pmatrix} \hat{u}_i & 1 \\ 0 & 1 \end{pmatrix}, \quad \text{as before,}$$

and the new relation

$$\begin{pmatrix} d_i & 1 \\ \sigma l_i & u_{i+1} - \sigma \end{pmatrix}\begin{pmatrix} 1 & -1 \\ l_i & d_i \end{pmatrix} = \begin{pmatrix} 1 & 0 \\ \hat{l}_i & d_{i+1} \end{pmatrix} \cdot det.$$

The last row yields

$$det = d_i + l_i = \hat{u}_i, \quad \text{as before,}$$

$$\hat{l}_i \cdot det = \hat{l}_i\hat{u}_i = \sigma l_i + (u_{i+1} - \sigma)l_i = u_{i+1}l_i, \quad \text{as before,}$$
$$d_{i+1} \cdot det = -\sigma l_i + (u_{i+1} - \sigma)d_i = u_{i+1}d_i - \sigma\hat{u}_i.$$

This is dqds(σ).

If one looks at the two matrices part way through the transformations $L \to \hat{U}$, $U - \sigma L^{-1} \to \hat{L}$, the singleton column in the second matrix (from Theorem 2) has disappeared and the relation of d_i to $(UL)^{-1}$ is more com-plicated.

Theorem 3   Consider L and U as described in Section 2. If U is invertible and $UL - \sigma I$ permits triangular factorization, with $\sigma \neq 0$, then the inter-mediate quantities d_i, $i = 1, \ldots, n$, generated by dqds(σ) applied to L and U satisfy

$$\frac{1 + \sigma[(\hat{U}UL)^{-1}]_{k,k-1}}{d_k + \sigma} = [(UL)^{-1}]_{k,k}.$$

We omit the proof.

9. Accuracy

The differential qd algorithms dqd and dqds are new to the scene of matrix computations. One feature that makes them attractive is that they seem to be more accurate than their rivals. In particular, in the positive case, all eigenvalues can be found to high relative accuracy as long as the shifts preserve positivity.

Let us begin with an extreme example.

Example 1. Take $n = 64$, $u_i = 1$, $i = 1, \ldots, 64$, $l_i = 2^{16} = 65536$, $i = 1, \ldots, 63$. Although $\det(LU) = 1$ the smallest eigenvalue is $\mathcal{O}(10^{-304})$.

In just 2 iterations dqd computed $\lambda_{\min}$ to full working precision. In contrast qd returns 0, a satisfactory answer relative to the matrix norm. Yet dqd does preserve the determinant to working accuracy, provided underflow and overflow are absent, while qd, LR and QR do not.

The reason for dqd's accuracy is that $\hat{u}_{64} = d_{64}$ reaches the correct tiny value through 63 multiplications and divisions. There are no subtractions.

A little extra notation is needed to describe the stability results compactly. When there is no need to distinguish l's from u's we follow Rutishauser and speak of a qd array

$$Z = \{u_1, l_1, u_2, l_2, \ldots, l_{n-1}, u_n\}.$$

The right unit for discussing relative errors is the ulp (1 unit in the last place held) since it avoids reference to the magnitudes of the numbers involved. In Example 1 the error in the computed eigenvalue $< \frac{1}{2}$ulp despite 2×63 divisions and multiplications.

Given Z the dqds algorithm in finite precision produces a representable output $\hat{Z}$. We write this $\hat{Z} = fl(\text{dqds}) \cdot Z$. Now we introduce two ideal qd arrays $\vec{Z}$ and $\check{Z}$ such that, in exact arithmetic, dqds with shift σ maps $\vec{Z}$ into $\check{Z}$. Moreover $\vec{Z}$ is a tiny relative perturbation of Z, and $\hat{Z}$ is a tiny relative perturbation of $\check{Z}$. See Figure 3.

The proof of the following result may be found in [2].

Theorem 4   In the absence of division by zero, underflow or overflow, the Z diagram commutes and, for all k, $\vec{u}_k$ ($\vec{l}_k$) differs from u_k (l_k) by 3 (1) ulps at most, and $\hat{u}_k$ ($\hat{l}_k$) differs from $\check{u}_k$ ($\check{l}_k$) by 2 (2) ulps, at most.

The proof is based on making small changes to Z and $\hat{Z}$ so that the *computed* sequence of d's is exact for $\vec{Z}$ and $\check{Z}$. There is no requirement of positivity so it is possible to have $\|\hat{Z}\| >>> \|Z\|$. Some people call this a *mixed stability* result because one had to perturb both input and output to get an exact dqds mapping. For example, such mixed accuracy results are the best that can be said about the trigonometric functions in computer systems; the output is within an ulp of the exact trigonometric function of a value within one ulp of the given argument.

Theorem 4 does not guarantee that dqds returns accurate eigenvalues in

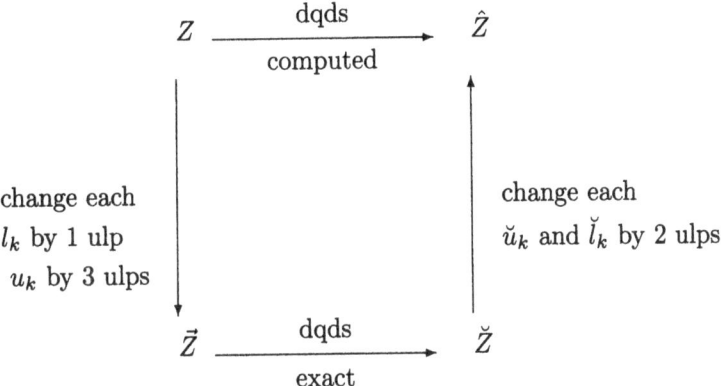

Fig. 3. Effects of roundoff

all cases, even when we only want errors to be small relative to $\|J\|$. We call such accuracy absolute rather than relative:

$$|\lambda_i - \hat{\lambda}_i| < \epsilon \|J\| \quad \text{versus} \quad |\lambda_i - \hat{\lambda}_i| < \epsilon \max\{|\lambda_i|, |\hat{\lambda}_i|\}.$$

In Fernando and Parlett (1994) a corollary to Theorem 4 establishes high relative accuracy for all eigenvalues computed by dqds *in the positive case.* The corollary is stated in terms of singular values but the algorithm computes the squares of those singular values, that is, the eigenvalues of a positive-definite matrix similar to a product LU.

Given these nice results it is natural to seek a conventional backward error analysis that says that $\hat{Z}$ is the exact output of dqds applied to some array $\overset{o}{Z}$, where either $\|Z - \overset{o}{Z}\| < M\epsilon\|Z\|$ or, better, $|u_i - \overset{o}{u}_i| < M\epsilon|u_i|$, $|l_i - \overset{o}{l}_i| < M\epsilon|l_i|$, for all i, which we write as $|Z - \overset{o}{Z}| < M\epsilon|Z|$. Here M is some reasonable constant.

This was one task given to Yao Yang for his doctoral research here and we quote here two of the results from his 1994 paper. Yang's first discovery was an unpleasant surprise. Even in the positive case it is not always true that $|Z - \overset{o}{Z}| < M\epsilon|Z|$. Here is an example.

Example 2. There is no small backward error for dqd.

$$n = 5, \ \sigma = 0, \text{ single precision: } 1 + 10^{-8} \to 1,$$
$$u = (1, 1, 1, 1, 1),$$
$$l = (10^4, 10^4, 10^4, 10^4, 0).$$

We omit tiny irrelevant terms in what follows.

$$\hat{u} = (10^4 + 1, 10^4, 10^4, 10^4, 10^{-16} - 10^{-20}),$$
$$\hat{u} \text{ exactly } (10^4 + 1, 10^4 + \mathbf{10^{-4}}, 10^4 + 10^{-8}, 10^4 + 10^{-12}, 10^{-16} - 10^{-20}),$$
$$\hat{l} = (1 - 10^{-4}, 1, 1, 1, 0),$$

$$\overset{o}{u} = (0, 1 - 10^{-4}, 1, 1, 1),$$
$$\overset{o}{l} = (10^4 + 1, 10^4, 10^4, 10^4, 0).$$

Thus $u_1 = 1$ must be changed to $\overset{o}{u}_1 = 0$ in order to have dqd· $\overset{o}{Z} = \hat{Z}$. The last 3 steps in computing $\overset{o}{Z}$ are instructive.

    1. $\overset{o}{u}_2 = \hat{u}_2 - \overset{o}{l}_2 + \hat{l}_1 = 10^4 - 10^4 + (1 - 10^{-4}) = 1 - 10^{-4}$

whereas the true $\hat{u}_2$ is $10^4 + 10^{-4}$ which would yield the ideal $\overset{o}{u}_2 = 1$.

    2. $\overset{o}{l}_1 = \hat{l}_1 \hat{u}_1 / \overset{o}{u}_2 = 10^4/(1 - 10^{-4}) = 10^4 + 1$ instead of 10^4.

    3. $\overset{o}{u}_1 = \hat{u}_1 - \overset{o}{l}_1 = (10^4 + 1) - (10^4 + 1) = 0$ instead of 1.

Note that $\overset{o}{u}_1 + \overset{o}{l}_1 = u_1 + l_1$ exactly!

This result shows why there was no backward error analysis in Fernando and Parlett (1994). Further investigation by Yang showed that the fault is in the formulation of the task, not in the algorithm. Recall from Section 2 that associated with any qd array Z are matrices L, U and their products J and J'.

    Here is Yang's second result.

Theorem 5   If dqds(σ) maps Z into $\hat{Z}$ in finite-precision arithmetic obeying (9.1) below and if both arrays are positive then there is a unique array $\overset{o}{Z}$ such that in exact arithmetic dqds(σ) maps $\overset{o}{Z}$ into $\hat{Z}$. Moreover the tridiagonal matrices J' and $\overset{o}{J}'$ associated with Z and $\overset{o}{Z}$ satisfy

$$|J' - \overset{o}{J}'| < 5\epsilon |J'|,$$

where ϵ is the roundoff unit.

    The inequality is interpreted element-by-element. This is a strong result and consistent with Theorem 4. The amplification factor 5 is a worst-case bound. Contemplation of the proof shows that in most cases the errors in executing dqds(σ) in finite precision can be accounted for by perturbing J's entries (not Z's) by 1 or 2 ulps.

    The strength of the result comes from the simplicity of the proof and what makes the proof simple is that, in exact arithmetic, dqds(σ) is equivalent to qds(σ) and qds brings in no intermediate quantities. Thus we may define $\overset{o}{Z}$ by $\overset{o}{Z} = $ qds$^{-1} \cdot fl($dqds$)Z$. Here fl denotes a result obtained with floating-point arithmetic.

    The diagram in Figure 4 illustrates the strategy.

    The invertibility of qds(σ) is proved by observing that if the output is positive and given then qds(σ) statements may be used in reverse order $(n, n-1, \ldots, 1)$ to recover the unique input.

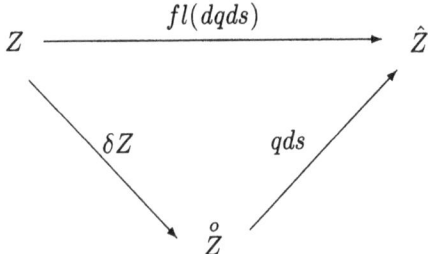

Fig. 4. Definition of $\overset{o}{Z}$

The model of arithmetic assumes the presence of a guard digit so that subtraction has the same relative accuracy as the other fundamental operations.

$$fl(a\square b) = (a\square b)(1 + \eta), \tag{9.1}$$

where $\eta \le \epsilon$, the roundoff unit (or arithmetic precision) which does not depend on a, b or $\square$, and $\square = \pm, *$ or $/$.

Proof of Theorem 5. First we write down the relationships satisfied by $\hat{Z}$. Subscripted variables γ, δ and ϵ denote roundoff quantities as needed by the model (9.1).

$fl(\text{dqds}) : d_1 = fl(u_1 - \sigma) = (u_1 - \sigma)(1 + \gamma_0');$

    for $i = 1, n - 1$

         $\hat{u}_i = fl(d_i + l_i) = (d_i + l_i)(1 + \epsilon_i)$

         $t_i = fl(u_{i+1}/\hat{u}_i) = u_{i+1}(1 + \delta_i)/\hat{u}_i, \; t_i$ should stay in a register

         $\hat{l}_i = fl(l_i * t_i) = l_i u_{i+1}(1 + \delta_i)(1 + \delta_i')/\hat{u}_i$

         $d_{i+1} = fl(d_i * t_i - \sigma) = (d_i u_{i+1}(1 + \delta_i)(1 + \gamma_i)/\hat{u}_i - \sigma)(1 + \gamma_i')$

    end for

         $\hat{u}_n = d_n.$

By definition $\text{qds}\cdot \overset{o}{Z} = \hat{Z}$, exactly, so with $\overset{o}{l}_0 = 0$,

qds :   **for** $i = 1, n - 1$

         $\hat{u}_i = \overset{o}{u}_i - \hat{l}_{i-1} + \overset{o}{l}_i - \sigma$

         $\hat{l}_i = \overset{o}{l}_i \overset{o}{u}_{i+1} / \hat{u}_i$

    end for

         $\hat{u}_n = \overset{o}{u}_n - \hat{l}_{n-1} - \sigma.$

Of course $\overset{o}{Z}$ is determined in the order $\overset{o}{u}_n, \overset{o}{l}_{n-1}, \overset{o}{u}_{n-1}, \overset{o}{l}_{n-2}, \dots, \overset{o}{u}_1$, but that is irrelevant here. Next we eliminate $\hat{u}_i$ and $\hat{l}_i$ using the two sets of equations from $fl(dqds)$ and qds. In what follows we omit terms that are $\mathcal{O}(\epsilon^2)$.

For $i = 1$

$$
\begin{aligned}
\overset{o}{u}_1 + \overset{o}{l}_1 - \sigma &= \hat{u}_1 = ((u_1 - \sigma)(1 + \gamma_0') + l_1)(1 + \epsilon_1) \\
&= u_1 + l_1 - \sigma + (u_1 - \sigma)(\gamma_0' + \epsilon_1) + \epsilon_1 l_1.
\end{aligned}
$$

In the positive case $\lambda_{\min}(J) \leq$ any 'pivot' in elimination, that is, $\sigma < \lambda_{\min}(J) \leq \min_j u_j$. Thus, in the positive case

$$
\left| \frac{\overset{o}{u}_1 + \overset{o}{l}_1 - u_1 - l_1}{u_1 + l_1} \right| < \frac{u_1 \cdot 2\epsilon + l_1 \epsilon}{u_1 + l_1} \leq 2\epsilon.
$$

For $2 \leq i < n$

$$
\begin{aligned}
\overset{o}{u}_i + \overset{o}{l}_i &= \hat{u}_i + \hat{l}_{i-1} + \sigma \\
&= (d_i + l_i)(1 + \epsilon_i) + l_{i-1} u_i (1 + \delta_{i-1})(1 + \delta_{i-1}')/\hat{u}_{i-1} + \sigma \\
&= d_{i-1} u_i (1 + \delta_{i-1})(1 + \gamma_{i-1})(1 + \gamma_{i-1}')(1 + \epsilon_i)/\hat{u}_{i-1} \\
&\quad - \sigma(1 + \gamma_{i-1}')(1 + \epsilon_i) \\
&\quad + l_i(1 + \epsilon_i) + l_{i-1} u_i (1 + \delta_{i-1})(1 + \delta_{i-1}')/\hat{u}_{i-1} + \sigma.
\end{aligned}
$$

Now we can combine the terms involving u_i, noting that

$$
\hat{u}_{i-1} = (d_{i-1} + l_{i-1})(1 + \epsilon_{i-1}).
$$

So, omitting terms of $\mathcal{O}(\epsilon^2)$, we have

$$
\begin{aligned}
\overset{o}{u}_i + \overset{o}{l}_i &= u_i + l_i + \frac{d_{i-1}}{d_{i-1} + l_{i-1}} u_i (\delta_{i-1} + \gamma_{i-1} + \gamma_{i-1}' + \epsilon_i - \epsilon_{i-1}) \\
&\quad + \frac{l_{i-1}}{d_{i-1} + l_{i-1}} u_i (\delta_{i-1} + \delta_{i-1}' - \epsilon_{i-1}) + \epsilon_i l_i - \sigma(\gamma_{i-1}' + \epsilon_i).
\end{aligned}
$$

Now use repeated terms like $\delta_{i-1} - \epsilon_{i-1}$ to simplify the roundoff terms:

$$
\begin{aligned}
\overset{o}{u}_i + \overset{o}{l}_i &= u_i + l_i + u_i(\delta_{i-1} - \epsilon_{i-1}) \\
&\quad + \frac{d_{i-1}}{d_{i-1} + l_{i-1}} u_i \gamma_{i-1} + \frac{l_{i-1}}{d_{i-1} + l_{i-1}} u_i \delta_{i-1}' \\
&\quad + \left(\frac{d_{i-1}}{d_{i-1} + l_{i-1}} u_i - \sigma \right)(\gamma_{i-1}' + \epsilon_i) + l_i \epsilon_i. \qquad (9.2)
\end{aligned}
$$

Thus, in the positive case, using $\sigma < u_i$,

$$
\begin{aligned}
| \overset{o}{u}_i + \overset{o}{l}_i - (u_i + l_i) | &< u_i 2\epsilon + u_i \epsilon + l_i \epsilon + \max\left\{ \frac{d_{i-1} u_i}{d_{i-1} + l_{i-1}}, \sigma \right\} 2\epsilon \\
&< u_i 2\epsilon + u_i \epsilon + l_i \epsilon + u_i 2\epsilon = u_i 5\epsilon + l_i \epsilon < 5\epsilon(u_i + l_i).
\end{aligned}
$$

For the subdiagonal entries we have, almost immediately,

$$\overset{o}{l_i}\overset{o}{u}_{i+1} = \hat{l}_i\hat{u}_i = l_iu_{i+1}(1+\delta_i)(1+\delta'_i), \tag{9.3}$$

so

$$\left|\frac{\overset{o}{l_i}\overset{o}{u}_{i+1} - l_iu_{i+1}}{l_iu_{i+1}}\right| < 2\epsilon$$

in all cases, positive or not. This completes the proof.   □

In the course of the proof the general case has been covered. Backward stability is guaranteed in neither a relative sense nor a norm sense. When there is element growth, that is, when $|d_i|$ or $|\hat{u}_i|$ or $|\hat{l}_i|$ greatly exceeds $|l_i|$ or $|u_i|$, then locally (in position i) $\overset{o}{J}'$ differs strongly from J'.

To state the result in terms of arrays treat $diag(M)$ as a linear array and define

$$\begin{aligned}
ones &= (1,1,\ldots,1),\\
\hat{l} &= (0,\hat{l}_1,\ldots,\hat{l}_{n-1}),\\
d &= (d_1,\ldots,d_n).
\end{aligned}$$

Corollary 3 In the general case, in the absence of under/overflow or divide by zero, it is element growth that governs $\overset{o}{J}' - J'$. Entry-by-entry,

$$\begin{aligned}
|diag(\overset{o}{J}') - diag(J')| &< \epsilon\{2|u| + |\sigma||ones| + |\hat{l}| + |\hat{u}| + 2|d|\},\\
|offdiag(\overset{o}{J}') - offdiag(J')| &< 2\epsilon|offdiag(J')|.
\end{aligned}$$

Proof. Relation (9.2) in the proof of Theorem 5 holds in the general case. By omitting $\mathcal{O}(\epsilon^2)$ terms we may undo the equation for d_i from *fl(dqds)*.

$$\begin{aligned}
\frac{d_{i-1}u_i}{d_{i-1}+l_{i-1}} &= \left(\frac{d_i}{1+\gamma'_{i-1}}+\sigma\right)(1+\epsilon_{i-1})/(1+\delta_{i-1})(1+\gamma_{i-1})\\
&= (d_i + \sigma - \gamma'_{i-1}d_i)(1 - \delta_{i-1} - \gamma_{i-1} + \epsilon_{i-1})\\
&= (d_i + \sigma)(1 + \mathcal{O}(\epsilon)),
\end{aligned}$$

Similarly we undo the equation for $\hat{l}_{i-1}$:

$$\begin{aligned}
\frac{l_{i-1}u_i}{d_{i-1}+l_{i-1}} &= \hat{l}_{i-1}(1+\epsilon_{i-1})/(1+\delta_{i-1})(1+\delta'_{i-1})\\
&= \hat{l}_{i-1}(1+\mathcal{O}(\epsilon)).
\end{aligned}$$

Now we can rewrite (9.2) in terms of $d_i, \hat{l}_{i-1}$ and $\hat{u}_i$.

$$\begin{aligned}
\overset{o}{l_i} + \overset{o}{u}_i &= u_i + l_i + u_i(\delta_{i-1} - \epsilon_{i-1}) + (d_i + \sigma)\gamma_{i-1} + \hat{l}_{i-1}\delta'_{i-1}\\
&\quad + d_i\gamma'_{i-1} + \epsilon_i(d_i + l_i).
\end{aligned}$$

So

$$| \overset{o}{u}_i + \overset{o}{l}_i - (u_i + l_i)| \; < \; |u_i| 2\epsilon + |d_i|\epsilon + |\sigma|\epsilon + |\hat{l}_{i-1}|\epsilon$$
$$+ |d_i|\epsilon + |\hat{u}_i|\epsilon,$$

omitting $\mathcal{O}(\epsilon^2)$ terms, as claimed in the corollary. The off-diagonal entries of $\overset{o}{J}$ do enjoy backward stability in a relative sense since the product rhombus rule is preserved to within 2 ulps as shown in (9.3) in the proof of Theorem 5.

□

10. Speed

All algorithms of LR type succeed or fail according to the sequence of shifts. The basic algorithms, when all shifts are zero, are just too slow. On the other hand zero is the natural shift when J is singular – unless it provokes element growth.

In practice all these algorithms find one or two eigenvalues at a time, deflate them from the matrix and proceed on the remaining submatrix. so the task, at each step, is to pursue these somewhat contradictory goals.

P1. The shift should approximate an eigenvalue, preferably a small one. The more accurate the better.

P2. The shift should not cost too much to compute.

P3. The shift should not lead to element growth in the transformation.

The positive case. Here P3 may be satisfied by always approximating the smallest eigenvalue from below. In this way positivity is preserved and dqds delivers high accuracy. Moreover the auxiliary quantities $\{d_i\}$ provide upper bounds on $\lambda_{\min}(J)$ that become very tight. In Theorem 3 the complicated quantity $[(\hat{U}UL)^{-1}]_{k,k-1}$ turns out to be positive and thus $\lambda_{\min}[J] < d_k + \sigma$. So,

$$\lambda_{\min}(J) < \min_j (d_j + \sigma).$$

With little extra expense dqds can return both the value of $\min_j d_j$ and the last index k at which the minimum occurs.

When $\sigma = 0$ The lower bound happens to be the Newton approximation to $\lambda_{\min}(J)$ from 0 for the characteristic polynomial of J. It is too expensive and too pessimistic to be a frequent choice except at the start of the algorithm.

The index k is useful in the selection of a cheap and realistic estimate of $\lambda_{\min}(J)$. We say that Z (and J) is in the *asymptotic regime* when $k = n - 1$ or $k = n$. At this point we remind the reader that all the algorithms we consider here may be run from the bottom of the matrix to the top, and so we may assume that $k > n/2$. In the asymptotic regime $\min_j d_j$ is an extremely good estimate of $\lambda_{\min}(J)$ but, of course, is a little too large.

A simple strategy that has satisfactory results in comparison with rival methods is to select a small subarray of Z surrounding row k and compute the Newton approximation to 0 for the polynomial associated with the subarray. This approximation is essentially

$$\left(\sum_{i=k-p}^{\min(n,k+p)} d_i^{-1} \right)^{-1}$$

for $p = 2$ or 3. This expression is modified appropriately when k is close to n. In the current implementation both Z and $\hat{Z}$ are available at the end of a transformation. Consequently, given d_k, it is easy to recompute neighboring d_i from $d_{i+1} = d_i u_{i+1}/(d_i + l_i) - \sigma$ going up or down.

Another strategy was proposed by Rutishauser (1960). If a shift σ causes qds (or dqds) to fail because $\hat{u}_n < 0$ but all $\hat{u}_i > 0$, $i < n$, then $\sigma + \hat{u}_n$ is an extremely good lower bound on $\lambda_{\min}(J)$, good to $\mathcal{O}(|\sigma - \lambda_{\min}|^3)$ asymptotically, see Fernando and Parlett (1994) for a new proof. So the bad transform is rejected and dqds is applied with the good shift.

In order to show that these qd algorithms are worth attention we quote some timing results from Fernando and Parlett (1994). On a test bed of several challenging cases of orders 20 to 100 a dqds code with the above shift strategy was between 4 and 11 times faster than LINPACK codes besides being more accurate. In more recent comparisons with code used in LAPACK (QR-based) routines the dqds program was, on the average, twice as fast.

It is likely that the current shifts will be replaced by better ones soon.

The symmetric indefinite case. Given a T that is symmetric, tridiagonal, but not positive definite we reduce it to the positive-definite case as follows. First compute the lower Gershgorin bound by

$$bound = \alpha_1 - |\beta_1|$$
$$\textbf{for } i = 2, n - 1$$
$$\qquad bound = \min\{bound, \alpha_i - |\beta_{i-1}| - |\beta_i|\}$$
$$\textbf{end for}$$
$$bound = \min\{bound, \alpha_n - |\beta_{n-1}|\}.$$

Next factor $T + bound \cdot I = LU$, with care, as follows:

$$u_1 = \alpha_1 + bound$$
$$\textbf{for } i = 2, n$$
$$\qquad l_{i-1} = \beta_i/u_{i-1}$$
$$\qquad u_i = (\max\{\alpha_i, bound\} - l_{i-1}) + \min\{\alpha_i, bound\}$$
$$\textbf{end for.}$$

It is because both *bound* and l_{i-1} are positive that we can use max and min to avoid computing $(big + little) - big$, the perennial danger when adding three quantities.

Now dqds may be applied to the positive case. Note that the eigenvectors are not altered by a shift. So, if eigenvectors are wanted, they may be computed as shown in Section 12 and the eigenvalues of T found afterwards by taking a Rayleigh quotient. This is to avoid subtracting *bound* from the computed eigenvalues that are very close to bound.

The general case. There are several open problems. We should expect to reject transforms from time to time when excessive element growth occurs. It is also possible to compute a dstqd transform (the stationary algorithm) at the same time as dqds for the same access to Z.

In other words, it is feasible to compute $\tilde{L}\tilde{U} = LU - \sigma I$ and $\hat{L}\hat{U} = UL - \sigma I$ at the same time. If σ is not too close to 0 then computation can proceed if either of the pairs $\tilde{L}, \tilde{U}$ and $\hat{L}, \hat{U}$ avoids element growth. It is also possible to apply dqds and dstqd to the reversal of Z, that is, $(u_n, l_{n-1}, u_{n-1}, \ldots)$. Our goal is to push $\min_j |d_j|$ to the closest end of the array.

When excessive growth occurs in $\hat{Z}$ it is not hard to evaluate the recurrence that governs the derivative of $\{p_i\}$ with respect to the shift. Here p_i is the characteristic polynominal of the top i by i submatrix of J. Given p'_j as well as p_j at a bad place one can calculate a new shift that will take the new p_j away from 0.

For each Z there is a bad set $Bad(Z)$ in $\mathbb{C}$ consisting of values that should not be used as shifts for qds. It would be useful to understand something of how $Bad(Z)$ changes under qds; that is, how does $Bad(\hat{Z})$ relate to $Bad(Z)$?

An alternative approach is to develop block versions of these algorithms.

11. Parallel Implementation

The algorithms qds and stqd seem to be irrevocably sequential in nature. In contrast the differential versions are less so.

Let us consider dqds from this point of view. The algorithm may be split into two parts.

Part 1. Compute $d = (d_1, \ldots, d_n)$ via
$$d_1 = u_1 - \sigma, \quad d_{i+1} = d_i u_{i+1}/(d_i + l_i) - \sigma, \quad i = 1, \ldots, n.$$
Part 2. As vector operations on l, u and d compute
$$\hat{u} = d + l, \quad l = (l_1, \ldots, l_{n-1}, 0),$$
$$\hat{l} = l * d^{\uparrow}/\hat{u},$$
where
$$d^{\uparrow} = (d_2, d_3, \ldots, d_n, 0).$$

Part 2 is ideal for vector or parallel processors.

It is interesting that Part 1 may, in principle, be executed in $\mathcal{O}(\log_2 n)$ time on a parallel processor but unfortunately the method seems to be unstable in finite precision. See the interesting paper of Mathias (1994). The technique is called 'parallel prefix' in computer science communities

The idea is to consider each d_i as a ratio p_i/q_i and rewrite the recurrence as

$$\frac{p_{i+1}}{q_{i+1}} = \frac{p_i(u_{i+1} - \sigma) + q_i l_i \sigma}{p_i + q_i l_i}$$

or

$$\begin{pmatrix} p_{i+1} \\ q_{i+1} \end{pmatrix} = \begin{pmatrix} u_{i+1} - \sigma & \sigma l_i \\ 1 & l_i \end{pmatrix} \begin{pmatrix} p_i \\ q_i \end{pmatrix} = M_i \begin{pmatrix} p_i \\ q_i \end{pmatrix}.$$

Note that each 2×2 matrix M_i is known a priori and we can start with $(p_1, q_1) = (u_1 - \sigma, 0)$. Consequently d_i is completely determined by column 1 of

$$N(i) = M_i M_{i-1} \cdots M_1.$$

The problem has been reduced to computing all the partial products indicated above. There is an intriguing way to compute the N's from the M's in $2(\log_2 n)$ parallel steps.

The pattern is indicated in the following diagram in Figure 5. This complicated algorithm can be written compactly in the following form.
Standard MATLAB notation is
$[i : j : k] = (i, i + j, i + 2j, \ldots, k); \quad [i : k] = [i : 1 : k].$
Let $p = \lceil \log_2 n \rceil$.

    Initialize $N(i) = M_i, \quad i = 1, 2, \ldots, n.$
    for $j = 2^{[1:p]}, \quad i = [j : j : n]; \quad N(i) = N(i) * N(i - j/2);$
    for $j = 2^{[p-1:-1:1]}, \quad i = [3 * j/2 : j : n]; \quad N(i) = N(i) * N(i - j/2).$

When $\sigma = 0$ the M's are all lower triangular and the task is simplified significantly and is well suited to implementation on the CM2 and CM5 massively parallel computers.

The strong potential for parallel implementation lies, not here, but at a coarser level in the computation of eigenvectors once the eigenvalues are known accurately. This is the topic of the next section.

12. Eigenvectors

Suppose that $Z (= L, U)$ is given along with an accurate approximation λ to an eigenvalue of $J = LU$. If λ were exact then

$$(LU - \lambda I)v = 0, \quad \|v\| = 1 \tag{12.1}$$

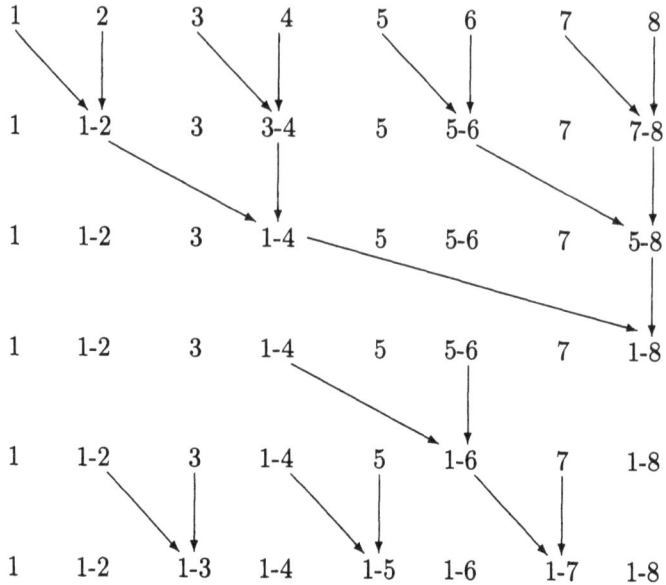

Fig. 5. Parallel prefix

is the equation defining a wanted eigenvector v. By applying the stationary qd algorithm $dstqd(\lambda)$ to Z we obtain, in the absence of breakdown, $\bar{Z}$ such that

$$LU - \lambda I = \bar{L}\bar{U}. \tag{12.2}$$

Since $\bar{L}$ is unit triangular it is invertible and it remains to solve $\bar{U}v = 0$, that is, $\bar{u}_i v_i + v_{i+1} = 0$, $i = 1, \ldots, n-1$. A little care must be exercised to avoid unnecessary overflows and underflows. Let k be the index of a maximal entry of v. Then

$$v_k = 1, \quad v_i = -v_{i+1}/\bar{u}_i, \quad i = k-1, \ldots, 1, \quad v_{j+1} = -v_j \bar{u}_j, \quad j = k, \ldots, n-1.$$

Note that $\bar{u}_n$ is not used and this is appropriate since $\bar{u}_n$ must vanish in exact arithmetic.

In practice there are two important changes to be made to the simple algorithm given above. First note that the matrix of interest is often not J but $\beta^{-1}J\beta$ for some diagonal β. So (12.2) changes to

$$\beta^{-1}LU\beta - \lambda I = (\beta^{-1}\bar{L}\beta)(\beta^{-1}\bar{U}\beta) \tag{12.3}$$

and

$$\beta^{-1}\bar{U}\beta = bidiag \begin{pmatrix} & \beta_1 & & \beta_2 & \bullet & & \beta_{n-1} & \\ \bar{u}_1 & & \bar{u}_2 & & \bullet & \bullet & & \bar{u}_n \end{pmatrix},$$

where

$$\beta = diag(1, \beta_1, \beta_1\beta_2, \ldots, \beta_1\beta_2 \cdots \beta_{n-1}).$$

So the bidiagonal system is

$$\bar{u}_i v_i + \beta_i v_{i+1} = 0, \quad i = 1, \ldots, n-1. \tag{12.4}$$

The second observation is that, even for extremely accurate approximations λ, the final value $\bar{u}_n$ is far from vanishing and the output from (12.4) or (12.3) is too often disappointing. Roundoff error spoils $\bar{U}$.

There is a remedy which is not complicated and appears to be new. In addition the new approach yields very good values for k, the index of a maximal or nearly maximal component of v. At the heart of dstqd is the t-recurrence:

$$t_{i+1} = t_i * l_i/(u_i + t_i) - \lambda \tag{12.5}$$

with $t_1 = -\lambda$. Once the t-vector is known $\bar{u} = (\bar{u}_1, \ldots, \bar{u}_n)$ and $\bar{l} = (\bar{l}_1, \ldots, \bar{l}_{n-1}, 0)$ may be found by vector operations: $\bar{u} = u + t$, $\bar{l} = u * l/\bar{u}$. The only thing that distinguishes a true eigenvalue λ from a non eigenvalue is that, in exact arithmetic, $0 = \bar{u}_n = u_n + t_n$. This gives a final value $t_n = -u_n$ and the 2-term recurrence may be solved in reverse order:

$$t_i = u_i / \left(\frac{l_i}{t_{i+1} + \lambda} - 1 \right) = \frac{u_i(t_{i+1} + \lambda)}{l_i - t_{i+1} - \lambda}. \tag{12.6}$$

In exact arithmetic with an exact λ the t-vectors from the forward and the backward passes will be the same but we use $\overset{o}{t}_i$ to denote the output of the backward pass. In practice t and and $\overset{o}{t}$ are not the same. We choose a k to satisfy

$$|t_k - \overset{o}{t}_k| = \min_j |t_j - \overset{o}{t}_j|.$$

Then we define $\bar{t} = (t_1, \ldots, t_k, \overset{o}{t}_{k+1}, \ldots, \overset{o}{t}_n)$, and compute an approximation x to v from

$$x_k = 1, \quad x_i = -\beta_i x_{i+1}/(t_i + u_i), \quad i = k-1, \ldots, 1, \tag{12.7}$$

$$x_{j+1} = -x_j(\overset{o}{t}_j + u_j)/\beta_j, \quad j = k, \ldots, n-1. \tag{12.8}$$

Note that t's are used going back from k, while $\overset{o}{t}$'s are used going forwards.

An attractive feature of (12.7) and (12.8) is that huge values of t_i and $\overset{o}{t}_j$, including ∞, do not impair the accuracy: From (12.6) $t_i = \infty$ implies $x_i = 0$ for $i = k-1, \ldots, 1$, and $\overset{o}{t}_{j+1} = \infty$ implies $\overset{o}{t}_j + u_j = 0$ implies $x_{j+1} = 0$, for $j = k, \ldots, n-1$. Wherever $x_i = 0$ the adjacent x-values are simply related by $\beta_{i-1}x_{i-1} + \beta_i x_{i+1} = 0$ when $\beta^{-1}LU\beta$ is symmetric and by $l_{i-1}u_{i-1}x_{i-1}/\beta_{i-1} + \beta_i x_{i+1} = 0$, in general. So, in the symmetric case, the full algorithm is: $x_k = 1$,

$$x_i = \begin{cases} -\beta_{i+1}x_{i+2}/\beta_i, & \text{if } x_{i+1} = 0, \\ -\beta_i x_{i+1}/(t_i + u_i), & \text{otherwise}, \quad i = k-1, \dots, 1, \end{cases}$$

$$x_{j+1} = \begin{cases} -\beta_{j-1}x_{j-1}/\beta_j, & \text{if } x_j = 0, \\ -x_j(\overset{o}{t}_j + u_j)/\beta_j, & \text{otherwise}, \quad j = k, \dots, n-1. \end{cases} \tag{12.9}$$

With (12.9) small entries in x are found by multiplication and division, not subtraction.

Let us now justify the choice of k. Suppose that all arithmetic operations are exact but, because λ is not an exact eigenvalue, the $\overset{o}{t}$-recurrence is not justified in using $\overset{o}{t}_n = -u_n$. Consequently there is truncation error and x is not an eigenvector. What can we say about x? If $T = \beta^{-1}LU\beta$ is symmetric then the residual

$$r = (T - \lambda I)x$$

vanishes in every component except the kth and at this position $|r_k| = |t_k - \overset{o}{t}_k|$. It is this result that justifies our choice of k. The assumption of symmetry is not essential but it simplifies the exposition. In what follows recall that $\beta_i^2 = l_i u_i$.

There are three cases. We consider r_j and use (12.9).

$$j < k. \qquad \beta_{j-1}x_{j-1} + (l_{j-1} + u_j - \lambda)x_j + \beta_j x_{j+1}$$

$$= x_j \left\{ -\frac{\beta_{j-1}^2}{\bar{u}_{j-1}} + l_{j-1} + u_j - \lambda - \bar{u}_j \right\}.$$

Now use (12.5):

$$r_j = x_j \left\{ -\frac{l_{j-1}u_{j-1}}{t_{j-1} + u_{j-1}} + l_{j-1} + u_j - \lambda - \left(\frac{t_{j-1}l_{j-1}}{t_{j-1} + u_{j-1}} - \lambda + u_j \right) \right\}$$

$$= x_j \cdot 0.$$

The case $j > k$ is similar but with $\overset{o}{t}_i$ replacing t_i and we omit it.

$$j = k. \qquad \beta_{k-1}x_{k-1} + (l_{k-1} + u_k - \lambda)x_k + \beta_k x_{k+1}$$

$$= x_k \left\{ -\frac{\beta_{k-1}^2}{\bar{u}_{k-1}} + l_{k-1} + u_k - \lambda - \bar{u}_k \right\}.$$

A little algebraic simplification together with (12.5) shows that

$$r_k = \left\{ \frac{t_{k-1}l_{k-1}}{t_{k-1} + u_{k-1}} - \lambda - \overset{o}{t}_k \right\} = t_k - \overset{o}{t}_k.$$

Roundoff errors make very little difference. More precisely it can be shown that with appropriately chosen 2 ulp perturbations to $L, U, t, \overset{o}{t}, x$, but not to λ, the equation corresponding to $(T - \lambda I)x = e_k(t_k - \overset{o}{t}_k)$ holds exactly. Here e_k is column k of I.

The significance of this result is that when λ is a very accurate approximation the quantities t_k and $\overset{o}{t}_k$ are quite often less than $\epsilon\|T\|$ and their difference is even smaller. Here ϵ is the precision of the arithmetic unit. Thus we compute vectors x whose residual norms $\|(T - \lambda I)x\|/\|x\|$ are significantly less than $\epsilon\|T\|$. This never happens with standard inverse iteration (i.e. TINVIT in the LINPACK package).

The cost for x from (12.9) appears to be $3n$ divisions (n for t, n for $\overset{o}{t}$, n for x) but for vector calculations dstqd may be rewritten so that the n divisions in (12.9) become n multiplications: $\bar{l}_i = \beta_i/\bar{u}_i$, $t_{i+1} = \bar{l}_i l_i t_i - \lambda$, $x_i = -\bar{l}_i x_{i-1}$. By way of comparison standard inverse iteration needs n calls for random numbers plus $2n$ divisions for a vector. So the new method is no slower than standard inverse iteration.

When two eigenvalues differ by about $\sqrt{\epsilon}\|LU\|$ the vectors x computed by (12.9) need to be refined by another step of inverse iteration in order to obtain eigenvectors orthogonal to working accuracy.

When m eigenvalues of T are very close (differing by n ulps, say) we can take care to pick m different values of k in order to try to produce outputs that are orthogonal. There is no need to perturb computed eigenvalues that are equal to working precision. This careful choice of k suffices in many but not all cases of clusters. Fortunately there is an extra modification of the method that takes care of the difficult cases by using appropriate submatrices of T. Since this modification is independent of qd algorithms it will not be described here, but see Parlett (1994).

The method described in this section is 'embarrassingly' parallel. Each processor is assigned a copy of Z and one or more eigenvalues. No communication is needed between processors.

13. Singular Values of Bidiagonals

Let

$$B = bidiag \begin{pmatrix} & b_1 & & b_2 & \bullet & & b_{n-1} & \\ a_1 & & a_2 & & \bullet & \bullet & & a_n \end{pmatrix}.$$

The goal is to find B's singular values $\sigma_1, \ldots, \sigma_n$. Since $\{\sigma_i^2\}$ is the eigenvalue set of $B^t B$ this is the eigenvalue problem of a positive-semidefinite symmetric tridiagonal with the constraint that $B^t B$ is not to be formed explicitly. Note that $B^t B$ is singular if, and only if, $a_i = 0$ for one or more i values. In these cases $B^t B$ is a direct sum of smaller tridiagonals and the 0 singular values may be deflated by applying qd or dqd independently to the appropriate submatrices as will be explained below. If some $b_i = 0$ the reduction to a direct sum is immediate and requires no arithmetic effort. Consequently there is no loss of generality in concentrating on the generic case: $a_i \neq 0$, $i = 1, \ldots, n$, $b_j \neq 0$, $j = 1, \ldots, n-1$.

We forsake symmetry and formally put $B^t B$ in the J-matrix format. Define

$$\Delta = diag(1, \pi_1, \pi_1 \pi_2, \ldots, \ldots, \pi_1 \cdots \pi_{n-1}), \quad \pi_i = a_i b_i.$$

It is readily verified that

$$\Delta B^t B \Delta^{-1} = J = LU,$$

where

$$L = bidiag \begin{pmatrix} 1 & 1 & \bullet & \bullet & 1 \\ & b_1^2 & b_2^2 & \bullet & b_{n-1}^2 \end{pmatrix},$$

$$U = bidiag \begin{pmatrix} & 1 & 1 & \bullet & 1 \\ a_1^2 & a_2^2 & \bullet & \bullet & a_n^2 \end{pmatrix},$$

and so we define $l_i = b_i^2$, $u_i = a_i^2$. Thus the singular-value problem leads to the positive case and dqds may be used to find the σ_i^2 in increasing order.

One begins with dqd (not dqds) in order to capture any tiny singular values and to make use of the lower and upper bounds in Corollary 2 to Theorem 2.

The only blemish is that the original data have been squared and thus the domain of application is smaller than for QR techniques since we must avoid overflow and underflow. However, there is an algorithm oqd described in Fernando and Parlett (1994) that has the same range as QR but the advantages of qd. The price that oqd pays for the extended range is that it must take a square root in the inner loop in the same way that QR techniques do. The response to this choice is easy: If exponent range permits then square up and use dqds, otherwise use oqd.

In order to find the right singular vector for σ_i^2 we apply the stationary algorithm dstqds to L, U with shift σ_i^2 to obtain $\tilde{L}$ and $\tilde{U}$ satisfying

$$\tilde{L}\tilde{U} = \Delta(B^t B - \sigma_i^2 I)\Delta^{-1}$$

whence

$$B^t B - \sigma_i^2 I (\Delta^{-1}\tilde{L}\Delta)(\Delta^{-1}\tilde{U}\Delta)$$

and the techniques of Section 12 may be envoked.

REFERENCES

J. Demmel and W. Kahan (1990), 'Accurate singular values of bidiagonal matrices', *SIAM J. Sci. Sta. Comput.* **11**, 873–912.

K. V. Fernando and B. N. Parlett (1994), 'Accurate singular values and differential qd algorithms', *Numerische Mathematik* **67**, 191–229.

P. Henrici (1958), 'The quotient-difference algorithm', *Nat. Bur. Standards Appl. Math. Series* **19**, 23–46.

R. Mathias (1994), 'The instability of parallel prefix matrix multiplication', preprint, Dept. of Mathematics, College of William and Mary, Williamsburg, VA 23187.

B. N. Parlett (1992), 'Reduction to tridiagonal form and minimal realizations', *SIAM J. on Matrix Analysis and Applications* **13**, 567–593.

B. N. Parlett (1994) 'Orthogonal eigenvectors for symmetric tridiagonals', to appear.

H. Rutishauser (1954), 'Der Quotienten-Differenzen-Algorithmus', *Z. Angew. Math. Phys.* **5**, 233–251.

H. Rutishauser (1958), 'Solution of eigenvalue problems with the LR-transformation', *Nat. Bur. Standards Appl. Math. Series* **49**, 47–81.

H. Rutishauser (1960), 'Uber eine kubisch konvergente Variante der LR-Transformation', *Z. Angew. Math. Mech.* **11**, 49–54.

H. Rutishauser (1990), *Lectures on Numerical Mathematics*, Birkhäuser, Boston.

Yao Yang (1994) 'Backward error analysis for dqds', submitted for publication.

For EU product safety concerns, contact us at Calle de José Abascal, 56–1°,
28003 Madrid, Spain or eugpsr@cambridge.org.

www.ingramcontent.com/pod-product-compliance
Ingram Content Group UK Ltd.
Pitfield, Milton Keynes, MK11 3LW, UK
UKHW060310090126

466816UK00021B/422